Handbook of Ground Water Development

Handbook of Ground Water Development

ROSCOE MOSS COMPANY
Los Angeles, California

WILEY

A Wiley-Interscience Publication

JOHN WILEY & SONS

New York   •   Chichester   •   Brisbane   •   Toronto   •   Singapore

This publication is designed to provide accurate and authoritative information in regard to the subject matter covered. While the information, analyses, suggestions, conclusions, and recommendations contained in this publication have been derived or compiled from sources believed to be reliable, the author and publisher make no guarantee as to, and assume no responsibility for, the correctness, sufficiency, or completeness of such information, analyses, suggestions, conclusions, or recommendations.

Library of Congress Cataloging in Publication Data:
Handbook of ground water development / Roscoe Moss Company.
      p.   cm.

   "A Wiley-Interscience publication."
   Includes bibliographies and index.
    1. Wells—Design and construction—Handbooks, manuals, etc.
  2. Water, Underground—Handbooks, manuals, etc. I. Roscoe Moss
Company.
TD407.H36 1989
627.1′14—dc19   89-30002
ISBN 0-471-85611-8   CIP

Printed in the United States of America

10 9 8 7 6 5 4 3 2

This book is dedicated to the memory of Roscoe Moss. Born in Rivera, California, in 1884, Mr. Moss joined the S.A. Clampett & Co., a pioneer water well drilling firm, in 1906. He became a partner in 1917 and by 1926 was the sole owner. In that year the company was incorporated as the Roscoe Moss Company and moved its headquarters to its present location in Los Angeles. Roscoe Moss remained active in the business until shortly before his death in 1977 at the age of 93.

Over the course of Mr. Moss' long business career, his company constructed thousands of high-capacity water wells, primarily in the southwestern United States and Hawaii. Many of the wells drilled in the early part of the century were for railroads and mines. Later they were required for agricultural development and municipal use, essential in areas where 50% of the water consumed comes from beneath the earth's surface. In the 1930s activities expanded to the eastern coast of the United States, and in the 1960s and 1970s ground water construction contracts were executed in Southeast Asia, Latin America, North Africa, and the Middle East.

Roscoe Moss' career began during a period of rapid economic development in the southwest. In this region where surface supplies are scarce, great quantities of ground water are found in the alluvial basins. To exploit the deep aquifers, new skills and advancements in drilling and pumping equipment were required. Roscoe Moss was an innovator, and his company was known as the best-equipped and most highly skilled contractor in the region. He made many contributions to improvements in the design, operation, repair, and maintenance of wells; casing and screen materials; and drilling equipment and technology.

Recognition of Roscoe Moss' expertise and contributions to the ground water development industry came in 1946 when he was selected by the governments of Great Britain and India to survey India and provide recommendations concerning the development of ground water in that country.

There has been a long-standing need for a comprehensive book on ground water development. The lack of sufficient texts may be because the industry is highly fragmented, and a number of technologies and sciences are needed for the understanding of the occurrence, extraction, replenishment, and proper exploitation of ground water. There are, however, numerous publications that concern one or a few aspects of this activity, but many are regional in scope or apply only to special conditions. To fill this gap, the Roscoe Moss Company sought out and enlisted the talents of distinguished experts in this multifaceted field and blended their knowledge with its 80 years of practical experience in water well design and construction. The result is this handbook which includes much valuable information not found in the literature but known to practioners in different segments of the industry.

This book is therefore written for use by all those involved in ground water development, be they designers, constructors, managers, or operators. It provides not only an overview of the subject, but it is sufficiently detailed to be useful to professionals. The range of treatment should particularly benefit students and newcomers to the industry.

The text is divided into three parts. In this way it traces the logical progression of the study of ground water from its origin through its development and exploitation. Part I deals primarily with the nature of ground water and where it can be found. Part II considers the parameters related to water well design and construction. Part III covers well and well field operation.

Although the emphasis is on high-capacity ground water-producing installations, most of the material applies to lower-yield wells. Although monitoring wells are not discussed specifically, the technologies of ground water development presented in this book are applicable, and those readers engaged in protection of this life-giving resource will find this publication useful.

ROSCOE MOSS JR.
GEORGE E. MOSS

Los Angeles, California

■■■■■■ ACKNOWLEDGMENTS

This wide-ranging book could not have been produced were it not for the generous help of a number of recognized experts. Dr. Joe Birman, formerly chairman of the Geology Department of Occidental College, wrote "Geologic Formations as Aquifers" and "Exploration for Ground Water." (Chapters 1 and 3). He was ably assisted by Barbara Brannan Birman. These two chapters provide valuable insights into these subjects, and the latter reflects Dr. Birman's extensive field experience.

Dr. Marcel Mougne authored "Geophysical Borehole Logging" (Chapter 4). Since this science primarily relates to petroleum exploration, the necessary modifications for water well applications required long and careful work. Dr. Mougne's patience and dedication to this difficult task is deeply appreciated.

Since the Roscoe Moss Company has been involved in the manufacture of water well casings and screens for over 60 years, we felt very qualified to write "Selection of Casing and Screen" (Chapter 11). "Well Design—General Considerations" (Chapter 6) was also written by the Roscoe Moss Company and reflects our over 80-year drilling experience in a wide variety of underground environments. "Drilling Systems" (Chapter 7) was primarily written by the Roscoe Moss Company, but with additional material contributed by Dr. John List, a faculty member of the California Institute of Technology. Dr. List also wrote "Stresses on Well Casing and Screen" (Chapter 9), with assistance from Dr. Gregory Gartrell, and he is the primary author of "Formation Stabilizer and Filter Pack" (Chapter 13). Chapter 9 presents material not published elsewhere and offers valuable information of great concern to water well designers and drilling contractors.

The first draft of "Drilling Fluid" (Chapter 8) was prepared by Dr. George Gray, whose unfortunate passing required the work to be completed by others. As with geophysical borehole logging, this chapter required the adaptation of techniques originally developed for use in oil well drilling to water wells. Sam Geffen of Baroid Corporation, an expert in both technologies, fulfilled the task superbly.

"Corrosion and Incrustation" (Chapter 10) in water wells is a subject of great importance. Although much has been published on this general topic, most of it is irrelevant to the water well environment. Expanding upon the knowledge of the Roscoe Moss Company, Dr. Gordon Treweek wrote a superb chapter on this complex subject. The late Jack Rossum, chief water quality chemist with the California Water Service Company for many years, aided Gordon with observations drawn from many years of practical experience. Since Gordon is an environmental chemist, he was able to describe and set forth the complicated reactions that accompany and cause corrosion and incrustation. The result is a chapter particularly useful to those involved in water well design.

"Cementing" (Chapter 12) was contributed by Dwight Smith, recently retired from the Halliburton Corporation. Again this effort was complicated because cementing techniques originated in the petroleum industry and material had to be edited and adapted for water wells. We are indebted to Dwight for his patience and persistence.

The techniques described in "Well Development" (Chapter 14) evolved from many years of practical experience. This chapter was first written by Jack Scheliga and later modified with assistance from Dr. John List and the staff of the Roscoe Moss Company.

Walt Webster, now retired, but formerly with Herkenhoff and Associates, Albuquerque wrote "Vertical Turbine Pumps" (Chapter 16) with substantial input from Jeptha Wade Jr., Mike Rossi, and Rob Guizetta, all of the California Water Service Co. Their dedication and enthusiasm is especially appreciated. This chapter was also carefully reviewed by John Dicmas, retired Chief Hydraulics Engineer for the Johnston Pump Co.

Although "Well and Pump Operation and Maintenance" (Chapter 17) was based on the experience of the Roscoe Moss Company, Paul Schreiber of the San Jose Water Company was a primary outside contributor. Carl Nuzman and Wayne Langley of Layne-Western Company added to the important area of well maintenance. Jacqueline S. Brophy of the Roscoe Moss Company furnished useful material during her review.

We are grateful to Dr. Scott Yoo for "Ground Water Quality and Contamination" (Chapter 18). Scott produced a work that presents an expert overview of this subject of great notoriety and importance today. Jack Rossum again supplied material from his many years of practical experience.

Special recognition must be given to the work of Dr. Dennis Williams. Dennis wrote five chapters, "Movement of Ground Water" (Chapter 2), "Hydraulics of Wells" (Chapter 5), "Well and Aquifer Evaluation from Pumping Tests" (Chapter 15), "Artificial Recharge" (Chapter 19), and "Ground Water Management" (Chapter 20). These chapters reflect the background and worldwide experience of a superb ground water hydrologist. His knowledge in this subject in part derived from study with two legendary teachers in the field, M. A. Hantush and C. E. Jacob. Building upon this background, Dennis himself advanced the frontiers of this science, most notably by his recent work on near-well turbulence. He has also been in the forefront of application of digital computer technology to solution of ground water mathematical equations and to management of this vital resource.

Dennis has been involved with this book from its conception throughout the entire seven-year period required for its completion. He executed a difficult task with great competence.

The scope and organization of this book were defined by Roscoe Moss Jr., and George E. Moss, and the major editing was also done by them. Indispensable processing of the multitudes of drafts was performed by Ofelia M. Quintero.

Glenn Spencer took drawings and sketches and, with his superlative skills, electronically transformed them into the lucid illustrations in this book.

The *Handbook of Ground Water Development* could never have been prepared without Kenneth G. Brown. Kenneth was the catalyst in the coordination of the many people involved. Drawing on his 40-year experience in water well construction in the United States and overseas with the Roscoe Moss Company, he processed and edited data into a form unique and useful to the industry. His doggedness, good humor, and patience were vital and irreplaceable. Without such dedicated people, fewer technical books would be written.

R. M. Jr.
G. E. M.

◼◼◼◼◼ ◼ CONTENTS

PART I

WATER IN THE GROUND

CHAPTER 1

Geologic Formations as Aquifers

1.1 INTRODUCTION

Water is an essential ingredient for all life forms on earth. Although generally abundant, it is not always available where needed, nor may it be of suitable quality. In many arid lands the only reliable supply lies beneath the earth's surface. In wetter areas, ground water may be superior in quality to surface supplies and also provide a reserve against drought. Because ground water is a hidden resource, knowledge of the conditions of its occurrence is essential in designing programs to utilize it efficiently.

Historical Perspective

Archeological evidence shows that ground water has been used since prehistoric times, especially in arid regions. Springs, for example, provided an easily available supply of freshwater, especially valuable during dry periods, and villages developed around them. Extensive references to the use of springs may be found in Genesis, Chapter 26, thought to be written as early as 2000 B.C.

Throughout the Middle East, tens of thousands of horizontal tunnels (*ghanats*) have been utilized for centuries and presently supply as much as 75% of all water used in Iran (1). *Ghanats* follow water table flow lines for distances as much as 20 mi, reaching depths in excess of 250 ft.

In ancient China water from hand-dug wells was raised using bucket and pulley devices. Cased wells were constructed by driving hollow bamboo rods into shallow saturated soils.

Hand-dug wells, probably the oldest method of developing ground water, are still in general use throughout the world. Production is limited as saturated materials can be penetrated only a few feet. Water is extracted from these shallow wells, using human, animal, or wind power.

The earliest methods of ground water development changed little until the advent of the industrial age. Not until the late 1800s, with development of mechanized techniques for drilling and bringing the water to the surface, could deeper, more productive wells be constructed.

Despite the early extensive use of ground water, little was understood of its origin. As a result of incomplete observation, Aristotle, and other ancient Greek and Roman natural philosophers, concluded that infiltration from pre-cipitation was insufficient to produce observed spring flow. They theorized that ground water was produced in the "roots of mountains," either by transformation of an assumed sub-terranean atmosphere or by conversion of seawater into freshwater in submarine canyons. This freshwater then rose to the surface discharging as springs.

Marcus Vitruvius (17 B.C.) introduced the present-day concept of ground water originating as precipitation and returning to earth in a continuous cycle. This theory gained little acceptance until propounded independently by Leonardo da Vinci and Bernard Palissy in the sixteenth century.

With the understanding of the "hydrologic cycle," the science of hydrology was born. Basic principles of artesian pressure and hydrostatics were formulated in the seventeenth and eighteenth centuries by Johann Becher, Bernardino Ramazzini, Antonio Vallisnieri, and others. Pierre Perrault, Edme Mariott, and Edmund Halley pioneered quantitative hydrology, measuring and correlating river flow, precipitation, and evaporation.

Development of theories of ground water storage and movement had to await the formulation of basic geological precepts. William Smith (1769–1839) was among the first to recognize the ground water storage potential of sedimentary strata.

The year 1856 marks the birth of ground water hydrology as a quantitative science. Henri Darcy, a French hydraulic engineer with the city of Dijon, published the results of a series of experiments on percolation of water through filter sands. Darcy showed that flow through sand is directly proportional to head difference and cross-sectional area. The constant of proportionality in Darcy's relation later became known as hydraulic conductivity, an important parameter in aquifer evaluation. The Darcy equation remains the basic expression describing flow through a confined aquifer. Jules Dupuit in 1863 developed a similar equation for ground water flow in a gravity drainage system (unconfined aquifers) and identified the curvilinear nature of flow near the well. Phillip Forcheimer, from 1886 through 1898, noted the analogy of ground water flow to heat flow and was the first to apply the LaPlace equation to solve problems in steady-state ground water flow.

Since the beginning of the twentieth century, the science of well hydraulics has been advanced by such investigators as Adolph and Gunther Thiem (father and son), Charles Theis, Hilton Cooper, C. E. Jacob, and Mahdi Hantush. They developed methods of determining aquifer characteristics through use of observations from pumping wells. Their methods are discussed in Chapters 5 and 10. Major contributions to ground water flow hydraulics were made by Oscar Meinzer on compressibility and elasticity of artesian aquifers, C. E. Jacob, who developed the general flow equation from fundamental principles, M. King Hubbert, who clarified the concept of hydraulic potential, and Mahdi Hantush, who, along with C. E. Jacob, developed the theory of leaky aquifers and made major contributions to well hydraulics.

Modern ground water investigators utilize computers to simulate ground water flow and contaminant migration so that ground water resources may be safely exploited. R. Allen Freeze, John Bredehoeft, Shlomo Neuman, Paul Witherspoon, Stavros Papadopoulos, and George Pinder are a few of the those who have contributed to further understanding of the ground water environment.

1.2   THE HYDROLOGIC CYCLE

Seventy percent of the surface of the earth is mantled in water. Over 90% of the world supply is found in the ocean, the source of virtually all water found upon or within the continental land masses. The processes that distribute the world's water supply are described collectively as the hydrologic cycle, shown in Figure 1.1.

At the ocean's surface, evaporation (powered by the energy of solar radiation) converts liquid water into water vapor. This vapor, which contains small amounts of dissolved material, is carried by air masses moving across the ocean. Approximately 3.6 million tons of water are evaporated annually from an average square mile of the ocean surface (2).

Moisture-laden air masses rise as they are heated by the sun's radiation. At higher elevations they expand and cool due to the lower pressure in the upper atmosphere. Minute particulate matter, such as combustion products, salt crystals, dust, and clay particles carried aloft by rising air currents, provide surfaces upon which water vapor is converted to liquid water or ice. These particles are referred to as condensation or freezing nuclei. Through this process, clouds composed of water droplets or ice crystals form. When individual droplets gain sufficient mass they accelerate downward by the force of gravity and fall as precipitation.

Interception and Infiltration

Some precipitation approaching the ground is intercepted and held on the surfaces of foliage, buildings, and so forth, by surface tension forces. After these surfaces have become saturated, interception occurs only when water is removed by evaporation or dripping.

Soil absorbs water because it is porous. The maximum rate at which a particular soil can absorb water is known as its "infiltration capacity." Infiltration capacity decreases as pore space become filled with water and ceases when the soil is fully saturated. Infiltration continues as water is removed from the pores by evaporation, lateral movement (interflow), or downward percolation.

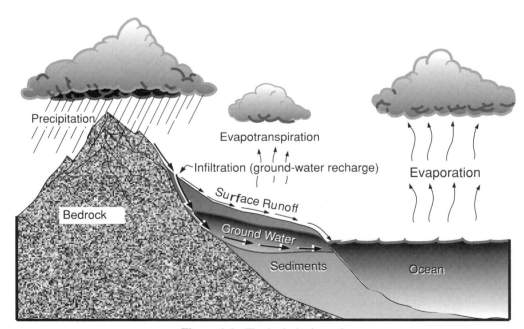

Figure 1.1. The hydrologic cycle.

Runoff, Depression Storage, and Interflow

Water that does not infiltrate into the soil or become stored in depressions flows down-gradient on the surface, initially moving for short distances in no particular channel as "sheet" or "overland flow." However, because flowing water possesses energy, it moves soil particles to form temporary rivulets. Water flowing in these rivulets reaches larger channels that eventually carry the runoff to a large water body (ocean or lake).

Some infiltrated water travels laterally through the shallow soil horizon, discharging where it intercepts a channel. This "interflow" delays the runoff due to the water traveling in longer flow paths around soil grains. Interflow will continue to discharge into channels for many hours after surface runoff has ceased (Figure 1.2). Interflow is distinguished from "base flow," which is the discharge from an aquifer to an adjoining channel with a lower water level.

Evapotranspiration

Depending upon climatic conditions such as temperature, humidity, and wind speed, most precipitation evaporates during and shortly after it occurs. Most of the remaining water percolates to the vegetation root zone and is absorbed through osmosis and discharged or transpired as water vapor through the plant leaves. The processes of evaporation and transpiration are known as "evapotranspiration." Evapotranspiration is also referred to as "consumptive use," because the water returns to the atmosphere as water vapor and is unavailable for further use at the earth's surface.

Recharge and Discharge

Typically, depending upon many factors, only a very small percentage of the infiltrated water percolates deeper. It ranges from possibly 10% to 20% in coarse alluvial deposits to very little in clayey soils. This percolating water moves

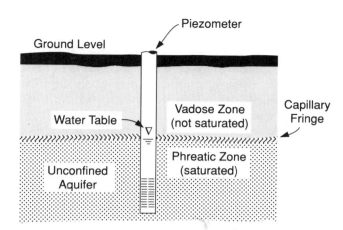

Figure 1.3. Piezometer installed to monitor water table elevation.

downward under the influence of gravity through a region defined as the "vadose zone." In this zone the pores between grains are only partially filled with water, some of which is bound to grains by surface tension and molecular forces.

Unbound water continues downward until it reaches the lower boundary of the vadose zone, known as the "capillary fringe." Here pore spaces are completely filled with water. The thickness of the capillary fringe varies from a few inches to several tens of feet, depending upon the nature of materials forming the zone. Coarse-grained deposits, which have large pore spaces and a low ratio of surface area to volume, have little or no capillary fringe. Materials composed primarily of fine particles have a large surface area to volume ratio and may have capillary fringes of 50 ft or more.

Water molecules are bound to grain surfaces in both the vadose zone and capillary fringe by a process known as surface retention. Here, water within the pore spaces of the capillary fringe is under tension at pressures that are less than atmospheric, and this zone is referred to as the "tension saturated zone" (3). Water held by molecular and surface tension forces does not drain by gravity and is said to be in "dead storage."

If not held in dead storage, the percolating water will eventually reach the zone of saturation, or the "phreatic zone." Water reaching the saturated zone constitutes ground water recharge. Ground water recharge may be natural or artificial, and it replaces ground water removed from storage by pumping or natural discharge such as springs.

A piezometer[1] installed into the saturated zone shows water standing at a level known as the "water table," or phreatic surface (Figure 1.3). The water table is the upper boundary of an unconfined aquifer.

Hydraulic head (potential) consists of both hydrostatic and gravitational potential. Ground water responds to decreasing head (hydraulic gradient) by moving slowly through

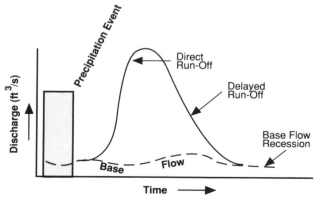

Figure 1.2. Stream hydrograph response to a precipitation event

---

1. A small-diameter pipe open or perforated at the bottom.

porous or fractured geologic materials. If not removed by pumping, the ground water may discharge as springs or seeps into channels, eventually reaching a body of surface water. Occasionally, ground water may come to rest in an area with no hydraulic gradient. It will remain there until either tectonic forces modify the geologic structure or it is removed by wells.

A system of branching streams collects drainage and carries it out of a basin. Such streams may flow perennially or intermittently. During and shortly after rainfall events, stream flow is composed primarily of surface runoff. During high flow some stream water infiltrates into stream banks, recharging the ground water. Such a stream is referred to as a "losing stream," and the process is known as "bank storage." After rainfall ends, stream stage diminishes and bank storage is released into the stream channel. The stream is then known as a "gaining stream."

Discharge of ground water into channels occurs whenever the head potential in the aquifer is higher than the elevation of the water surface in the channel, and the discharge may continue uniformly through the year. As shown in Figure 1.2, it constitutes the "base flow" of a stream. This is the component that the aquifer continues to contribute to stream flow after direct and delayed runoffs have ceased. At the low stream stage, the major component of flow is ground water discharge. Water flowing from an aquifer into a stream is known as "effluent seepage," and when flowing from the stream to an aquifer as "influent seepage."

1.3 GROUND WATER IN STORAGE

In the uppermost portion of the earth's crust, large volumes of porous and fractured materials store ground water. Although only a small fraction of the precipitation from any single storm may reach the saturated zone, the process has continued through geologic time, resulting in large reserves of subsurface water. Ground water storage represents the largest volume of freshwater available for consumption, as seen in the distribution of the earth's water resources (Table 1.1).

TABLE 1.1 Estimate of World Water Inventory

	Volume ($km^3 \times 10^6$)	Volume (%)
Oceans and seas	1370	93.77
Lakes, reservoirs, swamps, river channels	0.18	0.01
Soil moisture	0.07	0.01
Ground water	60	4
Icecaps and glaciers	30	2
Atmospheric water	0.01	<0.01
Biospheric water	0.01	<0.01

Source: Nace (4).

1.4 CONNATE WATER

Although virtually all ground water has been derived from geologically recent precipitation, a small amount may be the result of ancient events. Water trapped in sedimentary strata as they were formed is occasionally encountered. This water, known as "connate water," is normally highly mineralized because of the conditions during its deposition and its long residence in the formation. Connate water may discharge along fractures or faults, due to recent earth movement, or be extracted by wells penetrating these formations.

1.5 AQUIFERS

"Aquifers" are a formation or group of saturated geologic formations capable of storing and yielding freshwater in usable quantities. They are the target for all ground water exploration and development programs. The term aquifer can be ambiguous. In addition to its geological definition, it has an economic connotation. The same formation may be considered an aquifer in one location but not at another where an alternate formation can produce ground water at lower cost.

Saturated formations that store and yield little ground water are known as "aquicludes." This term has been largely supplanted by the term "aquitard." Some geohydrologists restrict the use of "aquitard" to formations whose yield capacities are significantly less than adjacent formations. Although an aquitard may not yield water economically, it can hold appreciable amounts of water.

Aquifer Characteristics

Good aquifers are characterized by relatively large values of storage, yield, and replenishment. The major determinant of aquifer storage is the amount of usable void space, or "effective porosity"; the major determinant of yield is the rate at which the formation allows ground water movement. Replenishment, also known as recharge, depends upon a number of factors such as precipitation and soil characteristics.

1.6 POROSITY

Total Porosity

The storage available in an aquifer is related to the void space that it contains (total porosity). Total porosity as a percentage is expressed as

$$n = \left(\frac{V_v}{V_T}\right) 100 \ ,$$

TABLE 1.2   Soil Classification, Total Porosities, and Effective Porosities of Well-sorted, Unconsolidated Formations

Material	Diameter (mm)	Total Porosity (%)	Effective Porosity (%)
Gravel			
Coarse	64.0–16.0	28	23
Medium	16.0–8.0	32	24
Fine	8.0–2.0	34	25
Sand			
Coarse	2.0–0.5	39	27
Medium	0.5–0.25	39	28
Fine	0.25–0.162	43	23
Silt	0.062–0.004	46	8
Clay	<0.004	42	3

Source: Morris and Johnson (5).

where

n = total porosity [%],
V_v = volume of void space in sample,
V_T = total volume of sample.

Unless individual voids are interconnected, they do not allow transmission of water. Some pyroclastic rocks such as pumice are an example of nonconnected pore spaces.

Effective Porosity

The portion of total interconnected pore space through which flow of water takes place is called the "effective porosity." Effective porosity is always less than total porosity, since not all interconnected pore space in a saturated material is available for flow. A portion of the voids is filled with water held in place against gravity by molecular and surface tension forces. The percentage of water that drains by gravity from a unit volume of material is the "drainable or active porosity," sometimes known as specific yield. Total porosity is therefore the sum of specific yield and surface retention. Table 1.2 lists values for total porosity and effective porosity for various materials.

Surface Retention. Individual water molecules are formed by an oxygen ion bonded between two hydrogen ions with a bond angle of about 105°. This imparts an unbalanced electrical charge to the molecule so that it is polarized with the oxygen ions electrically negative relative to the hydrogen ions. Porous media grains also bear unbalanced electrical charges on their surfaces. Attractive electrical forces, operating over very short distances strongly bind a layer of ground water a few molecules thick to grain surfaces. Because of this process of "surface retention," ground water firmly bound to grain surfaces cannot be removed by gravity. Surface retention is responsible for creating the capillary fringe.

The ratio of surface area to mass increases as particle size decreases. For a material of uniform specific gravity, mass and volume vary directly. The term "specific surface" is the ratio of surface area to volume of uniform spheres. Representative values of specific surface for porous media components are shown in Table 1.3.

Porosity in Aquifers

In aquifers there are three major types of porosity:

1. Spaces between grains in unconsolidated to moderately consolidated media.
2. Fractures in dense rocks.
3. Through-going tubes.

Porous media formations result from the deposition of individual grains eroded from rocks and transported by wind, water, or gravity. The void spaces are formed as the formation is deposited ("primary porosity").

Long after the rock is formed, fractures in rocks result from processes such as structural deformation and erosion ("secondary porosity").

Through-going tubes may represent primary porosity, such as lava tubes formed when a thread of molten lava is extruded from a solidified outer chamber. Tubes may also represent secondary porosity; for example, cavities formed

TABLE 1.3   Relationship of Surface Area per Unit Volume to Particle Size

Particle Diameter (mm)	Classification	Number of Particles	Surface Area per Volume (cm²)	Total Surface Area per cm³ (mm²)
10.	Medium gravel	1	3.14	3.14
1	Coarse sand	10^2	3.14×10^{-2}	31.4
0.1	Fine sand	10^6	3.14×10^{-4}	3.14×10^2
0.01	Silt	10^9	3.14×10^{-6}	3.14×10^3
0.001	Clay	10^{12}	3.14×10^{-8}	3.14×10^4

Sources: Todd (1) and Morris and Johnson (5).

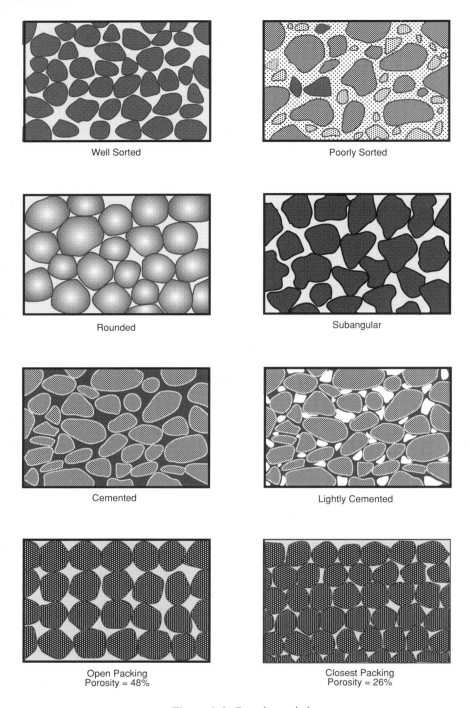

Well Sorted

Poorly Sorted

Rounded

Subangular

Cemented

Lightly Cemented

Open Packing
Porosity = 48%

Closest Packing
Porosity = 26%

Figure 1.4. Porosity variations.

in limestone from the solution of carbonates by ground water.

Porous Media. The factors controlling the amount of void space in a porous medium vary according to its type. In unconsolidated formations the primary determinants are shape, sorting, and packing of the grains (Figure 1.4). In

a consolidated (or semiconsolidated) formation of porous media, an additional factor is the extent to which pore spaces have become filled with cementing material. Formations of geologically older[2] porous media are usually cemented to

2. Older than Pleistocene or late Pliocene, or about a million to a few million years.

some degree. The cementing material may be calcium carbonate, silica, iron oxide, or sometimes clay. Since cementing greatly reduces porosity, older formations usually provide less ground water storage than younger ones.

Grain shape is measured in terms of sphericity and roundness. Sphericity is a three-dimensional aspect and describes how closely a particle approaches the shape of a sphere. Roundness is a two-dimensional aspect and relates to the presence or absence of angular corners. A cube has high sphericity and low roundness; a coin has low sphericity and high roundness.

Quartz and feldspar grains, which predominate in sand-size porous media, have high sphericity. Their roundness depends largely upon the distance they were transported prior to deposition; the longer the distance, the more corner abrasion and the higher the roundness. Clay particles, which predominate in shales, are platelike and have low sphericity.

"Grain packing" is the term describing the arrangement of individual grains. Spheres have the lowest ratio of surface to volume, whereas a circle has the lowest ratio of circumference to area. The most open type of packing of uniform spheres will result in a porosity of 48%, whereas the closest packing (rhombohedral) exhibits a porosity of 26%. Therefore, depending on packing, an aquifer composed of uniform-sized, spherical, well-rounded grains will have higher porosity than one composed of angular or flat particles.

Geologists refer to a porous medium whose grains are uniform in size as "well sorted." They refer to a porous medium exhibiting a wide range of grain sizes as "poorly sorted." If grain size varies, the smaller grains will fit into the pore spaces between larger grains, significantly reducing porosity. Paradoxically, in engineering practice a sample containing a wide range of grain sizes is termed "well graded."

Very fine-grained porous media may exhibit high porosity. However, because pore spaces are very tiny, specific surface is large relative to volume. Surface retention and frictional forces bind virtually all pore water, and ground water cannot move freely through the formation.

Clay particles are the smallest grains found in porous media formations. By definition, individual grains are less than 0.00015 in. in diameter. Clays have high porosity but, because of their minute size, very low permeability. The ratio of surface area to volume is very high, and pore spaces between grains are small. Dry clay will quickly absorb water (by capillary action) that binds to the grain surfaces, and the clay-water system becomes impermeable. Clays, which have a total porosity as high as 60% to 70%, surprisingly represent the most effective geological aquitards.

Fractured Rocks. Fracture and tubal porosities are characteristic of rocks sufficiently hard to maintain these openings. With few exceptions fractures occur long after the rock has been formed. In fractured media the determinants of porosity are the number and width of fractures per unit

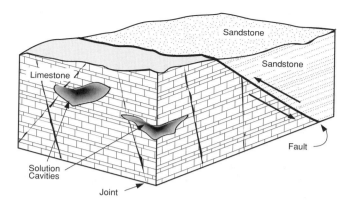

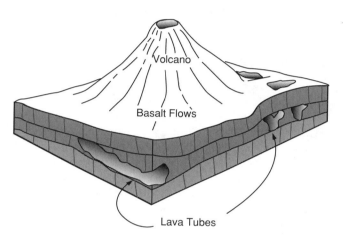

Figure 1.5. Fractures and tubes in rocks.

volume of rock, the degree to which they are interconnected, and the extent to which they may be filled with eroded or injected materials (Figure 1.5).

Fracture and tubal porosities tend to decline rapidly with increasing depth. As the pressure of the overburden increases, fractures close and tubes collapse. Nevertheless, in competent rock of low primary porosity, fractures may operate as conduits for ground water movement at great depth.

Permeability

Permeability is an intrinsic geometric property of the formation and is related to the ease with which a fluid can move through it. It is defined by the size and connections of the voids in the porous medium. For sand and silts, it is proportional to the square of material grain size.

1.7 HETEROGENEITY AND ANISOTROPY

An aquifer usually varies spatially in grain size and porosity due to different modes of deposition and stresses imparted after formation. Permeability therefore may differ significantly within an aquifer. Such a condition is referred to as "heterogeneity." If an aquifer shows little spatial variation in permeability, it is "homogeneous."

The term "isotropy" defines a condition in which ground water is able to move in any direction within the aquifer with equal ease. If the water may move with greater ease in one direction than another, the aquifer is "anisotropic." Subangular or subrounded grains are often deposited in a preferred orientation, having been transported by a moving fluid. This will cause horizontal anisotropy. A porous formation may be deposited over geologic time under varying conditions, resulting in a layered formation with permeability differing from layer to layer. Even if each layer is isotropic, the formation as a whole will be anisotropic. Some degree of heterogeneity and anisotropy characterize all aquifers.

Compressibility and Elasticity

All aquifers are elastic. Under an imposed stress of sufficient magnitude, they compress. If the stress is removed, they tend to return to their original volume. The modulus of elasticity expresses the elasticity of a material as the ratio of change in stress to change in strain. The ratio of the change in volume to change in stress is compressibility, the inverse of the modulus of elasticity.

A porous medium skeleton may compress through two mechanisms: the individual grains may decrease in volume, or the grains may be rearranged to a closer packing. Since individual grains are relatively incompressible, the result of an imposed stress is usually closer packing accompanied by a decrease of porosity. In a fractured rock aquifer, porosity decreases under stress through closure of fractures. In tubal aquifers, an imposed stress may cause collapse.

In many aquifers the response to a decrease in stress is not equal to the response to increase in stress. If an aquifer has compressed, it may not return to its original volume when stress is removed. For clays, the ratio of expansibility to compressibility is on the order of 1:10 (1). Extensive removal of ground water may lead to subsidence, formation of sinkholes, and structural damage to wells and surface installations.

1.8 FRACTURES: JOINTS AND FAULTS

All hard rocks contain fractures caused by tectonic or erosional processes after the rocks were formed. Joints are the most common fractures. They are simply cracks caused by the parting of rock masses. The only motion is perpendicular to the fracture plane. Faults are fractures along which the rocks have moved parallel to the fracture plane. Joints vary in size from submicroscopic to hundreds of feet in length; faults may range up to hundreds of miles.

The capacity of joints and faults to store and transmit water is highly variable and depends upon the type of stress, nature of the rock, depth of burial, and other factors. Rocks in which interconnected joints are only a short distance apart make reasonably good aquifers. Lava flows, in which joints are formed by extensive shattering of solidified lava from movement of underlying molten lava, may make excellent aquifers. This is especially true of recent lavas in which the joints have not yet collapsed or been filled with weathered material.

Joints are usually not random directionally but tend to occur in preferred orientations. This imparts anisotropy. They are more open on or close to the surface but become tightly closed at depth by the overburden pressure. Surface or near-surface open joints may become filled with products of weathering, restricting the movement of ground water. Consequently the interval at which jointed rock is effective in transmitting ground water generally occurs from a few tens of feet to a few thousand feet below the surface.

Faults, as shown in Figure 1.6, although similar in some respects to joint-type fractures, are less numerous and can extend for greater distances. Blocks of the earth's crust move laterally against each other, shattering rocks on each side of the fault. The wider the zone of shattering in consolidated formations, the more effectively the fault transmits ground water.

The intense grinding along a fault plane produces a fine-grained product called "gouge." This is often composed of clay-sized particles and is very impermeable. Therefore a fault is a permeable zone containing a central axis of impermeable gouge. The zone can vary in width from a few

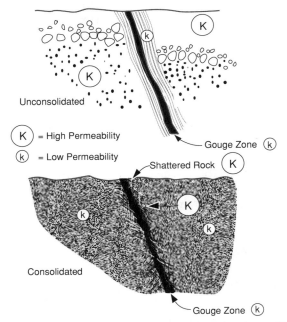

Figure 1.6. Effects of faulting in unconsolidated and consolidated formations.

TABLE 1.4 Representative Properties of Geologic Materials

Formation	Total Porosity (%)	Intrinsic Permeability (in.^{2})	Vertical Compressibility (ft^2/lb)
Permeable basalt	17	10^{-8}	
Fractured igneous and metamorphic rocks	5	10^{-9}	10^{-7}–10^{-9}
Limestones	30	10^{-9}	
Sandstones	35	10^{-9}	
Unfractured igneous and metamorphic rocks	0–10	10^{-10}	10^{-8}–10^{-10}
Shale	6	10^{-13}	
Unweathered clay	42	10^{-12}	10^{-5}–10^{-7}
Glacial till	6–34	10^{-9}	
Silt, loess	49	10^{-8}	10^{-6}–10^{-8}
Dune sand	45	10^{-8}	10^{-6}–10^{-8}
Clean sand	39–43	10^{-8}	10^{-6}–10^{-8}
Gravel	28–34	10^{-7}	10^{-7}–10^{-9}

feet to a few thousand feet in the case of the great plate boundary faults.

Although faults may operate as conduits for ground water flow in hard rocks, in unconsolidated formations they may act as barriers. Springs, oases, and marshes may occur on the up-gradient side of a fault that operates as a barrier. Although unconsolidated deposits cannot hold open fractures, they may be displaced on opposite sides of a fault, abutting a permeable formation against an impervious one.

1.9 DESCRIPTIVE PROPERTIES OF AQUIFERS

An aquifer as a water-bearing formation can be described by three basic physical properties: porosity, permeability,

and compressibility. Table 1.4 lists representative values for each property for common formations. It must be remembered that properties vary considerably within any given formation. However, the table is useful in comparing values among formations.

Aquifer Conditions

There are two major types of aquifers: confined and unconfined (Figure 1.7). Unconfined aquifers receive recharge directly from the overlying surface through percolation from precipitation or surface water. They usually exhibit a shallow water level. At the water table, water is at atmospheric pressure. Potential aquifer material extends to the ground surface, and the water table rises or falls in response to

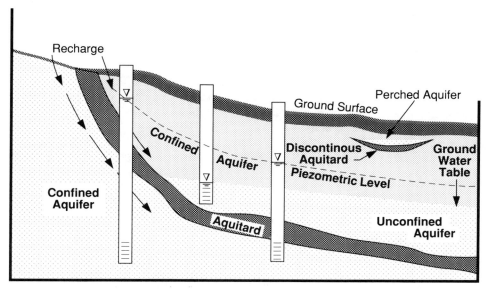

Figure 1.7. Types of aquifers.

changing patterns of precipitation, recharge, and discharge. When a well in an unconfined aquifer is pumped, the water level is lowered and gravity causes water to flow to the well. Formation materials in proximity to the well above the pumping water level are dewatered.

A confined aquifer is bounded both above and below by aquitards. It does not receive significant amounts of percolation from the overlying surface. The ground water within a confined aquifer is under a pressure that includes the sum of the weight of the atmosphere and the overburden. When a well penetrates a confined aquifer, the water level rises above the upper boundary of the aquifer. Such wells are artesian. In some instances the water may rise to the surface and flow (flowing artesian well).

Most confined aquifers are unconfined at their exposed edges and are recharged from percolation into this unconfined portion. Confined aquifers may also receive recharge through their bounding aquitards (semiconfined), especially when pressure changes are induced through pumping or injection. Recharge to an aquifer from or through an aquitard is termed "leakage." An aquifer system in which there is ground water communication through an aquitard is known as a "leaky aquifer system."

Because the water level in an artesian well stands above the upper boundary of the aquifer, pumping does not dewater the aquifer (unless the head is drawn below the top of the aquifer) but will reduce its pressure. This reduction increases the overburden load on the aquifer skeleton, resulting in compression of the aquifer.

A "perched aquifer" results when a lower permeability layer impedes the downward movement of water above it. Perched aquifers may occur above a confined or unconfined aquifer. They are usually very limited areally and may not receive sufficient recharge to support significant well production.

1.10 APPLICATION TO GEOLOGIC FORMATIONS

A formation can be assessed for its value as an aquifer by considering its storage, transmitting, and recharge potentials. Knowledge of the aquifer characteristics of porous media and other formations is of great importance in the study of ground water and its exploitation.

Porous Medium Aquifers

Porous medium aquifers are the most important sources of ground water due to their availability, accessibility, and productivity. Since most wells are drilled in porous media, their aquifer properties are best understood. The classic theories of ground water hydrology have been developed from observations and experimentation on porous media.

Weathering, tectonic, and other processes continually loosen rock. Fragments not carried aloft by winds are moved down-gradient by water or ice or occasionally gravitational force. Sediment deposited on land by running water is called "alluvium"; that deposited by wind is "aeolian."

Alluvium is by far the most common of the nonmarine porous media. Where channels are steep, flowing water has great energy and carrying capacity. Rock fragments collide with each other and are tumbled along the channel bottom, becoming more rounded and spherical through abrasion. Deposition of increasingly finer materials occurs as the gradient diminishes.

Stream velocity depends directly on flow volume. Variation of stage causes deposition of a succession of layers, each of which is well sorted, but with great variation of grain size from layer to layer. Individual layers range in thickness from a few inches to many feet.

Buried river sands and gravels are very prolific porous media aquifers. The layered deposits do not extend very far laterally from the original course of the river that deposited them. The extent of river deposits beyond the channel is the flood plain, deposited by the river at a stage high enough to overflow its banks. Flood plain materials are coarsest in the channel, becoming increasingly finer with distance from it. However, rivers that flow across softer materials tend to change course frequently because they erode their banks, resulting in more extensive channel and flood plain deposits. Because of the meandering and changing nature of the river channel through geologic time, it is often not possible to correlate formation samples among boreholes in alluvium, even when close together.

Buried river deposits have the shapes of valleys, long in the up-and-down stream direction and narrow across. They are trough-shaped at the bottom, relatively flat at the surface and pinched out at the edges. Coarse gravels are often found as the basal deposit.

It is common to find fine-grained layers (silts and clays) interbedded with coarser deposits, especially in the downstream area where the fine-grained materials originated as swamp, pond, or delta deposits. If they are extensive laterally, they may function as aquitards, confining some aquifers. River-deposited sediments therefore generally exhibit heterogeneity and anisotropy.

Aquifers deposited by a constant flowing river are in contact with and receive recharge from it. Although storage volume may be limited laterally, it is continuously replenished. River-deposited aquifers in arid regions often are not in direct contact with a constant flow, and recharge opportunities may be limited.

Alluvial Fan Sands and Gravels

In arid and semiarid regions, eroded rock is washed from the mountains during infrequent but intense precipitation events. The coarser materials are deposited against the mountains, with the finer grained extending outward. As erosion occurs within the range and deposition occurs beyond the range front, the gradient lessens. Succeeding flows swing back and forth creating a fan-shaped deposit below the range front, with the handle extending up into a canyon.

Streams that deposit alluvial fans have short total reaches and extremely variable flow. They are dry except during and after storms. As the fans build up they cover the retreating range front. Layers closer to the mountain front consequently can be expected to become coarser with increasing depth. During major storms slurries of clay, silt, and water may transport large rocks, many feet in diameter. Termed "debris flows," these are poorly sorted, tight, and may function as aquitards.

The coarse-grained upper edges of the fan are in an excellent position to receive orographically controlled precipitation, while the fine-grained deposits of clays and precipitated salts at the lower edge retard ground water outward flow. The ground water is usually unconfined near the mountain front and confined down-gradient. As a result alluvial fans operate as vast and efficient storage reservoirs in arid regions.

Alluvial fan deposits are heterogeneous, anisotropic, and composed of subrounded to subangular grains of a wide range of sizes. Wells may penetrate a number of aquifers separated by aquitards. Correlation of static water levels and lithologies from well to well may be difficult because of variation of deposition over short distances. It may be difficult to determine when a borehole has entered bedrock because large boulders may be buried within the fan.

Dune Sands

Dune sands are aeolian deposits that form in arid regions where topographic conditions cause a rapid decrease in the velocity of sediment-carrying winds. They consist of well-sorted, well-rounded, spherical sand grains and have high porosity and permeability. They are, however, limited in areal extent. Their high permeability prevents significant water storage since, as surface deposits, they quickly drain.

1.11 GLACIAL DEPOSITS

Glacial deposits formed during the Pleistocene Epoch are common in northern temperate regions throughout a wide range of elevations and at moderate-to-high elevations in the subtropics and tropics. These relatively young deposits are still unconsolidated, and some of them function as porous media aquifers. Ancient pre-Pleistocene glacial deposits are consolidated. If they provide water, they do so as bedrock-type aquifers.

Glacial deposits include all material deposited directly by ice or by meltwater streams derived from ice. The term "glacial drift" refers to all types of glacial deposits, regardless of manner of deposition. The materials deposited directly by ice are called "glacial till." Ridgelike or sheetlike deposits of glacial till are termed "moraines." Sediments deposited by glacial meltwater are referred to as "glacial outwash," "glaciofluvial deposits," or similar terms.

Glacial drift is widespread through large regions of North America and Europe. They were deposited by continental ice sheets centered over southeastern Canada and central Scandinavia. From these centers the ice extended radially, reaching southern New England and the central plains states in North America, and covering England and the northern European countries. As these continental glaciers expanded and contracted several times during the Pleistocene Epoch, they deposited vast sheets of rock material. Concurrently, alpine or valley glaciers formed in mountainous regions, following and deepening river valleys already in existence.

Glacial drift may consist of alternating layers of till and outwash, forming a series of aquitards and aquifers. Where several ages of glaciation have occurred, the intervals between episodes may have been sufficient to allow deep weathering of the till. The result is "gumbotil," a thick clay-rich layer of low permeability.

Layered silts and clays of low permeability are often found in glaciated areas. These deposits were created in the abundant lakes formed where water dammed against moraines or in hollows gouged out by ice. Glacial lake deposits are identical to those formed in nonglacial lakes.

Till

Moraines are till deposits that occur laterally along the valley-wall sides of valley glaciers. Terminal moraines were formed at the snouts of glaciers during times of little advance or retreat. In moraines where fine materials have not formed a significant part of the original deposit, or have been washed out by meltwater, local aquifers are created. In general, however, glacial till forms aquitards. The till is less permeable than outwash because it is poorly sorted and rich in silt and clay formed by the grinding of ice-borne sediment against bedrock. Examples are the ground moraines formed as discontinuous sheets of till left in the wake of retreating glaciers or plastered on underlying formations during ice advances.

Outwash

Outwash deposits may form good aquifers. Since they were deposited by running water, they have the hydraulic characteristics of stream sediments. Coarse texture is common because the meltwater from glaciers is abundant, and the climate that supports glacial development is usually wetter than in nonglacial periods. Because deposition is by running water, outwash deposits are well sorted and stratified.

The outwash deposits that spread from the edges of continental glaciers tend to be broad; those that were derived from valley glaciers are long and narrow, resembling river deposits. Outwash deposits are usually coarser grained than stream deposits because glaciers provide large volumes of water continuously.

"Eskers" are a special type of outwash formed by the filling of a crevasse or tunnel on or within the ice. Because they were deposited by high-velocity meltwater, they are generally very coarse grained, well sorted, and highly permeable. Eskers are shaped like railroad embankments and are distinctly linear. Most are several tens of feet wide, sinuous, and can extend for miles. They may not store significant amounts of water because of their high permeability, but if encased within less permeable deposits, they form excellent local aquifers.

"Loess" deposits are blankets of wind-blown silt associated with continental glaciation. They were formed outside the glaciated areas or in areas recently uncovered by retreating glaciers. Strong winds, abundant fine materials, and lack of vegetation to stabilize these materials have promoted the deposition of loess in vast regions. The deposits may be 50 to 100 feet thick, and they tend to stand in near-vertical bluffs when eroded. Although not good aquifers because of their fine-grained nature, they have formed rich soils throughout much of the American midwest and elsewhere.

1.12   BEDROCK AQUIFERS

Because most rock contains some water, the recognition and designation of bedrock as an aquifer is as much a local economic as a geohydrological decision. The term bedrock is relative. It usually refers to hard formations such as granite. In areas where hard formations do not occur, the term may be applied to weakly consolidated, early Pleistocene sediments when they are overlain by late Pleistocene or Holocene unconsolidated formations.

The competence or degree of consolidation of the bedrock controls the manner in which it stores and transmits water. Weakly consolidated rocks cannot support open fractures. Their value as an aquifer depends upon how closely they resemble porous media. As the degree of consolidation increases, permeability decreases. Fractured and cavernous rocks usually form aquifers only at relatively shallow depths, perhaps less than 2000 ft because, with increasing overburden pressure, fractures close and tunnels collapse.

The processes describing ground water flow in fractured rock have not been well studied until recently. It is usually assumed that if fractures are abundant, open, and interconnected, analytic methods developed for porous media can be applied. The assumption is also made that the aquifer is highly heterogeneous and anisotropic.

Sandstone, Siltstone, and Shale

Sandstone, siltstone, and shale are formed by the cementation of unconsolidated sands, silts, and clays. These porous media deposits have been compressed through burial and impregnated with cementing material.

Semiconsolidated sandstones may have sufficient primary porosity to transmit water as a porous media and may also be sufficiently competent to support open fractures. Sandstone aquifers are widespread, persist for long distances, and are the principle aquifers for many regions from the Colorado Plateau to Appalachia. Siltstone and shales rarely form good aquifers, even when extensively fractured.

Because deposits may grade into other deposits, aquifers may change from confined to unconfined over very short distances and stratigraphy may be complex. Siltstone and shales will function as aquitards in a sequence of lithified sedimentary deposits. Folding of lithified deposits may cause extensive fracturing along the axis of the folds, whereas faulting may offset beds, causing barriers to, or conduits for, ground water flow.

Carbonates, Limestone, and Dolomite

Much limestone and dolomite are chemically and biologically precipitated sedimentary rock formed in marine environments. Both are soluble in slightly acidic water. Limestone is composed primarily of calcium carbonate and is slightly more soluble than dolomite, a mixture of calcium and magnesium carbonate. The term "carbonate" is used for both types of formation. If a limestone incorporates considerable clastic material, it may have some primary porosity. If crystalline (composed of interlocking grains), little primary porosity is likely. Secondary porosity may result from the solution of calcite (calcium carbonate) along bedding planes. Some of these solution planes may enlarge over time to extensive caverns. Where thick limestone deposits prevail, such tunnels may be numerous, interconnected, and extend for several miles. If ground water withdrawals exceed recharge, tunnels may collapse, forming deep sinkholes.

Groundwater flow in limestone tunnels resembles open-channel flow. The aquifer may be confined, with discharge to the surface via springs issuing from solution cavities. Recharge to the system is rapid, so that potentiometric levels may rise and fall dramatically in response to climatic events. The aquifer may appear to be fairly homogeneous and isotropic over long distances if solution planes spread out in all directions.

Carbonates form major aquifers throughout the United States, especially in the southeast.

Plutonic and Metamorphic Rocks

Plutonic and some of the metamorphic rocks can support open fractures. Where fractures are numerous and interconnected, these rock types can function as aquifers. Plutonic rocks include granites, intermediate granitic types such as monzonite and granodiorite, and mafic and ultramafic types. Metamorphic rocks that function as aquifers are gneiss, massive quartzite, and some metavolcanics. Schist and slate are characteristically closely fractured, but these rocks are weak and usually the fractures have been closed. The weathered surface appearance of such rocks can be misleading because fractures that appear open at the surface are closed at shallow depth. Granite, which weathers easily in some climates, may exhibit a surficial zone in which fractures are filled with weathered detritus. At depth, fractures may be clean and open.

Because fractures in rocks, especially those caused by tectonic stresses, have preferred orientations, they are highly anisotropic. Stress-caused fractures may also vary in intensity and interconnectedness over short distances and exhibit heterogeneity. Recharge can be quite rapid if fractures are exposed at the surface. Granitic rocks have not been extensively utilized as aquifers but can contribute local water supplies in mountainous terrain.

1.13 VOLCANICS

Volcanic rocks range from extremely poor to some of the most productive aquifers known. Composition and age are the controlling factors. Rhyolitic and related lavas are the surface flow equivalents of the granitic intrusives. They may be extensively fractured but exhibit low permeability.

Basaltic lavas that form poor-to-moderate aquifers are geologically old or were chemically altered during deposition. In old basalts, fractures and lava tubes that might have formed during deposition have been closed by collapse and consolidation over time. Chemical alteration often occurs in marine basalts and andesites by the interaction of seawater during deposition. Fractures are filled with decomposition products and permeability is very low. Many marine lavas contain pillow structures, an important clue to underwater deposition.

It is important to distinguish pillow structure from cores and rinds developed in nonmarine lavas by spheroidal weathering. Nonmarine lavas that appear tight at the surface may contain abundant open fractures at shallow depth, below the zone of extensive weathering.

Young Quaternary basalts that still maintain open lava tubes and extensive, interconnected fractures formed during deposition are excellent aquifers. The *Aa* structure, which is best described as a deposit of rubble, occurs as congealed flow surfaces are broken up by the movement of liquid lava beneath, making it an excellent aquifer. Individual flows are tongue-shaped and are highly anisotropic and heterogeneous. Recharge into young lavas is very rapid.

Quaternary lavas form the principle aquifers of the Hawaiian Islands. Precipitation is orographically controlled, being in excess of 100 inches annually at higher elevations and 20 inches at sea level. Percolation through the volcanics is vertical until an ash bed is encountered. Ash beds formed of extruded exploded products form aquitards and perching structures for the percolating water. Below the ocean surface the less permeable lavas are saturated with seawater. The fresh percolating rainwater rests upon the saline water in a double convex lens known as the Ghyben-Herzberg lens.

In the Columbia Plateau of the northwestern United States, recent volcanics also provide extensive supplies of ground water.

REFERENCES

1. Todd, D. K. 1980. *Groundwater Hydrology.* J. Wiley and Sons, New York.
2. Chow, V. T. 1964. *Handbook of Applied Hydrology.* McGraw-Hill, New York.
3. Freeze, R. A., and J. A. Cherry. 1979. *Groundwater.* Prentice-Hall, Englewood Cliffs, NJ.
4. Nace, R. L., ed. 1971. "Scientific Framework of World Water Balance." UNESCO Technical Papers in Hydrology No. 7, 27 pp.
5. Morris, D. A., and A. I. Johnson. 1967. "Summary of Hydrologic and Physical Properties of Rock and Soil Materials." As analyzed by the Hydrologic Laboratories of the U.S. Geological Survey, 1948–60, Geological Survey Water-Supply Paper 1839-D.

READING LIST

Bennett, R. R. 1962. "Theory of Aquifer Tests." Flow net analysis by Ferris, J. G., et al. Geological Survey Water-Supply Paper, 1536-E.

Fetter, C. W., Jr. 1980. *Applied Geohydrology.* Charles E. Merrill Co., Columbus, OH.

Freeze, R. A., and P. Witherspoon. 1967. Theoretical analysis of regional groundwater flow, 2. Effect of water-table configuration and subsurface permeability variation. "Water Resources Research, No. 3," pp. 623–634.

Heath, R. C., and F. W. Turner. 1981. *Ground Water Hydrology.* Water Well Journal Publishing Co., Worthington, OH.

Hudson, A., and R. Nelson. 1982. *University Physics.* Harcourt Brace Jovanovich, New York, ch. 16.

Plummer, C. C., and D. McGeary. 1979. *Physical Geology.* Wm. C. Brown Co., Dubuque, IA, ch. 15.

Sun, R. J. 1986. "Regional Aquifer-System Analysis Program, Summary of Projects, 1978–84." U.S. Geological Survey, Circular 1002.

1984. *Drainage Manual.* U.S. Department of the Interior Bureau of Reclamation. U.S. Government Printing Office, Denver, CO.

1981. *Ground Water Manual.* U.S. Department of the Interior, Water and Power Resources Service. U.S. Government Printing Office, Denver.

1967. "Methods and Techniques of Ground-Water Investigation and Development." Water Resources Series No. 33. United Nations, New York.

1980. *National Handbook of Recommended Methods for Water-Data Acquisition*, ch. 2, Ground Water. U.S. Geological Survey.

1978. *National Handbook of Recommended Method for Water-Data Acquisition*, ch. 7, Basin Characteristics. U.S. Geological Survey.

CHAPTER 2

Movement of Ground Water

2.1 INTRODUCTION

Laws and principles governing ground water movement can be developed from the fundamental laws of physics. This chapter introduces these laws and applies them to ground water flow. Equations are developed that describe ground water flow in confined, unconfined, and semiconfined aquifers. At the conclusion of this chapter, several examples of common ground water flow problems are illustrated. In addition the equations produced are used to develop solutions for flow to wells (discussed in detail in Chapters 5 and 15).

2.2 FUNDAMENTAL LAWS AND PRINCIPLES

Laminar and Turbulent Flow

The direction and movement of ground water in porous media is determined by areas of recharge and discharge, and the system of interconnected passages comprising the microstructure of the porous matrix. The path followed by a sequence of water particles in steady flow is called a "streamline" and consists of a procession of similar particles, each overcoming frictional resistance to movement under a gentle but persistent hydraulic gradient.

Each streamline takes a slightly different path, diverging around individual grains and merging through narrow passages in the interstices. The rate of movement on each streamline also varies, speeding up when converging through constrictions and slowing down upon entering larger pore spaces. However, although the microscopic movement may appear highly varied, rotational, and tortuous, on a macroscopic scale the entire process when averaged over sets of adjacent streamlines seems perfectly orderly with the sets of streamlines moving in essentially parallel lines. Such flow is called "streamline or laminar flow" and is of fundamental importance to the understanding of ground water movement.

In turbulent flow, water particles move in irregular and rotational paths. Whereas laminar flow occurs at relatively low velocities, turbulent flow is characterized by high velocities. Flow in rivers and streams is generally turbulent, whereas ground water flow is generally laminar. Figure 2.1 illustrates both types of ground water flow.

Darcy's Law

Laminar flow relationships were first studied in capillary tubes by G. Hagen in 1839 and J. M. Poiseuille in 1846 (1, 2). They found that the rate of flow through these tubes varies directly as the hydraulic gradient. Later, in 1856, Henri Darcy verified this through experiments which led to his equation establishing the rate of flow through filter sands (3).

$$q = K \frac{(h_2 - h_1)}{L} \qquad (2.1)$$

where

q = the volume of water crossing a unit cross-sectional area per unit time $[LT^{-1}]$,

K = the factor of proportionality $[LT^{-1}]$,

h_2, h_1 = water heights above a reference level, measured by manometers terminated above and below the sand column $[L]$,

L = thickness of the sand (the length of the flow path) $[L]$.

The significance of this equation is that the passage of water through sands is proportional to the difference in pressure (head) and inversely proportional to the length of the flow path. This important relationship soon became known as Darcy's law and provided a fundamental basis for understanding movement of fluids through porous media. Figure 2.2 is a drawing of the original apparatus used by Darcy in his experiments.

Darcy's law is similar to Ohm's law, which describes the flow of electricity, and Fourier's law, which describes the flow of heat.

Reynolds' Number and the Range of Validity of Darcy's Law

Ground water flow is subject to both inertial and viscous forces. Its movement almost always occurs as laminar flow. However, turbulent flow may develop under special circumstances in the near-well zone as a result of high fluid velocities.

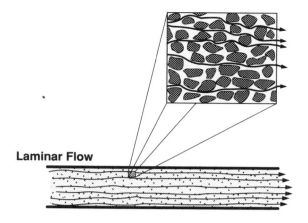

Laminar Flow

Turbulent Flow

Figure 2.1. Laminar and turbulent flow.

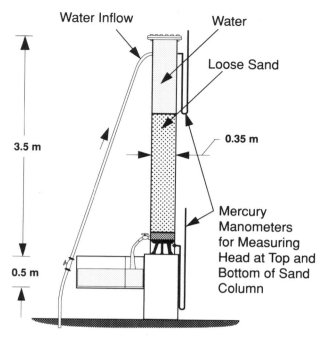

Figure 2.2. Original darcy apparatus (after Hubbert, 1956).

Osborne Reynolds in 1883 demonstrated the two modes of flow by a series of experiments involving various diameter glass tubes (4). The result of the experiments led to the definition of "critical velocity" (the velocity corresponding to the transition between laminar and turbulent flow) and to the establishment of a dimensionless ratio, known as the "Reynolds number," as an indicator of flow mode. This number may be expressed as

$$\text{Re} = \frac{\text{inertial forces}}{\text{viscous forces}} = \frac{vd}{\nu},$$

where

Re = the Reynolds number,
v = bulk (or Darcian flow) velocity $[LT^{-1}]$,
d = characteristic length (in porous media usually taken as the mean grain diameter)[1] $[L]$,
ν = kinematic viscosity $[L^2T^{-1}]$.

Reynolds number may be shown to be the ratio of the inertial force acting on the fluid per unit length of flow to the viscous force acting on the same fluid.

According to Hantush (5): In laminar flow through porous media, the inertial forces associated with even rapid changes

1. This diameter is such that if all the grains were of its size, the sand would transmit the same amount of fluid as it actually does. Hazen determined the mean grain diameter to be such that 10% of the natural sand (by weight) is of smaller grains and 90% is of larger grains (6).

in flow rates are much smaller than viscous forces and may be neglected in problems of practical interest. Thus ground water flow velocities are so small that viscous resistive forces predominate, with corresponding Reynolds numbers of 1 or less.

The transition from laminar to turbulent flow cannot be represented by any one value of the Reynolds number as defined above, due to variable factors such as porosity, grain shape, packing, and distribution. There is general agreement, however, that the laminar flow regime breaks down somewhere in the range of Reynolds' numbers of 1 to 10. When this occurs, Darcy's law is no longer applicable. Todd, however, summarizes experimental results of studies on flow through porous media that show ground water flow obeying Darcy's law for Reynolds numbers less than 1 and transitional flow for Reynolds numbers of 1 to 10 (7).

In practice, Darcy's law may be applied to flow conditions that exist when the Reynolds number is equal to or less than 10. Between 10 and 600–700 a state of partially turbulent flow is considered to exist, whereas above 600–700 fully turbulent flow is found (8). In Figure 2.3 the proportion of flow energy lost to friction is plotted as a function of Reynolds number for laminar and turbulent flow.

Bernoulli's Equation and Hydraulic Head

In 1738 two Swiss mathematicians and physicists, Daniel Bernoulli and his father John, developed a formula that has led to modern determinate hydraulics (9). This formula relates

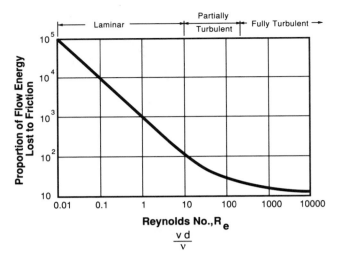

Figure 2.3. Flow regime versus reynolds number.

the velocity of running water with hydraulic head and is written

$$v = \sqrt{2gh} \, ,$$

where

v = flow velocity [LT^{-1}],
h = velocity head [L],
g = gravitational constant [LT^{-2}].

Daniel Bernoulli, making use of the principles of work and energy, further developed the relationship known as the "Bernoulli theorem" for steady flow of an incompressible fluid. This law expresses flow of water in conduits and states that if an incompressible fluid is in streamline flow, then the sum of the velocity head ($v^2/2g$), pressure head (p/γ), and elevation head (z) is equal to a constant.

This theorem is based on the concept that the total amount of kinetic and potential energy of a unit mass of fluid remains constant, or simply that the kinetic energy for unit mass gained by fluid falling a given distance is equal to the potential energy lost per unit mass. Algebraically, this is

$$\frac{v_1^2}{2g} + \frac{p_1}{\gamma} + z_1 = \frac{v_2^2}{2g} + \frac{p_2}{\gamma} + z_2 \, , \qquad (2.2)$$

where

v_1, v_2 = velocities at at two arbitrary flow sections [LT^{-1}],
g = gravitational constant [LT^{-2}],
γ = specific weight of water [FL^{-3}],
p_1, p_2 = fluid (hydraulic) pressure at sections 1 and 2 [FL^{-2}],

z_1, z_2 = gravitational potential (head) as measured above an arbitrary datum plane [L].

In ground water flow systems, velocities are usually so low that the velocity head ($v^2/2g$) can be neglected and Equation 2.2 reduces to

$$\frac{p_1}{\gamma} + z_1 = \frac{p_2}{\gamma} + z_2 \, . \qquad (2.3)$$

Equation 2.3 describes the total energy heads or fluid potential in a ground water flow system. This total head is called, the "potentiometric or hydraulic head" and is generally written

$$h(x, y, t) = \frac{p}{\gamma} + z \, , \qquad (2.4)$$

where

$h(x, y, t)$ = total potential (hydraulic head) of flow at any point (x, y) in the flow system at time (t) [L],
p/γ = hydrostatic pressure potential [L],
z = gravitational potential [L].

Figure 2.4 illustrates the concept of hydraulic head in ground water flow systems.

Hydraulic Conductivity

The parameter relating movement of fluid through porous media (the factor of proportionality in Darcy's flow equation, Equation 2.1), is known as "hydraulic conductivity." Hydraulic conductivity, or simply conductivity, may be expressed as

$$K = \frac{kg}{v} \, ,$$

where

K = hydraulic conductivity [LT^{-1}],
k = intrinsic permeability [L^2],
g = gravitational constant [LT^{-2}],
v = kinematic viscosity of the fluid [L^{2}T^{-1}].

The intrinsic permeability (k) is a property of the porous matrix only and is independent of the fluid moving through it. Intrinsic permeability is often written (7)

$$k = cd^2$$

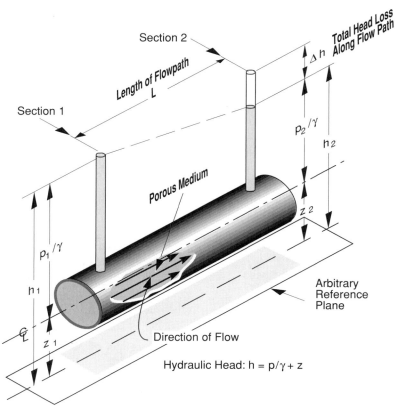

Figure 2.4. Illustration of hydraulic head in a ground water flow system.

where

k = intrinsic permeability [L^2],

c = a dimensionless constant controlled by various properties of the porous medium other than grain diameter (e.g., packing, distribution, and shape),

d = Mean grain diameter [L].

The hydraulic conductivity of different geologic materials varies widely, being greatest for materials with high effective porosity (e.g., sands and gravels) and lowest for silts and clayey materials. Table 2.1 shows relative hydraulic conductivity for different geologic materials.

TABLE 2.1 Hydraulic Conductivity for Various Classes of Geologic Materials

Material	Hydraulic Conductivity (gpd/ft²)
Coarse gravel	3,681
Medium gravel	6,626
Fine gravel	11,044
Coarse sand	1,104
Medium sand	295
Fine sand	61
Silt	2
Clay	0.005

Source: Todd, 1980.

Intrinsic permeability is usually measured in darcys (1 darcy $\simeq$ 1 μ^2), whereas the usual unit of hydraulic conductivity is in Meinzer units (gpd/ft², in honor of American hydrologist O. E. Meinzer). This unit is the number of gallons per day flowing through an aquifer section 1-ft thick and 1-mi wide under a hydraulic gradient of 1 ft per mile. Figure 2.5 illustrates the concept of hydraulic conductivity.

Differential Form of Darcy's Law

In an elemental cube of aquifer material, forces are in equilibrium and the velocity of flow in the s direction may be written (5)

$$v_s = -\frac{K\partial(p/\gamma + z)}{\partial s} \, . \qquad (2.5)$$

Differentiating the head in Equation 2.4 with respect to distance s yields

$$\frac{\partial h}{\partial s} = \frac{\partial(p/\gamma + z)}{\partial s} \, . \qquad (2.6)$$

Substituting Equation 2.6 into Equation 2.5 results in

$$v_s = -K\frac{\partial h}{\partial s} \, . \qquad (2.7)$$

The conventional units (Meinzer Units) are gpd/ft^2

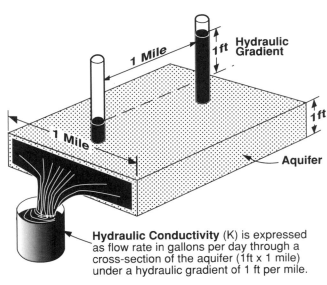

Hydraulic Conductivity (K) is expressed as flow rate in gallons per day through a cross-section of the aquifer (1ft x 1 mile) under a hydraulic gradient of 1 ft per mile.

Figure 2.5. Concept of hydraulic conductivity.

Equation 2.7 is the differential form of Darcy's law and states that the flow rate through porous media (in any direction) is proportional to the negative rate of change of head in that direction (the negative sign denotes that flow moves in the direction of decreasing head).

Ground Water Flow Velocity

Darcian or Bulk Velocity. The velocity expressed in Equation 2.7 assumes that flow occurs through the entire cross section of aquifer material without regard to pore space. This velocity therefore is a hypothetical velocity and represents the average velocity in the direction of decreasing uniform head. Use of Darcian velocity in flow equations is justified if one assumes that on a macroscopic scale, flow is orderly and the bulk of particles on the average move locally essentially in parallel paths.

Seepage Velocity. On a microscopic scale, ground water moves only through effective pore areas at a rate equal to "effective" or "seepage" velocity (9). Actual ground water flow therefore is faster than the average flow as represented by the Darcy velocity. The relation between Darcian and seepage velocity is stated as

$$v_e = \frac{v_d}{\theta} \; ,$$

where

v_e = seepage velocity [LT^{-1}],
v_d = Darcian velocity [LT^{-1}],
θ = effective porosity of material.

Ground water seepage velocities vary widely depending on aquifer materials and hydraulic gradients. Rates typically range between a few feet per year to several hundred feet per day (9).

2.3 DIFFERENTIAL EQUATIONS FOR GROUND WATER FLOW

The following section develops from fundamental concepts and principles the basic equations governing saturated flow in porous media. From these equations, solutions for flow to wells, drains, recharge basins, and so forth, can be mathematically derived.

Conservation of Matter and the Equation of Continuity

The fundamental principles of hydrodynamics are essentially reformulations of corresponding principles of mechanics, rearranged in such a way as to apply to flow of fluids. Although fluids are nonrigid systems, they are still subject to the law of conservation of matter, which states that fluid mass in any closed system can be neither created nor destroyed.

Specifically, this law may be restated as (10)

The net excess of mass flux, per unit of time, into or out of any infinitesimal volume element in the fluid system is exactly equal to the change per unit of time of the fluid density in the element multiplied by the free volume of the element

or more simply stated, inflow equals outflow plus storage change.

To illustrate this law, consider a representative cube of aquifer material with sides δx, δy, and δz and having a total volume V (see Figure 2.6).

The general equation for mass inflow is

Mass flow rate = fluid density × velocity

× cross-sectional area ,

or

Mass inflow in x-direction:

$$\rho v_x \delta y \delta z \; ,$$

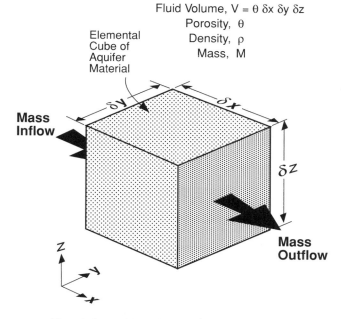

Fluid Volume, V = θ δx δy δz
Porosity, θ
Density, ρ
Mass, M

Elemental
Cube of
Aquifer
Material

**Mass
Inflow**

**Mass
Outflow**

Mass Inflow = Mass Outflow $\pm$ Change in Mass

Figure 2.6. Elemental Cube

Mass inflow in y-direction:

$$\rho v_y \delta x \delta z ,$$

Mass inflow in z-direction:

$$\rho v_z \delta x \delta y , \qquad (2.8)$$

where

ρ = fluid density [ML^{-3}],

v_x = fluid velocity (Darcian) in the x-direction [LT^{-1}].

Similarly, the general equation for mass outflow is stated as

Mass outflow = mass inflow

+ change in mass within the cube ,

or

Outflow in x-direction:

$$\left(\rho v_x + \frac{\partial(\rho v_x)}{\partial x} \delta x\right)\delta y \delta z ,$$

Outflow in y-direction:

$$\left(\rho v_y + \frac{\partial(\rho v_y)}{\partial y} \delta y\right)\delta x \delta z ,$$

Outflow in z-direction:

$$\left(\rho v_z + \frac{\partial(\rho v_z)}{\partial z} \delta z\right)\delta x \delta y . \qquad (2.9)$$

The total net inward flux mass in the cube is obtained by subtracting individual fluxes (inflow − outflow [Equations 2.8 − Equations 2.9]) and by adding together

Total net inward flux

$$= - \left(\frac{\partial(\rho v_x)}{\partial x} + \frac{\partial(\rho v_y)}{\partial y} + \frac{\partial(\rho v_z)}{\partial z}\right)V ,$$

where

$$V = \delta x \delta y \delta z \quad \text{(volume of cube)} [L^3] .$$

In accordance with the law of conservation of matter, the total net inward flux must equal change of mass with time. Quantitatively this is written

$$\left[\frac{\partial(\rho v_x)}{\partial x} + \frac{\partial(\rho v_y)}{\partial y} + \frac{\partial(\rho v_z)}{\partial z}\right]V = - \frac{\partial(M)}{\partial t} ,$$

where

M = mass in the small cube [M],

   = $\rho\theta\delta x \delta y \delta z$, = $\rho\theta V$

θ = effective porosity

so that

$$\frac{\partial(\rho v_x)}{\partial x} + \frac{\partial(\rho v_y)}{\partial y} + \frac{\partial(\rho v_z)}{\partial z} = - \frac{\partial(\rho\theta)}{\partial t} \qquad (2.10)$$

Equation 2.10 is a form of the law of conservation of matter and is also known as the "equation of continuity."

Compressibility and Elasticity

In unconfined aquifers, changes in water volume due to aquifer compressibility and elasticity are relatively unimportant compared to changes in water table elevation. However, in confined and semiconfined aquifers, compressibility and elasticity have a major impact on storage changes. If an aquifer is compressible and elastic, reduction of water pressure due to a pumping or flowing well relieves hydrostatic pressure against the confining beds. This results in increasing the load impressed on the aquifer skeleton and the aquifer compacts, reducing pore space. At the same time aquifer compaction is taking place the water volume is expanding because of the reduction in pressure. These two processes furnish all of the water produced from storage in a confined aquifer.

Aquifer elasticity was first noted by O. E. Meinzer in 1928 and has since been verified by both field and laboratory observations (11). Water level changes in wells (caused by pressure changes) due to atmospheric and tidal variations, land subsidence, and earthquakes are due to the elasticity of aquifers.

The following discussion describes the relationship between changes in hydraulic head and volume of water released from storage in confined aquifers.

Consider a small volume of soil V with a horizontal cross-sectional area A and vertical height δz (refer to Figure 2.6). The mass, M of water in this soil is given by

$$M = \rho \theta A \, \delta z \, , \qquad (2.11)$$

where

$$V = \theta A \, \delta z \, ,$$

$$A = \delta x \delta y \, .$$

This mass of fluid can change as a result of changes in density ρ, porosity θ, and height δz, each of which can occur as a result of a change in pressure within the soil volume. We wish to compute how much extra water will be stored in the soil volume V by an increase in the fluid pressure p by an amount dp. To accomplish this we form the logarithmic differential of Equation 2.11 to obtain:

$$\frac{dM}{M} = \frac{d\rho}{\rho} + \frac{d\theta}{\theta} + \frac{d(\delta z)}{\delta z} \, .$$

Also from Equation 2.11 we can write

$$\frac{dM}{V} = \theta d\rho + \rho d\theta + \frac{\rho\theta \, d(\delta z)}{\delta z} \, . \qquad (2.12)$$

This equation defines the mass of fluid released per unit volume of medium from an incremental change in the fluid density ($d\rho$), the porosity ($d\theta$), and the height of the volume element ($d(\delta z)$). We need to relate each of these to the pressure dp. First, the compressibility of water β is defined by

$$\frac{d\rho}{\rho} = +\beta \, dp \, , \qquad (2.13)$$

which states that a unit pressure increase will induce a density increase β per unit of density. For water, β is approximately 3.3×10^{-6} in²/lb.

To relate $d\theta$ to the pressure change dp we note that the volume of solid in the volume of medium V is approximately constant since it does not compress very easily so that

$$V_s = (1 - \theta) A \delta z \, .$$

Differentiating this logarithmically and using the result that $V_s =$ constant.

$$d\theta = (1 - \theta) \frac{d(\delta z)}{\delta z} \, . \qquad (2.14)$$

From Equations 2.12, 2.13, and 2.14 we have

$$\frac{dM}{V} = \rho\beta \, dp + \rho \frac{d(\delta z)}{\delta z} \, . \qquad (2.15)$$

Now we need only relate the expansion of the volume element to the pressure increase. An increase in the fluid pressure by an amount dp will approximately decrease the intergranular stress by an equal amount (see Figure 2.7). We now define a compressibility of the aquifer skeleton by σ so that

$$\frac{d(\delta z)}{\delta z} = +\sigma \, dp \, , \qquad (2.16)$$

then we have

$$\frac{dM}{V} = (\rho\beta\theta + \rho\sigma) \, dp \, . \qquad (2.17)$$

This equation states that an increase in the fluid pressure by an amount dp will increase the mass of water stored per unit volume by an amount

$$\rho(\beta\theta + \sigma) \, dp \, .$$

Conversely, if the pressure is reduced by an amount $+dp$, there will be released from each unit volume a mass of water

$$\rho(\beta\theta + \sigma) \, dp \, .$$

Since the pressure dp is related to piezometric head by the relationship

$$dp = \gamma \, dh \, . \qquad (2.18)$$

we have

$$\frac{dM}{V} = \rho(\beta\theta + \sigma) \, \gamma \, dh \, .$$

If ρ_0 is the density of water at standard temperature and pressure, then the volume of water released per unit volume of storage will be

$$\frac{d(M/\rho_0)}{V} = \left(\frac{\rho}{\rho_0}\right)(\beta\theta + \sigma) \, \gamma \, dh \, .$$

A decrease in water pressure
corresponds to an increase in
solid compressive stress of
the aquifer. The vertical dimension
is compressed by the amount a - b.

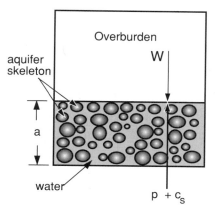
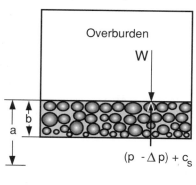

a.) System before decrease   b.) System after decrease in
in water pressure.   water pressure.

p = Pore Water Pressure
Δp = Decrease in Water Pressure
W = Weight of Overburden
c_s = Compressive stress of aquifer skeleton

Figure 2.7. Compressiblity of aquifers.

In general, the factor ρ/ρ_0 is very close to unity so the term on the right is written

$$\frac{dV}{V} = S_s \, dh \, , \qquad (2.19)$$

where

$$S_s = \gamma\beta\theta\left(1 + \frac{\sigma}{\beta\theta}\right) = \gamma\beta\theta + \gamma\sigma$$

and is defined as "specific storativity" of the aquifer. A typical range of specific storativities in alluvial aquifers would be from 1×10^{-4} to 1×10^{-9}/ft.

Specific Storativity

Specific storativity as defined by Jacob is the volume of water released from a unit volume of aquifer due to aquifer compression and water expansion under a unit decline in head (14). The fraction of total water released from storage derived from expansion of the water itself is given by

$$\theta\beta\gamma \, ,$$

and the fraction released due to aquifer compression is

$$\gamma\sigma \, ,$$

as shown in Figure 2.8.

Differential Equation for Flow in Confined Aquifers

If density (ρ) is assumed constant, Equation 2.10 becomes

$$\frac{\partial v_x}{\partial x} + \frac{\partial v_y}{\partial y} + \frac{\partial v_z}{\partial z} = -\frac{\partial(M/\rho V)}{\partial t} \, .$$

Furthermore it can be shown that when Equation 2.19 is substituted into the right-hand member of the preceding equation and considering density, mass and volume relationships, the following is obtained:

$$\frac{\partial v_x}{\partial x} + \frac{\partial v_y}{\partial y} + \frac{\partial v_z}{\partial z} = -S_s \frac{\partial h}{\partial t} \, . \qquad (2.20)$$

If Darcian velocities (Equation 2.7) are now substituted into Equation 2.20 (replacing the general direction s by specific directions x, y, and z); the following equation results:

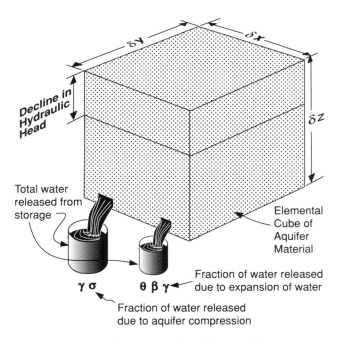

Figure 2.8. Concept of specific storativity.

$$K_x \frac{\partial^2 h}{\partial x^2} + K_y \frac{\partial^2 h}{\partial y^2} + K_z \frac{\partial^2 h}{\partial z^2} = S_s \frac{\partial h}{\partial t} \ . \quad (2.21)$$

If isotropic conditions are assumed (i.e., $K_x = K_y = K_z = K$), Equation 2.21 is simplified to

$$\frac{\partial^2 h}{\partial x^2} + \frac{\partial^2 h}{\partial y^2} + \frac{\partial^2 h}{\partial z^2} = \frac{S_s}{K} \frac{\partial h}{\partial t} \ . \quad (2.22)$$

Equation 2.22 is the general differential equation for flow in confined aquifers.

Approximate Differential Equation for Flow in an Unconfined Aquifer

Hantush has shown that for unconfined aquifers with horizontal bases, the approximate differential equation of flow in two dimensions may be written (5, 13)

$$\frac{\partial^2 h^2}{\partial x^2} + \frac{\partial^2 h^2}{\partial y^2} = \frac{\theta}{K\overline{D}} \frac{\partial h^2}{\partial t} \ , \quad (2.23)$$

where

$h = h(x, y, t) = $ average hydraulic head (L),
$\theta = $ effective porosity,
$\overline{D} = $ average saturated thickness (L).

Equation 2.23 is linear in h^2 and assumes that the average head (h) in a vertical section of the aquifer approximates the water table.

Differential Equation for Flow in Semiconfined Aquifers

Aquifers where flow takes place through one or more semipervious layers into or out of a main artesian aquifer are known as "leaky artesian" or semiconfined aquifers (see Figure 2.9).

The differential equation governing flow in anisotropic semiconfined aquifers as developed by Hantush is written (5, 13)

$$K_x \frac{\partial^2 h}{\partial x^2} + K_y \frac{\partial^2 h}{\partial y^2} + \frac{K'}{b'} \frac{(h_0 - h)}{b} = S_s \frac{\partial h}{\partial t} \quad (2.24)$$

For isotropic conditions, Equation 2.24 simplifies to

$$\frac{\partial^2 h}{\partial x^2} + \frac{\partial^2 h}{\partial y^2} + \frac{(h_0 - h)}{B^2} = \frac{S_s}{K} \frac{\partial h}{\partial t} \ , \quad (2.25)$$

where

B = leakage factor = $\sqrt{Kb/(K'/b')}$ [L],
K'/b' = leakance (coefficient of leakage) [T^{-1}],
h_0 = hydraulic head in the aquifer supplying leakage [L].

2.4 AQUIFER PARAMETERS

Physical properties that determine movement and storage of ground water in aquifer materials are called "aquifer parameters" (or formation parameters or formation constants). Aquifer parameters are integral components of all ground water flow and storage calculations, and vary widely due to the diversity of the geohydrologic environment.

Aquifer parameters may be grouped into:

• Parameters related to transmission of water through aquifers.

• Parameters related to gain or release of water from storage.

Transmissivity

The parameter describing how transmissive an aquifer is in moving water through pore spaces is called "transmissivity." Transmissivity is the product of hydraulic conductivity and the saturated aquifer thickness. Designated by the symbol T, transmissivity is defined as the rate of flow (gallons per

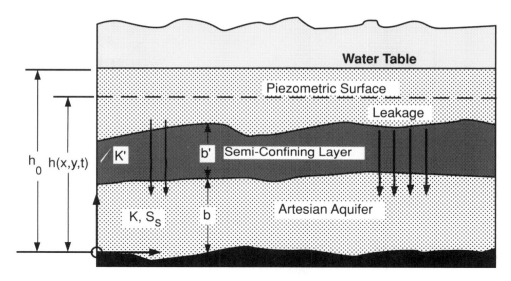

Figure 2.9. Leaky artesian aquifer system.

day) moving through the entire saturated thickness of an aquifer having a width of 1 mi under a hydraulic gradient of 1 ft per mile.

Transmissivity:

$$T = Kb \quad \text{(confined aquifers)} \; [\text{L}^2\text{T}^{-1}] \; ,$$

$$T = K\overline{D} \quad \text{(unconfined aquifers)} \; [\text{L}^2\text{T}^{-1}] \; ,$$

where

b = saturated thickness in the confined aquifer [L],
$\overline{D}$ = average saturated thickness in the unconfined aquifer [L].

The unit for transmissivity is the Meinzer unit of gpd/ft. Figure 2.10 illustrates transmissivity and its relationship to hydraulic conductivity for both confined and unconfined aquifers.

Storativity

Storativity is defined as the amount of water released or added to storage through a vertical column of the aquifer having a unit cross-sectional area, due to a unit amount of decline or increase in average hydraulic head.

For confined aquifers, storativity is the product of the saturated aquifer thickness and the specific storativity, and is due entirely to the compressibility and elasticity of the aquifer and the water. In unconfined systems, the volume of water released or taken into storage is due to dewatering or refilling, with only a very small part due to water and aquifer compressibility. Thus in unconfined aquifers the stor-

ativity for all practical purposes equals the effective porosity (specific yield) and represents the drainable pore volume. Storativity:

$$S = S_s b \quad \text{(confined aquifers)} \; ,$$

$$S = \theta + \overline{D}S_s \simeq S_y \quad \text{(unconfined aquifers, } S_s \to 0) \; ,$$

where

θ = effective porosity = S_y (specific yield),
S_s = specific storativity [L^{-1}].

Figure 2.11 illustrates the storativity parameter.

Leakance

The coefficient describing the ability of a semipervious layer (aquitard) to transmit vertical leakage is called "leakance," or the "coefficient of leakage." Leakance is defined as the rate of flow crossing a unit cross-sectional area of the aquifer/aquitard interface under a unit head differential measured between the top and bottom of the semipervious layer. Leakance is the quotient of the semipervious layer hydraulic conductivity and the layer thickness, and it is written

Leakance:

$$\frac{K'}{b'} \quad [\text{T}^{-1}] \; .$$

Figure 2.12 illustrates this concept.

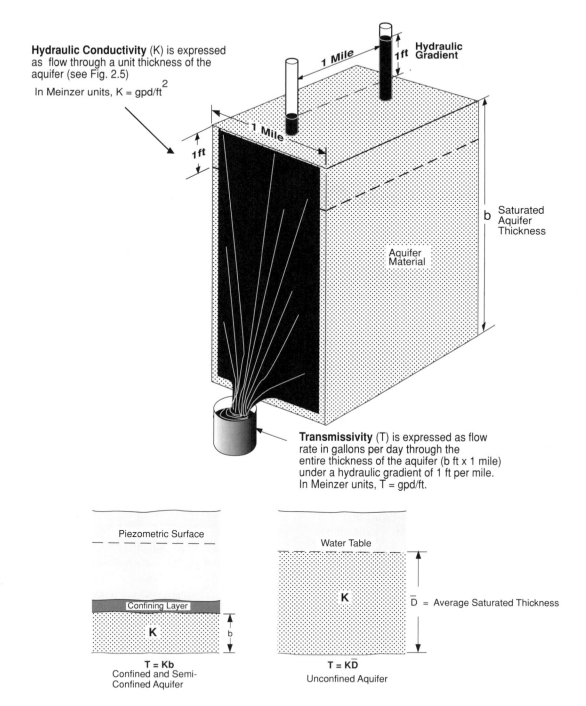

Figure 2.10. Transmissivity and hydraulic conductivity of aquifers.

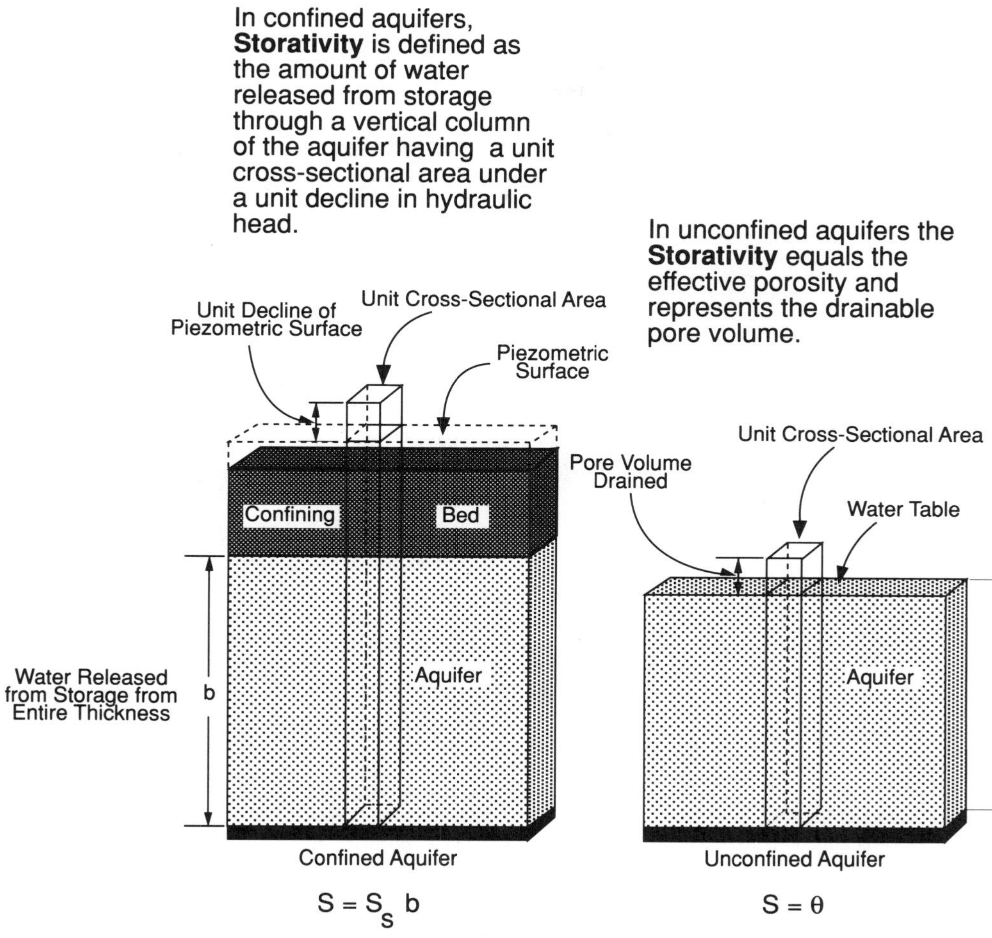
In confined aquifers, **Storativity** is defined as the amount of water released from storage through a vertical column of the aquifer having a unit cross-sectional area under a unit decline in hydraulic head.
In unconfined aquifers the **Storativity** equals the effective porosity and represents the drainable pore volume.
Unit Decline of Piezometric Surface
Unit Cross-Sectional Area
Piezometric Surface
Pore Volume Drained
Unit Cross-Sectional Area
Water Table
Confining
Bed
Aquifer
Aquifer
Water Released from Storage from Entire Thickness
b
D
Confined Aquifer
Unconfined Aquifer
S = S_S b
S = θ

Figure 2.11. Illustration of storativity.

Anisotropy

Due to wide variations in geologic depositional patterns and materials, considerable differences may be seen in the movement of ground water through aquifers in the same general area. These differences are attributed to anisotropy of the aquifer material resulting in horizontal and vertical variations in aquifer parameters. Typically, alluvial aquifers show strong anisotropic differences, and due to the nature of stratification, vertical flow is much lower than horizontal flow. In addition ground water flow in directions parallel to depositional patterns (perpendicular to mountain fronts) is generally greater than in the normal direction.

The average horizontal hydraulic conductivity ($\overline{K}_h$) can be calculated from

$$\overline{K}_h = \frac{K_1 b_1 + K_2 b_2 + \cdots + K_n b_n}{Z} \quad [LT^{-1}] ,$$

where

Z = the total saturated thickness and is expressed as $\Sigma_{i=1}^{n} b_i$ [L],

K_i = hydraulic conductivity of aquifer i having a saturated thickness, b_i [LT^{-1}].

For vertical flow, the equivalent average vertical conductivity between layers can be written

$$\overline{K} = \frac{Z}{(b_1/K_1) + (b_2/K_2) + \cdots + (b_n/K_n)} \quad [LT^{-1}] .$$

Figure 2.13 shows a stratified aquifer system.

2.5 APPLICATIONS OF GROUND WATER FLOW

This section illustrates a few practical applications of the material covered earlier in this chapter. In the development

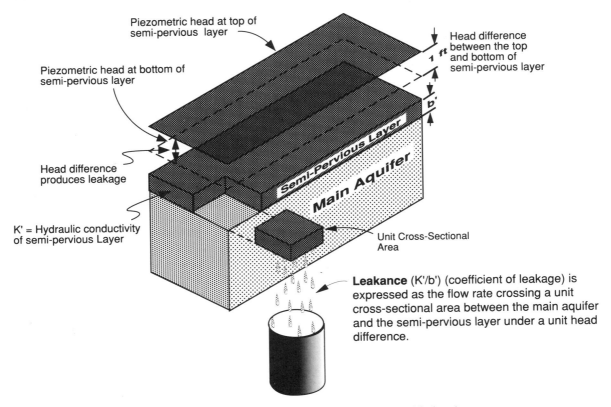

Figure 2.12. Illustration of leakance (coefficient of leakage).

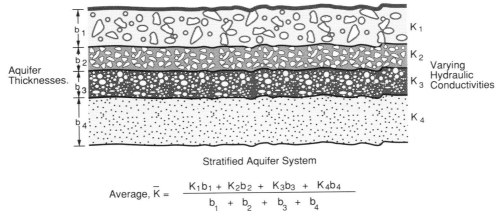

Figure 2.13. Vertical variation in hydraulic conductivity.

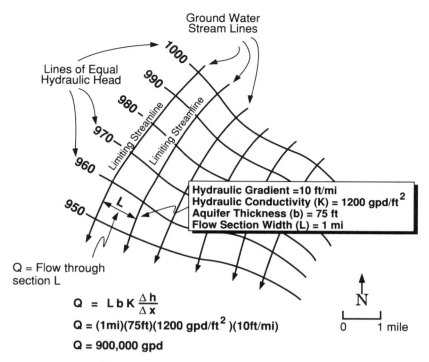

Figure 2.14. Ground water flow net example.

of the solution for the problem, field aquifer conditions are approximated by limiting spatial (x, y, z) and time (t) conditions, collectively forming "boundary conditions." The combination of these boundary conditions, together with the governing differential equation, constitutes the "boundary value problem."[2]

Solution of the particular boundary value problem is generally done by applying methods of transform calculus. However, simple cases can be solved directly through algebraic and integration procedures.

Steady-state conditions are distinguished from non-steady-state conditions by the absence of time as an independent variable in the governing equation.

Ground Water Flow Nets

Evaluation of ground water flow and storage changes may be calculated through use of "flow nets." A ground water flow net is composed of an orthogonal (mutually perpendicular) pair of curves. One curve represents ground water streamlines (flow lines) and the other curve the potential (hydraulic head).

2. Although the term "boundary conditions" in the definition of the boundary value problem includes hydrogeological boundaries (spatial) as well as time constraints, a distinction is often made for conditions of zero time, which are known as "initial conditions."

Ground water flow may be calculated between any two limiting streamlines from the following equation:

$$Q = LbK \frac{\Delta h}{\Delta x} ,$$

where

Q	=	flow [gpd],
L	=	distance between streamlines where flow is calculated [mi],
b	=	saturated aquifer thickness [ft],
K	=	hydraulic conductivity [gpd/ft²],
$\Delta h/\Delta x$	=	average hydraulic gradient at section L [ft/mi].

Figure 2.14 shows an example of the use of a ground water flow net to calculate subsurface flow (also refer to Appendix C).

Storativity may also be estimated using flow nets by measuring the difference in flow between any two sections of limiting streamlines and knowing changes in water level. Walton gives a formula for a calculation of storativity that assumes no leakage (15).

$$S = \frac{\Delta Q}{\Delta h A (2.1 \times 10^8)} ,$$

where

S = storativity of area A,

ΔQ = difference in flow crossing successive contour lines of hydraulic head between limiting streamlines [gpd],

A = area between limiting streamlines and successive water-level contours [mi^2].

Steady-State Flow between Two Drains

Consider the case of Figure 2.15 where flow between two line sources (drains) is occurring in an unconfined aquifer.

Since flow is steady state (constant) and in only one direction (x-direction), the governing equation (Equation 2.23) simplifies to

Governing equation:

$$\frac{\partial^2 h^2}{\partial x^2} = 0 \ .$$

Boundary conditions:

$$h(0) = h_1 \ ,$$

$$h(L) = h_2 \ .$$

Solution of the boundary value problem yields

$$h_1^2 - h^2 = \frac{(h_1^2 - h_2^2)x}{L} \ .$$

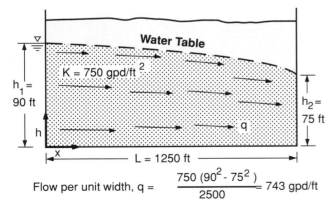

Flow per unit width, $q = \dfrac{750\,(90^2 - 75^2)}{2500} = 743 \text{ gpd/ft}$

Figure 2.15. Flow between two drains.

Solving for flow,

$$q = Av = -Kh\frac{\partial h}{\partial x} \ ,$$

$$\frac{h\partial h}{\partial x} = -\frac{h_1^2 - h_2^2}{2L} \ ,$$

$$q = \frac{K}{2L}(h_1^2 - h_2^2) \ ,$$

where

q = flow rate between drains per unit width [gpd/ft],

K = hydraulic conductivity [gpd/ft^2],

L = distance between drains [ft]

h_1, h_2 = heights of drains above reference [ft].

Nonsteady-State Flow to a Drain with a Semipervious Bed

Drains (or rivers or streams) in hydraulic continuity with artesian aquifers frequently have deposits of fine-grained materials (e.g., silt) lining the sides and bottom. This semipervious lining has an effect on the ground water flow and head distribution and must be accounted for in analyses. Consider the case shown in Figure 2.16.

A drain with a semipervious bed cuts through an artesian aquifer. Initially, the piezometric surface is flat and equals the water level in the drain. Later, the water level in the drain is instantaneously raised (e.g., due to irrigation supply changes), and water flows from the drain into the aquifer. The following example solves for the head distribution in the aquifer at any distance x (measured from the edge of the drain) and after any time t since the raising of the water level.

Equation 2.22, reduced to one direction, is written

$$\frac{\partial^2 h}{\partial x^2} = \frac{1}{\kappa}\frac{\partial h}{\partial t} \ ,$$

where

$\kappa = T/S$ = hydraulic diffusivity [L^{2}T^{-1}].

Initial conditions:

$$h(x, 0) = 0 \ .$$

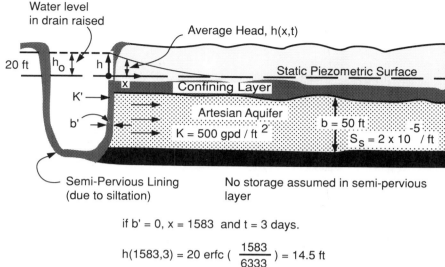

if b' = 0, x = 1583 and t = 3 days.

$$h(1583,3) = 20 \ \text{erfc} \left(\frac{1583}{6333} \right) = 14.5 \ \text{ft}$$

Figure 2.16. Flow near a drain with a semipervious bed.

Boundary condition:

$$h(\infty, t) = 0 \ ,$$

$$\frac{\partial h(0, t)}{\partial x} = - \frac{K'/b'}{K} [h_0 - h(0, t)] \quad \text{(discharge condition)} \ .$$

Hantush has shown that after applying Laplace transforms, the head distribution may be written (13)

$$h(x, t) = h_0 \left\{ -e^{x/a} e^{\kappa t/a^2} \ \text{erfc} \left(\frac{x}{\sqrt{4\kappa t}} + \frac{\sqrt{\kappa t}}{a} \right) + \text{erfc} \left(\frac{x}{\sqrt{4\kappa t}} \right) \right\} \ ,$$

(2.26)

where

$h(x, t)$	=	water level distribution in aquifer [L],
t	=	time since initial raising of water level [T],
K'/b'	=	leakance [T^{-1}],
a	=	$K/(K'/b')$ [L],
erfc	=	complementary error function (see Table 2.2).

For drains without semipervious linings ($b' = 0$ or $K' = \infty$), Equation 2.26 reduces to

$$h(x, t) = h_0 \ \text{erfc} \left(\frac{x}{\sqrt{4\kappa t}} \right) \ .$$

Length of Intruded Saltwater Wedge

Intrusion of seawater into fresh ground water reservoirs occurs in coastal aquifers in many areas of the world.

Knowledge of the length of the intruding "wedge" of saltwater and factors governing its movement are important to preservation of supplies. The following example calculates this length. Figure 2.17 shows an artesian aquifer in hydraulic continuity with the ocean.

Freshwater flows to the sea, discharging near the top of the aquifer. A wedge of saltwater extends inland beneath the freshwater to a distance (L). At any point on the interface (salt/fresh), the hydrostatic pressure of seawater ($\gamma_s z$) is balanced by the total pressure of freshwater [$\gamma_f (z + h)$].

Applying Darcy's law and continuity principles, the flow per unit width across any vertical section of the aquifer at a distance x may be calculated (5):

$$\frac{\partial (h - \alpha b')^2}{\partial x} = \frac{2\alpha q}{K} \ , \tag{2.27}$$

TABLE 2.2 The Error Function and Its Derivatives and Integrals

x	$e^{x^2}\,\text{erfc}\,x$	$4\pi^{-1/2}\,xe^{-x^2}$	$2\pi^{-1/2}\,e^{-x^2}$	$\text{erf}\,x$	$\text{erfc}\,x$	$2i\,\text{erfc}\,x$	$4i^2\,\text{erfc}\,x$	$6i^3\,\text{erfc}\,x$	$8i^4\,\text{erfc}\,x$	$10i^5\,\text{erfc}\,x$	$12i^6\,\text{erfc}\,x$
0	1.0	0	1.1284	0	1.0	1.1284	1.0	0.5642	0.25	0.0940	0.0313
0.05	0.9460	0.1126	1.1256	0.056372	0.943628	1.0312	0.8921	0.4933	0.2148	0.0795	0.0261
0.1	0.8965	0.2234	1.1172	0.112463	0.887537	0.9396	0.7936	0.4301	0.1841	0.0671	0.0217
0.15	0.8509	0.3310	1.1033	0.167996	0.832004	0.8537	0.7040	0.3740	0.1573	0.0564	0.0180
0.2	0.8090	0.4336	1.0841	0.222703	0.777297	0.7732	0.6227	0.3243	0.1341	0.0474	0.0149
0.25	0.7703	0.5300	1.0600	0.276326	0.723674	0.6982	0.5491	0.2805	0.1139	0.0396	0.0123
0.3	0.7346	0.6188	1.0313	0.328627	0.671373	0.6284	0.4828	0.2418	0.0965	0.0331	0.0101
0.35	0.7015	0.6988	0.9983	0.379382	0.620618	0.5639	0.4233	0.2079	0.0816	0.0275	0.0083
0.4	0.6708	0.7692	0.9615	0.428392	0.571608	0.5043	0.3699	0.1782	0.0687	0.0228	0.0068
0.45	0.6423	0.8294	0.9215	0.475482	0.524518	0.4495	0.3223	0.1522	0.0577	0.0189	0.0055
0.5	0.6157	0.8788	0.8788	0.520500	0.479500	0.3993	0.2799	0.1297	0.0484	0.0156	0.0045
0.55	0.5909	0.9172	0.8338	0.563323	0.436677	0.3535	0.2423	0.1101	0.0404	0.0128	0.0036
0.6	0.5678	0.9447	0.7872	0.603856	0.396144	0.3119	0.2090	0.0932	0.0336	0.0105	0.0029
0.65	0.5462	0.9614	0.7395	0.642029	0.357971	0.2742	0.1798	0.0787	0.0279	0.0086	0.0024
0.7	0.5259	0.9678	0.6913	0.677801	0.322199	0.2402	0.1541	0.0662	0.0231	0.0070	0.0019
0.75	0.5069	0.9644	0.6429	0.711156	0.288844	0.2097	0.1316	0.0555	0.0190	0.0057	0.0015
0.8	0.4891	0.9520	0.5950	0.742101	0.257899	0.1823	0.1120	0.0464	0.0156	0.0046	0.0012
0.85	0.4723	0.9314	0.5479	0.770668	0.229332	0.1580	0.0950	0.0386	0.0128	0.0037	0.0010
0.9	0.4565	0.9035	0.5020	0.796908	0.203092	0.1364	0.0803	0.0321	0.0104	0.0030	0.0008
0.95	0.4416	0.8695	0.4576	0.820891	0.179109	0.1173	0.0677	0.0265	0.0085	0.0024	0.0006
1.0	0.4276	0.8302	0.4151	0.842701	0.157299	0.1005	0.0568	0.0218	0.0069	0.0019	0.0005
1.1	0.4017	0.7403	0.3365	0.880205	0.119795	0.0729	0.0396	0.0147	0.0045	0.0012	0.0003
1.2	0.3785	0.6416	0.2673	0.910314	0.089686	0.0521	0.0272	0.0097	0.0029	0.0007	0.0002
1.3	0.3576	0.5413	0.2082	0.934008	0.065992	0.0366	0.0184	0.0063	0.0019	0.0004	0.0001
1.4	0.3387	0.4450	0.1589	0.952285	0.047715	0.0253	0.0122	0.0041	0.0011	0.0003	0.0001
1.5	0.3216	0.3568	0.1189	0.966105	0.033895	0.0172	0.0080	0.0026	0.0007	0.0002	
1.6	0.3060	0.2791	0.0872	0.976348	0.023652	0.0115	0.0052	0.0016	0.0004	0.0001	
1.7	0.2917	0.2132	0.0627	0.983790	0.016210	0.0076	0.0033	0.0010	0.0003		
1.8	0.2786	0.1591	0.0442	0.989091	0.010909	0.0049	0.0021	0.0006	0.0002		
1.9	0.2665	0.1160	0.0305	0.992790	0.007210	0.0031	0.0013	0.0003	0.0001		
2.0	0.2554	0.0827	0.0207	0.995322	0.004678	0.0020	0.0008	0.0002	0.0001		
2.1	0.2451	0.0576	0.0137	0.997021	0.002979	0.0012	0.0005	0.0001			
2.2	0.2356	0.0393	0.0089	0.998137	0.001863	0.0007	0.0003				
2.3	0.2267	0.0262	0.0057	0.998857	0.001143	0.0004	0.0002				
2.4	0.2185	0.0171	0.0036	0.999311	0.000689	0.0002	0.0001				
2.5	0.2108	0.0109	0.0022	0.999593	0.000407	0.0001					
2.6	0.2036	0.0068	0.0013	0.999764	0.000236	0.0001					
2.7	0.1969	0.0042	0.0008	0.999866	0.000134						
2.8	0.1905	0.0025	0.0004	0.999925	0.000075						
2.9	0.1846	0.0015	0.0003	0.999959	0.000041						
3.0	0.1790	0.0008	0.0001	0.999978	0.000022						

Source: Reprinted with permission from Carslaw and Jaeger (16).

where

L = length of saltwater wedge [ft],

q = freshwater flow to the ocean (per unit width of aquifer) [gpd/ft],

b' = distance between sea level and top of aquifer [ft],

z = distance between sea level and any point on interface [ft],

b = thickness of artesian aquifer [ft],

$\alpha = (\gamma_s - \gamma_f)/\gamma_f$,

γ_f = specific weight of freshwater = 62.4 lbs/ft^3,

γ_s = specific weight of seawater = 63.96 lbs/ft^3,

K = hydraulic conductivity [gpd/ft^2].

At $x = 0$:

$$h(0) = \alpha z(0) \simeq \alpha b' .$$

At $x = L$:

$$h(L) = \alpha z(L) = \alpha(b + b') .$$

Integrating Equation 2.27 gives

$$L = \frac{\alpha K b^2}{2q} .$$

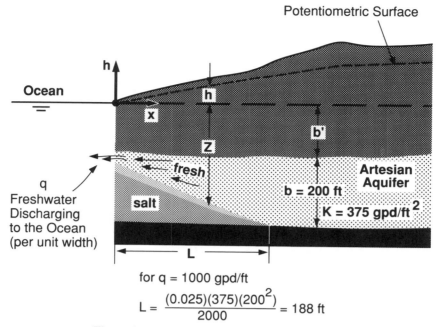

for q = 1000 gpd/ft

$$L = \frac{(0.025)(375)(200^2)}{2000} = 188 \text{ ft}$$

Figure 2.17. Length of intruding saltwater wedge.

REFERENCES

1. Hagen, G. 1839. *Ueber die Bewegung des Wassers in engen cylindrischen Rohren*. Vol. 46. Pogendorff Annalen.

2. Poiseuille, J. M. L. 1846. *Recherches experimentales sur le mouvement des liquides dans les tubes de tres petits diametres*. Vol 9. Acad. Sci., Paris.

3. Darcy, H. P. G. 1856. *Les fontaines publique de la ville de Dijon*. Paris.

4. Reynolds, O. 1883. "An Experimental Investigation of the Circumstances Which Determine Whether the Motion of Water Shall Be Direct or Sinuous and the Law of Resistance in Parallel Channels." *Roy. Soc., London Trans.* A174.

5. Hantush, M. S. 1964. "Hydraulics of Wells." In *Advances in Hydroscience*. Vol. 1. Academic Press, New York.

6. Hazen, A. 1893. "Some Physical Properties of Sands and Gravels with Special Reference to Their Use in Filtration." Massachusetts State Board of Health 24th Annual Report. Boston.

7. Todd, D. K. 1980. *Ground Water Hydrology*. J. Wiley and Sons, New York.

8. Hubbert, M. K. 1956. "Darcy's Law and the Field Equations of the Flow of Underground Fluids." *Trans. Amer. Inst. Min. and Metal. Engrs.* 207.

9. Meinzer, O. E. 1942. *Hydrology*. Dover, New York.

10. Muskat, M. 1946. *The Flow of Homogeneous Fluids through Porous Media*. J. W. Edwards, Inc., Ann Arbor, Mich.

11. Meinzer, O. E. 1928. "Compressibility and Elasticity of Artesian Aquifers." *Econ. Geol.* 23.

12. Jacob, C. E. 1950. "Engineering Hydraulics." In H. Rouse (ed.), *Flow of Ground Water*. J. Wiley and Sons, New York.

13. Hantush, M. S. 1965. Unpublished lecture notes taken by D. E. Williams. New Mexico Institute of Mining and Technology, Socorro, NM.

14. Jacob, C. E. 1966. Unpublished lecture notes taken by D. E. Williams. New Mexico Institute of Mining and Technology, Socorro, NM.

15. Walton, W. C. 1962. *Groundwater Resource Evaluation*. McGraw-Hill, New York.

16. Carslaw, H. S., and J. C. Jaeger. 1959. *Conduction of Heat in Solids*. Clarendon Press, Oxford.

CHAPTER 3

Exploration for Ground Water

3.1 INTRODUCTION

Since ground water is a hidden resource not amenable to direct observation, some level of exploration or assessment is usually necessary for the placement of an extraction system. However, the objective of an exploration program is not only to find water but also to provide a reliable supply meeting quantity and quality requirements. Data acquired during exploration can also form the basic information essential for the design of a ground water management program.

The scope and detail of an investigation depends upon its objectives, the size of the study area, and the record of previous exploration and development. In the case of an already developed basin, the investigation may consist simply of assembling information derived from nearby wells. Large development programs in areas where data are sparse may require use of a number of diverse and sometimes complex procedures to estimate the ground water potential.

Emphasis in this chapter is on exploration programs leading to development of ground water. However, the methodologies discussed are often applied to other ground water studies such as assessing the extent of aquifer contamination.

The Exploration Chart

Regardless of complexity, any thorough exploration project must answer three basic questions:

Where does the water come from?
Where does it go?
What is the nature of its geologic container?

Table 3.1 outlines the steps typically used to answer them. Each step must be justified by the preceding step, and decisions to alter, continue, or terminate the program are normally made at the end of each phase. Some steps may not be required, but the incorporation of data and program planning is a continuous process. Exploration concludes with an assessment of the quantity and quality of ground water economically and physically producible.

3.2 PROJECT DEFINITION

Prior to beginning a ground water investigation, the project objectives should be clearly defined in terms of total water demand and its required quality. The project boundary must be delineated. Finally, any constraints on the project must be identified and understood by the investigator. These constraints may be physical, legal, and/or economic.

Project Objectives

Frequently, contemplated development of an area hinges upon obtaining a specific quantity and quality of water, and the exploration project objective is to determine whether these needs can be met. Requirements should be clarified as continuous or intermittent. Some situations necessitate a complete water inventory and safe yield determination, whereas in others the only information necessary is to determine the quantity satisfying current demands. The concept of safe yield is detailed in Chapter 20.

When demand has not been quantified, other criteria may be used. For example:

- The needs of a single-family home can be satisfied by about 1 acre foot of water per year. A well producing 1 gallon of water per minute or 1440 gallons per day, in conjunction with a small storage tank, is sufficient for this demand. Its quality should meet state and federal drinking water requirements.
- Agricultural demand is based upon the amount of acreage to be irrigated, the project area's microclimate, and the specific crops. Local agricultural agencies can usually estimate the seasonal water requirements. Water quality requirements depend on crop type.

Project Boundary

The project boundary may be that boundary within which the owner has or can acquire control over the water resources. Alternatively, the project boundary may include the entire drainage basin that provides the ground water reserve. Usually

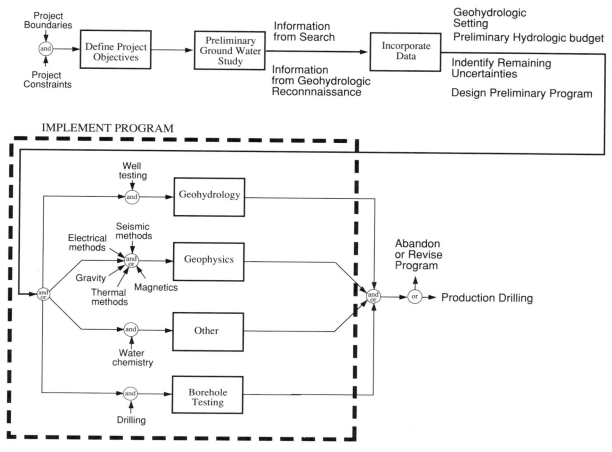

TABLE 3.1   The Exploration Chart

initial data accumulated cover a far larger area than that under control.

Project Constraints

There may be a number of constraints associated with the investigation. Physical constraints include limitations on the location and design of production wells. For example, in the case of domestic drinking water, consideration must be given to the existence of other facilities such as sanitary landfills, sewage treatment plants, or the presence of aquifers known to be contaminated. Existing distribution systems may dictate the location of new wells. Economic constraints include high costs for power or water treatment for water produced from deep pumping levels.

In some areas ground water may be controlled by state or local entities or by senior water-rights holders. Legislation may restrict extraction of ground water in an environmentally sensitive area (e.g., proximity to a wildlife refuge).

3.3   THE PRELIMINARY GROUND WATER STUDY

The objectives of the preliminary phase of a ground water exploration project are:

1. To define the project area geohydrologic setting.
2. To estimate the annual ground water recharge to the area.
3. To evaluate potential water-quality problems.
4. To plan fieldwork logistics.

Tasks performed during this phase always include an information search and on-site reconnaissance. Upon completion, a forecast of the scope and budget for the remainder of the project can be made.

Information Search

Reliable information produced by a careful search will almost always shorten the length and reduce the cost of a study by minimizing need for data developed in the field. Many sources are available to the ground water investigator.

Public agencies (federal, state, and local) have for many years routinely compiled data on surface and ground water resources. Their records may permit estimating the effect of long-term production rates on the ground water environment. Such estimates would not be possible with data developed during a relatively short-term study. Also useful are professional geological, geophysical, and hydrological organizations whose publications report both areal studies and applications of technology to ground-water problems.

Appendix A gives a short list of organizations likely to have useful information and, in some cases, indicates the type of information available. Although it is not possible to list individual state and local organizations, use of a directory will usually lead to the appropriate government agencies. Acquisition of a current report covering the area is useful as a first step because its bibliography, if comprehensive, may substantially reduce information source search time.

One important source of information may be the owner's own files. Geohydrologic investigations done on neighboring properties are also useful, although requests should be made for these reports only with the owner's permission.

In most states a permit must be obtained prior to drilling a water well or test hole, and a driller's report is filed after the well is constructed. This report includes the location, owner's and driller's name, and a description of lithologies encountered during drilling. It may also contain data on well construction and a record of tests performed. In some states the report is considered confidential and released only with the owner's permission.

3.4 GEOHYDROLOGIC SETTING

Regional Geohydrologic Setting

If the investigator has limited experience in the general project area, a review of a regional ground water study may be useful. The U.S. Geological survey has historically taken the lead in defining regional ground water systems in the United States, and continually updates this work. The latest edition by Heath, which builds upon the work of Thomas and Meinzer, uses criteria such as (1, 2, 3):

- Number and type of aquifers and aquifer-aquitard relationships.
- Type of porosity in dominant aquifer.
- Solubility of dominant aquifer matrix.
- Storativity and transmissivity of dominant aquifer.
- Recharge and discharge mechanisms of dominant aquifer.

Heath has divided the United States, including Alaska and Hawaii, into thirteen geographic ground water regions.

In addition a fourteenth, the alluvial valleys ground water region, though centered along the axis of the Mississippi River, includes elements throughout the coterminous United States. Appendix B lists these regions. Within a regional ground water system lie many local (usually porous media) aquifers that may provide significant amounts of ground water. These may include alluvial river deposits, glacial moraines and outwash deposits, and buried dune sands.

Another program useful in deriving information on the regional ground water setting is the Regional Aquifer–System Analysis Program (RASA) implemented by the U.S. Geologic Survey in 1978. The RASA criteria define a regional aquifer system as either a connected set of aquifers that act hydrologically as a single unit, or a set of independent aquifers that act similarly. The RASA ground water regions are shown in Figure 3.1.

Ongoing studies are made of these aquifer systems with results published as Professional Papers. Their master designation is PP813 plus a letter code designating the region from the map (Figure 3.1). These publications are indexed by the Geological Survey and by all major geologic and geohydrologic indexes. They contain comprehensive bibliographies, are usually relatively current, and are an excellent place to begin a geohydrologic information search.

Local Geohydrologic Setting

Any study area is actually a three-dimensional volume consisting of one or more aquifers separated or bounded by aquitards. Its depth may be defined as that below which no significant ground water occurs, ground water pumping cannot be sustained economically, or the quality of the ground water is unsuitable.

Within the local setting a convenient unit of study is the drainage basin in which all surface water drains toward a stream or the bottom of a closed valley such as a desert dry lake. On a suitably sized topographic map, the boundary is the line that is perpendicular to each maximum topographic contour that it crosses, as shown in Figure 3.2. The surface area of the basin may be calculated using a planimeter or by direct measurement using the map scale. A preliminary assumption that the ground water boundary coincides with the surface basin boundary is made.

The basin boundary may be considered a no-flow boundary because no flow crosses it. All precipitation falling within the boundary flows toward the center of the basin, whereas all precipitation falling outside the boundary flows into another basin. The geologic lower boundary of the study region, sometimes referred to as a hydrologic basement, may also be considered a no-flow boundary because no significant ground water flow crosses it.

A system of branching streams collects drainage and carries it out of basins (except for closed basins). Stream courses are marked on topographic maps. Depending upon

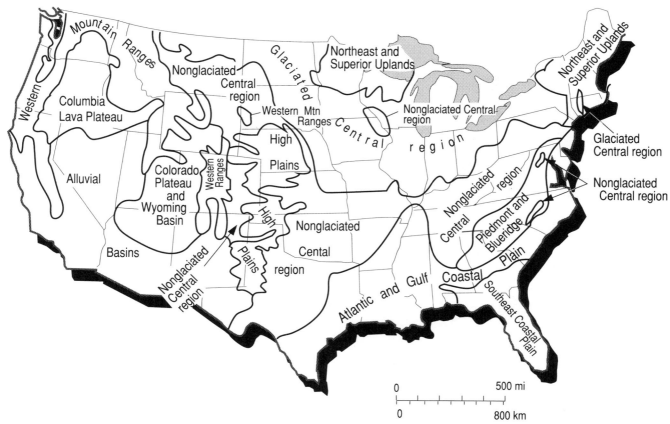

Figure 3.1. RASA ground water regions.

the size of the body and the map scale, positions of surface-water bodies fed or drained by streams, such as ponds and reservoirs, may be shown.

Aquifer-Aquitard Systems

The study area may include a multiaquifer system composed of two or more aquifers separated and bounded by aquitards. If there is little or no pumping within the area, it may be considered to be in a steady-state or equilibrium condition (heads within the aquifer do not change appreciably over time). With ground water pumping, leakance may allow aquitards to release significant amounts of water from storage into adjoining aquifers.

Geologic maps indicate the surficial extent of aquifer materials and their contact with bedrock and structures that control the movement of ground water. On standard geologic maps various shades of yellow are used to designate unconsolidated or slightly consolidated alluvial materials. The extent of such deposits can also be revealed on aerial photographs and, more precisely, by field mapping.

Major structural controls over ground water flow may also be shown on geologic maps as faults—solid where

mapped, slashed where inferred, and dotted where concealed. Study of satellite images sometimes reveals structural patterns that control ground water movement. These are shown as lineaments caused by soil or vegetation change too subtle to be seen at the surface or at lower altitudes.

Vertical cross sections may be shown on geologic maps. These are prepared using surface data such as geological structures formed by tectonic and erosional processes, and from borehole and geophysical data. Often the investigator is able to improve the accuracy of these cross sections with information gathered during the information search and reconnaissance mapping.

When the basin is large in relation to the study area, it is practical to isolate a portion of it for more detailed work. Boundaries should be established around the desired area parallel and perpendicular to the direction of inferred ground water flow.

Direction of Ground Water Movement and Hydraulic Gradient

A well inventory is prepared during the preliminary phase of the investigation. Essential data included are the surface

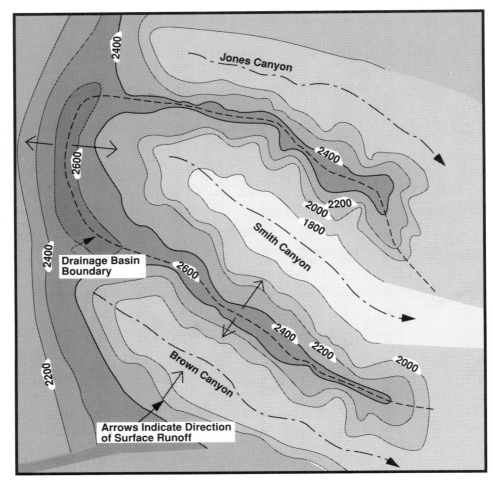

Figure 3.2. Drainage basins.

elevation of all boreholes, their total depths (which may be obtained by sounding), water levels, and screen or perforation schedules.

If no direct or indirect observation of the ground water elevation is available, the surface topography is assumed to reflect the direction of ground water flow. In areas of relatively low topographic relief, the surface elevation gradient may also be assumed to be the hydraulic gradient.

When water level data are available, they are plotted and contoured on a base map, usually topographic. The ground water flow lines are drawn at right angles to the water-level elevation contours in the direction of decreasing elevation, as shown in Figure 3.3. After the elevation contours have been drawn, the hydraulic gradient expressed as the decrease in water level elevation per unit of horizontal distance can be calculated at several different points on the map.

This procedure, though simple, can lead to erroneous conclusions because the potentiometric surface is neither static nor a simple plane. It moves and changes its profile in response to varying patterns of recharge, discharge, and

pumping. Unless all the water levels are measured over a short period of time, the contour map may be incorrect.

When more than one aquifer exists in the geohydrologic environment, each has its own potentiometric surface, direction of ground water flow, and hydraulic gradient. A potentiometric map prepared from measured water levels in more than one aquifer will usually reveal a complex surface, which should alert the investigator that a multiaquifer system exists. Sometimes the contouring of a potentiometric surface permits subdividing a ground water basin into several subbasins.

Some error is inherent in mapping a continuous surface from a fixed number of data points. In drawing water-level contours, interpolation must be performed between points. When data points are few and unequally spaced, only limited confidence can be placed in the resulting map. Extrapolation by contouring beyond the control points, as may occur with the use of computer-contouring programs, should therefore be interpreted with caution.

It should be noted that a ground water contour map may be used to construct a flow net (Appendix C) and, as such, can be used to estimate the magnitude of ground water flow. The Darcy equation shows that for constant flow, the hydraulic gradient bears an inverse relationship to the hydraulic conductivity. This suggests that areas where the hydraulic gradient is relatively low should be favorable for the location of wells (e.g., wider spacing of elevation contours infers areas of higher hydraulic conductivity).

The direction of movement of ground water may indicate the location of its recharge area. In hilly terrain it is desirable to locate a well within the area of main drainage rather than in areas of subsidiary drainage because recharge to the well will be greater and more reliable throughout extended dry periods.

Ground water contour maps reveal concealed barriers to ground water flow, such as faults (abrupt changes in water level) or shallow bedrock ridges. These linear features are indicated by zones of relatively large water level difference.

Contour maps sometimes show near-circular depressions that may be caused by pumping wells. An example of a ground water contour map is shown in Figure 3.3.

3.5   THE HYDROLOGIC BUDGET

If the information search justifies further study, the next step is to construct a simple hydrologic budget, using easily accessible rainfall data (Where does the water come from?), a topographic map (Where does it go?), and a generalized geologic map (What is the nature of its geologic container?). It quickly can be determined if the objectives in terms of the geohydrologic setting are reasonable. If so, a field reconnaissance visit is conducted. This will improve the geohydrologic perspective and provide the bulk of the

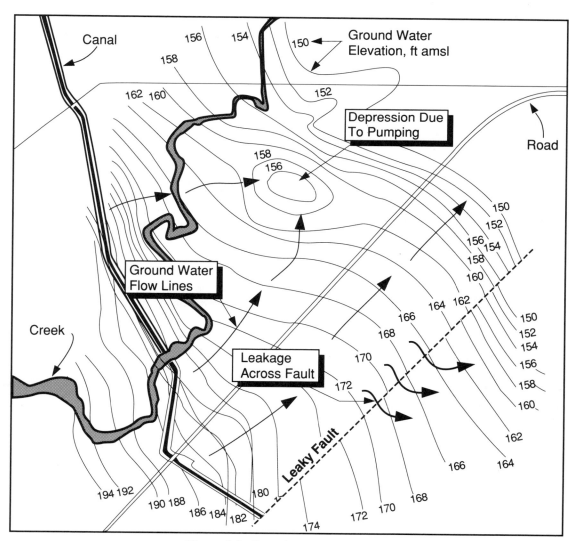

Figure 3.3. Ground water contour map.

conceptual solution to the problem. If not, the project must be either terminated or redefined.

A hydrologic budget is an accounting of the surface and ground water resources entering and leaving a specific geographic region. Since it is used to estimate the quantity of ground water that may be safely and economically withdrawn, every investigator and user of ground water should be familiar with the concept. Based upon the law of conservation of matter (Chapter 2), the hydrologic budget equates the total inflows with total outflows plus change in storage within the area and may be expressed in equation form as

$$P + SW(i) + GW(i) + So$$
$$= ET + SW(o) + GW(o) + Wd + \Delta S ,$$

where

P	=	precipitation,
$SW(i)$	=	surface-water inflow,
$GW(i)$	=	ground water inflow,
So	=	other sources (irrigation, spreading, interbasin transfers in),
ET	=	evapotranspiration,
$SW(o)$	=	surface-water outflow,
$GW(o)$	=	ground water outflow,
Wd	=	withdrawals (pumpage, interbasin transfers out),
ΔS	=	change in storage.

The ground water investigator is primarily concerned with estimating ground water inflow and outflow. For planning long-range needs, average annual values may be used for the variables. For predicting near-term needs, recent short-term values may be used. See Chapter 20 for an example of a ground water budget.

Total or partial hydrologic budgets (where not all of the foregoing terms are known) are used in many applications; some are:

- In agriculture, hydrologic budgets are used to predict the appropriate amounts and timing of irrigation applications. A significant variable is potential evapotranspiration, the amount of water that is evaporated from soil and transpired by plants assuming that there is no soil-moisture deficiency. A soil-moisture deficiency occurs when the plant does not receive all the water that can be transpired. Evapotranspiration depends upon climatic factors such as temperature, solar radiation, wind velocity, and relative humidity. In addition each crop has a unique water demand which varies according to the stage of its growth.
- Engineers use hydrologic budgets to develop forecasts of surface flows to design and operate water conveyance

and storage systems. They are particularly useful when the additional and sometimes competing goals of hydroelectric generation and flood control are considered. In arid climates where evaporation rates are high and surface-water loss significant, budget analysis may suggest that excess water be stored in alluvial basins through spreading or injection.

- Urban planners use hydrologic budgets for forecasting both near- and long-term water resource supplies. Near-term forecasts are used to allocate existing supplies, whereas long-term forecasts are used to plan major capital expenditures.
- Long-term hydrologic budgets are important in land-use planning, whether the target area is raw land or developed land undergoing conversion from one use to another. Vital decisions such as the project size and design of production and conveyance systems depend upon the hydrologic budget.

Calculation of a Hydrologic Budget

The basic calculation of a hydrologic budget is relatively simple, involving only the subtraction of total outflow from total inflow, plus or minus the change in storage within the study region. However, estimation of values for the individual variables within the equation may be difficult. Appendix C outlines methods of measuring these variables and calculations used.

An acceptable estimate of evapotranspiration may be especially difficult to derive since it includes water evaporated from water surfaces, soils, and other surfaces, as well as that transpired by vegetation. Many techniques are available to estimate potential evapotranspiration (PE). In humid areas this may present no problem. In arid areas the estimate may exceed precipitation by a factor of three or four, providing no disposition of precipitation by runoff or infiltration. In a study of an arid area, the budget value used is actual evapotranspiration (AE), the amount of water actually evapotranspired under conditions of moisture deficit. A preliminary estimate may be derived by calculating monthly PE and using the following considerations to derive AE:

1. In months when precipitation exceeds or equals PE, AE = PE.
2. When PE is greater than precipitation (i.e., no percolation), AE = precipitation plus a soil moisture increment stored from the previous month.

Arid areas with large topographic relief may also present difficulties in the estimation of precipitation. In desert basins long-term precipitation estimates of a few inches per year are common. However, their recharge areas, because of orographic effects, may receive in excess of 20 in. Consequently the study boundaries must encompass the recharge areas of confined aquifers, though they may lie far outside the area of interest.

Ground Water in Storage

Although the total amount of ground water stored within the aquifer or aquifer system is not an element of the hydrologic budget, it is essential to ground water investigators because it represents water available for use. However, often only a portion of ground water in storage may be exploited without creating such undesirable effects as seawater intrusion, land subsidence, or unacceptable deterioration of the ground water quality.

To derive an estimate for ground water in storage within a basin, it is necessary to multiply the volume of saturated aquifer materials by their average porosity. To obtain aquifer volume, it is necessary to know the basin boundary, the shape and thickness of the geological container, and the nature of the aquifer materials. Usually only a very rough estimate can be made during the preliminary phase of a ground water study.

Geohydrologic Reconnaissance

Field reconnaissance rarely takes more than a day or two. Tools include an automobile, a notebook, and a U.S. Geological Survey topographic map, preferably a 15-min rather than than a 7½-min quadrangle. Any geologic map used should be small scale and generalized.

Use of a four-wheel-drive vehicle, unless absolutely necessary, is not appropriate at this stage. It will entice time-consuming visits to a few remote localities, thereby reducing the general perspective that would otherwise be obtained by broad traverses along good roads throughout the entire area. Water wells are already in or will probably be located in easily accessible locations, and more water is found in broad flat valleys than along steep ridges.

For the same reasons, overflights are usually a waste of time. A competent investigator will learn more from a topographic map or aerial photographs than from a brief view from the air. Later, with specific questions in mind, airborne reconnaissance may be very valuable, but the investigator must learn how to direct his or her own reconnaissance.

A scan of the logistics base is worthwhile. Lodging, subsistence, availability of equipment, repair facilities, and supply of supplemental workers should be evaluated. The investigator must determine whether there are local people on whom to call for information and assistance. The confidentiality required by professional ethics must always be considered.

Local drilling contractors are good sources of information. Their concepts and terminology may differ from that of a geohydrologist, but they know where the water is, what its quality is apt to be, and, most important, the problems involved in its development.

Useful information may also be derived from selected local residents such as farmers, industrial plant managers, and sometimes long-resident water diviners. Paradoxically, the local diviner, whose proffered art rests on no scientifically established principles, may be an excellent observer.

Data collected during field reconnaissance are primarily areal. Details recorded in a field notebook are referenced to specific locations on a map. One major task is the mapping of the surficial extent of aquifer materials. Surface observations coupled with standard geologic mapping techniques such as the delineation and evaluation of outcrops, formation contacts, fractures, and, on layered materials, measurements of strike and dip are useful. Better understanding of the geohydrologic environment is aided by preparation of vertical cross sections from these data in conjunction with lithologic logs of existing drill holes. Lithological samples may be gathered and preserved for later, more detailed examination.

A well inventory is usually made during the reconnaissance visit. The location and surface elevation of each well found together with any other relevant information, such as owner, total depth, and current status—abandoned, inactive, or active—is recorded. Geohydrologic tasks such as water level, field water-quality measurements (temperature, conductivity, pH, dissolved oxygen, etc.), and collection of water samples for laboratory analysis may be accomplished. Frequently, however, wells are sealed for protection and special arrangements for access must be made. It may be worthwhile to repeat water level measurements during subsequent field visits to check for seasonal variations.

Stream courses should also be located on the field map, and an estimate of their flows made. This may be done by timing a float along a measured reach at a point where an approximation of cross-sectional area of the flow channel can be made. Mapping of natural surface water bodies, as well as associated seepage areas, may help to define a discharge zone or the location of barriers acting as natural dams to ground water flow.

Springs represent ground water discharge zones. They should be mapped, flows estimated, and samples taken for water-quality tests.

When the reconnaissance is completed, the data are compiled, analyzed, and collated with the data collected from the information search. The preliminary hydrologic budget is revised, allowing a more reliable opinion whether or not objectives can be reached. If the opinion is marginal or clearly positive, the investigator will design a phased program of investigation proceeding from the general to the specific.

Identify Remaining Uncertainties

The information search, field reconnaissance, and hydrologic budget analysis complete the preliminary phase of the exploration study, and those elements needing additional in-depth study are usually identified.

Options available are:

1. In a well-understood ground water environment with an apparently adequate potential resource, proceed to a program of production drilling and development.
2. If the potential ground water resource appears insufficient to meet project demand, investigate possible alternative sources to replace or augment the ground water resource. These may include utilization of surface water through the construction of a dam, water purchase, or project redesign to reduce demand.
3. If the potential resource appears adequate but the ground water environment is not well understood, complete the exploration study using appropriate geophysical or geohydrologic methods. The suitability of the various investigative methods is evaluated, and the most cost-effective program for providing the needed information is selected.

In a sense no geological or geophysical investigation is ever completed, since the results are only indicative and predictive. In the judgment of the investigator, sufficient work must be done either to justify the cost of exploration drilling or to dictate termination of the project. There are no hard and fast rules to define the course of exploration, except that the work should move in the direction of increasing costs for the information sought. The costs of geological and geophysical exploration must always be much lower than for exploration drilling. Otherwise drilling, which provides the only direct subsurface information, must be done instead.

Currently (1989), the costs of geological and geophysical exploration in other than single-family domestic well projects will range from a few thousand to a few tens of thousands, and rarely reach or exceed one hundred thousand dollars. The cost for a well that produces 1000 gpm or more ranges from several tens to a hundred thousand dollars or more. Even an exploration borehole is likely to be expensive because, unlike exploration drilling for ore deposits, an exploration borehole for water has to be completed into a test well.

3.6 GEOHYDROLOGIC METHODS OF INVESTIGATION

Testing of Existing Wells and Boreholes

When wells are located within the study area, many techniques of investigation are available that will yield valuable information. Wells may be geophysically measured using a variety of methods to determine the geohydrological character of the aquifer materials they penetrate. Geophysical borehole logging is discussed in detail in Chapter 4. Aquifer char-

acteristics can be found by conducting careful tests on pumping wells, as covered in Chapter 15.

Bail and Slug Tests

Bail and slug tests can be used to derive estimates of aquifer transmissivity. Both involve the measurement of residual drawdown, or water recovery, defined as the difference between the observed and the static water level at a given time. A bail test involves the instantaneous removal of a volume of water from a borehole, followed by timed measurements of water level recovery. A slug test involves the injection of a volume of water into a borehole, followed by timed monitoring of the water level decay.

Both methods are based upon the Theis equation, discussed in detail in Chapter 5, and have the same assumptions, including:

- The aquifer is homogeneous, isotropic, and of infinite areal extent.
- The well penetrates and receives water from the entire thickness of the aquifer.
- Water is removed from aquifer storage in immediate response to a drop in the piezometric surface or water table.

For bail tests the equation used is (4)

$$s' = \frac{V}{Tt}, \qquad (3.1)$$

where

s' = residual drawdown (drawdown measured after bailing stops) [L],
V = volume of water removed in one bail cycle [L^3],
T = Transmissivity [L^{2}T^{-1}],
t = length of time since bailer was removed, where t is sufficiently large [T].

When repeated bailing cycles are performed, the equation is modified as follows:

$$s' = \frac{1}{T} \left[\frac{V(1)}{t(1)} + \frac{V(2)}{t(2)} + \cdots + \frac{V(n)}{t(n)} \right], \qquad (3.2)$$

where $v(i)/t(i)$ identify each cycle of the bailer.

The slug test involves the injection of a slug of water into a well (5). An advantage is that it is not necessary to have bailing equipment available. It is based upon the same assumptions as the bail test.

When aquifers are shallow, small-diameter temporary wells are sometimes drilled for test purposes. However, results must take into consideration the scaling effects discussed in Chapter 15. This is an economic alternative to more indirect methods of exploration.

In utilizing data from bail and slug tests, it should be realized that these methods depart from standard pumping tests conducted under more controlled conditions. Less confidence can be placed in the values derived.

3.7 GEOPHYSICAL STUDIES

Geophysical studies involve measurements of basic physical parameters to infer subsurface conditions. Measurements may be made above the earth's surface by aircraft, at the earth's surface, or in boreholes.

Many geophysical techniques developed for oil exploration have been adapted to ground water exploration. Table 3.2 lists the potential contribution to exploration of most common techniques and estimates their cost in 1988 dollars. Costs include data reduction and interpretation.

Some geophysical techniques are passive, measuring naturally occurring phenomena. Among these are gravity studies that relate to the density of earth materials, magnetic studies that relate to their magnetic properties, and temperature studies. Others are electrical techniques that measure naturally occurring electromagnetic waves, such as spontaneous potential and telluric studies.

There are also active techniques in which the reaction of formations to a specific disturbance is measured. These include seismic studies that measure the response to a generated pressure wave and resistivity techniques that measure the electrical characteristics in response to the generation of an electrical field. Because both types are remote sensing, the information they generate is necessarily ambiguous. Moreover the field work and interpretation require a high level of technical knowledge and experience.

It is preferable to acquire geophysical data in proximity to logged wells and/or surface exposures of formations within the survey area to allow calibration with known lithologies.

Electrical Methods

There are a number of geophysical techniques that measure electrical properties of rocks in the earth's crust to predict subsurface conditions. Many geologic materials are relatively resistant to electrical current. When they are saturated with ground water, they are better conductors. Conductivity also increases with increasing ground water salinity. The most commonly used electrical method is the direct-current resistivity study in which a current is introduced into the ground and the potential induced by the current flow is measured.

The resistance, R, of a wire to electrical current depends upon its material, increases with increasing length of the wire, and decreases with increasing cross-sectional area of the wire so that

$$R = \frac{\lambda L}{A}, \tag{3.3}$$

where

R = resistance [ohm],
L = length of wire [L],
A = cross-sectional area [L^2],
λ = constant of proportionality, the material resistivity [ohm-L].

In electrical resistivity surveys, the length unit is the meter. Therefore the standard dimensional unit of resistivity is the ohm-meter.

Material resistivity may be derived from Ohm's law, which states that resistance is directly proportional to the potential difference across the resistance, V, and inversely proportional to the current, I, or

$$R = \frac{V}{I}, \tag{3.4}$$

where

R = resistance [ohm],
V = potential difference [v],
I = current [amp].

Combining these two equations and solving for λ,

$$\lambda = \frac{AV}{LI}. \tag{3.5}$$

The range of resistivity of geological materials is very broad. Clays exhibit very low resistivities. Other dry unconsolidated materials usually have much higher resistivities. if water is present in the pores of these same materials, their resistivities are significantly lowered. Non-water-bearing bedrock materials have very high resistivities.

Electrical resistivity equipment includes a generator and voltage-reading equipment. Electrodes may be driven copper stakes or surface electrodes (known as "pots" and containing copper sulfate) that are placed upon the ground. A geophysicist designs the array and operates the equipment, usually helped by one or more field assistants who move the electrodes.

Field measurements of resistivity are made by placing four electrodes at predetermined spacings. A direct or low-

TABLE 3.2 Geophysical Techniques Applied to Ground Water Exploration Projects

Method:	ELECTRICAL RESISTIVITY
Application:	To determine depth to layer boundaries, fresh/saline water interface, water in fractures
Field personnel requirements:	Geophysicist, technicians
Representative project size:	5 lines (1000′ ea)
Approximate dollars per unit:	$500–$1000/line
Method:	STREAMING POTENTIAL
Application:	To locate ground water flow zones
Field personnel requirements:	Geophysicist, technicians
Cost:	$500–$1000
Method:	SEISMIC REFRACTION
Application:	To determine depth to water table, number of and depth to layer boundaries and bedrock, dips of layers
Field personnel requirements:	Geophysicist, licensed blaster, field technician
Representative project size:	5 lines (2500 ft ea.)
Approximate dollars per unit:	$1000–$2000/line
Method:	GRAVITY
Application:	To determine relative depth to bedrock and thickness of alluvium within an area.
Field personnel requirements:	Geophysicist
Representative project size:	100 stations (readings)
Approximate dollars per unit:	$100/station
Method:	THERMAL SURVEY
Application:	To map high permeability zones for well location. To identify areas of shallow concealed bedrock and fault barriers
Field personnel requirements:	Geologist, drilling technician
Representative project size:	100 stations
Approximate dollars per unit:	$150/station
Method:	THERMAL LOGGING
Application:	To determine the depth and thickness of aquifers. To observe confined aquifer relationships
Field personnel requirements:	Geologist
Representative project size:	10 well logs
Approximate dollars per unit:	$250/log
Method:	MAGNETICS
Application:	To determine occurrence and relative depth to iron-rich bedrock
Field personnel requirements:	Geophysicist
Representative project size:	100 stations
Approximate dollars per unit:	$100/station

frequency alternating current is introduced via two of these electrodes, and the voltage difference is measured between the other two. There are two commonly used electrode spacings: the Wenner array, in which the spacing between electrodes is equal, and the Schlumberger, in which the spacing between the potential, or voltage, electrodes is much smaller than the spacing between the current electrodes (Figure 3.4). Other arrays such as dipole–dipole and bipole–bipole are becoming more popular. Mooney, Freeze and Cherry, and Fetter provide information on these methods (6, 7, 8).

Two techniques may be used in electrical resistivity surveying: sounding and profiling. In sounding, the electrodes are moved farther apart for successive readings. With increasing distance, the current will penetrate more deeply, and thus the potential difference reflects the electrical properties of materials at increasing depth.

In resistivity profiling, the spacing between electrodes is kept constant, and the array is moved from place to place over the study area. A resistivity profile is drawn to show the resistivity difference of an area at some constant depth.

Interpretation. The field data recorded include the magnitude of current applied, electrode spacings, and the potential reading at each electrode spacing. The potential reading depends upon the array of the electrodes. Therefore an array equation is used to convert the field reading to a value termed the apparent resistivity. These apparent resistivities are plotted on logarithmic paper against one-half the spacing of the outer, or current electrodes, $AB/2$, and the resulting curve is interpreted to derive the number of resistivity layers penetrated, the depth to the top of each layer, and the actual resistivity of each layer. The analysis of resistivity curves is complex because a curve can represent an infinite number of models. Interpretation has been greatly aided by the preparation of theoretical master curves against which field curves can be matched (9, 10).

Electrical resistivity surveys are most successful when they can be accomplished free of potential disturbances such as buried pipes, metal fences, and other structures that will distort the electrical field. When a point of known lithology, such as a borehole, lies within the survey area, a profile should be run close to it to provide a point of calibration. When bedrock is exposed, it is also worthwhile to obtain resistivity information on the bedrock to aid in data interpretation.

Use of Resistivity Surveys. Resistivity soundings and profiles have been used to determine the depths and thicknesses of saturated formations, and to investigate ground water occurrence in fractured bedrock. They are particularly appropriate in mapping the depth to freshwater–saline water interfaces and thus have been used extensively in saltwater intrusion problems and Ghyben-Herzberg conditions.

Streaming Potentials

The measurement of streaming potentials is a relatively new technique in ground water exploration. It is an inexpensive tool for helping to locate ground water flow zones and structures that control the ground water movement. A streaming effect occurs when a fluid is forced through a capillary or porous medium. As it flows, its electrons align and a small current is created. This induces an electrical field, called the "streaming potential," in the surrounding material (Chapter 4).

The instrumentation required to measure the electric field consists of a digital voltmeter, two nonpolarizing electrodes, and wire. A nonpolarizing electrode consists of a metal electrode submerged in a saturated solution of its own salt. Best results have been obtained using silver in silver chloride because of its short voltage-equilibrium period.

The electrodes are placed in a dipole array in the field. Measurement of the earth's voltage (in millivolts) is made using the voltmeter, and the electrodes are moved for the next reading. Necessary corrections are made for telluric signals (the earth's low-frequency current, about 1 mV/km,

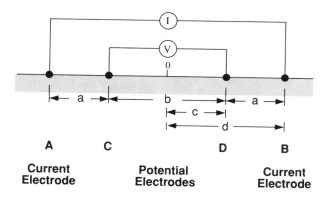

Wenner Array   (a=b)

$$\rho_a = 2\pi\, a\, \frac{V}{I}$$

When spacing a is increased, b is increased so that a = b.

Schlumberger Array $\left(c < \frac{d}{5}\right)$

$$\rho_a \text{ (apparent resistivity)} = \frac{\pi d^2}{c}\, \frac{V}{I}$$

Current electrode spacing (d) is increased for each measurement. To maintain a detectable voltage reading, the potential electrode spacing (c) is increased periodically. ρ_a = apparent resistivity.

Figure 3.4. Electrical resistivity arrays.

generated by variations in its magnetic field) and for electrode drift. Once corrected, the resulting values are drawn in profile or as a contour map of equipotentials and analyzed with respect to what is already known for the area. Although it is possible to map buried flow zones with this technique, it is not yet possible to find ground water flow rates.

Seismic Refraction Surveys

Compressional waves, such as those generated by sound or pressure, cause the materials through which they travel to oscillate in a direction parallel to the direction of the disturbance. This results in a wave train composed of intervals of compression and rarefaction, whose velocity is directly proportional to the material compressibility and inversely proportional to its density, or

$$v = \sqrt{\frac{B}{\rho}}, \qquad (3.6)$$

where

v = velocity $[LT^{-1}]$,
B = bulk modulus $[FL^{-2}]$,
ρ = density of the material $[ML^{-3}]$.

The bulk modulus, which measures the compressibility of a material in response to an imposed stress P, is

$$B = \frac{\Delta P}{\Delta v / v}. \qquad (3.7)$$

Although the density of earthen materials varies little, their compressibility varies by more than four to five orders of magnitude, depending upon porosity. Shallow unconsolidated porous media have relatively large porosity that decreases with increasing consolidation and depth. Unjointed crystalline rocks, on the other hand, essentially have no primary porosity. In general, the velocity of a seismic wave increases with increasing depth.

In a horizontal stratified model a sonic disturbance at or near the earth's surface will generate seismic waves that travel out in all directions from the source. For illustrative purposes, the actual paths followed from the energy source to a recording instrument, the geophone, are shown as rays drawn normal to the wavefront, as shown in Figure 3.5. One ray, $R1$, travels horizontally at constant velocity. Another, $R2$, travels downward until it reaches the upper boundary of a higher-velocity layer. It then traverses the boundary at the higher velocity until refracted back through the upper layer.

Depending upon the relative velocities of the layers and the distance of the geophone from the source, the refracted

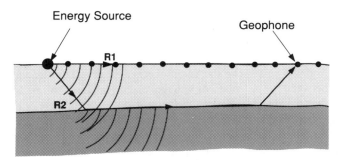

Figure 3.5. Seismic refraction survey.

ray may reach the geophone prior to the direct ray. In the survey several geophones are placed at increasing distances from the energy source. The distance is compared to the elapsed time between the generation of the disturbance and its first arrival at a geophone. A time–distance graph, similar to Figure 3.6, is prepared. From the slopes the velocities and depths of each layer from which energy has been refracted can be estimated.

A geophysicist is usually in charge of the survey and operates the equipment as it must be very precisely tuned in relation to background noise to obtain accurate recordings.

The energy for the seismic waves for short-distance seismic surveys may be a metal plate on the ground that is struck with a sledge hammer, or the dropping of heavy weights. Depending on the formation, penetration of the shock wave will be a few tens of feet. Recently, repeated hammer blows have been used with a stacker that cumulates the geophone

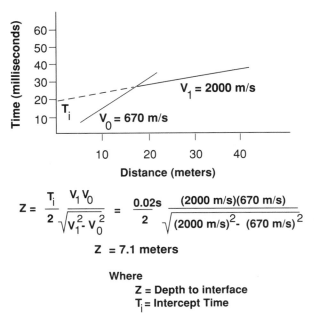

Figure 3.6. Time-distance graph.

response to each blow. A buried explosive charge can be used to achieve penetration of more than about one hundred feet.

The velocities calculated are apparent velocities and are equal to actual velocities if the layer boundaries are horizontal. For this reason seismic refraction profiles are usually run at both ends of a seismic profile. This is referred to as "back shooting." The time–distance curves are mirror images if the interfaces are horizontal. If they are not, back shooting allows the calculation of the inclination of the interface and the true layer velocities.

Interpretation. Information obtained through seismic refraction is used to interpret the lithology of each layer as well as the geometric configuration of the layer boundaries. Lithologic interpretations are made through reference to a table such as Table 3.3, which lists representative velocities for different types of geologic materials. Since many formations have similar seismic velocities, uncertainty is inherent in seismic refraction studies.

Use of Seismic Refraction Studies. Seismic refraction studies are used to estimate the thickness of unconsolidated materials, the top of the saturated zone, and the depth and configuration of the bedrock surface. However, this technique has one inherent disadvantage in that a lower-velocity layer cannot be detected beneath a higher-velocity layer. Since one or more field technicians are necessary to install geophones and to dig or drill holes for the explosives, seismic refraction surveys are relatively expensive.

Gravity Surveys

Gravitational acceleration varies slightly from point to point on the earth's surface because of small differences in the distance between the surface and the center of the earth, and because of slightly varying densities of the geological formations beneath the point of measurement. Since unconsolidated materials have a lower density than well-consolidated sedimentary formations and igneous rocks, the thickness of alluvium and the general configuration of the basin can be estimated through interpretation of gravity data.

The collection of gravity data is quite simple, but interpretation is time consuming and complex, requiring a number of corrections for latitude, elevation, and drift. Results give only gross information. Therefore gravity surveys are not often employed in geohydrologic investigations. However, since data have been published from a number of surveys performed by public agencies throughout the United States, it may be useful to acquire existing gravity data to aid in the analysis, especially for recognizing variations in depth to bedrock.

TABLE 3.3 Representative Seismic Velocities of Geological Formations

Formation	P-Wave Velocity (ft/s)
Topsoil (*u*)	330–1000
Topsoil (*s*)	4900–6500
Sand (*u*)	1000–1600
Sand (*s*)	4900–6500
Silt (*u*)	1600–3300
Silt (*s*)	4900–6500
Gravel (*u*)	1600–5000
Gravel (*s*)	4900–6500
Clay, till (*u*)	3300–7200
Clay, till (*s*)	5000–7200
Sandstone, conglomerate (*u*)	5000–8200
Shale	5900–14,700
Limestone	6500–19,600
Weathered, fractured granites, volcanics, metamorphics	3900–11,500
Unweathered granites, volcanics, metamorphics	13,100–32,300

Note: *u* = unsaturated; *s* = saturated.
Sources: D. K. Todd, 1980, *Groundwater Hydrology* 2nd ed., Wiley, New York.
J. J. Jakosky, 1950, *Exploration Geophysics*, 2nd ed., Times-Mirror Press, Los Angeles.

Thermal Surveys

Temperature is a basic property of any material. Heat, a form of energy, always flows from a region of high temperature to one of lower temperature. This process is known as conductive heat transfer. Since the interior of the earth is appreciably hotter than the surface, there is a constant outward flow of heat from the earth's interior toward the surface. Temperature increases with depth below surface. This is referred to as the geothermal gradient. At any location it is considered constant and is ordinarily between 1°C to 2°C per 100 ft of depth.

At shallow depths, the temperature is also affected by the atmospheric or ambient temperature. There are two major influential cycles: the diurnal, or day-to-night cycle, and the seasonal, or summer-to-winter cycle. The diurnal cycle penetrates only a few feet below the surface. The seasonal cycle is attenuated as depth increases and is no longer discernable beneath about 50 ft. Below this level, the temperature distribution is constant throughout time, unless influenced by flowing ground water.

In a porous medium through which lower-temperature ground water flows, heat is transferred from the porous medium to the ground water. This incrementally increases the temperature of the water and decreases the temperature of the surrounding materials. Since water has a high specific heat in relation to earthen materials, it is effective in absorbing and removing heat from the medium, a process combining conductive and advective heat transfer.

Because of these physical processes, as the permeability of porous media increases, ground water flow increases and advective heat transfer becomes more efficient. Above the zone of advective heat transfer, conductive heat flow adjusts the thermal configuration in the near-surface environment. Thermal surveys are used to map zones of higher transmissivity.

Field Methods. Thermistors—materials whose electrical resistance varies precisely with temperature—are emplaced at shallow depth below the surface (usually about 10 ft). Probe spacings through large basins are usually at mile or half-mile intervals, with the spacings later shortened to define specific areas of interest. They are located in a predetermined grid according to the objectives of the survey. This depth is below the penetration of the diurnal cycle but well within the region of the seasonal cycle. The probe holes are drilled with a power auger or a posthole digger. PVC pipes with removable caps may be installed for permanent access.

The probes are set in place and allowed to reach temperature equilibrium—a minimum of 24 hr after the holes are drilled, or 1 to 2 hr at any later reading. The sensors are read with a resistance bridge. The resistances are converted to temperatures through the use of calibration tables prepared for each sensor. At least two sets of readings are taken, a few days to a few weeks apart, in order to define the seasonal temperature drift. The measured temperatures are placed in base maps and contoured as isotherms, as shown in Figure 3.7.

The temperature configuration is interpreted with respect to the geohydrologic setting and all surface factors that may influence the temperature regime at the sensor depths. In the interpretation of a temperature survey, both the isothermal contours and the drift contours are used. Zones of higher permeability materials are characterized by curvilinear areas of lower temperature. Occasionally, linear zones of higher temperature are seen, frequently denoting deep ground water rising along faults. As with all geophysical methods, there are no unique solutions. Correct interpretations from a choice of alternatives depend directly on the investigators' understanding of the geohydrologic setting and experience with the use of thermal surveys.

Among the advantages of the thermal technique is that for ground water below the influence of the seasonal temperature wave, it responds only for water that is moving. This is an advantage because nonmoving ground water is likely to be of poor quality within tight materials. Another advantage is that the apex of the thermal anomaly is directly over the flow that causes the anomaly. This reduces the magnitude of any test drilling program. Also the cost of sensor emplacement is small, allowing large areas to be surveyed quickly and inexpensively.

There are important limitations. The thermal technique cannot provide quantitative values for depth to ground water nor the quantity of flow. Moreover the thermal effects of a shallow aquifer will conceal deeper, less active flow zones.

It is essential that the sensors be set well below the reach of the diurnal temperature wave. Most of the thermal effects produced by moving ground water are totally masked by the extremely complex thermal "noise" at the surface. For this reason ground-based or aerial infrared surveys are inadequate for detecting water other than that which reaches the surface, or within a few feet of the surface.

In general, the technique cannot be used in rugged terrain or in areas of extensively varying vegetation cover. A serious disadvantage is its deceptive simplicity. Inexperienced investigators are likely to make serious errors in interpretation, especially if the geohydrologic setting is not well understood. With these limitations the thermal technique is a very powerful tool in recognizing major ground water flow systems within about 500 ft of the surface.

Thermal Logging

An areal thermal survey is enhanced by down-hole thermal logs (if existing wells are available) that provide third-dimension information such as the number of aquifers, and their depths and thickness. A cable-mounted thermistor is used to record equilibrium temperature at selected depths. The vertical temperature profile reflects zones of greater horizontal groundwater flow and may also indicate crossflow, or flow through a borehole between confined aquifers due to pressure differences, as shown in Figure 3.8.

Equilibrium thermal logs are considered more useful than the continuous thermal logs sometimes run in a newly drilled borehole. This is due to the necessity of dissipating the heat generated during drilling, a process usually requiring several days.

Magnetic Surveys

Magnetic surveys are useful in helping determine the configuration of basement rocks containing slightly more iron-bearing mineral than overlying sediments. They consist of measurements of the intensity or change in intensity of the local magnetic field either at a number of ground stations or along one or more continuous aircraft traverses.

Two different types of magnetometers, fluxgate and proton, are commonly used. Two magnetometers of either type may be combined to form a gradiometer, an instrument capable of measuring directional components of the magnetic field.

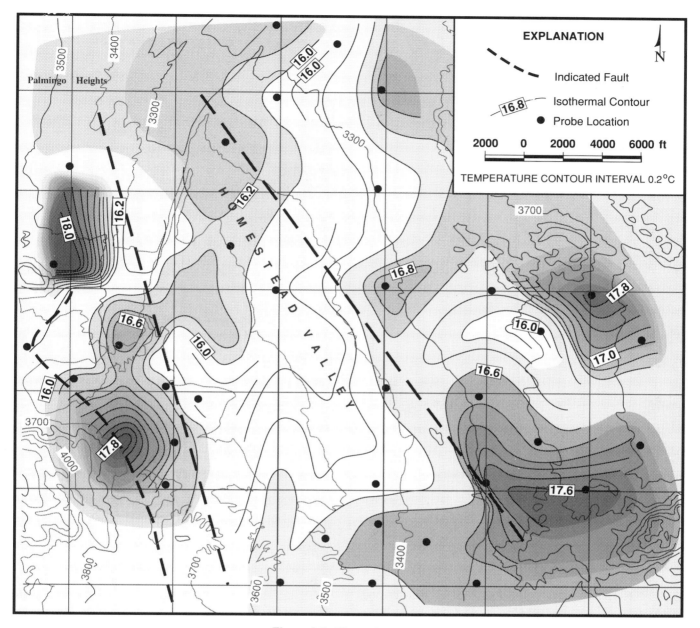

Figure 3.7. Thermal survey.

In the United States the vertical component of the magnetic field is more commonly measured than total field strength. The average intensity of the magnetic field in the United States is 50,000 nanotesla (1 nanotesla = 1 gamma).

Once measurements have been made they must be corrected for a number of factors, including secular (long-term) and diurnal variations in the earth's magnetic field, and for latitude and longitude. Measurements are also affected by magnetic storms. These storms are related to sunspot activity and have a recurrence interval of about 27 days. Surveys should

be halted during magnetic storms because of the rapid and extreme variations in the magnetic field intensity.

Due to the rapid decrease in intensity of induced magnetic field with distance, most interpretations of buried structures involving materials of similar magnetic susceptibility should be limited to those at shallow depths. Vertical offsets, even if very large, in material at depth in a large basin may produce only minor changes in magnetic intensity at the surface. Nevertheless, the distribution of lithologies with greatly dissimilar magnetic properties can often be inferred

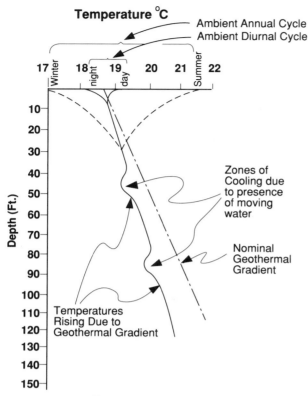

Figure 3.8. Thermal log.

from magnetic data, even when covered by a considerable thickness of sediments. The sediments generally have very low magnetic susceptibilities.

3.8 SYNTHESIS OF DATA

After all test and survey information has been gathered and analyzed, it is the ground water investigator's responsibility to derive a conceptual geohydrologic model that best fits the observed data.

From well records and static water level measurements, conclusions can be drawn about the number and types of aquifers. From the position of the static water level and the depths to various boundaries, estimates of aquifer thickness can be made. Estimates of hydraulic conductivity and aquifer thicknesses are used to predict transmissivities and potential aquifer yields. Aquifer thicknesses may also be used in conjunction with predicted zones of greater hydraulic conductivity to make well site selections. From surface resistivity studies, sampling programs, and other information, considerations of potential water-quality problems may be possible.

The investigator must also establish the confidence level of his or her predictions. If an area has been extensively

studied and already contains many wells, a production drilling program may be justified without first undertaking test drilling.

3.9 TEST DRILLING

The major objective of a test drilling program is information rather than water, and it should be designed accordingly. These programs are relatively expensive, and therefore boreholes are usually located at sites of predicted maximum ground water production potential. However, to test the hypothesized aquifer model in a large water-resource study, one or more test wells might be located at sites thought unfavorable.

The choice of drilling method in a test well program should consider the quality of drilling samples. Methods that use circulating freshwater−base drilling fluid to support the borehole are often economical but produce poor samples. In addition it is very difficult to determine the depth at which water is first encountered and the potential production through depth intervals. Cable-tool-drilled boreholes produce excellent formation samples, particularly when casing is carried co-incident with drilling. However, the technique is often slower and more costly. The recently introduced dual-tube drilling system generally affords a good compromise among information, samples, and cost but is limited to depths of approximately 1200 feet.

Drilling of small-diameter holes reduces the cost of a test drilling program. They should be large enough, however, to permit down-hole logging when desirable. If temporary casing is installed, it is possible to bail test. The test hole may also be preserved and used as an observation well during later test pumping of production wells.

Analysis of cuttings generated by direct or reverse rotary drilling requires care in the sampling operation and understanding of the limitations of the results obtained. To perform grain-size analysis, the samples must be washed to separate drilling fluid from natural formation. Unfortunately, in doing so, native clay may also be removed from the sample. The dried samples are then weighed and sifted through a set of standard sieves, and the portion retained by each sieve is weighed (mechanical grading analysis). A chart is prepared that shows grain diameter versus cumulative percent of sample passing through each sieve. Sieve analysis is used for selecting filter packs and screen openings (Chapter 13) and may also give a first-order estimate of hydraulic conductivity (see Appendix D).

REFERENCES

1. Heath, R. C. 1989. "Ground-Water Regions of the United States." U.S. Geological Survey Water-Supply Paper (in review).

2. Thomas, H. E. 1952. "Ground-Water Regions of the United States, Their Storage Facilities." U.S. 83rd Cong., House Interior and Insular Affairs Comm. "The Physical and Economic Foundation of Natural Resources." Vol. 3.

3. Meinzer, O. E. 1923. "The Occurrence of Ground Water in the United States." U.S. Geological Survey Water-Supply Paper 489.

4. Skibitzke, H. E. 1963. "Determination of the Coefficient of Transmissivity from Measurement of Residual Drawdown in a Bailed Well." In Ray Bentall (ed.), "Methods of Determining Permeability, Transmissibility and Drawdown." U.S. Geological Survey Water-Supply Paper 1536-1.

5. Ferris, J. G., D. B. Knowles, R. H. Brown, and R. W. Stallman. 1962. "Theory of Aquifer Tests." Geological Survey Water-Supply Paper 1536-E.

6. Mooney, H. 1977. *Handbook of Engineering Geophysics*. Bison Instruments, Minneapolis, MN.

7. Freeze, R. A., and J. A. Cherry. 1979. *Groundwater*. Prentice-Hall, Englewood Cliffs, NJ.

8. Fetter, C. W., Jr. 1980. *Applied Geohydrology*. Charles E. Merrill Co., Columbus, OH.

9. Orellano, E., and H. M. Mooney. 1966. *Master Table and Curves for Vertical Electrical Sounding over Layered Structures*. Interciecia, Madrid.

10. Zohdy, A. A. R. 1974. "Electrical Methods." In *Techniques of Water Resources Investigation of the United States Geological Survey*. Application of Surface Geophysics to Ground Water Investigations, bk. 2, ch. D1, U.S. Geological Survey.

READING LIST

Bennett, R. R. 1962. "Theory of Aquifer Tests," Flow net analysis by Ferris, J. G., et al. Geological Survey Water-Supply Paper, 1536-E.

Freeze, R. A., and P. Witherspoon. 1967. Theoretical analysis of regional groundwater flow, 2. Effect of water-table configuration and subsurface permeability variation. "Water Resources Research, No. 3," pp. 623–634.

Heath, R. C., and E. W. Turner. 1981. "Ground Water Hydrology." Water Well Journal Publishing Co., Worthington, OH.

Hudson, A., and R. Nelson. 1982. "University Physics." Harcourt Brace Jovanovich, New York, ch. 16.

Morris, D. A. and A. I. Johnson. 1967. "Summary of hydrologic and physical properties of Rock and Soil Materials" as analyzed by the Hydrologic Laboratories of the U. S. Geological Survey, 1948-60, Geological Survey Water-Supply Paper 1839-D.

Plummer, C. C. and D. McGeary. 1979. *Physical Geology*, Chapter 15, Wm. C. Brown Co., Dubuque, Iowa.

Sun, R. J. 1986. Regional Aquifer-System Analysis Program, Summary of Projects, 1978-84, U.S. Geological Survey, Circular 1002.

"Drainage Manual," 1984. U.S. Department of the Interior Bureau of Reclamation, Government Printing Office, Denver, CO.

"Ground Water Manual," 1981. U.S. Department of the Interior, Water and Power Resources Service, Government Printing Office, Denver.

"Methods and Techniques of Ground-Water Investigation and Development," 1967. Water Resources Series No. 33, United Nations, New York.

National Handbook of Recommended Methods for Water-Data Acquisition, 1980. Chapter 2, Ground Water. U.S. Geological Survey.

National Handbook of Recommended Method for Water-Data Acquisition, 1978. Chapter 7, Basin Characteristics. U.S. Geological Survey.

CHAPTER 4

Geophysical Borehole Logging

4.1 INTRODUCTION

Prior to development of geophysical well-logging methods, down-hole information on water wells was limited to lithologic logs. Although formation sampling is still a vital component of the well construction process, geophysical borehole logging provides additional information independently and through correlation with formation samples.

Lord Kelvin performed the first borehole log in 1869 when he measured the earth's temperature in a shallow hole. The Schlumberger brothers recorded the first electrical resistivity log (*E*-log) at Pechelbron, France, in 1927. Since then more than 50 geophysically derived well logging devices have been introduced, based on various principles including electrical, nuclear, and chemical.

Since geophysical borehole logging was developed for use in oil and gas exploration, there is relatively little published information available on log interpretation for freshwater development. Unfortunately, the methods for oil and gas do not directly apply because of the large differences in the ionic makeup and resistivities of interstitial waters between shallow freshwater and deep hydrocarbon-bearing zones. Nevertheless, experience demonstrates that geophysical borehole logging can be used to evaluate aquifers, although the parameters of interest in water wells (yield and water quality) are different from those important to oil and gas exploration. This chapter discusses the basic principles of geophysical borehole logging and their use in evaluation of freshwater zones. Figure 4.1 classifies the principal geophysical logging systems applicable to ground water exploration. Although neutron and density logs are used in ground water studies outside the United States, they are not discussed because federal regulations generally prohibit their use.

Functions of Well Logging Surveys

The primary functions of geophysical water well logging are to obtain information on formation character, yield potential, and water quality. These are obtained indirectly through measurement or calculation of:

1. Formation electrical resistivity (water-yielding properties).
2. True resistivity of formation water (water quality).

3. Formation porosity (water-yielding properties).
4. Formation clay content.

Other information obtainable through down-hole logging devices includes:

1. Variations in borehole diameter.
2. Formation temperature.
3. Fluid velocity in the well.

Figure 4.2 is a block diagram of typical geophysical well-logging equipment.

4.2 FORMATION RESISTIVITY AND CONDUCTIVITY

Resistivity logging measures the resistance of formations to the flow of electricity. These measurements are corrected to estimate the true formation resistivity, and through interpretation formation characteristics are derived. Resistivity and conductivity are reciprocals of each other:

$$\sigma = \frac{1}{\rho},$$

where

σ = conductivity,
ρ = resistivity.

The unit of conductivity used in geophysical borehole logging is the mho/m, which is the conductance of a cube, one meter on a side, drawing one ampere of current under one volt applied between two opposite faces of the cube. The unit of resistivity logging is the ohm-m.

For comparison, resistivity for various electrolytic solutions are listed in Table 4.1.

Sources of Formation Conductivity

Geologic formations conduct electricity through interconnecting pore spaces filled with formation water which acts

Log Type or Device	Uses: Open Hole (O) or Cased Hole (C)	Lithology	Bed Thickness	Clay & Shale Content	Total Porosity	Effective Porosity	Chemical & Physical Properties	Source & Movement of Water in Well	Cementing	Well Construction	Casing Corrosion	Hole Diameter
Spontaneous Potential	O	X	X	X			X					
Short Normal	O	X	X	X								
Long Normal	O	X	X	X	X	X						
Lateral (6 ft)	O	X		X								
Caliper Log	O											X
Caliper Log	C									X	X	X
Gamma Ray Log	O	X	X	X								
Gamma Ray Log	C	X	X	X								
Sonic Log	O	X			X							
Sonic Log	C	X								X		
Temperature Log	O							X				
Temperature Log	C							X	X			
Flowmeter Log	O											
Flowmeter Log	C							X				
Directional Survey	O											
Directional Survey	C											

Figure 4.1. Summary of logging systems used in water wells.

as an electrolyte. In most cases the matrix surrounding the pores is essentially nonconductive, thus conductivity of a given volume of porous formation will be lower than an equal volume of formation water. Pure rock matrix containing no water would have no conductivity (or infinite resistivity). Formation-water conductivity is due to the presence of ions of salts dissolved in the water and is primarily a function of two factors:

1. The salt concentration in the solution [ppm].
2. The chemistry of the salt, or the nature of the ions.

The chart of Figure 4.3 shows the conductivity of various salt solutions versus concentration at 64°F.

4.3 PRINCIPLES OF RESISTIVITY MEASUREMENT

Single-Point Resistivity Logs

The single-point logging method uses a single down-hole electrode, with electric current supplied from the surface through an insulated conductor. Current flows from the electrode into the formation, returns to the surface, and flows back to the current generator through a grounded electrode (usually placed in the mud pit or clamped to the surface or conductor casing). This is illustrated in Figure 4.4

Depth of investigation is a direct function of the down-hole electrode radius. If the electrode is spherical, the radial depth of penetration is approximately 10 times the electrode radius. Use of a large-diameter electrode reduces the drilling fluid effect. The selection of the size of the current electrode is a compromise between depth of investigation and vertical resolution.[1] The main shortcomings of the single point tool are the lack of depth of investigation and excessive influence of borehole drilling fluid and diameter, particularly with borehole enlargement.

Normal Resistivity Log

The normal device, or two electrode system, is shown in Figure 4.5. A current of constant value is passed between electrodes A and B. Electrode A is the current-emitting electrode, and electrode B is the current return electrode. The voltage or potential difference is measured between electrodes M and N.

The equation describing electrode potential is

$$V_M = \frac{\rho I}{4\pi AM} ,$$

where

V_M = electrode M potential [v],

ρ = formation resistivity [ohm-m],

1. The ability to distinguish between formation layers, usually due to their varying resistivity. The greater the vertical resolution, the greater the ability to distinguish individual layers.

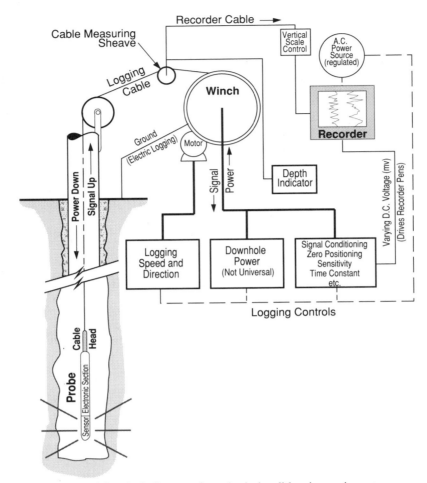

Figure 4.2. Block diagram of geophysical well-logging equipment.

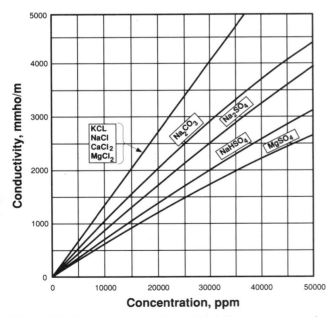

Figure 4.3. Conductivity of various salt solutions vs. concentration at 64° F.

TABLE 4.1 Resistivities of Electrolytic Solutions (at 77°F)

Electrolyte	Resistivity (ohm-m)
Brine	0.04
Seawater	0.2
Brackish water	0.2–0.5
Drilling fluid	0.04–5.0
Tap water	7–15
Distilled water	>100

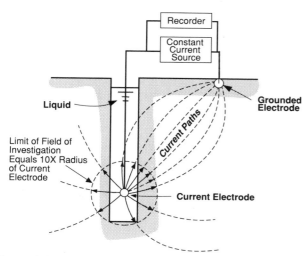

Figure 4.4. Single-point resistance logging system (courtesy of Welenco).

I = electrode A current [amp],

AM = electrode spacing [m].

If I and AM are kept constant, V_M is proportional to ρ. When logging, measurements consist of a continuous recording of voltage V_M that varies with formation resistivity ρ. The normal log is more indicative of formation resistivity than the single point since use of two electrodes permits resistivity measurements of the formation in the vicinity of the probe.

Depth of penetration increases directly with electrode spacing. Consequently vertical resolution decreases with increasing electrode spacing. The extent of depth of investigation is clarified by Figure 4.6. This figure represents a graph of the variation of measured voltage, V_M, of the normal device compared to the distance from the probe's A electrode. The unit of distance is spacing distance AM. Fifty percent of the measured voltage originates in a shell of formation of thickness $2AM$, and 75% in a shell of thickness $4AM$. The contribution of the formation to signal voltage diminishes rapidly beyond $4AM$.

Functions of the Normal Curve. In order to provide two different depths of investigation, normal AM spacings have been standardized at 16 in. (short normal) and 64 in. (long normal). They offer the following information:

1. Determination of bed boundaries and thickness.
2. Measurement of the apparent resistivity (R_a) of the invaded zone (see Figure 4.10.) from the short-normal reading and of the noninvaded zone from the long-normal reading.
3. Qualitative information pertaining to formation porosity and permeability.

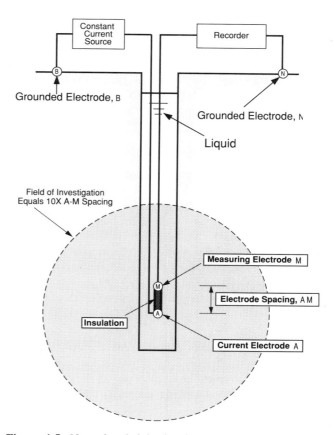

Figure 4.5. Normal resistivity logging system (courtesy of Welenco).

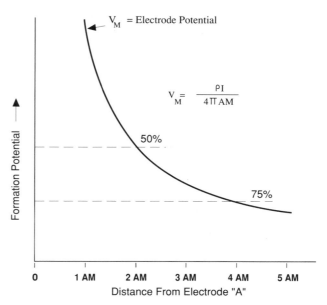

Figure 4.6. Depth of investigation of normal device (courtesy of M. L. Mougne).

Environmental Effects on Resistivity Measurements

Temperature. Resistivity of the electrolyte decreases with increase in formation temperature. Figure 4.7 expresses resistivity of sodium chloride solutions as a function of solution temperature, and ion concentrations in parts per million (ppm).

Drilling Fluid Resistivity and Borehole Diameter. Figure 4.8 shows the distortion of current pattern caused by drilling fluid resistivity. If the drilling fluid in the borehole is a better conductor than the formation (true in most cases), electric current flows more readily in the borehole. In addition current concentration will be greater in a larger borehole. This is a masking influence which affects apparent measured resistivity (R_a) values, making it more difficult to differentiate zones of high resistivity.

Table 4.2 lists the resistivities of drilling fluids of varying salinity.

Bed Thickness. Bed thickness affects the difference between apparent resistivity (R_a) and true formation resistivity (R_o) since measurement error increases as the ratio of bed thickness to electrode spacing decreases. When a bed is thinner than the spacing of the normal device and lies between two low resistivity clay beds, R_a will be lower than the clays resistivity. This negative anomaly, known as a "reversal," is shown in Figure 4.9.

When the thickness of a bed is equal to the spacing (critical spacing), the recording is almost flat. If bed thickness is several times spacing, R_a equals R_o. Normal resistivity curves always show resistive beds thinner than they actually are by an amount equal to the spacing, with half of this error at each boundary. These relationships assume no drilling fluid invasion.

Drilling Fluid Invasion. Figure 4.10 represents an invasion resistivity profile encountered in a typical permeable formation. In freshwater aquifers the filtrate[2] will mix with pore water, resulting in a flushed zone. The resistivity of this flushed zone will differ from R_o.

Filtrate invasion depth will be influenced by effective porosity. However, although in many freshwater aquifers, such as unconsolidated sands, effective porosity is relatively high, depth of invasion is shallow. The low hydrostatic head encountered in water wells further reduces the extent of

---

2. The liquid portion of the drilling fluid that passes through the filter cake into the formation adjacent to the borehole.

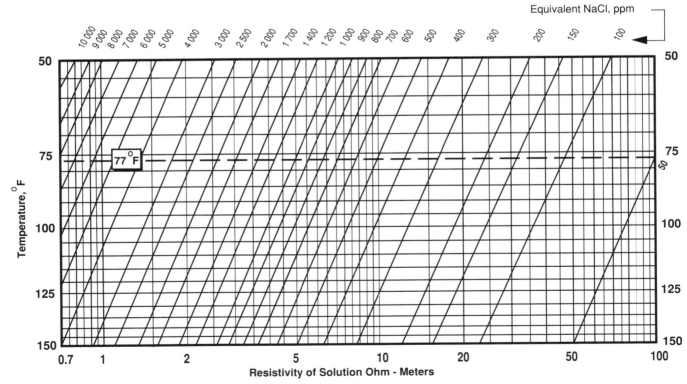

Figure 4.7. Resistivity of water as a function of salinity and temperature. Salinities are in terms of NaCl concentration (courtesy of Schlumberger Corp.).

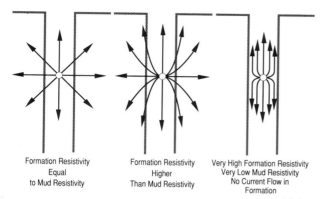

Figure 4.8. Distortion of current pattern by drilling fluid resistivity (courtesy of M. L. Mougne).

invasion. Therefore, with reasonable drilling fluid control, a normal curve will usually provide a measurement very close to the true resistivity (R_o) of thick beds.

Analysis of Normal Resistivity Logs

Although there are many spacings available for measuring the resistivity of the invaded zone, the one most used is the 16-in. short normal (AM) resistivity curve. The 16-in. curve is shown on the electric log in Figure 4.11 as the solid curve

TABLE 4.2 **Resistivity of Drilling Fluids (at 77°F)**

Drilling-Fluid Classification	Resistivity (ohm-m)
Very fresh	>2
Fresh	0.5–2
Brackish	0.2–0.5
Salt	0.04–0.2

in the center column. It is scaled in ohm-m. Resistivity increases with increasing deflection to the right.

Although the variables listed below can affect the measured log resistivity, R_a, the 64-in. long normal usually records values of R_a close to R_o when depth of invasion is less than depth of investigation and bed thickness. The 64-in. curve is shown on Figure 4.11 as the dotted curve in the center column. R_a must be corrected to obtain R_o, as will be described. As discussed, R_a is a function of:

- True formation resistivity, R_o
- Borehole diameter, dh
- Mud resistivity, R_m
- Electrode spacing, AM
- Bed thickness, th
- Invasion parameters (borehole diameter, flushed zone fluid, etc.)

Neglecting invasion for purposes of analysis, the dimensionless parameters R_o/R_m, AM/d, and th/dh can be formulated. Resistivity departure curves for an AM spacing of 16 in. are shown in Figure 4.12. Figure 4.13 shows similar departure curves for an AM spacing of 64 in. These correction charts allow the determination of R_o from R_a and R_m. They apply to bed thicknesses more than 3.5 times the electrode spacing when drilling for water using freshwater-base drilling fluid having low water loss properties. R_m is measured with an ohm-cell (resistivity or conductivity meter) from a sample of the drilling fluid.

R_w Determined from True Formation Resistivity

Another important parameter to be determined is R_w, the resistance of the interstitial water. In freshwater aquifers R_w, corrected for temperature, correlates with water quality. The conventional method for estimating water quality from R_w using *SP* logs is discussed later in this chapter.

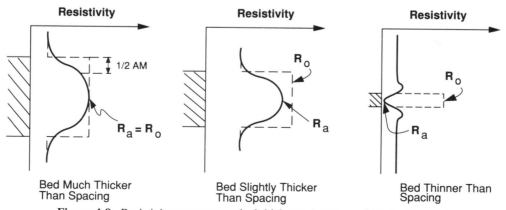

Figure 4.9. Resistivity responses to bed thickness (courtesy of M. L. Mougne).

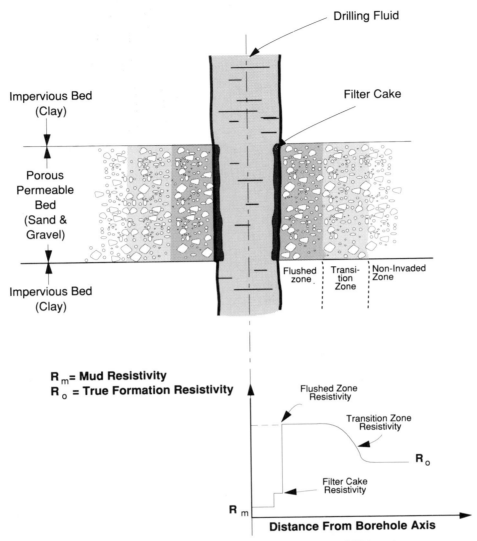

Figure 4.10. Invasion resistivity profile (courtesy of Welenco).

Alternate Method of R_w Determination. In the case where the *SP* curve is unreliable (very unusual for water wells), an alternate method is available. This method assumes a formation factor, *F*, that is constant for a given formation. R_w is obtained by dividing true formation resistivity, R_o, by F:[3]

$$R_w = \frac{R_o}{F} . \tag{4.1}$$

F values are based on empirically derived data obtained from field studies and are fairly constant in an unconsolidated,

clean, consistent sand in any local area. *F* values can be used to find R_w, as long as the formation sands are clean and similar in gradation. As shown in Figure 4.14, the presence of clay and silt in sand formations reduces the amplitude of the resistivity deflection, since these materials are good conductors.

It has been found empirically that *F* varies in freshwater sands in different areas, not only with R_w but also with porosity and grain size. The general relationship is

$$F = \frac{a}{\varnothing^m} ,$$

where

a = an empirical constant related to the formation,

---

3. This relationship states that (for a nonconducting matrix) the formation resistivity is proportional to the pore water resistivity. With clayey sands the matrix can be conductive, thus altering the relationship.

BOREHOLE AND DRILLING FLUID DATA

Depth: 0 to 712 ft
Conductor casing: 30" Diameter—depth 50'
Bit size: 15"
Fluid type: Fresh water with bentonite
Source of sample: Mud pit

Rm at measured temperature:11.1 ohm-m @ 75°F
Rmf at measured temperature: 9.8 ohm-m @ 75°F
Rm @ BHT (Bottom Hole Temp.): 9.0 ohm-m @ 89°F
Time since circulation: 2.0 hours

Max recorded temperature: 88.7°F

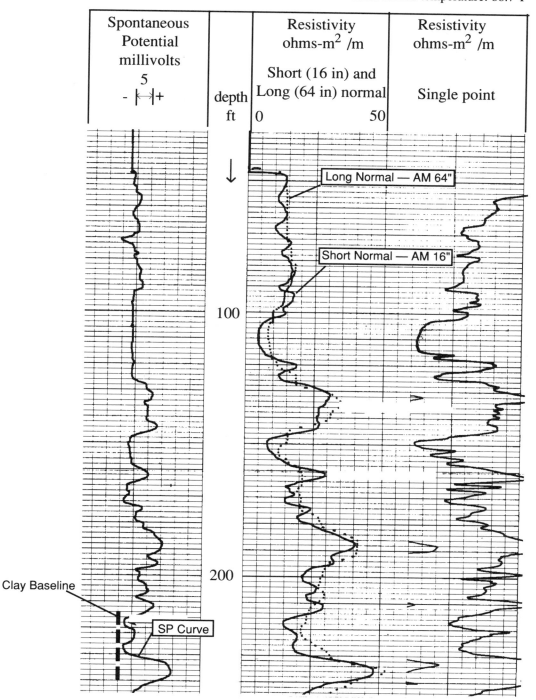

Figure 4.11. SP and resistivity logs (courtesy of Welenco).

(*Figure continues on p. 60.*)

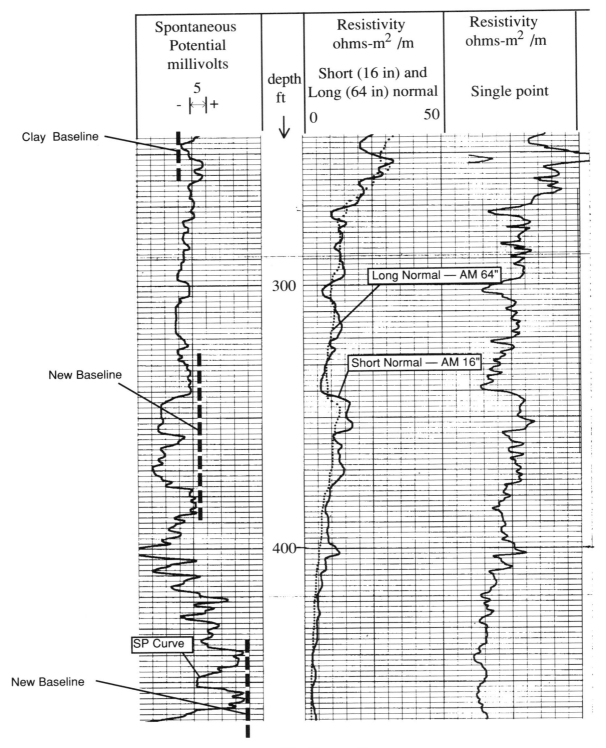

Figure 4.11. (*Continued*)

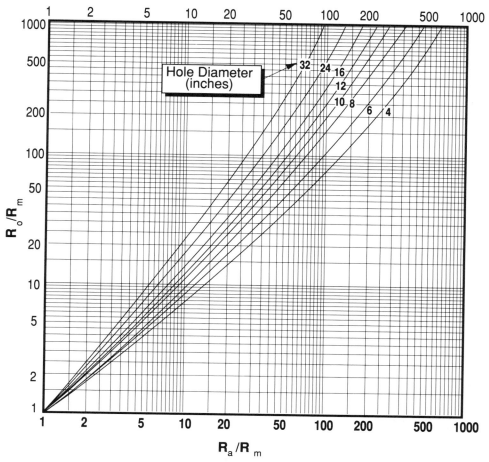

Figure 4.12. Resistivity departure curves, 16″ normal readings (courtesy of Gearhart Industries and Welenco).

$\varnothing$ = total porosity (expressed as a fraction)

m = cementation factor.

Values of formation factors can be derived by standardizing a to a value of one by varying m. Figure 4.15 shows relationships between porosity, formation factor, and m, with a equal to one (1, 2). These relationships must be found from empirical data for a specific basin. The "Humble" curve may be used in the absence of precise lithological data for semiconsolidated sand. This empirically derived variation of the standard formula for F is given as

$$F = \frac{0.62}{\varnothing^{2.15}} \ .$$

However, in freshwater aquifers the relationship between F and porosity is not always constant. The value of F, given a stable matrix configuration (constant porosity, pore geometry, and chemical composition), becomes a function of

R_w (Equation 4.1). Research has shown that for freshwater in unconsolidated sands, computed formation factors are appreciably lower than described by the F versus porosity relationship commonly used in petroleum industry log interpretations.

Surface Conductance

In their investigations of sands, Hill and Milburn found it necessary to qualify F by the R_w used in the determination (2). They confirmed that higher values of R_w tended to produce lower values of F. When the formation factor is an abnormally low value (4–6), normal relationships between formation factor and permeability or grain size may be "masked" due to surface conductance (ionic double layer) effects. When this phenomenon is suspected, locally derived curves such as Figure 4.17 must be used. The magnitude of surface conductance is related to the ion concentration of the water. As electrolytic concentration increases, the

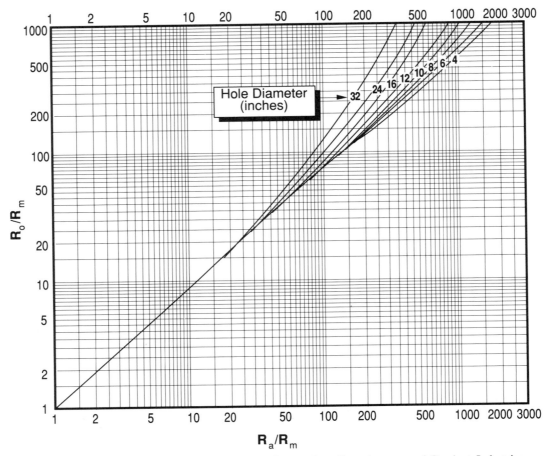

Figure 4.13. Resistivity departure curves, 64″ normal readings (courtesy of Gearhart Industries and Welenco).

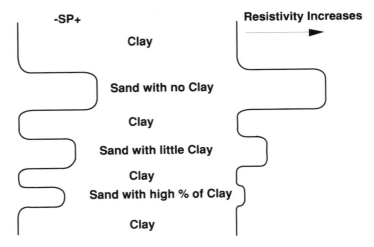

Figure 4.14. Effect of increasing clay content on SP and resistivity logs (courtesy of M. L. Mougne).

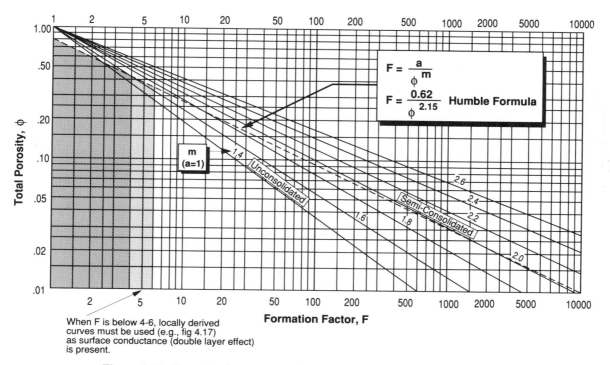

Figure 4.15. Formation factor—porosity relationships (courtesy of Welenco).

magnitude of surfaces increases up to the saturation point. In the low-conductivity, low-electrolytic concentration environment of a freshwater sand, even this increase is significant. Figure 4.16 illustrates this effect.

A more significant factor affecting the magnitude of surface conductance is the surface exposed. Total surface conductance is a direct function of the surface exposed to the water. This is significant because the surface area of sands is related to both grain size and permeability.

Grain Size versus Formation Factor

Formation factor is a direct function of grain size when the ion concentration in the water is very low. Figure 4.17 gives an example of this relationship (3).

Grain Size versus Hydraulic Conductivity versus Formation Factor

The relationship between grain size, hydraulic conductivity, and formation factor shown in Figure 4.18 is very important in evaluating aquifers from log data (3). In summation, R_o increases with higher water quality, coarser grain size, and greater formation permeability. Thus for freshwater aquifers, the best zones are indicated by the highest resistivities.

4.4 LATERAL RESISTIVITY DEVICE

The lateral tool is a three-electrode device comprising two voltage-measuring electrodes (M, N) and a current electrode (A), as illustrated in Figure 4.19. The voltage measuring electrodes are close to each other but remotely located from the current electrode.

The effective measuring point of the lateral device is midway between the potential electrodes M and N and is labeled O. The nominal spacing of the device is the distance from this midpoint O, to the current electrode A and is called the AO spacing. Standard AO spacing for freshwater borehole surveys is 6 ft.

The lateral device measures the resistivity of a small shell of material far out in the formation. The material nearer the borehole is averaged into the global measurement but its contribution is extremely small. The dashed circles in Figure 4.19 show the "equipotential spheres," the tool measures.

Characteristic Responses

There are characteristic responses from lateral logs. The response shown in Figure 4.20a, occurs in a thick high resistivity bed (several times the AO spacing) between two beds of low resistivity. A false indication of low resistivity

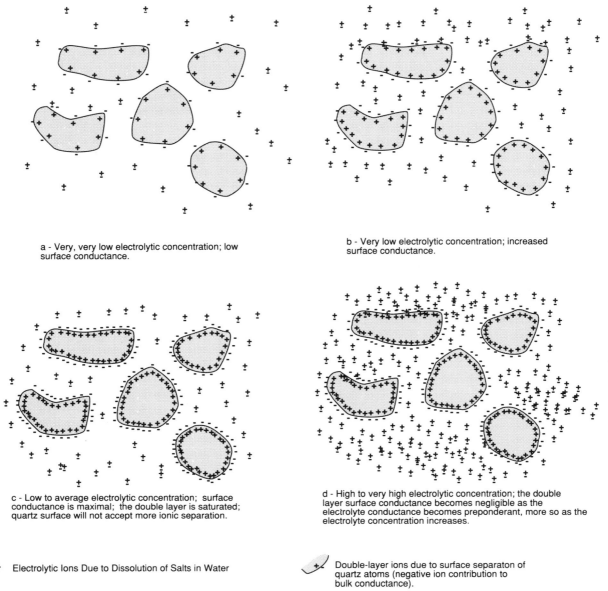

a - Very, very low electrolytic concentration; low surface conductance.

b - Very low electrolytic concentration; increased surface conductance.

c - Low to average electrolytic concentration; surface conductance is maximal; the double layer is saturated; quartz surface will not accept more ionic separation.

d - High to very high electrolytic concentration; the double layer surface conductance becomes negligible as the electrolyte conductance becomes preponderant, more so as the electrolyte concentration increases.

± Electrolytic Ions Due to Dissolution of Salts in Water

Double-layer ions due to surface separaton of quartz atoms (negative ion contribution to bulk conductance).

Total Conductance = Electrolytic Conductance + Surface Conductance

Figure 4.16. Double layer effect on quartz grains immersed in an electrolyte (courtesy of M. L. Mougne).

is recorded in the upper part of the bed equal to the AO spacing. R_a increases to a value that is greater than the true resistivity, R_o, from the bottom of this interval to the bottom of the bed. An acceptable procedure for finding R_o from the lateral curve in thick, resistive beds involves choosing the indicated value as shown in Figure 4.20a. Note that the lateral resistivity curve is not symmetrical, whereas the normal curves are symmetrical.

With thinner beds the peak deflection of the lateral curve more closely approximates R_o. When bed thickness is 1½

times the AO spacing, R_o is measured at a point ⅔ of the distance out on the slope, below a distance AO from the top of the bed. This is illustrated in Figure 4.20b.

When bed thickness decreases to 1⅓ times the AO spacing the peak response may be chosen as R_o, as shown in Figure 4.20c. When thickness equals the AO spacing, the lateral device records its minimum response. Although R_o can only be estimated at this "critical thickness," R_a is approximately ¼ to ⅓ of R_o. The critical thickness response is shown in Figure 4.20d.

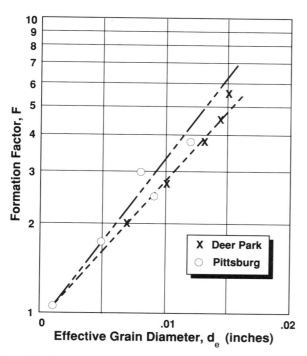

Figure 4.17. Formation factor vs. grain size (after Alger, 1966).

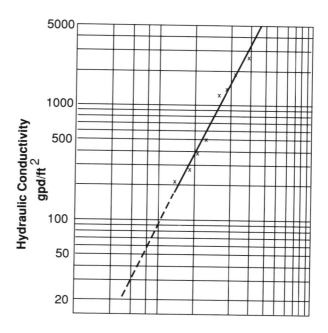

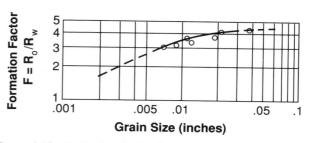

Figure 4.18. Grain size, hydraulic conductivity, and formation factor (after Alger, 1966).

Thin Bed Response. With thickness less than the *AO* spacing, R_a increases, although it is less than R_o. With bed thicknesses of ¼ to ½ the *AO* spacing, R_o is estimated by multiplying the peak value in the bed by the resistivity of the adjacent clay and dividing this product by the minimum value below the thin bed. Figure 4.20*e*, illustrates this procedure.

Response below Thin Beds. Other phenomena are associated with the lateral device. For beds thinner than the *AO* spacing, the lateral device falsely indicates a very low resistivity for a distance below the bed equal to the difference between the *AO* spacing and the bed thickness. Because the lateral device cannot indicate high resistivity in this zone, it is known as the "blind or dead zone."

At the bottom of the dead zone when the current electrode *A* is entering the top of the thin, resistive bed above, R_a begins to increase, and continues to increase until current electrode *A* leaves the bottom of the thin resistive bed. This increase in R_a does not necessarily represent a formation resistivity change but is caused by the current electrode *A* passing through the thin resistive bed. This false indication of increased resistivity is known as a "reflection peak," as shown in Figure 4.20*f*.

In sequences of heterogeneous beds, the response may become so confusing as to be of little value. Nevertheless, the lateral curve does have several advantages. In extremely

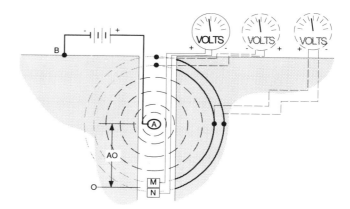

Figure 4.19. Lateral device (courtesy of Welenco).

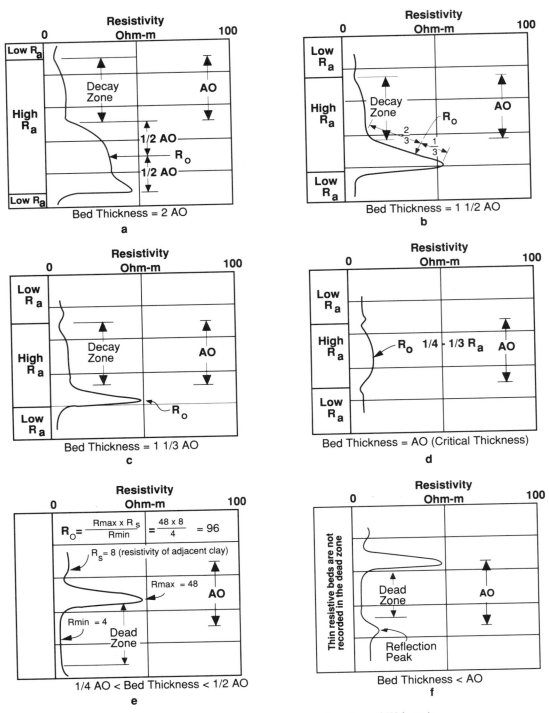

Figure 4.20. Lateral log responses (courtesy of Welenco).

thick beds it yields a relatively uninvaded value of R_o and is useful in estimating the extent of invasion when compared with the normal curves. Some approximate true formation resistivity values for thin-bed analysis can be estimated from the features of the lateral curve(s).

4.5 SPONTANEOUS POTENTIAL LOGGING

There are three connected media needed to generate a *SP* log:

1. A permeable bed containing water.
2. A clay or shale bed.
3. A borehole filled with freshwater-base drilling fluid containing bentonite or other clay material.

Voltage is measured between the point of investigation in the borehole and ground surface, as shown in Figure 4.21. Figure 4.22 shows the current generated within the three media. Figure 4.23*a* reflects a condition when formation water is less resistive (saltier) than the drilling fluid. In Figure 4.23*b* the formation water is more resistive (fresher). The larger the resistivity difference, the larger is the magnitude of the potential. If the drilling fluid and formation water resistivities are equal, no current lines are generated (4,5).

In *SP* generation two voltages are present:

1. A membrane potential across clay boundaries (E_m) [mv].
2. A liquid-junction potential at the filtrate formation-water interface (E_j) [mv].

Figure 4.22 indicates the voltage polarity resulting from the ion exchanges and the corresponding equivalent circuits.

Membrane Potential (E_m)

As shown in Figure 4.22, positive cations accumulate at the clay-mud interface, while the sand-clay interface remains charged with negative anions. Due to high salt-concentration differences, more cations flow across the clay surface than from the mud.

The membrane potential in millivolts is expressed as follows:

$$E_m = -59.2 \ \log_{10}\left(\frac{R_{mf}}{R_w}\right) \quad \text{at } 77°F \ ,$$

where

R_{mf} = resistivity of the mud filtrate [ohm-m].

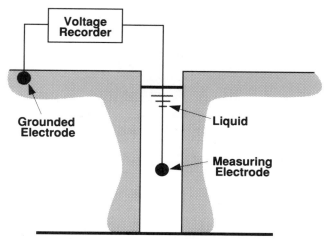

Figure 4.21. Simplified SP circuit (courtesy of Welenco).

Liquid-Junction Potential (E_j)

Figure 4.22 also illustrates the electrical effects at the junction of the flushed zone with the uninvaded zone water. Through the phenomenon of diffusion, ions migrate from the solution with the highest concentration to the lowest. Due to relative ionic mobilities, a negative layer is created at the junction of the filtrate. This liquid-junction potential is expressed as follows:

$$E_j = -11.5 \ \log_{10}\left(\frac{R_{mf}}{R_w}\right) \quad \text{at } 77°F \ .$$

Total *SP* Potential

Figure 4.22 combines both potentials ($E_m + E_j$) to give the total *SP* signal:

$$SP = -70.7 \ \log_{10}\left(\frac{R_{mf}}{R_w}\right) \quad \text{at } 77°F \ .$$

In making the combination, it is assumed that the drilling fluid has a lower sodium chloride concentration than the formation water, or that $R_{mf} > R_w$, which is the most general condition encountered, particularly in oil-well logging. If, however, the drilling fluid has more sodium chloride, polarity is reversed and current flows in the opposite direction.

If $R_{mf} < R_w$, then

$$SP = +70.7 \ \log_{10}\left(\frac{R_{mf}}{R_w}\right) \quad \text{at } 77°F \ .$$

This indicates a reversal in the deflection on the *SP* curve.

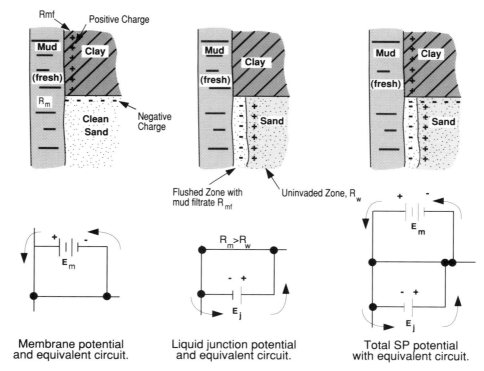

Figure 4.22. Electrochemical SP sources.

Streaming Potential

An electromotive force, the streaming potential or electro-filtration as described in Chapter 3, appears across a permeable medium through which an electrolyte is being flushed. It vanishes in the absence of flow (i.e., hydrostatic equilibrium). In wells drilled with circulating fluid having poor filtration properties (high water loss), the electrofiltration potential may be continuously generated and will mask the *SP* signal. The *SP* curve should not be used under such conditions.

Functions of the *SP*

The spontaneous potential curve provides the following information:

1. Delineation of clay beds.
2. Delineation of permeable beds.
3. Lithologic correlation.
4. Qualitative information pertaining to formation water resistivity (R_w) compared to mud resistivity (R_m).
5. Calculation of R_w from *SP* deflection, R_m, and formation temperature. As stated R_w, with temperature correction, allows determination of relative water quality.

The *SP* deflection in millivolts is given by the formula:

$$SP = -70.7 \left(\frac{460 + T}{537} \right) \log_{10} \left(\frac{R_{mf}}{R_w} \right) , \quad (4.2)$$

where

T = formation temperature, degrees Fahrenheit.

Because the relationship between R_w and *SP* varies in accordance with the type of ion in the formation water, it is convenient to consider the water resistivity determined from Equation 4.2 an equivalent water resistivity, R_{we}.

Figure 4.24 represents a family of curves for the variations of the ratio R_{mf}/R_w as a function of *SP* deflection and formation temperature. R_{we} can be calculated if R_{mf} is measured from a filtrate sample. Since *SP* deflection is measured at formation temperature, the surface R_{mf} value must be corrected to formation temperature, using Figure 4.7. If a filter press is not available for obtaining mud filtrate for testing, Figure 4.25 can be used to find R_{mf} from R_m.

Environmental Effects on *SP* Measurement

The two main environmental factors that affect the amplitude of the *SP* curve are bed thickness with respect to borehole diameter and bed resistivity with respect to drilling-fluid resistivity. Bed thickness is defined by the distance between

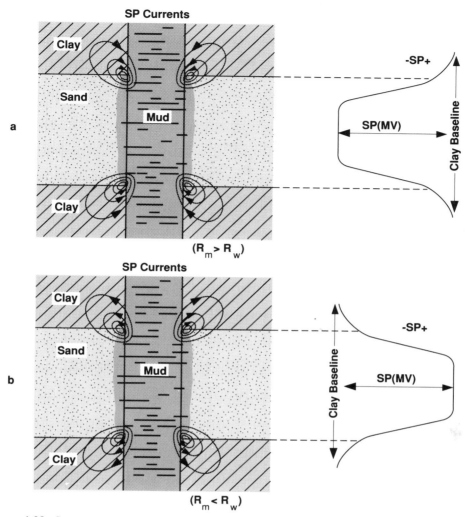

Figure 4.23. Spontaneous potential for different formation and drilling fluid resistivities (courtesy of M. L. Mougne).

the inflection points of the *SP* curve at the upper and lower boundaries. When the ratio of bed thickness to borehole diameter is high, *SP* deflection reaches maximum amplitude. This is called "static *SP*" (*SSP*). When this ratio is low the *SP* deflection is lessened and requires correction.

When the ratio of formation resistivity to drilling-fluid resistivity is low, static *SP* is established. When this ratio is large, *SP* deflection is reduced and requires correction. *SP* correction for bed thickness can be obtained from Figure 4.26. The corrected *SP* is used to find R_{we} from Figure 4.24.

The *SP* curve is affected by other environmental factors, including:

1. Filtrate invasion, which reduces the amplitude of *SP* signals (only when very large invasion diameters are encountered).

2. The presence of clay particles in sands, which reduces the amplitude of the *SP* signals in proportion to their fractional volume (discussed later in "The Reduction of the *SP* Deflection in Clayey Sands").

Effect of Other Salts

The preceding analysis applies when sodium chloride is present in both the drilling fluid and formation water. This is true in most cases when drilling for oil or gas in deep sediments. However, in shallow drilling for freshwater, there may be other dissolved salts, and sodium chloride may not be present. As developed by Alger, in this case both the membrane and liquid-junction potential differences will be generated in terms of ion concentration and the nature of the ions present in the solutions (drilling fluid and formation

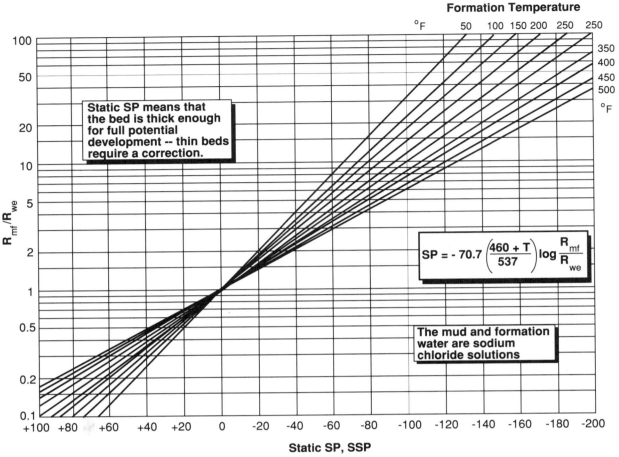

Figure 4.24. Variations of R_{mf}/R_{we} with temperature and static SP.

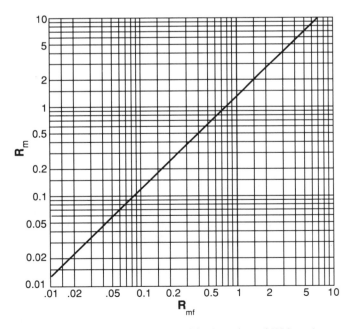

Figure 4.25. $R_m - R_{mf}$ relationship (courtesy of Welenco).

water) (3). For instance, the divalent cations magnesium and calcium strongly affect *SP* potentials. Gondouin et al. show that for such cases, the magnitude of the electrochemical *SP* is as follows (6):

$$SP = -K \log \frac{a_w}{a_{mf}} \qquad (4.3)$$

and

$$SP = -K \log \frac{(a_{Na} + \sqrt{a_{Ca} + a_{Mg}})_w}{(a_{Na})_{mf}}, \qquad (4.4)$$

where a_{Na}, a_{Ca}, and a_{Mg} are, respectively, the solution activities due to sodium, calcium, and magnesium ions, a_w and a_{mf} are aquifer water and drilling-fluid filtrate solutions, and K is a function of the formation temperature.

In Equation 4.4 the filtrate is considered to act as a NaCl solution. There are cases where divalent ions are present in significant concentrations. However, through base exchange in clay additives or shales, their concentration is often reduced

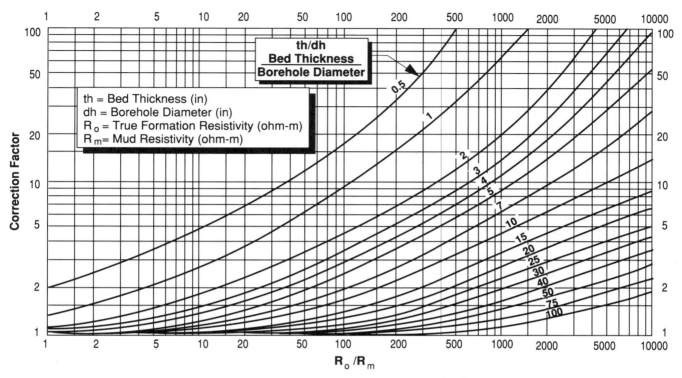

Figure 4.26. SP correction for bed thickness (courtesy of Welenco).

from that of the makeup water. Thus, in the absence of information to the contrary, it is customary to consider the filtrate as a NaCl solution. For NaCl solutions, the relationship between activity and resistivity is shown on Figure 4.27.

When considered a NaCl solution, the activity of the filtrate used in the denominator of Equation 4.4 may be found when R_{mf} is known. However, the relationship in Figure 4.27 is correct only for a solution temperature of 77°F. This corresponds with the 25°C used as a standard for water well computations. For R_{mf} values at other temperatures, Figure 4.7 can be used to convert to standard temperature conditions.

The relationship between cation concentration and activity for dilute solutions is given in Figure 4.28. In the figure the solid line represents the relationship for Na ions, and the dashed line for Ca and Mg ions. The total water activity, for use in Equation 4.4, is the sum of $a_{Na} + \sqrt{a_{Ca} + a_{Mg}}$. Thus activities can be computed from solution cation concentrations.

Bicarbonate. In many freshwaters the predominant anion is bicarbonate, HCO_3. If the accompanying cation is Na^+, the activity of the solution, and thus the effect on an *SP* curve, is usually close to that for a NaCl solution having the same Na^+ concentration (6). That is, the activity of the solution is primarily dependent on the cation concentration. However, the resistivities of NaCl and $NaHCO_3$ solutions (of the same Na^+ concentrations) are different. The HCO_3

ion contributes only 27% as much conductivity as an equal weight of Cl ion. The R_w of an $NaHCO_3$ solution is 1.75 times greater than R_w of an NaCl solution having the same Na^+ concentration.

Other Ions. The relationships between concentration and resistivity for other ions also differ from that of NaCl. Listed in Table 4.3 are the multipliers required to convert concentrations of commonly encountered ions to equivalent NaCl concentrations for R_w determination (3).

Water Quality from *SP*

The foregoing discussion of divalent cations and bicarbonates shows that use of the *SP* for water-quality determinations requires some knowledge of the ion content of the aquifer water. The activities of the various cations differ and cause *SP* response to depart from that for NaCl solutions. Also the relationships between concentration and resistivity vary depending on the type of ions in solution.

The following example (from Alger 1966) illustrates the application of water-quality data to find R_w (3). Table 4.4 presents the ion concentrations from a water sample and, through application of the appropriate multipliers, the conversion to equivalent NaCl concentration.

The NaCl total-equivalent concentration for this example is extremely low and falls off the chart of Figure 4.7 (used to convert NaCl concentration to R_w). However, for dilute

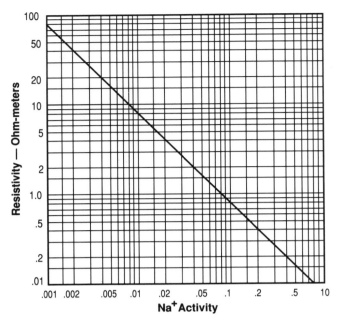

Figure 4.27. Relationship between sodium ion activity and resistivity for a sodium chloride solution at 77° F (after Alger, 1966).

TABLE 4.3   Conversion of Common Ions into Sodium Chloride Equivalent

Ion Type	Symbol	Multiplying Factor k
Sodium	Na^+	1.0
Calcium	CA^{2+}	0.95
Magnesium	Mg^{2+}	2.0
Chlorine	Cl^-	1.0
Carbonate	CO_3^{2-}	1.26
Bicarbonate	HCO_3^-	0.27
Sulfate	SO_4^{2-}	0.5

TABLE 4.4   Converting Common Ions from a Water Sample into NaCl Equivalent

Ion	Concentration (ppm)	Multiplier	NaCl Equivalent (ppm)
Na	10	1.0	10
Ca	3	0.95	2.85
Mg	0.8	2.0	1.6
Cl	6	1.0	6
SO₄	0	0.5	0
CO₃	0	1.26	0
HCO₃	29.3	0.27	7.9
Total equivalent NaCl			28.35 ppm

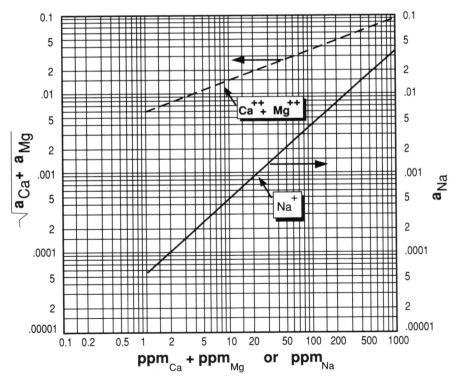

Figure 4.28. Cation concentration *vs.* activities for dilute solutions at 77°F (after Alger, 1966).

solutions, concentration and resistivity are inversely proportional. Therefore we can compute R_w from the chart by extrapolation where a concentration of 50 ppm corresponds with an R_w of 100 ohm-m at 77°F:

$$R_w = 100\left(\frac{50}{28.35}\right) = 176 \text{ ohm-m} \ .$$

Using Equation 4.3 and the cation concentrations obtained from chemical analysis of the water, the *SSP* that should be observed opposite this sand can be computed.

From Figure 4.28, 10 ppm Na gives $a_{Na} = 0.00048$, and

3.8 ppm Ca + Mg gives $\sqrt{a_{Ca} + Mg} = 0.010$.

Thus

$$a_w = 0.01048 \ .$$

From Figure 4.27, $R_{mf} = 68$ ohm-m at 77°F (as measured) gives $a_{mf} = 0.0012$.
Then

$$SP = -71 \log \frac{a_w}{a_{mf}}$$
$$= -71 \log \frac{0.01048}{0.0012}$$
$$= -66 \text{ mv} \ .$$

Thus the computations confirm the measured SP, as shown in Figure 4.29. It was only necessary to properly account for the types and concentrations of ions present.

Derivation of R_w can be done only after chemical analysis of water samples collected in the specific local area. The curve R_w versus R_{we} is then drawn for that particular area and can be used to establish R_w "in situ" from the *SP* log.

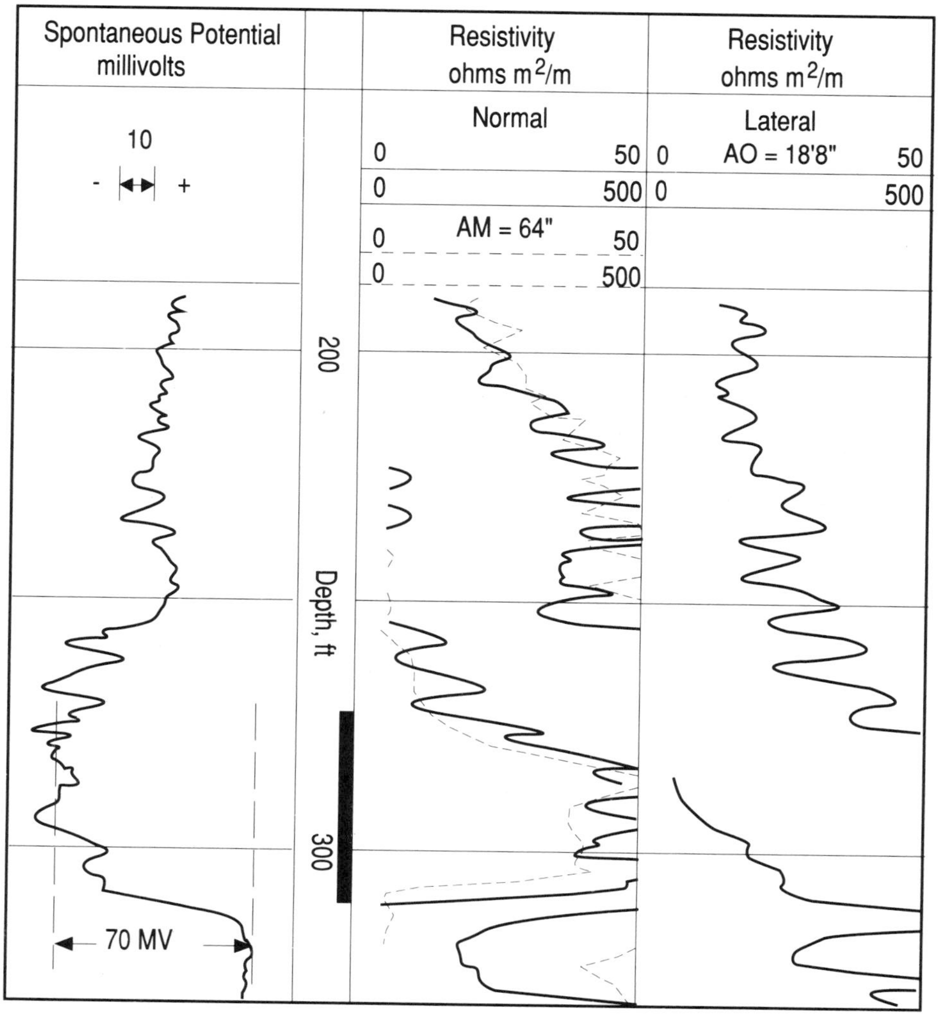

Figure 4.29. Electric log, recorded in a shallow East Texas water well (after Alger, 1966).

Chemical analysis also permits the evaluation of the total dissolved solids (TDS) in terms of R_w.

This approach can be used for the determination of R_w versus R_{we} from a few discrete water samples because, in a given area, the ion assemblage is fairly constant with regard to the nature and relative percentage of the ions, even though the total concentration may vary quite widely.

Figure 4.30 shows examples of response curves of R_w versus R_{we}. The sodium chloride (NaCl) curve has a 45° slope so that R_w equals R_{we} at all points. The sodium bicarbonate (NaHCO$_3$) and calcium chloride (CaCl$_2$) curves are also presented. The CaCl$_2$ curve represents the upper departure from sodium chloride solutions likely to be found in nature.

An empirical curve compiled by Gondouin et al. that uses laboratory and *SP* results from a large number of wells drilled in the United States, Venezuela, and Indonesia is also drawn on Figure 4.30 (6). This curve may be used as a first approximation, but ideally each geographical area should use its own empirical data.

Summary of Method for Deriving R_w from the *SP* Signal

In summary, the way to establish R_w versus R_{we} is as follows:

1. Measure R_{mf} at the surface from a filtrate sample, or measure R_m and derive R_{mf} from the chart of Figure 4.25. This is R_{mf} at ambient temperature or approximately 77°F.
2. Calculate R_{mf} at formation temperature, using the chart of Figure 4.7 and the borehole temperature gradient chart Figure 4.31.
3. Correct the *SP* log deflection for bed thickness and formation resistivity, if required (Figure 4.26).
4. Derive R_{we} from the *SP* log deflection and the chart of Figure 4.24. This is R_{we} at formation temperature.
5. Derive R_{we} at ambient (77°F), using the chart of Figure 4.7.
6. Evaluate R_w from a locally derived R_w vs. R_{we} curve (Figure 4.30.)

SP Baseline Shifts

In many instances, particularly in shallow wells, the clay baseline[4] will shift. It is believed that the nature of the clays causes these shifts. Under such conditions the *SP* deflection is taken with reference to the overlying and underlying clay beds. Examples of such shifts appear in Figure 4.11.

---

4. The clay baseline is drawn by joining the points of minimum resistivity on the *SP* log.

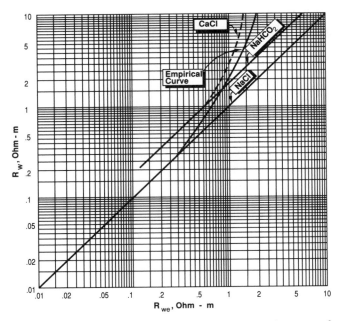

Figure 4.30. R_w vs. R_{we} for fresh formation waters (courtesy of M. L. Mougne).

The Reduction of the *SP* Deflection in Clayey Sands

Compared to the *SP* signal in clean formations (static *SP*), the *SP* curve in clayey formations undergoes a reduction in signal amplitude designated as pseudostatic *SP*. The reduction ratio is conventionally called α. Figure 4.32 shows qualitatively how clay will reduce the *SP* deflection from clean sand *SSP*.

Water-Quality Evaluation from Spontaneous Potential Measurements

Calculation of total dissolved solids (TDS), chloride, and other ion content can be made from the *SP* curve when the predominant ions are known. In general, the *SP* curve constitutes the best source of data for making quality estimates. This may not always be true, however. In some cases the normal resistivity curve may supply more useful information, especially in areas where water quality data are available. Ion concentration (ppm) for individual salts can be found from R_w by using Figure 4.33.

When there are several salts in solution, one point of the curve (TDS vs. resistivity) must be found through chemical analysis and resistivity calculation or through measurement. Once this point is established, another point on the curve can be calculated by using the equivalent NaCl resistivity for another arbitrary concentration, from Figure 4.33. Then the curve (TDS vs. resistivity) for this particular solution can be traced on a log-log grid, as in Figure 4.33. (Note that it was assumed that the ion assemblage is uniform even

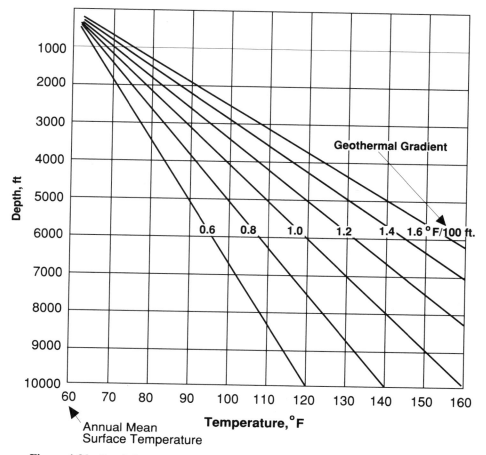

Figure 4.31. Borehole temperature gradient chart (courtesy of Schlumberger Corp.).

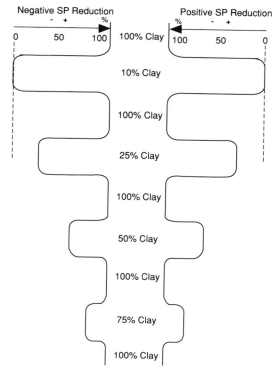

Figure 4.32. SP reduction due to clayey sands (courtesy of M. L. Mougne).

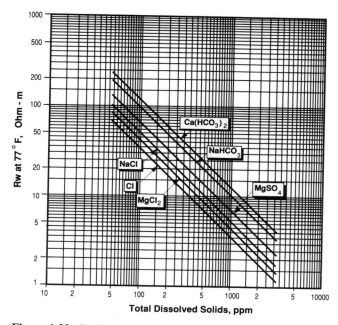

Figure 4.33. Resistivity vs. total dissolved solids for various electrolyte solutions (courtesy of M. L. Mougne).

though the TDS may vary widely in that geographical area (7).

Variations in ion assemblage refers to variations in percentage of some ions present with respect to the other ions. With additional chemical analyses the method described earlier can be extended to such cases in which the variations are minor. The larger the variations, the more chemical analyses are required.

Water-Quality Estimate from Resistivity Measurements

Resistivity measurements are measurements of formation water resistivities if the influence of the matrix is a constant, which is true for most clean sand aquifers. Porosity (which is matrix derived) affects formation conductivity since more pores result in higher conductivity. As previously discussed, if porosity is fairly constant, F is fairly constant, and formation resistivity is a function of formation water resistivity. Therefore a measurement of R_o yields R_w. Water quality is determined by using the methodology set forth in *SP* analysis.

4.6 POROSITY MEASURING TECHNIQUES

Sonic Logging

The sonic log was originally intended as an aid to the interpretation of seismic data. However, it has become widely used as a porosity log in water well surveys. Sonic logging measures the velocity of sound through various subsurface materials. Table 4.5 shows typical ranges.

The sonic log is very sensitive for high porosities, but measurements are strongly affected by borehole caving and diameter irregularities. A fairly sophisticated instrumentation and measuring technique is required, as illustrated in Figure 4.34.

Extensive research has established that sound-transit time in porous rocks is a direct linear function of formation porosity.

TABLE 4.5 Sound Velocities through Various Materials

	Velocity (ft/sec)	Transit Time (μsec/ft)
Cement (cured)	12,000	83.3
Dolomite	23,000	43.5
Granite	19,700	50.7
Limestone	21,000	47.6
Quartz	18,900[a]	52.9
Steel	20,000	57.0
Sandstones	11,500–16,000	62.5–87.0
Shales	7000–17,000	142.8–58.8
Water	4800	208.3

[a] Arithmetic average of values along axes (Wyllie et al. 1956).

When logging, a toroidal transducer (transmitter) emits a burst of ultrasonic energy at its natural vibrational frequency of about 20 kilohertz (KHz). The rays of interest are those that hit the borehole wall at their critical angle of incidence, follow a path along the wall, and are refracted into another transducer (receiver) located at some distance from the transmitter.

Snell's laws of optics are used to model the transmission of acoustical energy in solid and liquid media. The elapsed time, or "transit time," between the emission of the burst of ultrasonic energy and the instant of energization of the receiver is the sum of the travel times in the drilling fluid and the formation. The formation-only transit time can be derived mathematically since it is a function of spacing, drilling-fluid transit time, and probe and borehole diameters.

The probe consists of a multitransducer-carrying cylindrical mandrel. In order to avoid direct transmission of sound from transmitter to receiver through the mandrel material, a low velocity (high transit time) material must be used together with transmitter-receiver acoustic isolation. Effects of variations in borehole diameter and sonde tilt are eliminated by the use of "borehole-compensated sonic tools" with which compensation is obtained by placing two alternately pulsed sonic transmitters above and below two pairs of sonic receivers.

The equipment for powering the transducer (or transducers) is located on top of the mandrel. Signal waveforms are transmitted through the logging line to the surface for processing. The probe must be centered to achieve symmetrical transit times and avoid errors due to "wave interference."

The sonic logging system is relatively free from borehole environmental effects, except those due to drilling-fluid invasion. However, the presence of shale and clay impurities affects measurements. As a rule the depth of investigation of sonic probes is about one-half the wavelength of the wave traveling in the formation.

Formation transit times range from 50 to 200 μsec per foot. The minimum scale value of 40 μsec per foot is close to the 57 μsec per foot figure for transit time in steel, which can be verified in the conductor casing as an indication of equipment working order. Other scale ranges can be used according to local geology.

The following linear relationship exists between the transit time measured on the log curve and the total formation porosity:

$$\emptyset = \frac{t - t_{ma}}{t_f - t_{ma}}$$

or, to keep numerator and denominator positive,

$$\emptyset = \frac{t_{ma} - t}{t_{ma} - t_f}, \tag{4.5}$$

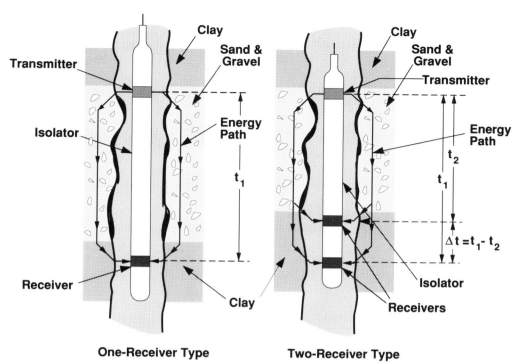

One-Receiver Type   **Two-Receiver Type**

Figure 4.34. Sonic logging system.

where

$\emptyset$ = total porosity,

t = log transit time reading [μsec/ft],

t_{ma} = matrix transit time [μsec/ft],

t_{f} = water transit time [μsec/ft].

Equation 4.5 applies to compacted sands or carbonate formations. In unconsolidated formations Equation 4.5 gives values higher than the actual porosity, so a correction is necessary. The degree of consolidation can be estimated from the transit time values of adjacent clays.

Sonic Correction. The corrected sonic porosity is calculated as follows:[5]

$$\emptyset_{sonic} = \left(\frac{t_{log} - t_{ma}}{t_{f} - t_{ma}} \times \frac{100}{t_{sh}} \right) \\ - V_{sh}\left(\frac{t_{sh} - t_{ma}}{t_{f} - t_{ma}} \right), \tag{4.6}$$

where

t_{log} = log sonic transit time [μsec/ft],

---

5. The terms in the last expression of Equation 4.6 can be reversed to keep numerator and denominator positive.

t_{ma} = matrix sonic transit time [μsec/ft] (Table 4.5),

t_{f} = formation fluid sonic transit time [μsec/ft] (Table 4.5),

t_{sh} = clay sonic transit time [μsec/ft] (measured from adjacent clay beds),

V_{sh} = clay volume fraction of the bulk formation volume (derived from a clay indicator such as the gamma ray log).

4.7 GAMMA RAY LOGGING

Gamma ray surveys consist of measurements of natural radioactivity of the formations encountered. The three elements contributing to formation radioactivity are thorium, uranium, and potassium. Gamma ray radiations are quickly absorbed. However, they can penetrate casing and/or drilling fluid to reach the detector (9). The gamma ray logging device detects radiation and is used in conjunction with other logs to determine formation lithology.

Instrumentation

Gamma ray logging relies on a device incorporating a sodium iodine crystal coupled to a photomultiplier tube (8). When exposed to the irradiating radioactive medium, the crystal detects and counts radiation. This device is shown in Figure 4.35. Geiger-Müller tubes are also used in this application.

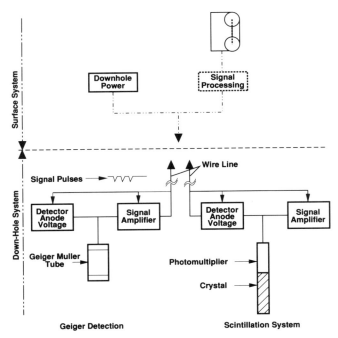

Figure 4.35. Gamma-ray logging system (courtesy of Schlumberger Corp).

Depth of Investigation

The depth of investigation is largely controlled by the absorption capability of the formation, which is directly proportional to its density. Fortunately, radiations emitted at the borehole wall and within the near-well zone pass through a comparatively low density medium drilling fluid, and the gamma ray flux undergoes little absorption in reaching the detector. In average density formation, the depth of investigation is about 1 ft (9).

Environmental Effects

The two basic borehole environmental effects are

1. Drilling-fluid density. With high density there is higher absorption and lower detected gamma radiation.
2. Borehole diameter. Absorption is greater as the diameter gets larger and fewer rays are detected (9).

Gamma Ray Log Interpretation

The gamma ray curve is widely used for geological correlations, particularly in cased wells where the steel would short-circuit electric log currents. The highest radiation levels correspond to pure clays, whereas the lowest are found in formations with no clay. Intermediate radiation levels correspond to variable amounts of clayey impurities found in the rock matrix (Figure 4.36).

The log analysis for water wells often requires an evaluation of the clay or shale content of the formations. For a proper evaluation, the chemical composition of the formation should be uniform within the zone being investigated. When this favorable condition exists, the gamma ray may be scaled in clay percentage by taking 100% at the maximum radiation response and 0% at its minimum, and various percentages in between these limits.

This scaling can not be applied to a well drilled through several zones of different age and chemistry. In this case individual scaling should be established for each type of

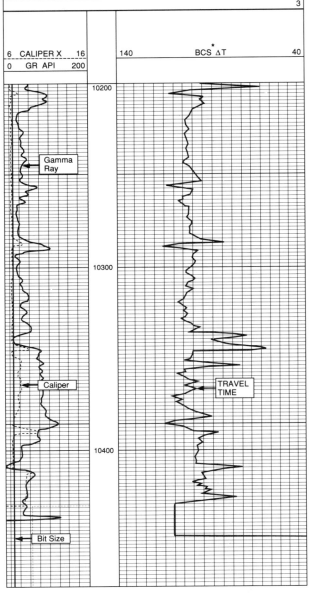

Figure 4.36. Typical gamma ray, caliper and sonic log.

sediment. Geological information obtained from cuttings and examination of all available logs usually define zones of homogeneous deposition.

Evaluation of Shale or Clay Volume Percentage from a Gamma Ray Log

The following algorithms provide reliable estimates of clay volume percentage V_{sh} as a function of the gamma ray index I_{GR} (10):[6]

$$V_{sh} = 0.33[2^{(2 \times I_{GR})} - 1.0] \quad \text{(older rocks)} ,$$

$$V_{sh} = 0.083[2^{(3.7 \times I_{GR})} - 1.0] \quad \text{(tertiary rocks)} ,$$

$$I_{GR} = \frac{GR_{log} - GR_{min}}{GR_{max} - GR_{min}} ,$$

where

$$GR_{max} = \text{clay},$$
$$GR_{min} = \text{clean sand}.$$

In some unconsolidated sediments, radioactive levels vary only slightly between formations. In this case the log will have insufficient detail to assist in formation evaluation.

4.8   OTHER LOGGING DEVICES

Caliper Log

Except in the case of drilling very hard rocks, borehole diameters are not equal to nominal bit sizes. Caliper logs measure the borehole diameter by the use of two or more arms that are mechanically linked to a precision potentiometer in the tool body. Their movement biases an electronic circuit so that changes in diameter are converted to pulses or dc signals transmitted to the surface for recording (Figure 4.36). A caliper log can establish the amount of borehole erosion, the presence of swelling clays, diameter reduction due to mud-cake deposition, and the presence of fractures or caverns. The volume of cement or gravel envelope material required can be calculated from the log. The caliper log is also used to determine corrections for borehole diameter effects in *SP* logging and to interpret the resistivity log.

Caliper tools are varied in design. Bow-spring-type calipers have the advantage of centering any probes that are simultaneously run, whereas finger-type calipers have very high vertical resolution and can detect very small borehole enlargement.

6. There are exceptions to the usefulness of these algorithms. In rare cases clays are not radioactive. Most volcanic compounds generate gamma radiation.

Flowmeter Logging

The continuous-spinner flowmeter incorporates an impeller that is rotated by a moving fluid. A series of magnets in the instrument and on the shaft of the rotating impeller generate electrical pulses that are transmitted through a logging line to surface equipment. These pulses are translated into a spinner-shaft rotation rate that is converted into fluid velocity and recorded. Fluid volume passing the instrument can be calculated with appropriate calibration curves.

A limitation of the continuous-spinner flowmeter is that the impeller does not rotate until the fluid velocity past the instrument reaches or exceeds a rate known as "threshold velocity." After the threshold velocity is exceeded, the speed of rotation increases linearly with fluid velocity, although rates only slightly exceeding threshold velocity may be unpredictable. If the threshold velocity is not reached, the instrument may be lowered or raised at a known rate that exceeds it. Spinner-shaft rotation greater than that due to instrument movement translates into fluid velocity in the well.

The problem of the threshold velocity can be minimized by use of a basket-type flowmeter, which incorporates a collapsible basket or funnel controlled from the surface. In operation, it is lowered with the funnel closed until a depth is reached where a measurement is desired, and the funnel is then opened. This concentrates the flow, increasing even low water velocity sufficiently to activate the propeller meter.

Figure 4.37 shows an injection well with two perforated zones. A flowmeter survey was made at an uprun at 50 ft per minute. After the calibration was done, a percentage of flow scale was superimposed on the log. This facilitates the calculation of the percentage of water entering each zone.

The flowmeter is useful for measuring the vertical fluid movement in wells completed in multiple aquifers with different piezometric levels. Head difference will result in flow between the aquifers. A flowmeter is also often run in a pumping well to measure production from the various aquifers penetrated, and the incremental increase in production through the length of screen.

Temperature Logging

Temperature logging continuously measures and records temperatures throughout the borehole (see Chapter 3). Most temperature logging tools are based on high resistance servoconductor sensing elements. Temperature logging is usually done while slowly lowering the tool into the borehole in order to minimize temperature influence from the tool and cable and to avoid disturbing the drilling fluid. This log is used for water-level location, cement detection, location of loss-of-circulation zones, and measuring formation temperature for electric-log interpretation.

The borehole fluid temperature is equal to the formation temperature under equilibrium conditions. A plot of depth

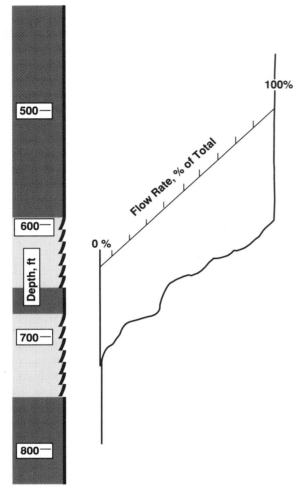

Figure 4.37. Flowmeter survey (courtesy of Welenco).

versus temperature ideally shows a constant slope that duplicates the geothermal gradient. Departures from the constant gradient usually are caused by fluid migrations inside the borehole. High temperature anomalies found at depth are associated with highly mineralized waters.

4.9 LOG INTERPRETATION AND APPLICATION TO WATER WELL COMPLETION

Making the electric-log interpretation is more an art than a science. Good results can only be achieved with empirical data. Varying environmental conditions can produce identical effects on the *SP* and resistivity curves.

Reliable log data and interpretation are also dependent upon an appropriate drilling-fluid program. Low water loss will reduce invasion and promote the recording of a usable *SP*. Drilling without bentonite will make the *SP* log interpretation very difficult.

A qualitative analysis should precede any quantitative analysis. This evaluation will define zones unworthy of exploitation and thus will assist in preventing the completion of an unproductive well. Particular attention should be given to high positive *SP* deflections. The greater the deflection, the greater is the aquifer-water resistivity, and the lower is the salt concentration.

Sequential Evaluation Method for Freshwater Wells

The following discussion explains the use of an algorithm that can be used to analyze commonly run geophysical logs in water wells (i.e., *SP*, 16 in. and 64 in. normal, and 6-ft lateral). When properly applied, the algorithm will yield quantitative information on aquifers' upper and lower boundaries, hydraulic conductivity, and water quality (total dissolved solids).

To illustrate the procedure, log data from a borehole drilled through alluvial aquifers in Southern California are used (Figure 4.11). In this example the general qualitative analysis is as follows:

Top aquifer	125–272 ft, clean to slightly clayey sands, freshwater bearing (positive *SP* and high resistivities). *SP* and resistivity reductions due to clay are noted at 125, 148, 180, and 198 ft. The departure between the long and short normal curves reflects a freshwater zone of lower conductivity than the drilling fluid.
Middle aquifer	272–377 ft, brackish water sands (negative *SP*).
Lowest zone	377 ft to bottom, saltwater sands, interbedded with clays (highly negative *SP*). Note the low resistivity with baseline shift. The long normal log reflects the high conductivity formation fluids, and the short normal log reflects the lower conductivity invaded zone.

The cleanest sand at 230–242 ft is selected for analysis.

Table 4.6 itemizes the steps used in the algorithm with the values obtained from the log example.

Step 1. Determination of Clay Baseline on *SP* Log

The first step in the analysis is to determine the clay baseline from which measurements of *SP* deflections will be made. The clay baseline is drawn on the *SP* log (Figure 4.11)

TABLE 4.6   Sequential Evaluation of Resistivity and *SP* Logs

Step	Description		Figures Used	Results
1	Determination of clay baseline	4.11	*SP* and resistivity logs	
2	Determination of bed boundaries and thickness	4.11	*SP* and resistivity logs	230–242 ft (12 ft)
3	Calculate formation temperature	4.11		84°F
4	Calculate mud resistivity at formation temperature	4.7	Salinity-resistivity chart	$R_m = 10$ ohm-m @ 84°F
5	Measurement of apparent resistivity (R_a)	4.11	64 in. normal resistivity log	$R_a = 52$ ohm-m @ 84°F
6	Calculate true formation resistivity (R_o)	4.13	Borehole departure curve	$R_o = 50$ ohm-m @ 84°F
7	Calculate resistivity of the mud filtrate (R_{mf})	4.7, 4.25	R_m vs R_{mf}	$R_{mf} = 8.6$ ohm-m @ 84°F
8	Measurement of the *SP* signal	4.11	*SP* log	$SP = +18$ mv
9	Correct SP to static value (*SSP*)	4.26	*SP* correction for bed thickness	$SSP = +18$ mv
10	Find (R_{we})	4.7, 4.24	R_{mf}/R_{we} vs *SSP*	$R_{we} = 15.36$ ohm-m @ 84°F
11	Find resistivity of interstitial water (R_w)	4.7, 4.30	R_w vs R_{we} for freshwater	$R_w = 26.3$ ohm-m @ 84°F
12	Find total dissolved solids (TDS)	4.33	R_w vs TDS at 77°F	TDS = 345 ppm
13	Calculate formation factor (F)		$F = R_o/R_w$	$F = 1.9$
14	Calculate porosity, grain size, and hydraulic conductivity	4.18	F vs. grain size and hydraulic conductivity	grain size = 0.0026 in. hydraulic conductivity = 23 gpd/ft^2

and is identified as being opposite zones of low resistivity. As can be seen on the log, the baseline is not constant but changes significantly due to hydrogeologic heterogeneities affecting the membrane or liquid-junction potential.

Step 2. Determination of Bed Boundaries and Thickness from the *SP* Log

The upper and lower bed boundaries are located from the inflection points of the *SP* curve. In the example, the boundaries are 230 and 242 ft, respectively (from Figure 4.11). This zone is confirmed by the deflection on the normal log.

Note that deflection of the *SP* signal is measured relative to the clay baseline. The direction of the maximum SP deflection is dependent on the relationship of the drilling-fluid and formation-water salinities (Figure 4.23).

The bed thickness is the difference between the upper and lower boundary depths:

$$\text{Bed thickness} = 242 - 230 = 12 \text{ ft} .$$

Step 3. Calculate Formation Temperature at Depth

When the bottom-hole fluid temperature is available, calculate the temperature at the depth of interest (assumed to be 236 ft) by

$$T_{236} = T_{bot} - \frac{t_d - 236}{100} °F ,$$

where

t_d = total depth,

T_{bot} = temperature at total depth measured by a maximum reading thermometer as 88.7°F.

Assume a 1° per 100 ft gradient from the surface temperature (as measured from the drilling-fluid temperature):

$$T_{236} = 88.7 - \frac{710 - 236}{100} = 84°F .$$

In most wells less than 1000 ft deep, the bottom-hole temperature is not a critical measurement in calculation of TDS, if variations between circulated drilling-fluid temperature and formation temperature are less than 10°. Generally, the circulated drilling-fluid temperature is used as the formation temperature.

Step 4. Calculate Mud Resistivity at Formation Temperature

The mud resistivity R_m at the formation temperature of 88° is calculated from Figure 4.7:

$R_m = 11.1$ ohm-m   at 75°F (measured) ,
$R_m = 10.0$ ohm-m   at 84°F (from Figure 4.7) .

Step 5. Measurement of Apparent Resistivity (R_a) from the Resistivity Log

The 64-in. normal resistivity log is chosen for the example since this log generally does not reflect invasion (i.e., R_a is closest to R_o). Other resistivity logs could also be analyzed using the same procedure (e.g., 16-in. normal and 6-ft lateral). Measured directly from Figure 4.11,

$$R_a = 52 \text{ ohm-m} \quad \text{at 236-ft depth.}$$

Step 6. Calculate True Formation Resistivity (R_o)

The true resistivity of a formation saturated with water can be calculated from Figure 4.13. For $R_a/R_m = 52/10 = 5.2$, and a borehole diameter of 15 in., the ratio R_o/R_m (Figure 4.13) is 5.0. The true resistivity may be calculated as

$$R_o = R_m \times 5.0 = 10.0 \times 5.0$$
$$= 50 \text{ ohm-m} \quad \text{at 84°F} .$$

Step 7. Calculate Resistivity of the Mud Filtrate (R_{mf})

$$R_m = 11.1 \text{ ohm-m} \quad \text{at 75°F (measured)} ,$$

$$R_{mf} = 8.0 \text{ ohm-m} \quad \text{(Figure 4.25)} ,[7]$$

$$R_{mf} = 9.8 \text{ ohm-m} \quad \text{(measured with a filter press at the well site)} ,$$

$$R_{mf} = 8.6 \text{ ohm-m} \quad \text{at 84°F (Figure 4.7) based on } R_{mf} \text{ of 9.8 ohm-m at 75°F} .$$

The formation temperature of the zone of interest (from Step 3) was calculated to be 84°F.

Step 8. Measurement of the *SP* Signal

The *SP* signal is the maximum deflection (either positive or negative depending on mud/formation salinity factors described in Step 2) measured from the clay baseline. For the example zone between 230 and 242, the *SP* measurement is +18 mV.

Step 9. Correct *SP* to Static Value

Due to thin beds the observed *SP* signal may be less than the static *SP*, and therefore must be corrected. Figure 4.26 may be used to correct the measured *SP* signal to the static value (*SSP*) as follows:

Bed thickness (th) = 144 in.,
Borehole diameter (dh) = 15 in.,
$th/dh = 9.6$,

R_o/R_m at 84°F = 5,
Correction factor (from Figure 4.26) = 1,
$SSP = SP \times$ correction factor $= +18 \text{ mV} \times 1 = +18$ mV.

Step 10. Find R_{we}

The equivalent resistivity of the interstitial water in the example assumes only NaCl in solution. It may be calculated directly from Equation 4.2 or from a chart. Using the equation

$$SSP = -K \log \frac{R_{mf}}{R_{we}}$$

or

$$\frac{R_{mf}}{R_{we}} = 10^a = 0.5607 ,$$

where

$$a = \frac{SSP}{-70.7[(460 + T)/537]} = \frac{+18}{-70.7[(460 + 84)/537]}$$
$$= -0.2513$$

Using Figure 4.24 with a static *SP* of +18 mV, and a formation temperature of 84°F, $R_{mf}/R_{we} = 0.56$,

$$R_{we} \text{ at 84°F} = \frac{R_{mf}}{0.56} = 15.36 \text{ ohm-m,}$$

R_{we} at 77°F (standard lab temperature)
$$= 16.8 \text{ ohm-m} \quad \text{(from Figure 4.7).}$$

Step 11. Find Resistivity of Interstitial Water (R_w)

Figure 4.30 is used to determine true formation water resistivity, R_w. If the predominant ion of the water is known from nearby wells, the specific curve is used. If nothing is known regarding ground water quality in the area, then

---

7. High mud resistivities may not yield accurate mud filtrate resistivities when Figure 4.25 is used since the chemicals of the mud solid phase become important with respect to the filtrate conductivity. In such cases Figure 4.25 is only used if R_{mf} cannot be measured at the surface.

the empirical curve is used. The empirical curve is made up of a composite of water samples from all over the world.

In the example the predominant ion is known to be sodium bicarbonate ($NaHCO_3$):

$$R_w = 29 \text{ ohm-m} \quad \text{at } 77°F \text{ (Figure 4.30)} ,$$

$$R_w = 26.3 \text{ ohm-m} \quad \text{at } 84°F \text{ (Figure 4.7)} .$$

Step 12. Find Total Dissolved Solids (TDS)

$$TDS = 345 \quad \text{(from Figure 4.33)} .$$

Step 13. Calculate Formation Factor F

$$F = \frac{R_o}{R_w} \quad \text{at } 84°F = \frac{50}{26.3} = 1.9 .$$

Step 14. Calculate Grain Size and Hydraulic Conductivity

If Figure 4.15 is used to convert the formation factor to porosity, an unrealistically high value (>60%) is obtained. This is due to the double layer effect. Therefore locally derived curves such as Figure 4.18 must be used. These curves of formation factor versus grain size are obtained in the area based on sieve analyses of samples taken at various depths.

From Figure 4.18,

$$\text{Grain size} = 0.0026 \text{ in.}$$

This translates into a hydraulic conductivity of 23 gpd/ft^2.

REFERENCES

1. Jones, P. H., and T. B. Buford. 1951. "Electrical Logging Applied to Ground-Water Exploration." *Geophysics* 16, 1.

2. Hill, H. J., and J. D. Milburn. 1956. "Effect of Clay and Water Salinity on Electrochemical Behavior of Reservoir Rocks." *J. Pet. Tech.* 8, 3.

3. Alger, R. P. 1966. "Interpretation of Electric Logs in Freshwater Wells in Unconsolidated Formations." Trans. SPWLA Seventh Annual Logging Symposium, Tulsa.

4. Wyllie, M. R. J. 1949. "A Quantitative Analysis of the Electrochemical Component of the SP Curve." *J. Pet. Tech.* (January).

5. Doll, H. G. 1948. "The SP Log, Theoretical Analysis and Principles of Interpretation." *Trans. AIME* 179.

6. Gondouin, M., M.P. Tixier, and G. L. Simard. 1957. "An Experimental Study on the Influence of the Chemical Composition of Electrolytes on the SP Curve." *J. Pet. Tech.* 9, 2.

7. Sarma, V. V., and B. V. Rao. 1962. "Variations of Electrical Resistivity of River Sands, Calcite, and Quartz Powders with Water Content." *Geophysics* 27, 4.

8. Knoll, G. F. 1979. *Radiation Detection and Measurement.* J. Wiley and Sons, New York

9. Tittman, J. 1956. "Radiation Logging: Physical Principles." University of Kansas Petroleum Engineering Conference, April.

10. Dresser. 1983. Log Interpretation Charts. Dresser Industries.

READING LIST

Denson, K. H., A. Shindala, and C. D. Fenn. 1968. "Permeability of Sand with Dispersed Clay Particles." *Water Resources Research* 4, 6.

Doll, H. G. 1950. "The SP Log in Shaly Sands." *Trans. AIME* 189.

Dunlap, H. F., H. L. Bilhartz, E. Schuler, and C. R. Bailey. 1949. "The Relation Between Electrical Resistivity and Brine Saturation in Reservoir Rocks." *J. Pet. Tech.* 1, 10.

Fons, L. C., E. Johns, and M. L. Mougne. 1970. "New Way to Evaluate Shaly Formations." *Oil and Gas J.* 3 (August).

Guyod, H., and L. E. Shane. 1969. "Introduction to Geophysical Well Logging, Accoustical Logging." Vol. 1. Guyod, Houston.

Kolesh, F. P., R. J. Schwartz, W. B. Wall, and R. L. Morris. 1965. "A New Approach to Sonic Logging and Other Acoustic Measurements." *J. Pet. Tech.* 17, 3.

Porter, C. R., and J. E. Carothers (Phillips Petroleum Co). 1970. "Formation Factor Relation Derived from Well Log Data," *SPWLA* (May).

Tittman, J., and J. S. Wahl. 1965. "The Physical Foundations of Formations Density Logging (Gamma-Gamma)." *Geophysics* (April).

Tixier, M. P., R. P. Alger, and C. A. Doh. 1959. "Sonic Logging." *J. Pet. Tech.* 11, 5.

Turcan, A. N., Jr. 1962. "Estimating Water Quality from Electrical Logs." USGS Prof. Paper 450-C, Art. 116.

Urban, F., H. L. White, and E. A. Strassner. 1935. "Contribution to the Theory of Surface Conductivity at Solid-Solution Interfaces." *J. Phys. Chem.* 39.

Wahl, J. S., J. Tittman, and C. W. Johnstone. 1964. "The Dual Spacing Formation Density Log." *J. Pet. Tech.* (December).

Winsauer, W. O., and W. M. McCardell. 1953. "Ionic Double Layer Conductivity in Reservoir Rocks." *J. Pet. Tech.* 5, 5.

Wyllie, M. R. J., A. R. Gregory, and G. H. F. Gardner. 1956. "Elastic Wave Velocities in Heterogeneous and Porous Media." *Geophysics* 21, 1.

Wyllie, M. R. J., A. R. Gregory, and G. H. F. Gardner. 1958. "An Experimental Investigation of Factors Affecting Elastic Wave Velocities in Porous Media." *Geophysics* 23, 3.

Schlumberger Well Surveying Corporation. 1949. *Resistivity Departure Curves.* Document No. 3.

Gearhard-Owen. 1977. *Formation Evaluation Handbook.*

CHAPTER 5

Hydraulics of Wells

5.1 INTRODUCTION

Knowledge of well hydraulics is essential to proper well design and construction, and optimum performance of the well/aquifer system. Part I, Chapter 2, developed from fundamental laws of physics the general equations for ground water flow. These equations describe flow of ground water in porous media for three basic aquifer types:

- Confined systems.
- Unconfined systems.
- Semiconfined or leaky systems.

This chapter develops solutions for the most commonly encountered field conditions for ground water flowing in and around water wells. The solutions are equations for drawdown as a function of various aquifer, well, and operational parameters. The many steps necessary to develop the equations are omitted, although the complete mathematical derivations can be found in the references.

Equations are developed for drawdown for the three basic porous-media aquifer types and two flow conditions, nonsteady (time varying) and steady state (equilibrium). The equations are subject to a specific set of mathematical initial and boundary conditions that must be considered in their application. Numerical values for parameters required to apply the equations may be found by the methods discussed in Chapter 15. The equations presented here are consistent with the field methods presented in that chapter.

Basic Assumptions

Many investigators have derived useful solutions for flow to wells by considering ideal flow through porous media modified by assumed initial and boundary conditions. Although the solutions are based upon simplified assumptions and approximations of field conditions, they have been nevertheless proved sufficiently accurate to promote good ground water planning and well design (1). Unless otherwise indicated, the assumptions listed here are appropriate for the applications described in this chapter. These assumptions are:

- The aquifer is composed of porous media, with ground water flow obeying Darcy's law.
- The velocity of flow is proportional to the tangent of the hydraulic gradient, and not the sine as is actually the case.
- The aquifer has a horizontal base and uniform thickness and is homogeneous, isotropic, and of infinite areal extent.
- The pumping well penetrates the complete thickness of the aquifer and receives water from the entire thickness by horizontal flow.
- Water is released from storage instantaneously in response to decline in head.
- The well diameter is sufficiently small that the volume of water removed from the well bore during pumping is negligible.
- Prior to pumping, the initial water level is horizontal.
- The well is pumped at a constant discharge rate.

Differential Equations for Ground Water Flow

The differential equations for ground water flow developed in Chapter 2 are solved here for the three basic aquifer types.

Confined Aquifers. For confined aquifers the governing differential equation for nonsteady-state flow is (2)

$$K_x \frac{\partial^2 h}{\partial x^2} + K_y \frac{\partial^2 h}{\partial y^2} + K_z \frac{\partial^2 h}{\partial z^2} = S_s \frac{\partial h}{\partial t} , \quad (5.1)$$

where

h	=	hydraulic head (L)
K_x, K_y, K_z	=	hydraulic conductivity in x-, y-, and z- directions [LT^{-1}],
S_s	=	specific storativity [L^{-1}],
t	=	time [T].

Incorporating two of the basic assumptions (horizontal flow in an isotropic aquifer) and transforming into polar coordinates, Equation 5.1 written in terms of drawdown (see Figure 5.1) becomes

$$\frac{\partial^2 s}{\partial r^2} + \frac{1}{r}\frac{\partial s}{\partial r} = \frac{1}{\kappa}\frac{\partial s}{\partial t} \qquad (5.2)$$

where

s = drawdown $(h_0 - h)$ [L],
r = radial distance from well [L],

t = time since pumping began [T],
κ = hydraulic diffusivity, T/S [L^2T^{-1}],
T = transmissivity [L^2T^{-1}],
S = storativity.

For steady-state flow, Equation 5.2 reduces to

$$\frac{\partial^2 s}{\partial r^2} + \frac{1}{r}\frac{\partial s}{\partial r} = 0 \ .$$

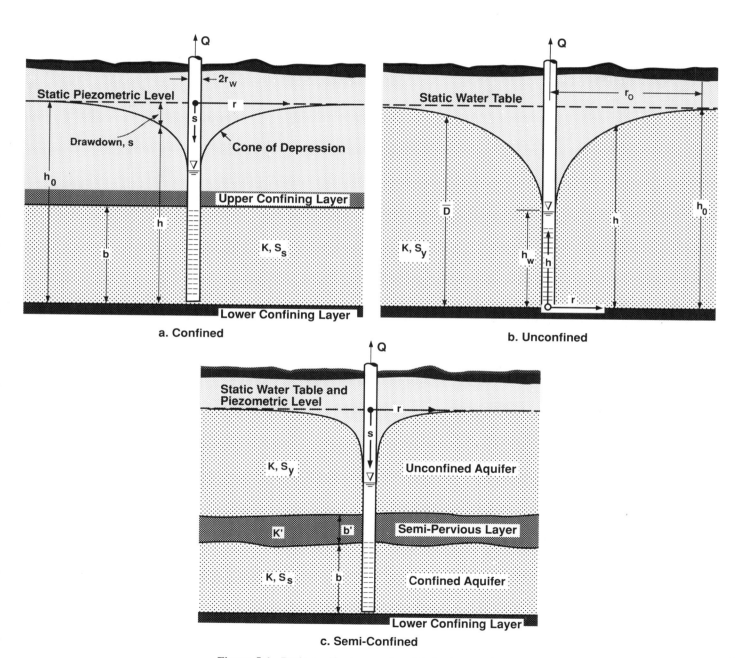

Figure 5.1. Basic aquifer types referred to in the equations.

Unconfined Aquifers. For nonsteady horizontal flow in an unconfined isotropic aquifer, the differential equation of flow is

$$\frac{\partial^2 h^2}{\partial x^2} + \frac{\partial^2 h^2}{\partial y^2} = \frac{\theta}{\overline{D}K} \frac{\partial h^2}{\partial t} ,$$

where

$\overline{D}$ = average depth of saturation [L],

θ = effective porosity,

K = hydraulic conductivity [LT^{-1}].

In polar coordinates,

$$\frac{\partial^2 h^2}{\partial r^2} + \frac{1}{r} \frac{\partial h^2}{\partial r} = \frac{\theta}{\overline{D}K} \frac{\partial h^2}{\partial t} .$$

For steady-state flow,

$$\frac{\partial^2 h^2}{\partial r^2} + \frac{1}{r} \frac{\partial h^2}{\partial r} = 0 .$$

Semiconfined Aquifers. The differential equation for nonsteady flow in a semiconfined (leaky) aquifer in terms of drawdown and polar coordinates is given as

$$\frac{\partial^2 s}{\partial r^2} + \frac{1}{r} \frac{\partial s}{\partial r} - \frac{s}{B^2} = \frac{1}{\kappa} \frac{\partial s}{\partial t} ,$$

where

B = leakage factor [L].

For steady-state flow,

$$\frac{\partial^2 s}{\partial r^2} + \frac{1}{r} \frac{\partial s}{\partial r} - \frac{s}{B^2} = 0 .$$

5.2 EQUATIONS FOR AQUIFER DRAWDOWN

In the following discussions, solutions are presented in terms of the governing equation and applicable initial and boundary conditions. The convention of polar coordinates is used and solutions for drawdown presented in terms of radial distance to the "cone of depression" from the pumping well and time since start of pumping. The cone of depression surrounding a pumping well is the free surface of gravity drainage in an unconfined aquifer, and the pressure surface (piezometric surface) in a confined or semiconfined aquifer. This cone, logarithmic in form, extends radially outward from the well,

the extent and shape being a function of aquifer parameters, pumping rate, and time.

Nonsteady-State Flow in a Confined Aquifer

The Theis Equation. In 1935 C. V. Theis, noting similarities between heat flow and ground water flow, developed the nonsteady-state equation for ground water flow in confined aquifers (2). This equation and its derivatives are the most often applied equations in transient ground water hydraulics.

Governing Equation (Equation 5.2; also See Figure 5.1a)

$$\frac{\partial^2 s}{\partial r^2} + \frac{1}{r} \frac{\partial s}{\partial r} = \frac{1}{\kappa} \frac{\partial s}{\partial t}$$

Initial Condition

$s(r, 0) = 0$ (drawdown at any radial distance at time, $t = 0$) .

Boundary Conditions for s(r, t)

$s(\infty, t) = 0$ (drawdown at an infinite distance for any time) .

$$r_w \frac{\partial s(r_w, t)}{\partial r} = -\frac{Q}{2\pi Kb}$$ (discharge condition at the well) .

Solution for Drawdown

$$s(r, t) = \frac{Q}{4\pi T} W(u) \quad \text{(Theis equation)} , \quad (5.3)$$

where

$s(r, t)$ = drawdown at distance (r) at time (t) after start of pumping [L],

Q = discharge rate [L^{3}T^{-1}],

$W(u)$ = well function of Theis (see Table 15.2.)

Specifically,

$$W(u) = \int_u^\infty \frac{e^{-y}}{y} \partial y$$

$$= -\gamma - \log_e u + u - \frac{u^2}{2 \cdot 2!} + \frac{u^3}{3 \cdot 3!}$$

$$+ \cdots + \frac{u^n}{n \cdot n!} ,$$

where

$$\gamma = \text{Euler's constant} = 0.5772,$$

$$u = \frac{r^2 S}{4Tt} .$$

In conventional units,

$$s(r, t) = \frac{114.6Q}{T} W(u) \qquad (5.4)$$

and

$$u = \frac{1.87r^2 S}{Tt} ,$$

where

t	=	time since pumping started [days],
$s(r, t)$	=	drawdown [ft],
T	=	transmissivity [gpd/ft],
S	=	storativity.
Q	=	discharge rate [gpm]

Jacob's Equation.

In its general form the Theis equation is less convenient to use, requiring access to tables of the Well function. Jacob in 1946 noted that for small values of $u \leq 0.05$, the well function may be approximated by (3)

$$W(u) \simeq -\gamma - \log_e u \simeq \log_e \left(\frac{e^{-\gamma} 4Tt}{r^2 S} \right) . \qquad (5.5)$$

Substituting Equation 5.5 into Equation 5.3 results in Jacob's approximation to the Theis equation:

$$s(r, t) = \frac{Q}{4\pi T} \log_e \left(\frac{2.25Tt}{r^2 S} \right) ;$$

in conventional units,

$$s(r, t) = \frac{264Q}{T} \log \left(\frac{0.3Tt}{r^2 S} \right) , \qquad (5.6)$$

where parameter units are the same as defined for Equation 5.4.

Because the fundamental assumption for Jacob's solution requires $u \leq 0.05$, the range of validity of Jacob's equation requires time $t > 5r^2 S/T$ or, in conventional units, $t > 37.4r^2 S/T$ (days). Figure 5.2 illustrates the range of validity of Jacob's equation.

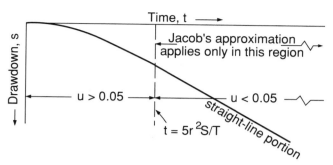

Figure 5.2. Range of validity of Jacob's equation.

Steady-State Flow in Confined Aquifers—Thiem's Equation.

Solutions for drawdown in confined aquifers under steady-state flow conditions are described by

Governing Equation (See Figure 5.1a)

$$\frac{\partial^2 h}{\partial r^2} + \frac{1}{r} \frac{\partial h}{\partial r} = 0 .$$

The solution as given by Thiem was solved with the following conditions (4):

Initial Conditions

Steady-state flow.

Boundary Conditions

$$h(r_0) = h_0 ,$$

$$r_w \frac{\partial h(r_w)}{\partial r} = -\frac{Q}{2\pi Kb} \quad \text{(discharge condition}$$

$$\text{at the well)} ,$$

where

h_0 = hydraulic head where zero drawdown occurs [L],
r_0 = radial distance measured from the center of the well where zero drawdown occurs [L].
r_w = radius of well [L].

Solution for Drawdown

$$s(r) = h_0 - h(r) = \frac{Q}{2\pi T} \log_e \left(\frac{r_0}{r} \right) ;$$

in conventional units,

$$s(r) = \frac{528Q}{T} \log \left(\frac{r_0}{r} \right) . \qquad (5.7)$$

Equation 5.7 is known as the equilibrium or Thiem equation.

Nonsteady-State Flow in Unconfined Aquifers

Without Delayed Yield. In unconfined aquifers with no delayed yield effects, ground water flow in the vicinity of the pumping well is identical to flow in a confined aquifer. Thus the Theis or Jacob equation may be applied within limitations of the basic assumptions and after replacing the drawdown (*s*) with

$$s' = s - \frac{s^2}{2h_0} .$$

With Delayed Yield—Boulton's Equation. Water produced by a pumping well in unconfined aquifers is derived from both gravity drainage and aquifer compaction/water expansion. However, gravity drainage does not usually occur instantaneously with the decline in head, and wells typically show an effect of "delayed yield."

Boulton pointed out that delayed yield is a form of recharge, which can be seen by examining the time-drawdown curve (5). This curve shows three distinct components, the first of which occurs for a short period after start of pumping before gravity drainage is effective. During this period the unconfined aquifer behaves in the same way as a confined aquifer with water being released instantaneously from storage by compaction of the aquifer and expansion of the water.

The second period shows the effects of recharge as replenishment takes place by gravity drainage from the effective pore spaces above the cone of depression.

The third segment of the curve again closely conforms to the confined aquifer condition as equilibrium is established

between the rate of gravity drainage and the rate of decline of the water table. This segment may begin several minutes to several days after start of pumping. Figure 5.3 illustrates this phenomenon.

Solutions for drawdown in unconfined aquifers showing delayed yield effects are described by

Governing Equation

$$\frac{\partial^2 \phi}{\partial r^2} + \frac{1}{r} \frac{\partial \phi}{\partial r} + \frac{\partial^2 \phi}{\partial z^2} = \frac{S_s}{K} \frac{\partial \phi}{\partial t} ,$$

where

$\phi = \phi(r, z) = p/\gamma + z$ (hydraulic head [L]; see Figure 5.4), where γ is the specific weight of water [FL^{-3}] and p is the hydrostatic pressure [FL^{-2}],

$z =$ gravity potential [L].

Initial Condition

$$\phi(r, z, 0) = h_0 .$$

Boundary Conditions

$$\phi(\infty, z, t) = h_0 ,$$

$$\frac{\partial \phi(r, 0, t)}{\partial z} = 0 ,$$

$$Q = 2\pi r_w K \int_0^{D_w} \left[\frac{\partial \phi(r_w, z, t)}{\partial r} \right] dz .$$

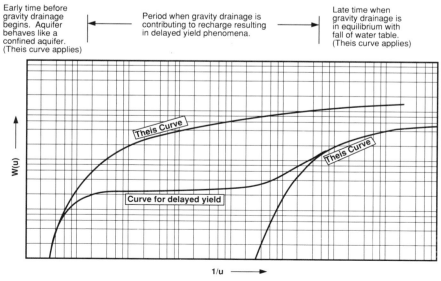

Early time before gravity drainage begins. Aquifer behaves like a confined aquifer. (Theis curve applies)

Period when gravity drainage is contributing to recharge resulting in delayed yield phenomena.

Late time when gravity drainage is in equilibrium with fall of water table. (Theis curve applies)

Theis Curve

Theis Curve

Curve for delayed yield

W(u)

1/u

Figure 5.3. Effect of delayed yield in unconfined aquifers.

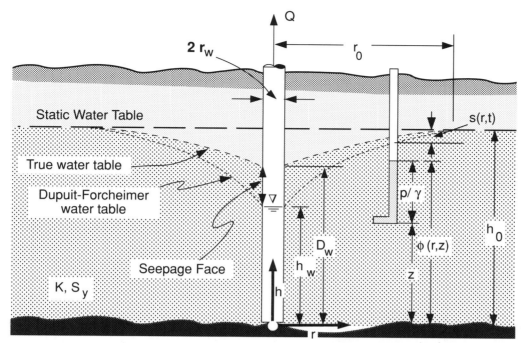

Figure 5.4. Flow to a well in an unconfined aquifer.

A solution for this problem is not easily found. However, Boulton proposed that if $h_0 - h_w < 0.5h_0$, the solution is analogous to that for a leaky artesian aquifer (5, 7),

Solution for Drawdown

$$s(r, t) = \frac{114.6Q}{K\overline{D}} W(u_{AY}, r/B) ,$$

where

$s(r, t)$	=	drawdown [ft],
r	=	distance from pumping well [ft],
t	=	time since pumping began [days],
Q	=	pumping rate [gpm],
B	=	drainage factor [ft],
K	=	hydraulic conductivity [gpd/ft^2],
$\overline{D}$	=	average saturated thickness
	$\simeq$	$(h_0 + h_w)/2$ [ft],
$W(u_{AY}, r/B)$	=	well function of Boulton,
u_A	=	$1.87r^2S/Tt$ (small values of t),
u_Y	=	$1.87r^2S_y/Tt$ (large values of t),
S_y	=	specific yield (effective porosity).

Steady-State Flow in Unconfined Aquifers—The Dupuit–Forcheimer Equation

Dupuit's Assumption of Horizontal Flow. In 1863 Dupuit assumed that if in gravity flow systems with horizontal

bases the slope of the hydraulic gradient is low, the hydraulic gradient is the same at all points in a vertical section, resulting in horizontal flow. Under this assumption the slope is equal to the tangent of the angle of inclination and not the sine (as is actually the case). Dupuit's assumptions therefore do not account for the curvilinear nature of flow in the immediate vicinity of the well and indiscriminate use of the following equation will lead to overestimation of drawdown (see Figure 5.4).

Investigations have shown that the following equation yields accurate results if r is sufficiently large so that the curvilinear effects are negligible (i.e., $r > 1.5h_0$) (9):

Governing Equation

$$\frac{\partial^2 h^2}{\partial r^2} + \frac{1}{r} \frac{\partial h^2}{\partial r} = 0 .$$

Initial Condition

Steady-state flow.

Boundary Conditions

$$h(r_o) = h_0 ,$$

$$h(r_w) = h_w .$$

The solution for discharge results in the Dupuit–Forcheimer equation:

$$Q = \frac{K(h_0^2 - h_w^2)}{1055 \log(r_0/r_w)} \quad . \qquad (5.8)$$

Equation 5.8 may be converted to drawdown in the general case, $s(r) = h_0 - h(r)$:

$$s(r) - \frac{s(r)^2}{2h_0} = \frac{528Q}{Kh_0} \log\left(\frac{r_0}{r}\right) , \qquad (5.9)$$

where all units are as previously defined.

As can be seen, Equation 5.9 is identical to Equation 5.7 (Thiem's equation for confined aquifers) if drawdown in the Thiem equation is replaced by $s - s^2/2h_0$ and T by Kh_0, (6).

Nonsteady-State Flow in Semiconfined Aquifers— The Hantush–Jacob Equation

An artesian aquifer is commonly confined above and below by semipervious materials (silt and clay). Water ponded above the aquifer, or recharged to an underlying aquifer, supplies the "leakage" induced by lowering the hydraulic head in the artesian aquifer.

The actual flow system created by a pumping well in such a system is three-dimensional. The flow, however, may be treated as two dimensional with certain assumptions (10, 11). These assumptions are:

- The head in the layer supplying the leakage is constant.
- The hydraulic conductivity contrast between the semipervious upper and lower layers and the artesian aquifer is great, so that flow is vertical in the semiconfining beds and horizontal in the artesian aquifer.
- Storage in the semipervious layers is neglected.

Based on these assumptions, an equation for nonsteady flow to a well in the vicinity of an infinite leaky aquifer was developed by Hantush and Jacob 1955; (also see Figure 5.1c) (12):

$$\frac{\partial^2 s}{\partial r^2} + \frac{1}{r}\frac{\partial s}{\partial r} - \frac{s}{B^2} = \frac{1}{\kappa}\frac{\partial s}{\partial t} \quad .$$

Initial Condition

$$s(r, 0) = 0 \quad .$$

Boundary Conditions

$$s(\infty, t) = 0 \quad ,$$

$$r_w \frac{\partial s}{\partial r}(r_w, t) = -\frac{Q}{2\pi T} \quad .$$

Solution for Drawdown in Conventional Units

$$s(r, t) = \frac{264Q}{T} W(u, r/B) \quad \text{Hantush–Jacob equation}$$

where

$W(u, r/B)$ = well function for leaky aquifers (see Table 15.2),

u = $1.87r^2 S/Tt$,

B = leakage factor = $\sqrt{\dfrac{Kb}{(K'/b')}}$ [ft],

K'/b' = leakance (coefficient of leakage), which is a characteristic of the semipervious layer and a measure of the ability of this layer to transmit vertical leakage (6).

Steady-State Flow in Semiconfined Aquifers

Governing Equation

$$\frac{\partial^2 s}{\partial r^2} + \frac{1}{r}\frac{\partial s}{\partial r} - \frac{s}{B^2} = 0 \quad .$$

Initial Condition

Steady-state flow.

Boundary Conditions

$$s(\infty) = 0 \quad ,$$

$$r_w \frac{\partial s}{\partial r}(r_w) = -\frac{Q}{2\pi T} \quad .$$

Solution for Drawdown in Conventional Units

$$s(r) = \frac{229Q}{T} K_0\left(\frac{r}{B}\right) ,$$

where

$K_0(r/B)$ = modified Bessel function of the second kind and zero order.

5.3 APPLICATIONS AND SPECIAL WELL/ AQUIFER CONDITIONS

Partially Penetrating Wells

Wells that are screened in less than the full thickness of the aquifer penetrated are called "partially penetrating" wells. Unlike flow to fully penetrating wells, which is essentially horizontal and parallel to bedding planes, flow to partially penetrating wells is three dimensional. Consequently the drawdown observed in a partially penetrating well is greater. Drawdown is a function of the length and position of the screen in the aquifer of the pumping and any observation wells (see Figure 5.5).

Experience has shown that when $r > 1.5b$, the effect of partial penetration is negligible and equations for fully penetrating wells may be used (6). If this is not the case, the following solution applies.

Governing Equation

$$\frac{\partial^2 s}{\partial r^2} + \frac{1}{r}\frac{\partial s}{\partial r} + \frac{\partial^2 s}{\partial z^2} = \frac{1}{\kappa}\frac{\partial s}{\partial t} \ .$$

Initial Condition

$$s(r, z, 0) = 0 \ .$$

Boundary Conditions

$$s(\infty, z, t) = 0 \ ,$$

$$\lim_{r \to 0}\left[(D - D')r\,\frac{\partial s}{\partial r}\right] = 0 \quad \text{for } 0 < z < D' \ , \quad \text{(above screen)}$$

$$= -\frac{Q}{2\pi K_r} \quad \text{for } D' < z < D \ , \quad \text{(in screen)}$$

$$= 0 \quad \text{for } D < z < b \ . \quad \text{(below screen)}$$

For relatively long time periods,

$$t > \frac{b^2 S_s}{2K_z} \ .$$

The solution for drawdown is (13).

Solution for Drawdown

$$s(r, z, t) = \frac{114.6Q}{T}\left\{W(u) + f\left[\frac{r}{b}, \frac{D}{b}, \frac{D'}{b}, \frac{z}{b}\right]\right\} \tag{5.10}$$

where

$$f = \frac{4b}{\pi(D - D')}\sum_{n=1}^{\infty}\left(\frac{1}{n}\right)K_0\left(\frac{\pi n r}{b}\right)\cos\left(\frac{\pi n z}{b}\right)$$

$$\times \left[\sin\left(\frac{\pi n D}{b}\right) - \sin\left(\frac{\pi n D'}{b}\right)\right] \ ,$$

$W(u)$ = well function of Theis,
K_0 = modified Bessel function of second kind, zero order.

All angles are in radians.

Principle of Superposition

Because the differential equation of ground water flow is linear in the dependent variable (head or drawdown), a linear combination of solutions is also a solution (6). This principle, known as "superposition," is a powerful mathematical tool and is widely applied to problems in ground water hydraulics. For example, in a well field consisting of n discharging artesian wells, the drawdown distribution around each well separately satisfies the solution to the boundary value problem.

If several wells are pumping (or recharging) in the same aquifer, the drawdown at any point in the aquifer is the net sum of the individual drawdowns as if each well were operating alone. Assuming that the boundary condition at the well face (discharge condition) is not influenced by other wells, the drawdown distribution may be written

$$s(x, y, t) = \frac{114.6}{\sqrt{T_x T_y}}\sum_{i=1}^{n} Q_i W$$

$$\times \left\{\frac{1.87S}{t}\left[\frac{(x - x_i)^2}{T_x} + \frac{(y - y_i)^2}{T_y}\right]\right\} \ , \tag{5.11}$$

where

$s(x, y, t)$ = drawdown at a point (x, y) in the field at time (t) [ft],
Q_i = discharge of ith well in the field [gpm],
T_x, T_y = transmissivities in the x and y directions, respectively [gpd/ft],
t = time since start of pumping [days],
x_i, y_i = coordinates of the ith well [ft].

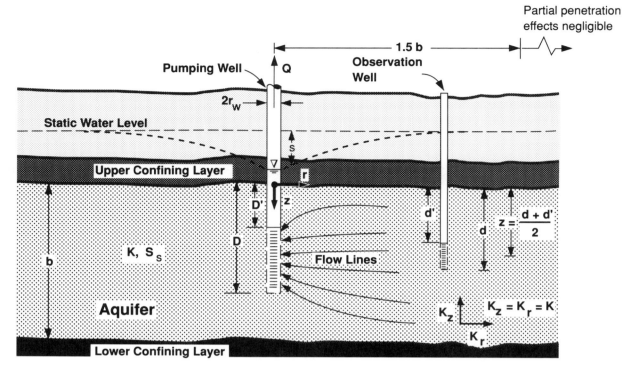

Figure 5.5. Flow to a well partially penetrating a confined aquifer.

Figure 5.6 diagrammatically shows the principle of superposition.

Image Well Theory

A special application of superposition is the method of images. In this method the effect of a hydrologic boundary such as a stream or fault is mathematically replaced by a system of imaginary "image" wells. These wells are arranged in such a fashion as to transform the complex system of well and boundaries into a simple multiple-well field that can be solved using Equation 5.11.

Full and Partial Strength Image Wells. Solution of problems using image wells involves adding or subtracting hydraulic effects (due to boundaries or other wells). Because interference at any point in the flow system is only a function of the image well discharge and distance (assuming constant time and aquifer parameters), a full strength image well at a distance of 1000 ft might have the same interference effect as a partial strength image well placed at 350 ft. This concept allows simulation of boundary conditions, ranging from impervious and semipervious barriers to constant head.

Hydrologic Boundaries

Recharge Boundary. Figure 5.7 shows a constant head condition as a result of the hydraulic connection between a stream and a well. Because of the constant recharge from the stream, drawdowns are less on the stream side of the well.

An equivalent hydraulic model for the stream–well system may be simulated by placing a recharging image well an equal distance away on the other side of the stream. The system may now be analyzed, assuming an infinite aquifer system containing two wells, a real discharging well, and an imaginary recharging well.

Impermeable Boundary. In cases where wells are located close to impermeable boundaries such as bedrock or impervious faults, the resultant cone of depression closest to the boundary is skewed deeper and flatter than on the other side (Figure 5.8). The equivalent hydraulic system for an infinite aquifer is created by placing a discharging image well an equal distance away from the impermeable boundary.

Leaky Boundary. In some cases the boundary may not be completely impermeable, and a nearby well will receive some recharge through leakage. Common examples are alluvial faults with semipervious fault planes or streams with semipervious beds.

Faults transecting alluvial aquifers frequently have a semipervious "gouge" layer in the vicinity of the plane of faulting (14). This layer transmits water through the fault zone, depending on the leakance and the hydraulic gradient across the zone. Leaky faults exist in many areas of the world and

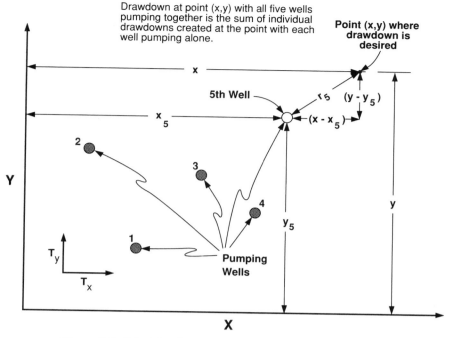

Figure 5.6. Principle of superposition applied to a five-well field.

are characterized by noticeable water-level differences across the fault (Figure 5.9). In some cases the upstream water level is close enough to ground surface to support phreatophyte growth.

The equivalent hydraulic system for an infinite aquifer containing a well near a leaky fault is simulated using a "partial strength image well." As shown in Figure 5.9, the pumping image well is placed an equal distance from the fault on the opposite side. A first-order estimate of the amount of induced leakage (q_L) may be made by knowledge of the water-level differential (Δh) across the fault, namely

$$q_L = L\overline{D}\,\frac{(K'/b')\Delta h}{192.5}\;,$$

where

q_L = induced leakage per length of the fault [gpm],
L = length as measured along fault face [ft],
$\overline{D}$ = average depth of saturated thickness [ft],
K'/b' = leakance of the fault gouge zone [day^{-1}],
Δh = water level difference across fault [ft].

Flow in Nonporous-Media Aquifers

Aquifers such as cavernous limestone, basalt, or fractured bedrock having no primary porosity (all ground water flow occurs through secondary porosity) do not obey conventional laws governing flow through porous media. Jenkins and

Prentice offer a procedure covering linear flow to a well penetrating a fracture with hydraulic conductivity many times higher than the surrounding aquifer (15). Their solution for drawdown requires the following assumptions:

- The aquifer is infinite, homogeneous, isotropic, and confined, containing a long, finite, fully penetrating vertical fracture having an infinitesimal width.
- The fracture has no storage capacity.
- Resistance to flow within the fracture is negligible.
- Flow within the fracture is laminar and linear toward the pumped well.
- Flow within the aquifer is laminar and linear toward the plane of the fracture, and the flow through the fracture zone is constant and uniform along the entire length of the fracture.

Based on these assumptions, the solution for drawdown can be written

$$s(x,\,t) = \frac{720Q}{LT}\left[\sqrt{\frac{0.17Tt}{S}} - x\right],$$

where

$s(x,\,t)$ = drawdown at perpendicular distance x from the fracture at time (t) [ft],
L = length of fracture [ft],

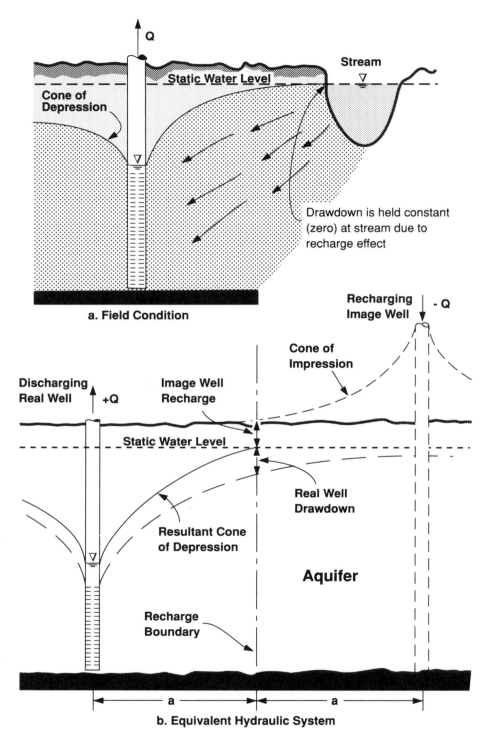

Figure 5.7. Well pumping near a recharge boundary.

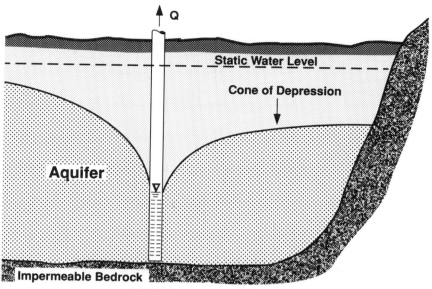

Q

Static Water Level

Cone of Depression

Aquifer

Impermeable Bedrock

a. Field Condition

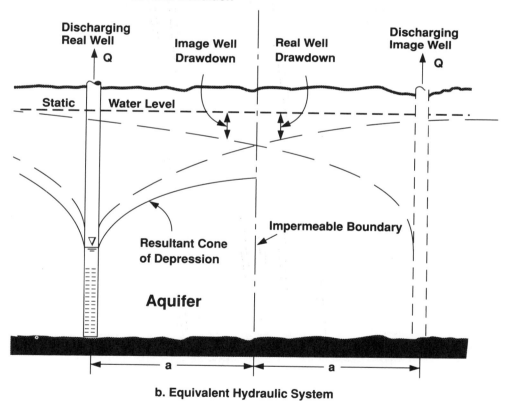

Discharging
Real Well
Q

Image Well
Drawdown

Real Well
Drawdown

Discharging
Image Well
Q

Static   Water Level

Resultant Cone
of Depression

Impermeable Boundary

Aquifer

a

a

b. Equivalent Hydraulic System

Figure 5.8. Well pumping near an impermeable boundary.

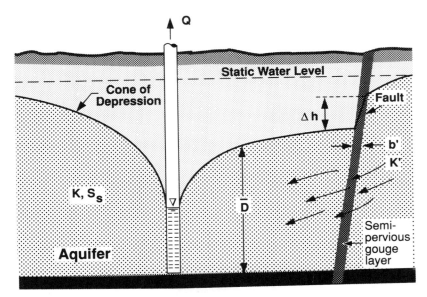

a. Field Condition

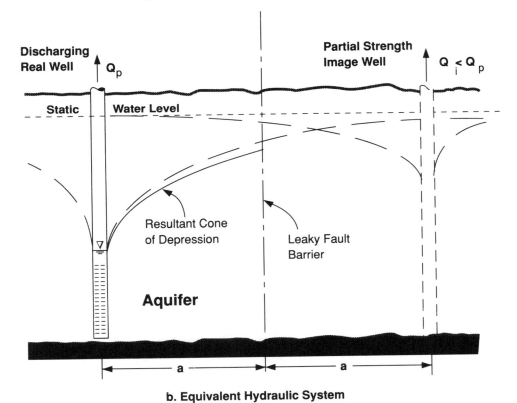

b. Equivalent Hydraulic System

Figure 5.9. Well pumping near a leaky barrier.

T = transmissivity [gpd/ft],

t = time since pumping started [days],

Q = discharge rate [gpm],

S = storativity.

Fracture theory is applied to wells completed in bedrock and other consolidated formations where fractures form the only conduits for ground water movement.

Miscellaneous

Vertical flow in the filter pack and the effect of well screen length on the specific capacity and drawdown are discussed in Chapter 11.

5.4 WELL AND NEAR-WELL CHARACTERISTICS

Laminar and Turbulent Flow

As discussed in Chapter 2, with the flow of ground water in porous media, streamlines move in essentially parallel layers. Although the microscopic movement may appear highly varied, rotational, and tortuous, on a macroscopic scale the entire process is perfectly orderly, with each layer sliding parallel with respect to its neighbor.

As flow velocity increases above a critical level, the flow becomes turbulent, and secondary irregular motions and velocity fluctuations are superimposed on the average flow. Turbulent flow is characterized by eddies or whirls that absorb kinetic energy and increase the fluid friction transfer of pressure energy to heat.

Turbulence occurs in and near the vicinity of a pumping well when inertial forces predominate over viscous forces. The transition between completely laminar and completely turbulent flow in porous media consists of the gradual dissemination of turbulence throughout the pores. Thus the first deviations from Darcy's law correspond to the beginning of appreciable eddy losses in the larger pores. These localized regions of turbulence spread to the smaller pores as velocity increases.

The following equations are based on Darcy's law and the Darcy–Weisbach formula, and apply equally to uniform aquifers and filter packs (16). Darcy's law shows that for laminar flow, the hydraulic gradient $\partial h/\partial r$ may be expressed as

$$\frac{\partial h}{\partial r} = \frac{a_1 v^2 \text{Re}}{gkd} . \qquad (5.12)$$

Similarly, for turbulent flow, the gradient is

$$\frac{\partial h}{\partial r} = \frac{a_2 v^2 \text{Re}^2}{gkd} \qquad (5.13)$$

where

Re = Reynolds number = vd/v,

v = flow velocity [LT^{-1}]

v = kinematic viscosity [L^{2}T^{-1}] (for water = 1.233 × 10^{-5} ft^2/sec),

d = mean grain diameter (assumed = 50% passing),

k = intrinsic permeability of filter zone [L^2],

a_1, a_2 = constants (a_1 = 1 for laminar flow, a_2 = 1/Re$_c$),

Re$_c$ = critical Reynolds number (the constant a_2 is equal to the reciprocal of the critical Reynolds number, Re$_c$, and is primarily a function of grain shape, packing, and distribution),

g = gravitational constant [LT^{-2}].

At the point of transition from laminar to turbulent flow (defined as the "critical point" and denoted by the subscript c), the hydraulic gradients are equal and Equations 5.12 and 5.13 may be equated, resulting in

$$\left[\frac{\partial h}{\partial r}\right]_c = a_1 \frac{v^2}{gkd} \text{Re}_c = a_2 \frac{v^2}{gkd} \text{Re}_c^2 .$$

This relationship was confirmed for several different filter pack materials by measuring the hydraulic gradient (dh/dr) at varying distances from the well for different flow rates (16, 17). The velocity was then calculated at each of the radial distances, and the Reynolds number was computed. Results from these tests are graphically shown in Figure 5.10. The critical Reynolds number interpreted from these plots is approximately equal to 30. This value was derived by observation of the general range where the slope of the curves changed from 1 (pure laminar flow) to 2 (pure turbulent flow).

Critical Radius

The critical radius is defined as that distance, measured from the center of the well, where flow changes from predominately turbulent to mostly laminar (Figure 5.11). Based on the foregoing mathematical analysis and experimental results, a critical radius may be calculated for any given set of flow conditions. Within the critical radius, flow is turbulent (or partially turbulent), with corresponding head losses varying as some exponential power of the velocity (approaching 2 for fully turbulent flow). Assuming a critical Reynolds number of 30, the critical radius may be calculated from

$$r_c = 0.9587 \frac{(Q/L)d}{\theta} \simeq \frac{(Q/L)d}{\theta} ,$$

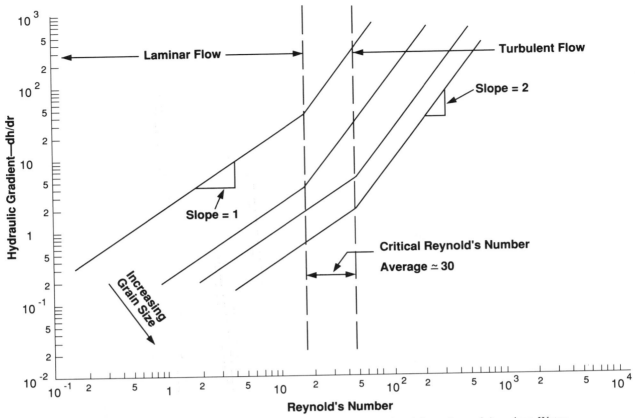

Figure 5.10. Hydraulic gradient *vs.* Reynold's Number (reprinted from *Journal* American Water Works Association, Vol. 77, No. 9 (September 1985) by permission. Copyright © 1985, American Water Works Association).

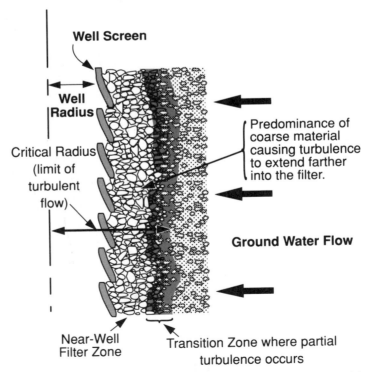

Figure 5.11. Near-well turbulence showing critical radius (reprinted from *Journal* American Water Works Association, Vol. 77, No. 9 (September 1985) by permission. Copyright © 1985, American Water Works Association).

where

r_c = critical radius [in.],

Q/L = specific aquifer discharge [gpm/ft],

Q = discharge rate [gpm],

L = length of screen [ft],

d = mean grain diameter [in.],

θ = effective porosity of material [fraction].

Figure 5.12 is a plot of specific aquifer discharge versus critical radius for typical filter zone materials.

Drawdown in and around a Pumping Well

The drawdown in a pumping well is the sum of various individual head-loss terms. Some of these terms are the result of natural processes and behave according to relevant physical laws. Others are the result of a formation damage or an excessive pumping rate, and may be minimized by using proper well design and construction. This section identifies these losses and shows their relationships to well performance.

The components of drawdown in a production well may be expressed as

$$s = ds + ds' + ds'' + ds''' + \text{minor losses} ,$$

where

s = total drawdown measured in the well [L],

ds = head loss in the aquifer (formation loss) [L],

ds' = head loss in the damage zone [L],

ds'' = turbulent loss in the filter zone [L],

ds''' = well losses [L].

See also Figure 5.13.

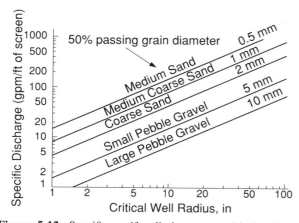

Figure 5.12. Specific aquifer discharge *vs.* critical well radius (reprinted from *Journal* American Water Works Association, Vol. 77, No. 9 (September 1985) by permission. Copyright © 1985, American Water Works Association).

Aquifer Loss. The head loss measured at the interface between the aquifer and the damage zone is known as "aquifer or formation loss." For steady-state conditions, aquifer loss may be expressed as

$$ds = BQ , \qquad (5.14)$$

where

ds = head loss in the aquifer [ft],

T = transmissivity [gpd/ft],

B = $(528/T) \log (r_0/r_a)$,

r_a = distance to aquifer/damage zone interface from the center of the well [ft],

r_0 = distance from center of well to point of zero drawdown [ft].

Head Loss through the Damage Zone. The damage zone consists of finely ground drilling debris, filter cake, drilling fluid, or other material whose hydraulic conductivity is considerably less than that of the aquifer. Its extent depends upon well construction and development procedures. Head loss through the damage zone is generally laminar and may be expressed in a manner similar to aquifer loss:

$$ds' = B'Q ,$$

where

ds' = head loss through the damage zone [ft],

B' = $(528/K'b) \log (r_a/r_g)$,

K' = hydraulic conductivity of damage zone [gpd/ft^2],

r_g = radial distance to inner edge of damage zone [ft].

If the well is fully developed, the damage zone component is minimal and can be ignored.

Turbulent Losses in the Filter Zone. Turbulent flow losses, which may be a significant component of the total drawdown, occur if the critical radius exceeds the nominal well radius. These losses do not obey Darcy's law and vary exponentially with the flow velocity. The transition between purely laminar and purely turbulent flow varies in any given situation. In the zone of partial turbulence, the exponent of the head-loss velocity term varies between 1 (for purely laminar flow) and 2 (for purely turbulent flow). The existence and degree of these losses depends upon filter material, screen length, and pumping rate.

The turbulent filter zone losses may be written as

$$ds'' = B'' Q^n ,$$

where

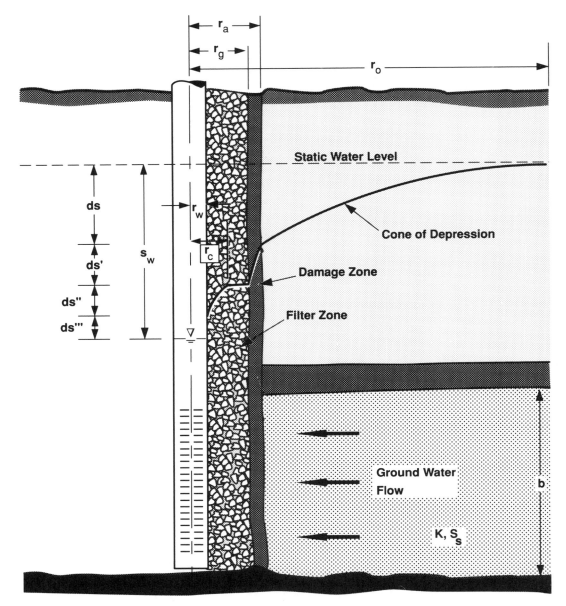

Figure 5.13. Drawdown components in and around a pumping well (reprinted from *Journal* American Water Works Association, Vol. 77, No. 9 (September 1985) by permission. Copyright © 1985, American Water Works Association).

ds'' = fully turbulent head losses in the near-well filter zone [ft],

B'' = turbulent filter zone loss coefficient [ft/gpm^n],

n = exponent varying between 1 and 2.

Well Losses. Head losses associated with the entrance of water through the well screen and the axial flow toward the pump intake are known as "well losses." These losses are caused by turbulent flow conditions and vary as the square of the velocity. Well losses can be expressed as

$$ds''' = CQ^2 \ , \qquad (5.15)$$

where

ds''' = well losses [ft],

C = well-loss coefficient [ft/gpm^2],

Q = discharge rate [gpm].

Minor Losses. In addition to the major drawdown components, several minor losses occur. Laminar losses through the filter zone and head losses associated with a restricted inlet area (convergence losses) are typical minor losses. Because these losses are generally much smaller than other losses, they are neglected in the total well drawdown equation.

Boulton showed that for long vertical slots spaced equally around the circumference of the well, the convergence head loss is closely given by (18)

$$h_L = \frac{528Q}{NLK} \log\left(\frac{2}{1 - \cos \alpha\pi}\right) \qquad (5.16)$$

where

h_L = head loss due to convergence [ft],
Q = discharge rate [gpm],
N = number of vertical slots around the circumference of the screen,
L = length of screen [ft],
α = slot/width ratio (width of slot divided by distance between centers).

The example given in Figure 5.14 shows the calculation of convergence losses in a typical well completed with a vertical machine-slotted screen in an alluvial aquifer. Convergence losses are insignificant for this type of screen and are even less for most other types.

Pumped Well Efficiency

The performance of a pumping well may be thought of in terms of a machine doing work, with the efficiency of the machine defined as the ratio of the work output to the work input.

The work input to the machine (well) consists of all drawdown (head-loss) components necessary to produce the particular flow rate. The work output from the machine is the minimum aquifer drawdown required to produce flow in the aquifer supplying the well. Because of aquifer damage, less than ideal well design, construction, or operation, the drawdown inside the well (work input to the machine) is always more than the drawdown in the aquifer (work output by the machine), with resulting efficiency being less than 100%.

In well hydraulics, well efficiency is defined as the ratio of actual specific capacity to the theoretical specific capacity. Since actual specific capacity is related to drawdown in the well and theoretical specific capacity to drawdown in the aquifer, well efficiency may be stated as the ratio of drawdown in the aquifer to drawdown in the well.

For a fully developed well (i.e., no damage zone losses), and neglecting filter zone and minor losses, well efficiency may be written

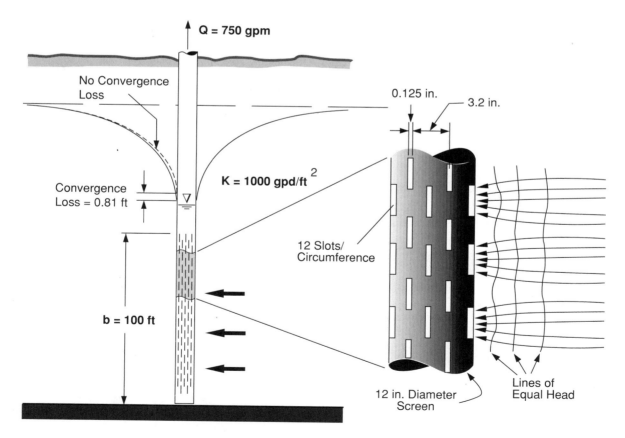

Figure 5.14. Example of convergence losses.

$$E = \frac{100}{1 + (CQ/B)} ,$$

where

E = well efficiency,

Q = discharge rate [gpm],

B = formation loss coefficient [ft/gpm],

C = well loss coefficient [ft/gpm^2].

In a 100% efficient well, the total dynamic head required to lift water to the initial (nonpumping level) would equal the total resistive (laminar) losses in the aquifer. This ideal well rarely exists in practice, with the exception of unscreened wells in fractured rock aquifers or cavernous limestone. Typically, pumping wells develop turbulent losses through and near the screen, resulting in a loss of efficiency (Figure 5.15).

Well efficiencies considerably less than 100% may result from a combination of damage-zone laminar losses and near-well turbulent flow losses. Short lengths of screen combined with high discharge rates may increase turbulent flow in the filter pack. Also improper construction or poor filter pack and screen design can result in partially or completely plugged screen apertures. Head losses in the vicinity of the aperture

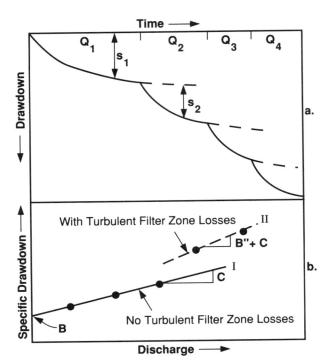

Figure 5.16. Step-drawdown test analysis (reprinted from *Journal American Water Works Association*, Vol. 77, No. 9 (September 1985) by permission. Copyright © 1985, American Water Works Association).

and through the screen are then extremely high and well efficiencies very low.

Determination of Well Efficiency—Step-Drawdown Testing.

Evaluation of efficiency, degree of development, and calculation of some types of individual head losses may be made by well and aquifer tests. In the step-drawdown method, multiple-discharge rates (steps) are run on the pumping well. At least three steps are necessary, and if turbulence in the filter zone is suspected, as many as six steps may be required (see Figure 5.16a).

In a fully developed well (no damage-zone head-loss component) penetrating the complete thickness of the aquifer, the drawdown in the well may be written

$$s = BQ + B''Q^n + CQ^2 .$$

This equation may be rewritten in terms of specific drawdown s/Q, assuming fully turbulent flow in the filter zone (i.e., $n = 2$):

$$\frac{s}{Q} = B + (B'' + C)Q .$$

From Figure 5.16b, it can be seen that the formation loss coefficient B may be found from the zero intercept of the best-fit straight line on a plot of specific drawdown versus discharge. During low discharge steps near-well turbulence

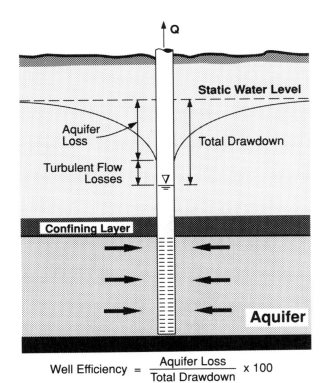

Figure 5.15. Concept of pumped well efficiency.

is not significant ($B'' = 0$), and the only turbulent losses are well losses. The slope of the best-fit straight line through these points will be equal to C. During higher discharge rates, however, near-well turbulent flow losses may be significant and will be reflected by a change (increase) in the slope of the line (see line II, Figure 5.16b). The slope of the best-fit straight line through the higher discharge points is equal to $B'' + C$.

In most wells, screen lengths are sufficiently long and the critical radius does not exceed the nominal radius. When all points of the step drawdown test lie on a straight line, this is confirmed. An example for calculation of minimum screen length (and diameter) to avoid losses from an excessive critical radius is given in Chapter 11.

Once the constants B, B'', and C are determined, the well efficiency may be calculated for the complete range of discharge as

No Near-Well Turbulence

$$E_I = \frac{100}{1 + (CQ/B)} . \qquad (5.17)$$

Near-Well Fully Turbulent Flow Present

$$E_{II} = \frac{100}{1 + (B'' + C)Q/B} \qquad (5.18)$$

in which E_I = well efficiency during lower discharge steps when no near-well turbulence is evident, and E_{II} = well efficiency during the higher discharge steps when near-well turbulent flow losses are present (Figure 5.17).

Identification of Damage Zone Losses. The foregoing procedure applies to fully developed wells with no damage-zone losses. Analysis of step-drawdown data on a well with damage-zone losses results in a lower formation loss coefficient

B than the true value. Step-drawdown testing by itself is not adequate to differentiate between types of laminar losses in the near-well zone.

If damage-zone head losses are suspected, they may be identified using an interference test in conjunction with the standard drawdown test. An interference test provides distance-drawdown data from observation wells located near the pumping well (preferably within several hundred feet) (see Chapter 15). These data may be then extrapolated logarithmically to the outer edge of the filter zone. The extrapolated drawdown at the aquifer-filter zone contact represents the true formation loss at this point.

The difference between the apparent formation loss, as calculated from step-drawdown testing, and the true formation loss represents damage-zone laminar losses (Figure 5.18).

Effective Well Radius

Another indication of the degree of well development is to calculate the increase in hydraulic conductivity in the near-well zone. A method of accomplishing this involves the concept of effective well radius.

As defined by Jacob, effective well radius is that distance measured radially from the axis of the well at the point of actual drawdown and extending to the theoretical drawdown (19). It indicates the effectiveness of well development on increasing the hydraulic conductivity in the immediate vicinity of the well (see Figure 5.19). The effective well radius may be calculated from step-drawdown test data once aquifer transmissivity and storativity are known.

The following equation can be used to evaluate well development by comparing effective radius (r_e) with nominal radius (r_w). When $r_e > r_w$ hydraulic conductivity in the near-well zone is greater than the surrounding aquifer, reflecting proper development,

$$r_e = \sqrt{\frac{0.3Tt^*}{S} \, 10^{-BT/264}} , \qquad (5.19)$$

where

$B = B(r_e, t^*)$ = formation loss coefficient determined from step-drawdown testing [ft/gpm],

t^* = time since the beginning of each new discharge step [days].

Specific Capacity

The ratio of the production rate or yield of a well to the drawdown required to produce that yield is known as "specific capacity." Specific capacity is expressed in units of discharge/length (commonly gpm/ft) and is a useful index of well performance.

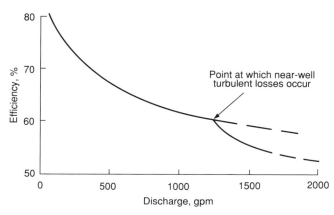

Figure 5.17. Well efficiency showing the effect of near-well turbulence.

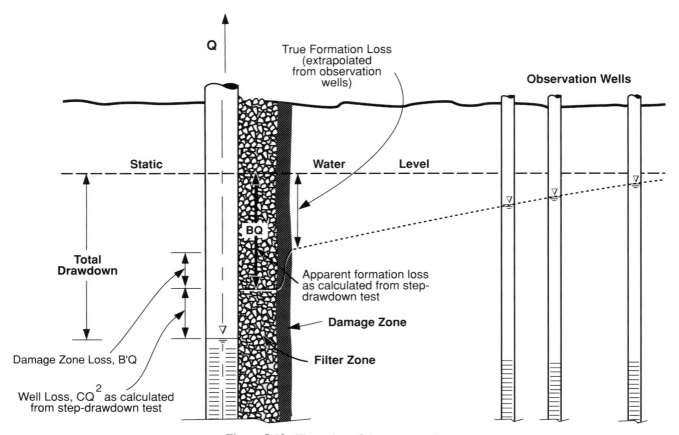

Figure 5.18. Illustration of damage-zone loss.

Specific capacity should be used with knowledge of its sensitivity or variability with discharge rate, time, and aquifer type. From Equation 5.6 and substituting r_w for r and assuming only aquifer and well losses.

$$s_w = \frac{264Q}{T} \log\left(\frac{0.3\,Tt}{r_w^2\,S}\right) + CQ^2$$

$$\frac{Q}{s_w} = \frac{1}{(264/T)\log(0.3Tt/r_w^2 S) + CQ},$$

(5.20)

where

r_w = well radius [ft],

s_w = total drawdown measured in the well [ft].

Other terms as previously defined.

Equation 5.20 shows that specific capacity is a function of well discharge (Q) and time (t). Figure 5.20 illustrates this relationship. For steady-state flow, specific capacity may be written

$$\frac{Q}{s_w} = \frac{1}{(528/T)\log(r_0/r) + CQ}.$$

(5.21)

Also see Equation 5.7 and Figure 5.1 for clarification of terms.

The decrease of specific capacity with time is due to the expanding cone of depression. The decrease with increasing discharge is due to the presence of turbulent losses.

Specific Loss. Well efficiency tests have been shown to be useful in assessing the condition of a well. But because the formation loss coefficient (B) is a function of aquifer type, comparison of well efficiencies between wells in different geologic environments may be misleading.

To overcome this, the effect of the different aquifer types is removed, and comparisons are made only between turbulent head-loss terms (well losses and near-well turbulent losses). This latter index, defined as "specific loss" (SL), is expressed as (16)

$$SL = CQ \quad \text{(no near-well turbulence)}.$$

For cases where near-well turbulence is present, the specific loss can be written

$$SL = (B'' + C)Q \quad \text{(near-well turbulent losses present)}.$$

Figure 5.21 gives an example of two wells completed in dissimilar geohydrologic environments. As demonstrated, specific loss comparisons are meaningful where comparison of well efficiencies is not.

Well Yield and Diameter

The following example illustrates the effect of screen diameter on well yield by comparison of specific capacities for two different well diameters. For a 12-in.-diameter well in an aquifer with a transmissivity (Kb) of 100,000 gpd/ft, specific capacity can be calculated from Equation 5.21, assuming no well losses ($C = 0$) and also that $r = r_{\mathrm{w}}$, $r_0 = 1050$ ft, and $Q/s = 57$ gpm/ft. If the well screen diameter were 24 in., the specific capacity would be $Q/s = 63$ gpm/ft. Therefore doubling the diameter increases specific capacity only 10%.

Special Consideration for Small-Diameter Wells

As noted so far in this chapter, and in Chapter 11, screen diameter is not a major factor in theoretical specific capacity, but the results of pumping tests from small-diameter test wells with well losses must be corrected to obtain estimates of well losses in production wells. The reason for this is demonstrated in the following example, using field data from an 8-in. test well.

The drawdown equation in a fully developed well with no filter or minor losses may be expressed as

$$s = BQ + CQ^2 \, ,$$

Step-drawdown testing on the 8-in. well resulted in

$$B = 0.05 \text{ ft/gpm} \, ,$$

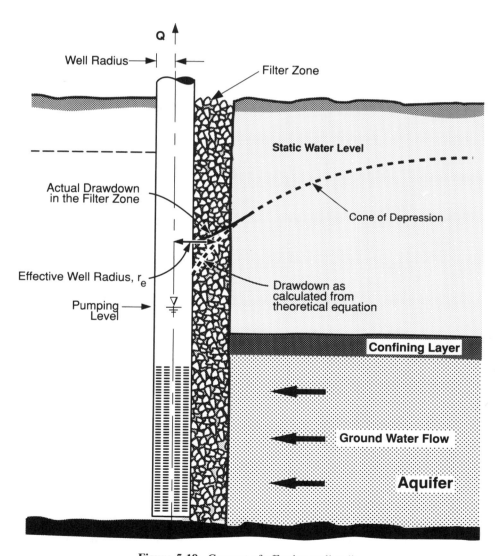

Figure 5.19. Concept of effective well radius.

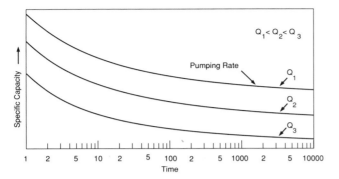

Figure 5.20. Variation of specific capacity with time and pumping rate.

$$C = 0.0003672 \text{ ft/gpm}^2 \ ,$$

and

$$s = 0.05Q + 0.0003672Q^2 \text{ ft} \ .$$

From the continuity equation,

$$Q = AV \ ,$$

where

A = area open to flow ($\pi \times$ diameter $\times$ screen length $\times$ percent open area) [L^2],

V = screen entrance velocity of the water [LT^{-1}].

Well losses may be expressed as

$$CQ^2 = C(AV)^2 \ . \tag{5.22}$$

Combining the turbulent head-loss component, $KV^2/2g$ from Bernoulli's equation, with Equation 5.22, the following relation is obtained:

$$\frac{K}{2g} = CA^2 \ ,$$

where

K = frictional loss coefficient = assumed constant.

An estimate may now be made for the well loss coefficient for a 16-in.-diameter production well based on results from the 8-in.-diameter test well.

$$\frac{K}{2g} = \text{constant} = C_8 A_8^2 = C_{16} A_{16}^2 \ .$$

Assuming the same percentage open area of screen, the well loss coefficient for the production well is C_{16} estimated to be

$$C_{16} = C_8 \left(\frac{d_8}{d_{16}}\right)^2 = 0.0003672 \times 0.25$$

$$= 0.000092 \text{ ft/gpm}^2$$

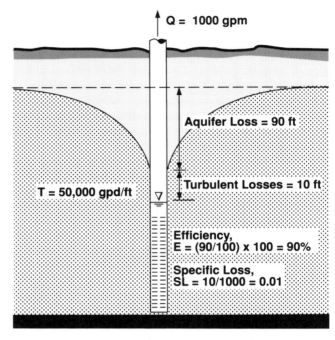

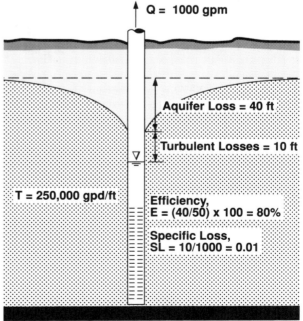

Figure 5.21. Examples of specific loss (reprinted from *Journal American Water Works Association*, Vol. 77, No. 9 (September 1985) by permission. Copyright © 1985, American Water Works Association).

This value is approximately 25% of C_8.

Although this is only a first-order estimate, the usefulness of this analysis becomes important in the drawdown estimate for a production well. Assuming a production rate of 750 gpm for the 16-in. well and using the C_{16} value, the drawdown estimate would be

$$s = 0.05 \times 750 + 0.000092 \times 750^2 = 89 \text{ ft} .$$

This is significantly less than the 244 ft predicted from the 8-in.-diameter test well. Based on the preceding analysis, it was decided that approximately 1000 gpm could be developed in the area with reasonable drawdowns.

Note that although the formation loss coefficient decreases with increasing effective radius, the change is small:

$$B_{16} = B_8 + \frac{528}{T} \log \left(\frac{r_8}{r_{16}} \right)$$

Assuming a transmissivity of 50,000 gpd/ft,

$$B_{16} = 0.05 + \frac{528}{50,000} \log \left(\frac{4}{8} \right) = 0.047 \text{ ft/gpm} .$$

REFERENCES

1. Maasland, D. E. L., and M. W. Bittinger. 1963. "Proceedings of the Symposium on Transient Ground Water Hydraulics." CSU, Fort Collins, CO.

2. Theis, C. V. 1935. "The Relation between the Lowering of the Piezometric Surface and the Rate and Duration of Discharge of a Well Using Ground-Water Storage." *Amer. Geophys. Union Trans.*, 16th Ann. Mtg., Pt. 2, pp. 519–524.

3. Cooper, H. H., and C. E. Jacob. 1946. "A Generalized Graphical Method for Evaluating Formation Constants and Summarizing Well Field History." *Amer. Geophys. Union Trans.* 27, pp. 526–534.

4. Thiem, A. 1906. *Hydrologische Methoden*. Gebhardt, Leipzig.

5. Boulton, N. S. 1954. *Unsteady Radial Flow to a Pumped Well Allowing for Delayed Yield from Storage*. Vol. 2. Intern. Assoc. Sci. Hydrology, Gen. Assembly, Rome, Publ. 37.

6. Hantush, M. S. 1964. "Hydraulics of Wells." In *Advances in Hydroscience*. Vol. 1. Academic Press, New York.

7. Boulton, N. S. 1963. "Analysis of Data from Non-Equilibrium Pumping Tests Allowing for Delayed Yield from Storage." *Proc. Inst. Civ. Eng.* 26, pp. 469–482.

8. Dupuit, J. 1863. *Etudes theoriques et pratiques sur le mouvement des eaus dans les canaux decouvert et a travers les terraines permeables*. 2d ed. Dunot, Paris.

9. Hantush, M. S. 1962. "On the Validity of the Dupuit–Forcheimer Well-Discharge Formula." *J. Geophys. Res.* 67.

10. Hantush, M. S., and C. E. Jacob. 1955. "Steady Three-Dimensional Flow to a Well in a Two-Layered Aquifer." *Amer. Geophys. Union Trans.* 36, pp. 286–292.

11. Hantush, M. S. 1960. "Modification of the Theory of Leaky Aquifers." *J. Geophys. Res.* 65, 11.

12. Hantush, M. S., and C. E. Jacob. 1955. "Non-Steady Radial Flow in an Infinite Leaky Aquifer." *Amer. Geophys. Union Trans.* 36, pp. 286–292.

13. Hantush, M. S. 1957. "Non-Steady flow to a Well Partially Penetrating an Infinite Leaky Aquifer." Iraqi Scientific Soc., p. 10.

14. Williams, D. E. 1970. "Use of Alluvial Faults in the Storage and Retention of Ground Water." *Ground Water*, 8, 5.

15. Jenkens, D. N., and J. K. Prentice. 1982. "Theory for Aquifer Test Analysis in Fractured Rocks under Linear (Non-Radial) Flow Conditions." *Ground Water* (January–February).

16. Williams, D. E. 1985. "Modern Techniques in Well Design." Am. Water Works Assoc., September.

17. Williams, D. E. 1981. "The Well/Aquifer Model, Initial Test Results." Roscoe Moss Company, Los Angeles.

18. Boulton, N. S. 1947. Discussion in C. E. Jacob, "Drawdown Test to Determine Effective Radius of Artesian Wells." *Trans. Amer. Soc. Civil Engrs.* 112, pp. 1047–1070.

19. Jacob, C. E. 1947. "Drawdown Test to Determine Effective Radius of Artesian Well." *Trans. Amer. Soc. Civil Engrs.* 112, pp. 1047–1070.

PART II

EXPLOITATION

CHAPTER 6

Well Design—General Considerations

6.1 INTRODUCTION

A successful well is a culmination of an iterative process ranging from considerations of well purpose and hydrogeology to operation and maintenance procedures after the well has been placed in service. The synthesis of this process is the well design. The design in ground water developed areas usually evolves from, or is highly influenced by, years of local contractor and operator experience.

Declining water tables and more stringent water-quality requirements increasingly necessitate exploitation of deeper and more complex aquifers. Deeper conductor casings and sealing to eliminate unusable ground waters and contamination have become commonplace. These and other changing requirements, coupled with new technologies, have increased well design complexity.

Many factors involved in well design are detailed in Chapters 6 through 15. This chapter discusses general design aspects applicable to most wells.

6.2 DESIGN OBJECTIVES

The primary function of a water well is to provide a conduit from the aquifers to the surface. This conduit must be designed to

1. Match discharge required to pumping plant and aquifer characteristics.
2. Achieve designed production rates with maximum well efficiency.
3. Produce acceptable quality water and protect it from contamination.
4. Maximize well life commensurate with cost effectiveness.

6.3 PRELIMINARY EVALUATION

A preliminary evaluation of all relevant data is part of the well design process. In a particular end use, operational parameters and hydrogeologic considerations are crucial.

6.4 END USE

Although high construction and performance standards are always required, end use will influence a number of design parameters, including well life, sand production standards, and acceptable water quality.

6.5 OPERATIONAL REQUIREMENTS AND THEIR RELATIONSHIP TO THE SYSTEM

Maximum demand and the pumping schedule make up part of the design criteria for the pumping plant. This in turn influences well design. A decision must be made as to how the ground water supply is to be utilized. Is the system totally dependent upon wells or is there supplemental water available from another source? Is there an adequate ground water supply convenient to the system that can be developed in a straightforward way? Is there a known history of water-quality problems that indicates that the well(s) may have to be located at some distance from the area where the water is to be delivered, or can the well(s) be located in the area of consumption?

Custom and practice have an influence on water-system design and operation. Some agencies prefer to pump water from a well field to a reservoir and transport it through pipelines to the point of use. In other cases, when the ground water supply is sufficiently well known, the well may be located at the point of demand and discharged directly into the distribution system.

There are certain advantages to a central well field system. The well field can be designed in conjunction with the amount of storage available. Water treatment, if needed, is simpler at a single location than at multiple locations. However, experience has shown that individual wells, if properly designed and constructed, offer a protected supply coupled with their own aboveground reservoirs. Strategic location of wells throughout a distribution system is common practice.

6.6 SITE SELECTION

Site or alternative site characteristics must be analyzed in order to select drilling method and well design. Typical site considerations include the following.

Hydrogeologic Evaluation

In well-site investigation it is necessary to evaluate and quantify the ground water available for exploitation. The methods for doing this are covered in Chapter 3.

General Hydrogeology of the Area

General site hydrogeology may affect the well construction method, and therefore site size and well design. Location of the individual well site(s) can be made based on information from the history of nearby wells, in addition to knowledge of local hydrogeologic characteristics (see Chapter 3). Operating cycles, pumping rates, and well interference as they affect pumping lift and total dynamic head are important factors in proper well spacing.

Water-Quality Considerations

The quality of ground water is controlled primarily by contact with the geologic formations through which it passes, but in some cases people may contaminate it. When the site considered is located above a known or suspected contaminated aquifer system, special design considerations are necessary. These require knowledge of both the areal and vertical extent of poor-quality water and the effect of proposed production rates on induced migration of contaminates. Selective zone testing may be required to locate acceptable aquifers. The well design should incorporate the sealing of contaminated zones.

Site Geography

Geographical factors affecting well design include site topography and drainage.

Site Area and Configuration

Although as little as 60 × 100 ft of working area is required for construction and pump and well maintenance, there may be other site considerations. Size and configuration may be affected by local conditions such as minimum lot size, setbacks, and proximity to power lines, sewers, and septic tanks. Health regulations generally require at least 50-ft separation from a sewer line, and 100 to 150 ft from present or future septic tanks and drainage fields. It is important to consider the proximity of oil transmission lines. In the case of a domestic water supply a large area may be needed for a reservoir and/or water-treatment equipment.

The additional cost of a site larger than the minimum necessary may be recovered in construction cost savings and through a more easily maintained production facility. A larger site offers room for drilling a replacement well, often a valuable consideration in urban areas. An irregular or narrow site may make placement of drill rig, pits, auxiliary equipment, gravel, casing and cuttings storage, and so on, difficult and expensive. In a few instances a limited working area may dictate a drilling method, since as a rule, cable tool equipment requires less working area than rotary.

Site Access

Ease of access facilitates well servicing. It is rarely wise to subdivide a well site since higher costs of well and pump servicing during the life of the well may exceed any gain from the property sale. The value of any site diminishes if future development restricts accessibility. Present or planned overhead high-tension lines must be considered in well location. Too frequently they constitute a hazard to servicing. Buildings housing the well pump, piping, controls, and so on, must be designed for easy removal, or with openings sufficiently large for access to the well by service equipment. Discharge lines and electrical and other equipment, should be placed so that equipment can be freely erected and operated.

Local Ordinances

A complete review of regulations from both public and private agencies is necessary because they may affect the design of the well (e.g., sanitary seal, maximum pumping rates, or environmental constraints).

Deed covenants and restrictions and zoning requirements must be reviewed. A title check may reveal a lien of covenants and conditions on the selected property. Covenants that are generally intended to provide uniformity of development within a residential area may preclude the construction of a water well. A title check also reveals easements that would interfere with construction. Mineral reservations that permit the exploitation of resources, such as oil or gas, could affect the well operation.

Zoning regulations normally provide for the granting of use permits to a utility. This recognizes the fact that such use is needed for urban development.

Drilling Water Availability

Availability of drilling water is an important factor in selecting the well construction method and design. Both rotary drilling techniques require large quantities of potable water (typically up to several hundred gallons per minute for reverse circulation) during the drilling process. Even large quantities must be available during mud thinning and other completion and development operations. Cable tool and air drilling operations normally require considerably less water (as low as 50 to 100 gal per hour for cable tool).

Handling Surplus Drilling Fluid

For rotary drilled wells, the process of drilling fluid thinning and conditioning prior to and throughout graveling operations significantly increases fluid volume discharge, and the site should be sufficiently large for storage of this material. If conditions permit, any excess can be allowed to flow to a location where the test pump water will eventually dilute and dissipate the mud. The only acceptable disposal method on some urban sites is the use of vacuum trucks for transport to a liquid disposal dump, a very expensive alternative.

Development and Test Water Disposal

High-capacity wells require the production of large quantities of water during development and test pumping operations, particularly when step-drawdown and maximum safe pumping rate tests are made. Disposal facilities must accommodate pumping up to 150% of the anticipated permanent production rate. In many cases as much as 50 acre-ft of water will be pumped during a short period of time. Drilling debris, which cannot be separated readily from the pumped water and fine formation material will also be produced, sometimes in considerable quantity. This material cannot be readily separated.

Authorities controlling urban disposal facilities, such as storm drains, do not always permit large quantities of dirty water to enter these systems. Consequently disposal facilities should be investigated before selecting a well site in an urban area. Even in agricultural locations, the production of large quantities of ground water may tax the natural drainage capacity.

Noise Levels at Site

Most well construction operations are noisy, and some phases cannot be limited to daylight hours or to scheduled intervals without seriously compromising operating efficiency and final well performance. Cable tool operations are much less critical than rotary in this respect and can be conducted on a single-shift basis. Rotary drilling loses much efficiency if continuous operation is curtailed. Shutting down at critical junctures may put the whole drilling operation at risk and jeopardize successful completion. Sound-control procedures are available, but their implementation can be expensive, particularly if stringently low sound levels are required at the perimeter of the drill site.

6.7 WELL COMPONENTS

A completed well installation will usually include the following:

1. Conductor casing (cemented in place) to protect the well from surface contamination.

2. Pump housing casing to protect the pump, to support the borehole, and to provide a conduit for transmission of water from the well intake.

3. Well intake that can be open borehole, well screen, or casing perforated in place.

4. Pumping plant and surface appurtenances.

A schematic drawing showing a typical well installation is shown in Figure 6.1.

6.8 COMPLETION

The nature of the producing formations will determine the type of completion: naturally developed, gravel envelope, or open borehole. Gravel envelope wells are further classified into those that are completed either with filter pack or with formation stabilizers. Gravel envelope function and design are covered in Chapter 13. Some consolidated rock formations do not deteriorate with time, and wells drilled in them can be completed with an open hole below the pump housing casing.

6.9 DRILLING METHOD

Knowledge of the attributes and characteristics of each drilling system is essential to the well-design process. Drilling systems may be dependent on well design and are usually associated with certain combinations of well design and formation characteristics. Details are covered in Chapter 7.

6.10 CRITERIA FOR SELECTION OF WELL DESIGNS

Standard well designs and variations of them have evolved over the years according to the ground water environment, demand requirements, and equipment capabilities. Nine of the standard designs are described in this chapter. A tenth, the "casing path well," though not of general application, has been added because of its success in irrigated regions with perched water conditions.

Complete Gravel Envelope

The most common well design in gravel envelope construction is shown in Figure 6.2: gravel envelope well with gravel envelope to surface. This type of well has been constructed to depths of over 2000 ft. Advantages of this design include simplicity, ease of construction and maintenance, and low initial cost.

In this design, the conductor casing is set and cemented to stabilize the upper formation while drilling, to seal the

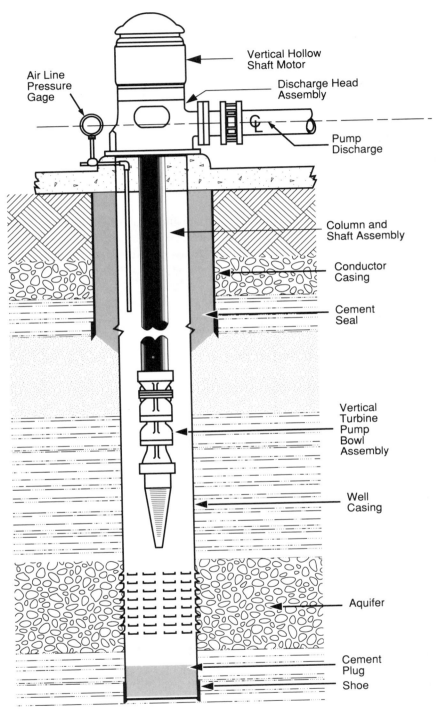

Figure 6.1. Typical well installation.

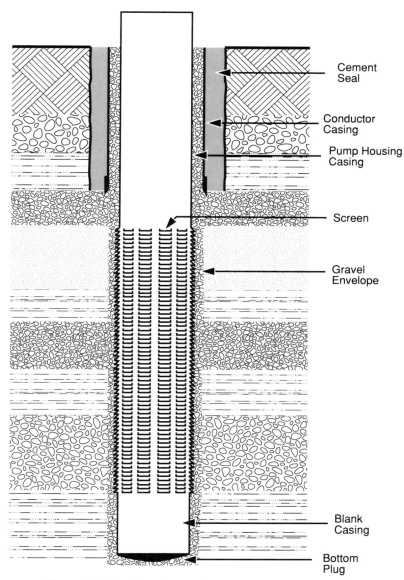

Figure 6.2. Gravel envelope well with gravel envelope to surface.

well against contamination, and to support the suspended weight of casing and screen. The annular space above the screen serves as a reservoir for gravel. Pump housing casing is installed to the deepest anticipated pump setting depth. The well screen usually extends continuously below the pump housing casing to a short length of blank casing that serves as a settlement reservoir.

With gravel envelope wells, unless there is a requirement to prevent production from a specific zone, experience confirms the merit of screening the entire well below the pump housing casing to within 20 ft of total depth. Where there are alluvial aquifers from meandering stream flow deposition, drillers and geophysical borehole logs do not always accurately define aquifer and aquitard boundaries, particularly if the aquifers are in thin lenses. Wells completed with intermittent screens have frequently proved less productive than nearby wells fully screened. Gravel placement and well development are far more difficult to execute through alternate sections of screen and blank casing. This is due to the tendency of gravel to bridge at the top of a screen section and the difficulty of removing drilling fluid and debris from the gravel behind blank casing. Construction cost savings by selectively screening the well only in apparently productive areas are seldom warranted. The annulus between the borehole and the casing and screen is filled to the ground surface with either a formation stabilizer or a selected filter pack material (Chapter 13). Wells deeper than 1000 ft, or using a stainless steel screen, may be reduced in diameter, preferably at the juncture of the pump housing casing and screen. However, the casing and screen are installed together in one continuous operation. Complete gravel envelope wells are normally drilled by the direct or reverse rotary method and in unconsolidated alluvial formations.

Partial Gravel Envelope

A variation of the complete gravel envelope design is shown in Figure 6.3: gravel envelope well with cemented pump housing casing and gravel replenished with feed line. In this design the entire length of pump housing casing is cemented

in place. A permanent gravel feed line is installed to permit replenishment.

A cemented conductor casing is needed to stabilize the upper formation during drilling and to support the casing and screen during installation, graveling, and cementing. The sanitary seal is provided by the cemented pump housing casing. The gravel envelope should extend at least 20 ft above the top of the well screen to serve as a reservoir. A minimum of 20 ft of blank casing is placed below the screen. The casing and screen are installed together in one string. Partial gravel envelope wells are drilled in unconsolidated alluvial formations using either rotary method.

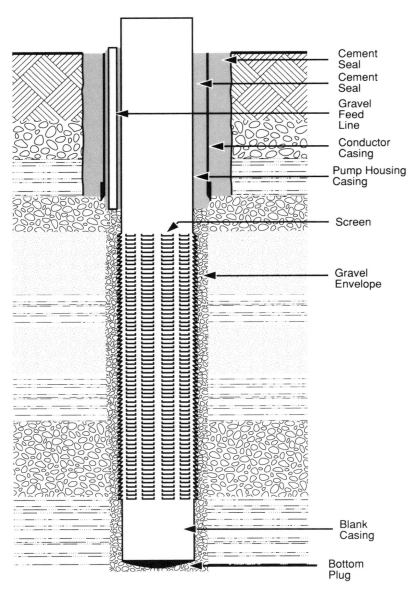

Figure 6.3. Gravel envelope well with cemented pump housing casing and gravel replenished with feed line.

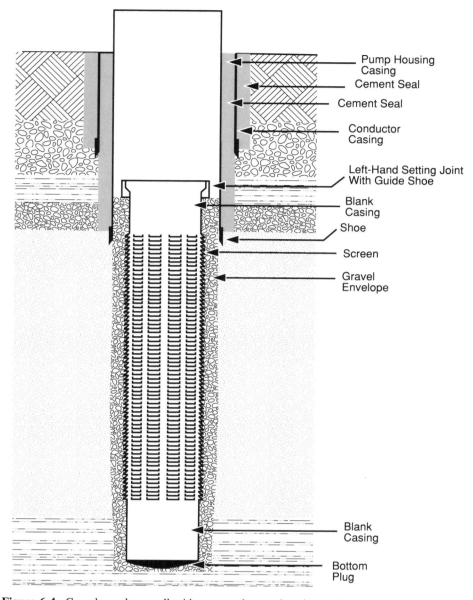

Figure 6.4. Gravel envelope well with cemented pump housing casing and telescoped screen.

The cement seal protects the pump housing casing against external corrosion. This design is particularly useful in situations where undesirable deeper zones must be sealed off. The installation and sealing of large-diameter conductor casing to those depths is much more expensive.

Telescoped Well Screen

Figure 6.4 illustrates another variation: gravel envelope well with cemented pump housing casing and telescoped well screen. The well screen is installed by telescoping through the pump housing casing. A threaded-and-coupled pipe connected to a left-hand back-off joint on top of the screen is used for placement. The screen is held in suspension until the gravel envelope is placed. The screened borehole diameter is limited by the inside diameter of the pump housing casing, necessitating a large reduction in the screen's diameter (6 to 8 in.). The pump housing casing is cemented in place, satisfying the requirements for a sanitary seal. A minimum of 20 ft of blank casing is connected to the top of the screen, and this casing laps into the pump housing casing. Cable tool or rotary drilling is appropriate for this design, depending upon the formations to be encountered.

Telescoped Well Screen in Underreamed Borehole

Figure 6.5—gravel envelope well with cemented pump housing casing and telescoped screen in underreamed borehole—is identical to the preceding design, except that the borehole is underreamed below the pump housing casing. This allows a screen diameter that is 4 in. less than the inside diameter of the pump housing casing. Underreaming is performed with either a hydraulic or mechanical underreamer. These tools can enlarge the borehole up to 16 in. greater than the inside diameter of the pump housing casing.

This design calls for the direct rotary method in deep, well-defined aquifers that consist primarily of fine sand.

Multiple-Zone Completion

Figure 6.6—gravel envelope well with isolated multiple-zone completions—illustrates a well design used when the screening involves two or more aquifers separated by a well-defined aquiclude(s). The annulus opposite the aquicludes is cemented, and the envelope material is placed in the screened aquifer sections. Unless gravel feed lines are installed (necessitating a very large borehole), gravel installation must be carefully executed because there is no way to replenish it. This design is rarely used, and usually only in deep wells where aquifers are separated by thick aquicludes.

As with other gravel envelope wells, a conductor casing is set and cemented to satisfy health and construction re-

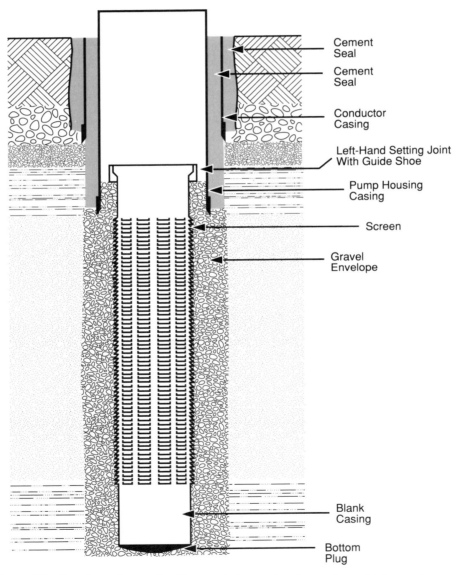

Cement Seal

Cement Seal

Conductor Casing

Left-Hand Setting Joint With Guide Shoe

Pump Housing Casing

Screen

Gravel Envelope

Blank Casing

Bottom Plug

Figure 6.5. Gravel envelope well with cemented pump housing casing and telescoped screen in underreamed borehole.

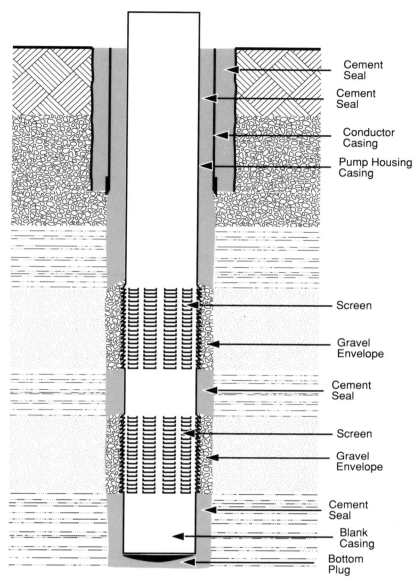

Figure 6.6. Gravel envelope well with isolated multiple-zone completions.

quirements. Drilling is by reverse or direct rotary equipment, depending upon formation and depth. Because of potential problems with envelope settlement during well development and operation, multiple-zone completions are not specified if an alternative design is acceptable.

Casing Path Well

The casing path well[1] was developed to ameliorate the problems occurring when pumping levels are below the top of the screen, particularly when caused by perched water con-

ditions. When an upper aquifer is separated from lower producing zones by an aquiclude extending over a broad area, pumping from the lower aquifers does not affect the upper zone. If the upper aquifer is screened and the pumping water level is below the top of the screen, water will cascade into the well, carrying air with it. This problem can also occur in wells with wide fluctuations in static and pumping levels caused by seasonal rainfall or varying surface water conditions.

Cascading water can entrain air if any part of the screen or the perforations are above the pumping water level. This phenomenon takes place when water entering the well above the pumping level impinges on the pump column and breaks up into spray. As the spray settles, air is entrained. This

1. The casing path well design is patented.

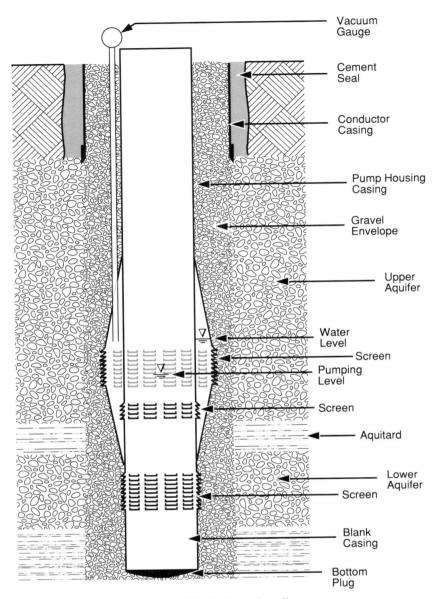

Figure 6.7. Casing path well.

air–water mixture is very tenacious, and no process of separating the air and water before it is pumped has proved to be satisfactory.

Pumping an air–water mixture is costly. The pumped air displaces water so that far less water is produced. Cavitation, which is described in Chapter 16, also occurs as this mixture passes the pump impeller.

It is difficult to predict when air entrainment will occur, but two conditions must exist for it to happen: some screen or perforated casing must be above the pumping water level, and the supplying aquifer must be prolific. There have been many instances of wells producing harmful quantities of air when the screen is as little as 6 in. above the pumping water level. Remedies such as throttling the pump discharge or lining the exposed screen solve the problem but usually result in a dramatic production loss.

As illustrated in Figure 6.7, the upper aquifer can be exploited with a well screen placed coaxially outside the lower portion of the pump housing casing. At top and bottom the screen has tapered reducing cones that are welded to the casing. The connections must be airtight. Lower aquifer(s) are screened conventionally. A short length of screen is placed at the bottom of the double section, allowing water produced from the upper zone to enter the pump housing casing. A 2-in. pipe with a vacuum gauge is placed through the well annulus to the upper well screen.

The water initially pumped from the upper aquifer rapidly exhausts the air between the screen and casing, creating a partial vacuum in the annulus and preventing further air entrainment. Under the conditions described earlier, casing path wells are frequently much better producers than conventional wells screened only in the lower aquifers.

It is important to recognize that this design does not permit upper aquifer development or redevelopment other than by pumping, and consolidation of the pack can be difficult. For these reasons this well design is probably best suited to areas where the gravel envelope functions primarily as a formation stabilizer (Chapter 13) and aquifers consist of coarse unconsolidated alluvial material. The reverse circulation rotary method using water without additives best meets construction requirements.

6.11 NATURALLY DEVELOPED WELLS

There are two categories of naturally developed wells. In the first, casing is installed to total depth, and selected sections are perforated in place to form a screen. In the second, casing is set, and a screen is telescoped through and installed below it.

Naturally Developed Well with Well Casing Perforated in Place

Figure 6.8—naturally developed well, with casing perforated in place—depicts this first design. This type of well is drilled by the cable tool method with casing carried more or less coincident with drilling by driving or jacking. Perforations are made in place opposite strata whose character permits natural development. Advantages of this well design include the ability to obtain very reliable water quality and formation samples, selective aquifer completion, and use of cable tools to drill formations difficult to penetrate by other methods. The conductor casing is installed to the depth required.

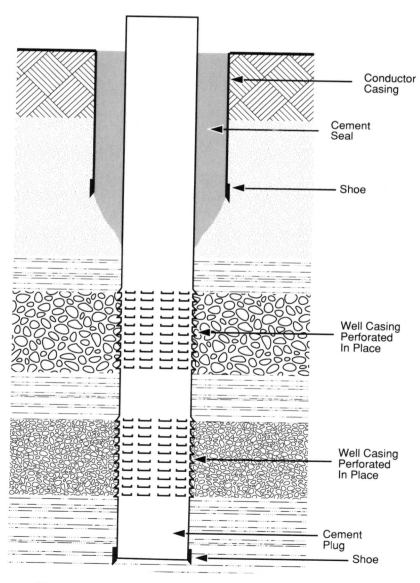

Conductor Casing

Cement Seal

Shoe

Well Casing Perforated In Place

Well Casing Perforated In Place

Cement Plug

Shoe

Figure 6.8. Naturally developed well with casing perforated in place.

Naturally Developed Well with Screen Set in Open Hole, Bailed Down, or Installed by the Pull-Back Method

In Figure 6.9—naturally developed well with screen set in open hole, bailed down, or installed by the pull-back method—the pump housing casing is installed at the top of the aquifer(s) to be exploited. In one procedure, an open hole is drilled below the pump housing casing (with the hole kept open if it is unstable by filling the borehole and casing to ground surface with drilling fluid). Then the screen is telescoped through the pump housing casing into the open hole. A packer is placed on the screen section so that the formation material cannot be pumped into the well. In another, the telescoping screen is placed in the pump housing casing

and installed into the aquifer by bailing or washing down. In a third procedure, the well is drilled and cased through the aquifer, the screen is set, and the casing is pulled back to expose the screen. The conductor casing is placed to the depth required and sealed by cementing between the two casings.

All of these techniques are used with cable tool equipment and are generally limited to shallow wells and short screen lengths.

Wells in Water-Bearing Rock

Figure 6.10—well with open hole completion in water-bearing rock—and Figure 6.11—well completed in water-bearing rock with formation stabilizer around screen—depict

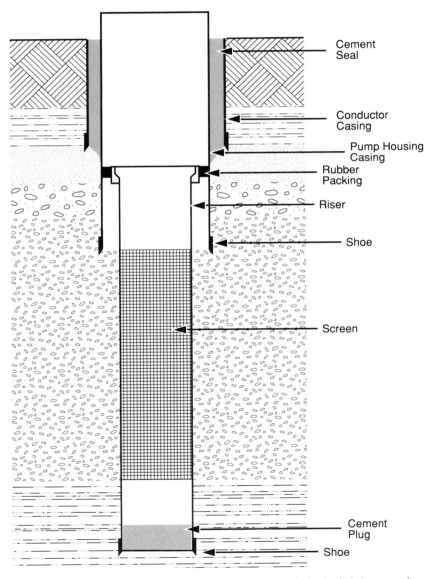

Figure 6.9. Naturally developed well with screen set in open hole, bailed down, or installed by the pull-back method.

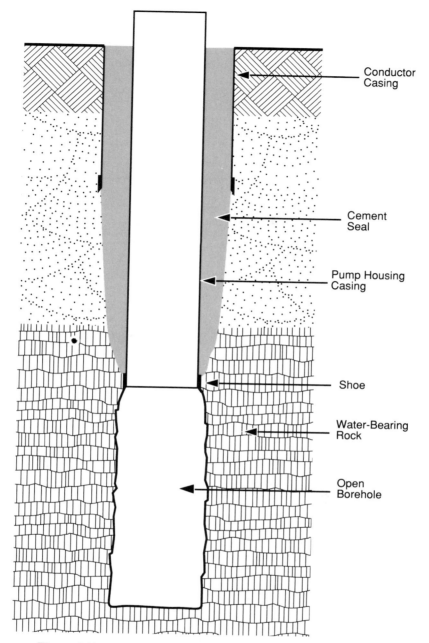

Figure 6.10. Well with open hole completion in water-bearing rock.

designs for wells completed in high producing zones in rock, such as basalt or limestone. As shown in Figure 6.10, the pump housing casing is normally set at the top of the producing zone, and an open hole is drilled below this point. Production comes from the unscreened open hole.

Semiconsolidated sandstone aquifers overlaid by well-consolidated aquicludes can be developed and produced in an open borehole. This procedure usually requires substantial development time, up to hundreds of pumping hours. The developed borehole diameter must be large enough to reduce velocities below the level required to transport sand into the pump housing casing.

If the aquifer cannot be stabilized, a well screen is installed with a filter pack (Figure 6.11). In highly fractured rock formations a well screen is used without a filter pack to protect the borehole from caving.

Pump housing casing is cemented to the depth required to seal off unwanted surface waters. Because of the nature

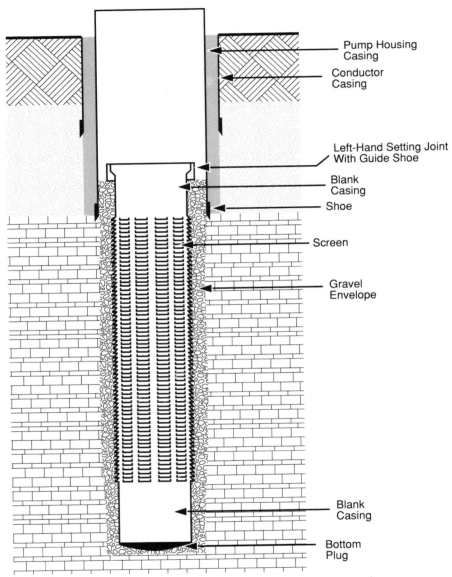

Pump Housing
Casing

Conductor
Casing

Left-Hand Setting Joint
With Guide Shoe

Blank
Casing

Shoe

Screen

Gravel
Envelope

Blank
Casing

Bottom
Plug

Figure 6.11. Well completed in water-bearing rock with formation stabilizer around screen.

of these formations, it may be necessary to cement its entire length.

6.12 WELL DEPTH

The total well depth is a function of the anticipated well production and the aquifer(s) being exploited. It must provide sufficient aquifer penetration to achieve the desired production and specific capacity.

6.13 DEPTH OF CONDUCTOR CASING

The depth of conductor casing is normally established by regulatory agencies and the character of the upper formations. But the usual depth is not less than 50 ft.

6.14 DEPTH OF PUMP HOUSING CASING

The pump setting is determined by pump and well characteristics and the discharge requirements. The pump housing casing must accommodate not only the designed pump setting but also any projected deeper setting due to a lowered water table, interference from other wells, or increased production requirements. Failure to make these provisions may require a change in pump design and/or setting the pump into the screen section. A pump set into a screen sometimes causes serious problems with well production and longevity (Chapter 16).

6.15 CRITERIA FOR CASING DIAMETER

Conductor Casing

The inside diameter of the conductor casing must be adequate to permit the passage of drilling tools (generally 2 in. of clearance is allowed).

Pump Housing Casing

The inside diameter of the pump housing casing is a function of the pump's diameter. The pump-bowl diameter is established by production requirements, the type of pump, and the design rpm. With line-shaft turbine pump assemblies, the maximum diameter is nearly always the outside diameter of the bowl assembly. For a submersible pump, the largest diameter is the bowl-assembly diameter plus the power cable. Occasionally, the controlling diameter is the column pipe collar, or with high horsepower units, the submersible motor.

The inside diameter of the pump housing casing must be at least 2 in. larger than the largest pump component in order for it to compensate for possible well misalignment or any future change in the type or capacity of pumping equipment. This additional clearance allows diameter-restricting well repairs without pump-bowl changes and loss of production.

Table 6.1 shows the inside diameter of the casing as a function of the maximum pump diameter and pumping rate. For well casing, the flow rate should be less than 5 ft/sec to avoid excessive head loss.

TABLE 6.1   Minimum Casing ID vs. Nominal Bowl Diameter

Nominal Bowl Diameter (in.)	Operating Pump Speed (RPM)	Discharge Rate (gpm)	Minimum Casing ID (in.)
8	3500	200–1200	10
	1800	100–600	
	1200	160–400	
10	1800	200–1500	12
	1200	370–670	
12	1800	400–2300	14
	1200	250–1500	
14	1800	1000–4500	16
	1200	700–3000	
16	1800	2000–5200	18
	1200	1300–3400	
18	1800	3200–4100	20
	1200	2200–4000	
	900	2800–3000	
20	1200	3100–4400	24
	900	2300–3600	
22	1200	7500	24
	900	5600	

Note: More clearance may be required for submersible pumps.

Intermediate Casing

Intermediate casings may be installed in deep rotary drilled wells to facilitate completion through loss of circulation zones, hydrating shales, and other problem formations. A 4-in. clearance is required to permit passage of each string with a minimum 20-ft lap.

Special Considerations for Cable Tool Drilling

In alluvium and some other formations, the cost of cable tool drilling is relatively independent of the borehole diameter. Efficient drilling of cobbles, boulders, or basalt necessitates the use of heavy large-diameter tools that require casing with a diameter much larger than that of the pump.

Since cable tool wells are often drilled under difficult and unpredictable conditions, the ability to drive or jack casing in a single string to total depth in a deep well may be limited by the borehole wall's friction. Depending upon the nature of the formation and method of placement, a single string of casing can usually be installed 500 to 1000 ft. Nominal diameter reductions are 4 in. with a lap of 10 to 20 ft, although a 2-in. reduction can be made using a $\frac{7}{8}$-in. drive shoe.

6.16 CRITERIA FOR SCREEN DIAMETER

The hydraulic criteria for the screen diameter are discussed in detail in Chapter 11. In general, the screen diameter is determined by the diameter of the pump housing casing which may differ according to well type:

1. Complete gravel envelope well. The screen diameter is preferably the same as the pump housing casing diameter. If a reduction is made, it is not more than 4 in., and the maximum axial flow velocity of 5 ft per second must not be exceeded.
2. Partial gravel envelope well. The relationship is the same as for a complete gravel envelope well.
3. Telescoped well screens are normally 4 in. smaller in diameter than the inside diameter of the pump housing casing.
4. Casing path well. The lower well screen is the same diameter as the pump housing casing. The outer well screen is at least 4 in. larger than the pump housing casing.
5. Naturally developed well with casing perforated in place. If the well is completed in one casing string, the screen is the same diameter as the pump housing casing. Reductions, if required, are in increments of 2 to 4 in.
6. Naturally developed well with the screen set in the open hole, bailed down, or installed by the pull-back

method. Screens to be installed in the open hole require up to 4-in. clearance, except for machined vertical slotted screens which can be driven into open hole with 2-in. clearance.

6.17 CRITERIA FOR BOREHOLE DIAMETER

Borehole diameters are a function of casing and screen diameters. Details regarding borehole diameter selection are covered elsewhere but can be summarized as follows:

1. For direct rotary drilling of high-capacity wells, the pilot bore is usually in the range of $9\frac{7}{8}$ to 15 in.
2. The borehole diameter for gravel envelope wells is 8 to 12 in. larger than the nominal screen or casing diameter.
3. The borehole diameter for casing to be sealed is approximately 4 to 6 in. larger than the nominal diameter of the casing.
4. The final borehole diameter for naturally developed wells is the same as the outside diameter of the screen or the perforated casing.
5. The borehole diameter for casing to be installed with a minimum of clearance should be 2 to 4 in. larger than the casing outside diameter.

6.18 PLUMBNESS AND ALIGNMENT

Although plumbness and alignment are not normally thought of as components of well design, geometry requirements are frequently cited in well-design specifications for very important reasons. The two primary factors influencing the pump life are the amount of sand pumped and freedom of operation. The latter depends on well geometry. Satisfactory well geometry requires consideration of two elements: plumbness and alignment (straightness). Alignment is the more critical element.

Little thought is usually given to the effect of alignment. Often a well-drilling specification will require that a 40-ft rigid dummy slightly smaller in diameter than the inside diameter of the pump housing casing pass freely through the well to the anticipated permanent pump setting. It is not difficult to visualize a well with a dog-leg or S-curve that allows the passage of the dummy but will severely bind a pump column and a line shaft. Pumps have operated for long periods in straight wells considerably out of plumb, whereas a crooked but relatively plumb well will cause premature bearing and shaft failure.

Even if the operation of a submersible pump is planned, alignment should be carefully specified and checked at the completion of the well construction. This is to prevent cable damage, and consideration must be given to the possibility that a line-shaft pump may eventually replace the submersible pump.

6.19 PROCEDURE FOR CHECKING WELL PLUMBNESS AND ALIGNMENT

Well plumbness and alignment should be measured to the depth of anticipated pump setting or the bottom of the pump housing casing (caging). A wire cage lowered from a surface tripod is used for this purpose. Most cages are fabricated from spring steel wire and are approximately circular in their vertical cross section. The diameter of the cage is varied by tightening a nut on a threaded rod that forms its vertical axis. The cage is adjusted to be slightly smaller in diameter than the inside diameter (ID) of the pump housing casing. A bail, fabricated on the threaded rod, connects to a wire line that suspends the cage in the well. This line runs over a pulley in a tripod positioned to center the cage in the top of the casing. The tripod legs are adjusted to place the pulley a selected distance above the top of the casing and to center the cage in the casing at the surface.

Two steel bars, attached to each other at right angles, are set on the casing for measurement of the movements of the wire line from the center of the well. The deviations are measured at 10-ft intervals in two planes (for convenience labeled $N-S$, $E-W$) at right angles.

6.20 CALCULATION OF WELL DEVIATION

The deviation of the well at any depth from a vertical line centered at the surface is calculated from the following equation:

$$X = \frac{D(H + h)}{h},$$ (6.1)

where

X = well deviation at any given depth [in.],
D = distance the line moves from the center of the casing [in.],
H = distance to the top of the cage from the top of casing [ft],
h = distance from the center of the tripod pulley to the top of the casing [ft].

6.21 PLOTTING WELL DEVIATION

After caging, a plot is made of the resulting deviations from vertical for both the $N-S$ and $E-W$ readings. The pump

housing casing is scaled on the deviation plot. After the casing is plotted, a line is drawn from the center of the well at the surface to the center of the well to the depth that well alignment is to be maintained according to specifications. The well casing should not be closer to the plotted center line than the amount shown in the following list:

8 in. for 24 in.-ID well casing
7 in. for 20 in.-ID well casing
6 in. for 18 in.-ID well casing
5 in. for 16 in.-ID well casing
4 in. for 14 in.-ID well casing
3 in. for 12 in.-ID well casing
2 in. for 10 in.-ID well casing

Figure 6.12 is an example of a deviation plot. Figure 6.13 illustrates the alignment standard.

The usual standard for plumbness allows 6 in. out of plumb for every 100 ft of well depth to the specified depth. Some engineers feel that 6 in. per 100 ft is excessive and

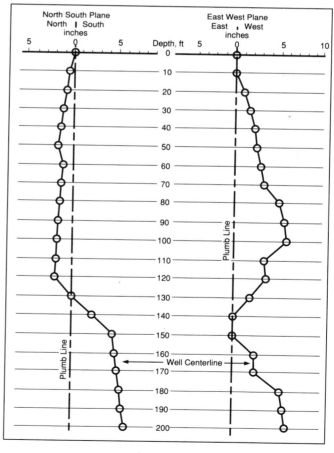

Figure 6.12. Deviation plot.

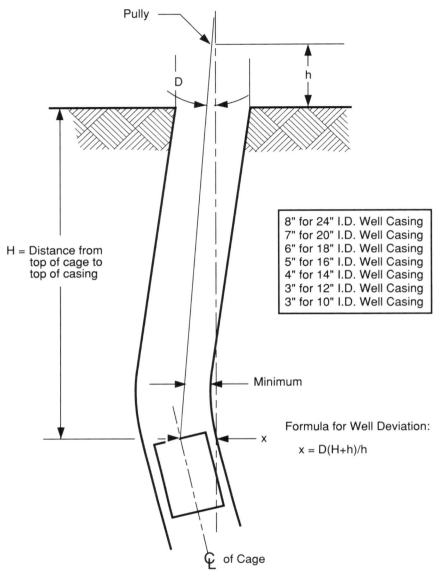

Pully

D

h

H = Distance from
top of cage to
top of casing

8" for 24" I.D. Well Casing
7" for 20" I.D. Well Casing
6" for 18" I.D. Well Casing
5" for 16" I.D. Well Casing
4" for 14" I.D. Well Casing
3" for 12" I.D. Well Casing
3" for 10" I.D. Well Casing

Minimum

Formula for Well Deviation:

$$x = D(H+h)/h$$

x

℄ of Cage

Figure 6.13. Alignment standard

allow only 3 in. per 100 ft, although there is little benefit in insisting on this more rigid requirement.

Conformity to these standards ensures freedom of installation and operation of line-shaft and submersible pumps. Other standards with recommended testing procedures are in AWWA A 100-84.

6.22 WELL PAD

A concrete well pad seals the surface from contamination. It is designed with a slight slope to drain water away from the well. The pad also functions as a working platform and supports any enclosure. A typical pump foundation for deep well settings is 24-in. thick and rises 12 in. above the pad. It must be 40 to 48 in. square to enclose the gravel feed pipe (for gravel envelope wells) and sounding pipe. A typical well pad with pump foundation design is shown as Figure 6.14. In some jurisdictions, pads are subject to health department standards.

6.23 GRAVEL FEED LINE

The gravel feed line is 3 to 4 in. in diameter and enters the conductor casing approximately 10 ft below ground level.

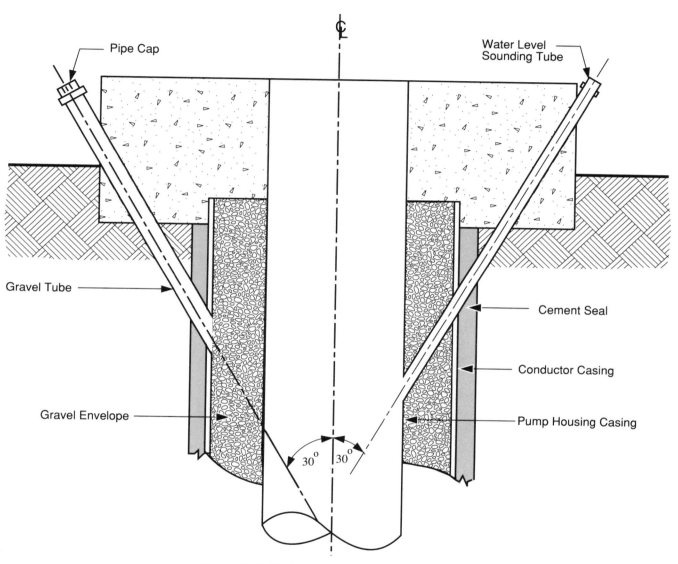

Figure 6.14. Typical pump foundation and well pad.

It is used to check and replenish the gravel in gravel envelope wells.

In designs where the upper portion of the annulus is cemented, a minimum of 4-in.-diameter gravel feed line is installed through the cement seal into the gravel envelope. It passes through the conductor casing below the ground surface, extends above the pump pad or foundation, and is protected with a threaded cap.

6.24 SOUNDING PIPE

A sounding pipe is normally incorporated in any well design to allow checking of water level and well fill. It is usually 2 in. in diameter, follows the annulus (gravel envelope wells), and enters the well casing below the pump setting. This arrangement eliminates the possibility of lines entangling around the pump. The sounding pipe enters the casing at a shallow depth in naturally developed wells that have no open annulus. At the surface it is protected by a threaded end cap.

A 4-in. sounding pipe is occasionally used to accommodate the newer smaller-diameter TV inspection equipment. However, there are not many situations in which a well can be adequately inspected without withdrawing the pump. Since a larger-diameter sounding pipe is more difficult to install, it may not be worthwhile. Reduced clearance between casing and borehole may introduce bending stresses which in combination with tensile stresses may put the casing and screen at risk (see Chapter 9).

6.25 DISINFECTION

An excellent discussion of methods for disinfection of new wells, gravel envelopes, and permanent well equipment is found in AWWA standard C 654-87.

CHAPTER 7

Drilling Systems

7.1 INTRODUCTION

Methods for extracting ground water have been developed over several thousand years. They include dug wells, ghanats, water tunnels, and vertical borehole drilling systems. All of these methods are in use today, but the borehole drilling method is by far the most common. The application of vertical borehole drilling systems to water well construction depends upon geologic and hydrologic conditions, general practice in the area, available tools and equipment, economic considerations, and final well design.

There are two basic drilling systems used in most water well construction: hydraulic rotary and cable tool, with a number of variations. This chapter introduces the fundamentals of drilling technology. An understanding of these principles is required for the proper design and construction of water wells. Special emphasis is given to the hydraulics of drilling fluids, the maintenance of borehole alignment, and the dynamics of cable tool drilling.

7.2 ROTARY DRILLING

The drilling of boreholes by the hydraulic rotary method requires a drill bit, a system for rotating the bit, the means for controlling bit pressure on the formation, and a medium for removing the material displaced by the bit.

Direct Hydraulic Rotary: System Components

With this system a circulating fluid is pumped down the rotating drill string, through the bit, and up the annulus between the drill string and borehole wall to the surface where the transported cuttings are removed. The rig and basic components of the system are shown in Figure 7.1.

Derrick. The portable derrick, or mast, must safely carry the expected loads encountered in drilling and completion of wells of diameter and depth for which the rig manufacturer specifies the equipment. The derrick is rated for its collapse resistance to vertical loading. The largest dead load imposed will normally be the casing string, whereas pulling on a stuck drill pipe or casing will probably be the maximum dynamic load. To allow for contingencies, the rated derrick capacity should be at least twice the anticipated drill string or casing loads.

Power Plant. The power plant(s) is used principally for rotating the drill string, hoisting, and circulating drilling fluid. It may be necessary to perform any of these jobs separately or a combination of them simultaneously, making flexibility a principal requirement. Internal combustion engines are used on water well drilling rigs, with minimum capacities of 100 hp per 1000 ft of rig-design depth. Horsepower requirements for drilling-fluid system operation are discussed later.

Draw-works. The draw-works, or hoist, includes the control center from which the driller operates the rig. The principal components of the draw-works are the hoisting drums, cat-heads, brakes, and clutches. Draw-works are designated by a horsepower and depth rating. The depth rating must specify the size drill pipe to which the rating pertains. The draw-works horsepower output required for hoisting is

$$\text{HP} = \frac{WVn}{33,000e},$$

where

W = hook load [lb],

Vn = hoisting velocity of the traveling block [ft/min],

e = hook to draw-works efficiency [commonly 80%–90%].

Drilling Line. The drilling line operates from the draw-works' main drum through the crown sheaves on the mast and the traveling block. Usually the dead line is returned to the rig structure near the draw-works. The wire line most commonly used for this service is 6 × 19 construction, fiber core plow steel cable. Premium lines with an independent wire-rope center are used for high load requirements.

Traveling Block. The traveling block and hook are part of the hoisting system, along with the mast, draw-works, crown block, elevators, drill line, and weight indicator.

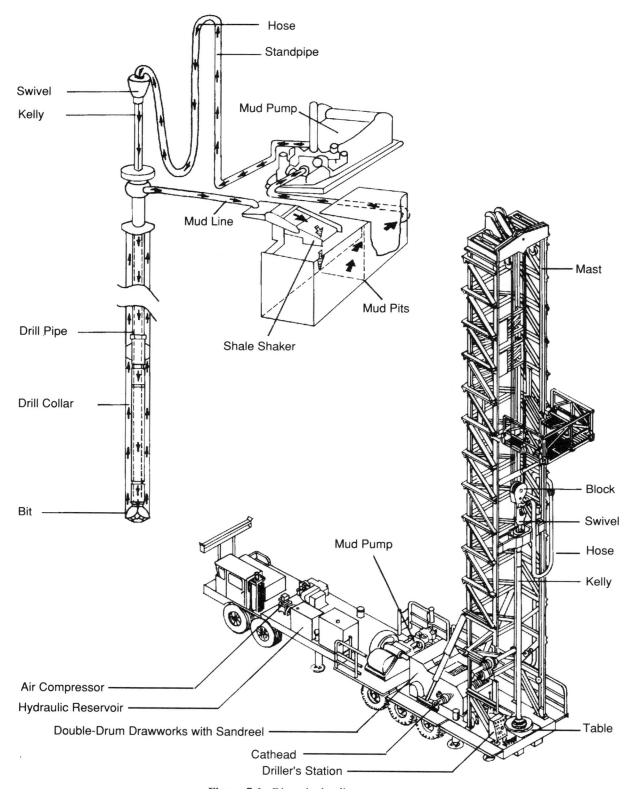

Figure 7.1. Direct hydraulic rotary system.

Circulating System. Mud pumps usually are positive displacement, two- or three-cylinder double-acting (pumping from both sides of the piston) types with pistons driven from a crank shaft. The diameter of the suction line from the mud pit or tank should be at least 1½ times that of the mud pump liners to ensure complete cylinder filling during each pumping stroke. Use of auxiliary centrifugal pumps for maintaining positive intake-side pressure materially enhances pump efficiency and permits running at speeds up to 50% greater than the customary 40 to 50 rpm.

The mud-pump discharge line connects to a vertical standpipe mounted on the rig mast. The rotary hose, attached to the upper end of the standpipe, conducts the drilling fluid to the swivel. The swivel carries the weight of the drill string and allows passage of the drilling fluid into the kelly and drill string while they rotate.

The drilling fluid is pumped through the drill string and bit, and returns the drill cuttings to the surface in the annulus between the drill string and borehole wall. The drill cuttings may be separated from the fluid by a vibrating screen, called a "shaker." Finer particles passing the shaker are removed by hydrocyclones. Solids may also settle in the mud pit(s) or tank(s), or be recirculated. The pit(s) or tank(s) is also used to add material to the drilling fluid to control its properties.

Drill String. The drill string is the major system component used in the actual process of drilling the borehole. It includes the kelly, drill pipe, drill collars, stabilizers, and bits, as illustrated in Figure 7.2.

The kelly, a hollow length of alloy-steel square bar stock, is suspended from the swivel and is designed primarily to transmit the torque from the rotary table to the drill string. It is turned as its square section slides through bushings in the rotary table so that the rotating drill string may be lowered or raised during drilling operations.

The major length of the drill string consists of the drill pipe, which connects the kelly to the drill collars. The drill pipe is manufactured from hot-rolled pierced seamless tubing and is connected by tool joints. Range 2 drill pipe, which has an average length of 30 ft, is commonly used with larger rigs.

The lower section of the drill string consists of drill collars and, in some cases, stabilizers. The purpose of drill collars, each typically about 30 to 40 ft in length, is to furnish weight and stiffness in the bottom portion of the drill string. Depending upon rig capacity, hole diameter, and depth, drill collars used in water well drilling range from 6 to 11 in. in diameter.

During drilling, the drill pipe should be in tension, since drill pipe is essentially a tube of medium wall thickness with little resistance to buckling under compression. Therefore the total drill collar weight is controlled by the weight to be carried on the drill bit, which in turn is a function of hole diameter, bit type, and the formation being drilled.

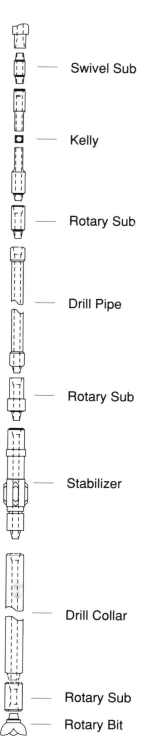

Swivel Sub

Kelly

Rotary Sub

Drill Pipe

Rotary Sub

Stabilizer

Drill Collar

Rotary Sub

Rotary Bit

Figure 7.2. Hydraulic rotary drill string (courtesy of Hydraulic Institute).

Generally, not more than 70% of the effective drill collar weight should be applied to the bit, since excessive weight may cause hole deviation in some formations. This subject is examined in more detail in Appendix E.

Drill Bit. The drill bit is attached to the lower end of the drill collars and penetrates the formation as it is rotated at the bottom of the hole. Drilling fluid pumped through the drill string is discharged through ports or jet nozzles in the bit, cooling it and cleaning both the bit and hole bottom. Strata lying horizontally can typically be drilled faster and with greater bit weight than inclined beds where the maintenance of hole alignment necessitates less bit weight. The type of bit used is governed primarily by formation characteristics. Softer clays are drilled very effectively with drag bits, whereas rolling-cutter bits are used for harder formations.

The simplest of drag bits is the fishtail type with two blades spaced at 180°. Drag bits having three or four blades are also used, but the cutting edges are usually fingered. Tungsten carbide inserts or hard-surfacing minimizes wear. Many soft formations drilled with drag bits essentially are drilled by the jetting action of the drilling fluid. In this case the penetration rate is related to the hydraulic horsepower at the bit.

Rolling-cutter (roller-rock) bits are the most common type for direct rotary use and are classified as tri-cone or cross section. The teeth are hard surfaced (usually with tungsten carbide) for longer life. The rolling cutters rotate independently, turning as the drill string is rotated by virtue of their contact with the bottom of the hole.

Different types of roller bits are used, depending on the formation to be penetrated. Those used in the drilling of unconsolidated alluvia or other soft formations are designed with broad teeth, which are thin in cross section, long, and widely spaced. They are placed on offset or skewed cutters so that they gouge, scrape, or pry out the formation. The teeth of adjacent cutters intermesh so that self-cleaning occurs. Some teeth may be deleted to allow release of large cuttings.

Hard formations are dense, high strength, and abrasive. They can only be drilled by crushing, namely, by applying sufficient weight on the bit to cause the material to fracture or fail. Hard formation bits have short teeth protected by tungsten carbide on the gage surfaces. Sintered tungsten carbide inserts are used in lieu of steel teeth for extremely hard formations.

Proper bit-type selection is an important factor in efficient drilling. In unconsolidated alluvial formations, drilling-fluid hydraulics and nozzle design must be considered, since excessive weight will cause the bit to founder due to formation material packing around the cones and between the teeth. The addition of jet nozzles with corresponding higher fluid horsepower to soft-formation bits can markedly improve performance. Better cleaning of the teeth and hole bottom permits the application of additional weight before a new foundering point is reached. However, considerable experience is required to define the economic limit of increasing hydraulic horsepower.

Limitations on available bit weight and cuttings removal preclude the drilling of a large borehole in one stage with direct circulation rotary, and hole openers are required to achieve final diameter. The final diameters of high-capacity gravel-envelope wells range from 20 to 48 in. A 9⅞ to 15-in. pilot hole is typically drilled first, and then enlarged in 8- to 12-in. increments.

Analysis of System Hydraulics

The efficiency of direct rotary drilling methods depends upon the use of a properly designed drilling fluid to maintain the borehole and remove cuttings. Composition and use of drilling fluids is covered in Chapter 8.

The key factor in the design of the circulation system is the velocity necessary for the particular fluid to remove cuttings. With the flow rate established, the selection of pumping equipment for liquids or compressors for air is defined by the pressure losses associated with circulation.

Annular Flow Velocities. Much laboratory and theoretical work has been done to develop relationships for the slip velocity of cuttings suspended in drilling fluids. However, the annular velocity used in design calculations is still based mainly on operating experience, since the important factors of gradation and shape of cuttings are a function of drill bit design, load on the bit, and formation being drilled.

Annular flow velocities of 100 to 200 ft/min are typical with direct rotary drilling. The lower velocities are used in slow drilling of a hard rock formation, and the higher velocities for drilling in unconsolidated alluvium where rate of penetration is rapid. These velocities require very high circulation flow rates with larger-diameter boreholes. Reverse circulation rotary drilling has a major advantage in this situation, since cuttings are lifted inside the smaller cross-sectional area drill pipe, maintaining high return velocities. Nevertheless, the principles of analysis of both fluid circulating systems are the same.

Reynolds Number. In considering power requirements, it is necessary to determine whether the fluid motion is laminar or turbulent. In laminar motion the fluid and drill cuttings travel in smooth trajectories parallel to the pipe walls. In turbulent flow, motions are chaotic, and cuttings mix uniformly across the flow cross-sectional area. The measure of whether a flow is laminar or turbulent is a non-dimensional parameter known as the Reynolds number (discussed in Chapters 2 and 5). The Reynolds number at any particular flow location is defined as the product of mean fluid velocity and a transverse dimension of the flow area, divided by the fluid kinematic viscosity. The number defines the local ratio of inertial force in the flow to the viscous force, so that a low Reynolds number implies that viscosity has a dominant influence on the fluid motion. Thus the Reynolds number for pipe flow is defined as

$$\text{Re} = \frac{\rho U d}{\mu},$$

where

Re = Reynolds number,
ρ = fluid density [slugs/ft^3],
U = mean fluid velocity [ft/sec],
d = pipe diameter [ft],
μ = dynamic viscosity [slugs/ft sec],

For Reynolds numbers in excess of 2000, the flow will always be turbulent and the frictional losses should be calculated according to turbulent flow laws, as will be described.

In common drilling practice the units of fluid density are lb/U.S. gal, velocity units are ft/sec, the pipe diameter is quoted in in., and the viscosity[1] is in centipoise (cp). The Reynolds number formula in these units is:

$$\text{Re} = 927 \, \frac{\rho(\text{lb/gal}) U(\text{ft/sec}) d(\text{in.})}{\mu(\text{cp})}.$$

Gel Strength. Another important drilling-fluid property significantly influencing the power required for circulation is gel strength (see Chapter 8). Fluids with gel properties are described as "Bingham plastic," and those without them are referred to as "Newtonian."

The gel strength of drilling fluid is a major factor in the shear stress necessary to start fluid movement. For a Bingham plastic drilling fluid that is stationary in a pipe, limiting shear stress is reflected by the pressure level necessary to initiate fluid motion. This is defined for a drilling fluid as the Bingham yield point and is generally expressed in lb/100 ft^2. Newtonian fluid viscosity is the proportionality coefficient between stress applied and rate of strain produced in the liquid. Calculation of system pressure losses must take into account whether the flow is laminar or turbulent and whether the fluid is Newtonian or Bingham plastic.

Calculation of Required Pumping Pressure. Since drilling fluid starts at and returns to the surface, the required pumping pressure is simply the sum of pressure losses through system components. These components include the following:

1. Pump suction and discharge piping, standpipe, hose, and swivel.
2. Drill pipe.
3. Drill collars and stabilizer.
4. Drill bit ports or nozzles.
5. Annulus between drill string and borehole.
6. Surface piping and discharge.

The computation of pressure losses is identical in rotary and reverse rotary drilling, although the flow direction must be considered. First, dimensions of the system are established, especially the largest diameter of the borehole, along with the properties of the drilling fluid (density, viscosity, and Bingham yield point). The flow rate necessary to give the 100 to 200 ft/min return velocity needed to lift the cuttings is then computed. Using this flow rate, the mean-flow velocities through each of the system components are calculated. With velocity, diameter, density, and viscosity known, it can be determined whether the flow is laminar or turbulent. For Newtonian fluids, this means verifying whether or not the Reynolds number exceeds 2000. For Bingham plastic fluids, it is accomplished by finding the Reynolds number using an equivalent Newtonian dynamic viscosity as found from the relationship

$$\mu_c = \mu_p + \frac{d Y_p}{6 U}$$

where

μ_c = equivalent Newtonian dynamic viscosity [ML^{-1}T^{-1}],
μ_p = plastic viscosity [ML^{-1}T^{-1}],
Y_p = Bingham yield point [ML^{-1}T^{-2}],
d = flow diameter [L],
U = mean velocity [LT^{-1}].

For units commonly employed within the drilling industry, the formula is

$$\mu_c(\text{cp}) = \mu_p(\text{cp}) + \frac{6.6 Y_p(\text{lb/100 ft}^2) d(\text{in.})}{U(\text{ft/sec})}.$$

The factor 6.6 is empirically reduced to 5.0 in some literature.

For laminar flow, pressure losses in surface piping and the drill string are calculated using the relationship

---

1. The units of viscosity can be confusing. In English units, dynamic viscosity, μ, is given in lb-sec/ft^2, which are identical to slugs/ft-sec. However, the most commonly used unit is the centipoise (cp), which is a metric unit of one-hundredth of a poise (whose actual units are 10^{-1} kg/m-sec). Centipoise units are converted to English units by dividing centipoise by 47,900 to obtain lb-sec/ft^2. The dynamic viscosity of water is about 1 cp. Kinematic viscosity (ν) is dynamic viscosity divided by fluid density. The English unit is ft^2/sec and the metric unit is m^2/sec. However, kinematic viscosity is usually enumerated in Stokes units, which are 10^{-4}/m^2/sec.

$$\Delta p = \frac{32\mu_c LU}{d^3},$$

where

Δp = pressure drop $[ML^{-1}T^{-2}]$,
L = pipe length $[L]$.

In the commonly used units of drilling practice, this becomes

$$\Delta p(\text{lb/in.}^2) = \frac{\mu(\text{cp})L(\text{ft})U(\text{ft/sec})}{1497d^3}.$$

For turbulent (Newtonian) flow the Darcy-Weisbach friction factor f is used in the relationship

$$\Delta p = \frac{\gamma f L U^2}{2gd},$$

where

γ = specific fluid weight $[ML^{-2}T^{-2}]$,
g = gravitational acceleration $[LT^{-2}]$.

For common units, the formula is

$$\Delta p(\text{lb/in.}^2) = \frac{\rho(\text{lb/gal})L(\text{ft})U^2(\text{ft/sec})}{103.3d(\text{in.})}.$$

The Darcy-Weisbach friction factor is a function of the relative roughness of the pipe material and the Reynolds number and is generally presented in a Moody diagram, illustrated in Figure 7.3.

In chemical and petroleum engineering the friction factor (f) used is the Fanning friction factor, given by

$$4f_{\text{Fanning}} = f_{\text{Darcy-Weisbach}}.$$

For flow through the annulus, an equivalent diameter is used in the pressure calculation formulas. It is the difference in diameters between the borehole and the drill string.

The pressure drop across the bit nozzles is computed using a discharge coefficient for the nozzle C_D so that

$$\Delta p = \frac{\gamma V_D^2}{2gC_D^2},$$

where

V_D = nominal nozzle velocity $[LT^{-1}]$,
γ = fluid specific weight $[ML^{-2}T^{-2}]$.

In common units this formula is

$$\Delta p(\text{lb/in.}^2) = \frac{\rho(\text{lb/gal})V_D^2(\text{ft/sec})}{1240C_D^2}.$$

For simple round holes $C_D = 0.8$, whereas for jet nozzles $C_D = 0.95$. The pressure drop is the same for all nozzles of the same diameter and nozzle velocity, regardless of the number of nozzles.

The minor losses associated with elbows, valves, tees, or other fixtures, can be accumulated using standard pressure-loss tables, such as provided by Crane Co. or the Hydraulics Institute. Charts and curves for pressure drop–flow rate relationships in typical rotary drilling systems are available from equipment manufacturers.

The required hydraulic horsepower for the system can then be estimated from the total accumulated pressure losses, $\Delta P(\text{lb/in.}^2)$ and flow $Q(\text{gpm})$ from the formula

$$\text{HP} = \frac{\Delta P(\text{lb/in.}^2)Q(\text{gpm})}{1715}.$$

The required engine horsepower is the hydraulic horsepower divided by the hydraulic and mechanical efficiencies of the pump, usually about 0.77.

Advantages and Disadvantages of the Conventional Direct Hydraulic-Rotary System

The direct rotary drilling system has a number of advantages under many water well drilling conditions. These include:

1. Ability to drill and maintain open borehole in a wide variety of formations to depths in excess of those required for water wells.
2. Relatively high penetration rates.
3. Ability to drill small diameter, low-cost borehole for formation sampling and geophysical logging. This information leads to the final well design. In most cases the pilot borehole is used for this purpose.
4. Ability to drill and maintain open borehole facilitates installation of casing and screen, cementing, and installation of gravel envelope.
5. Low cost of well construction in soft, unconsolidated alluvium, particularly with depths greater than 1000 ft.
6. Combination of small-diameter pilot borehole with appropriate diameter and length of drill collars that ensures good alignment in soft unconsolidated formations. For analysis of factors causing borehole alignment deviations, see Appendix E.
7. A wide variety of casing and screen designs and material able to be installed in open boreholes.

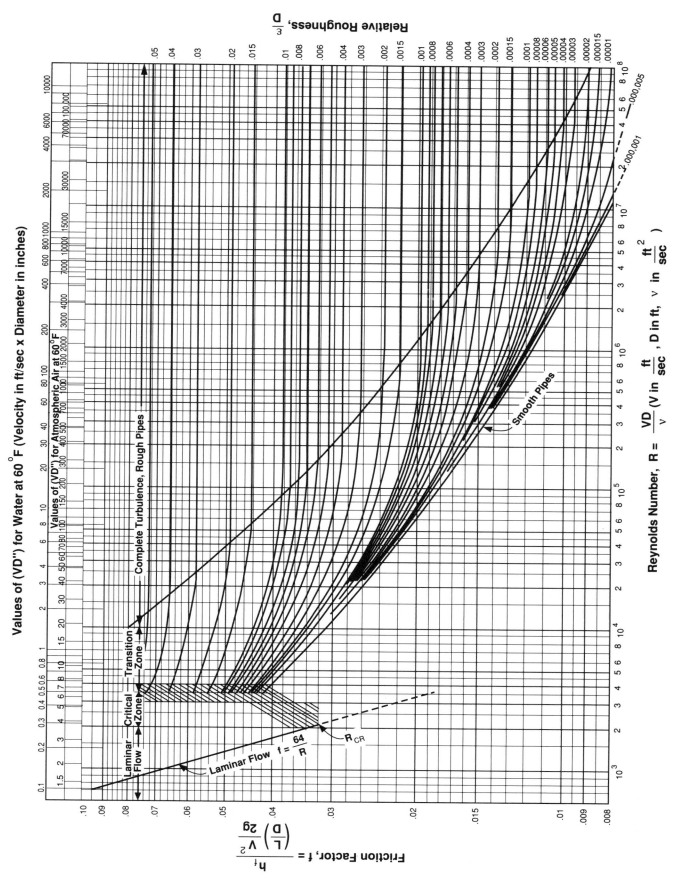

Figure 7.3. Moody diagram (courtesy of the Hydraulic Institute).

137

Disadvantages include:

1. A more complex drilling system than with cable tool.
2. Relatively high equipment capital cost.
3. Higher bit cost, particularly in hard formations.
4. Engineering and control of drilling-fluid properties critical to well logging, completion, and development.
5. High noise levels that create operating problems in urban areas.
6. Greater daily operating cost.
7. Relatively high makeup water requirements.
8. Relatively high equipment transportation cost.
9. High cost of drilling karstic formations.

Air Drilling

Drilling with air or with a combination of air and additives is feasible where borehole stability is not a problem and the relatively low water infiltration can be removed, as described in Chapter 8.

Rotary-type compressors are used for drilling shallow holes in consolidated rock. These machines are capable of producing 600 to 1200 cfm. But their upper pressure limit of 125 psi can be inadequate, and their volumetric efficiency is very low at maximum pressure. Five hundred to 750 cfm two-stage positive-displacement compressors having working pressures of 250 to 350 psi are more satisfactory, particularly when submergence requires higher pressures.

The compressor horsepower requirement for air drilling is fixed by the velocity of air necessary to lift the cuttings and the pressure losses in the air circulation system. Computation of the pressure losses follows the same basic principles presented for liquid-drilling systems, except that calculation of nozzle losses will, in most cases, need to account for the compressibility of the air. If so, air expansion through the bit nozzles will change the temperature down the hole. The velocity of flow in the circulating system is defined by both the temperature and the pressure of the air within that element. For detailed calculations, the *Air and Gas Drilling Manual* by Lyons should be consulted (1).

The required air velocity to raise cuttings depends on the fall velocity of cutting particles. Experience indicates that removal of particles up to $\frac{3}{8}$ in. in size requires annular velocities of 2000 to 4000 ft/min. An average practical velocity is 3000 ft/min.

In addition to the basic requirement of lifting the drill cuttings, air is used to enhance drilling capability by directing jets at the cutting heads to increase penetration rates. This subject was addressed by Angel, who produced tables of the standard air volume required (2). With the flow rate defined, the compressor horsepower can be estimated from the adiabatic compression of the air according to the formula:

$$HP = \frac{QP_0}{229e_v}\left(\frac{k}{k-1}\right)\left[\left(\frac{P_d}{P_0}\right)^{k/k-1} - 1\right]$$

where

e_v = volumetric efficiency of compressor,

k = specific heat ratio for dry air = 1.4,

P_0 = suction pressure (varies with altitude) [lb/in.^{2} absolute],

Q = standard intake volume flow of air [SCF/min],

P_d = discharge pressure from the compressor [lb/in.^{2} absolute].

Ratios of P_d/P_0 greater than 6 may require a second-stage compressor. Estimation of P_d depends on losses in the circulating air system, which are very dependent on the water content of the air when operating in wet conditions (3). Drilling to 1000 ft will need pressures of 250 psi, requiring two-stage compressors, with each stage having a pressure ratio of 4:1.

Air-Foam Drilling. Air drilling in large-diameter boreholes entails enormous air volume flow rates to maintain the minimum required annular velocity. As described in the next chapter, proprietary foaming agents are mixed with air and water, and sometimes bentonite or polymers, to produce a foam that can remove cuttings at relatively low velocities. Thus velocities can be reduced from those necessary for dry-air drilling, and water demands are much lower than are required for a liquid circulation system.

Use of air-foam drilling systems is discussed in some detail by Driscoll, and part of the material presented here on this topic is from his work (4). Design of an air-foam drilling-fluid system is not straightforward because of the two-phase flow aspect, with each phase reacting differently to changes in temperature and pressure. At the bottom of the hole the air fraction is compressed, but it expands as the foam rises up the borehole. Heat transfer in the upflowing borehole tends to be minimal, so this aspect of the calculation is generally ignored. The ideal gas law is used to determine volumetric change in the gas fraction of the foam after it leaves the bit, with temperature usually assumed constant. The gas-volume fraction at any point is therefore inversely proportional to pressure, so that effective foam density is much greater at the bottom of the borehole. Since pressure is reduced as foam rises, the gas expands and fluid velocity must increase. Roughly speaking, if pressure at the bottom of the borehole is 5 atm, surface flow rate must be five times that at the bottom of the hole.

The key variable in design of a foam drilling-fluid circulation system is the liquid-volume fraction (LVF) of the foam, since this controls the ability of the foam to lift drill cuttings. Experiments have shown that an optimum lift ca-

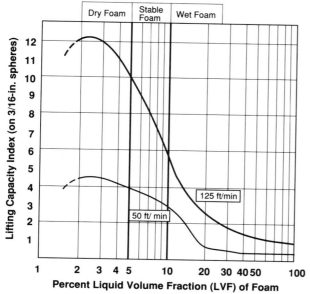

Figure 7.4. Liquid-volume fraction and lift (after Drilchem).

pacity is attained at a LVF of 2% based on the free air volume (i.e., the volume of air at standard atmospheric pressure), as shown in Figure 7.4. If the LVF is 2% at atmospheric pressure at the top of the borehole, then at the bottom it will be larger than 2%, since the volume of air is compressed by the borehole pressure. The bottom hole pressure is determined by the total weight of drilling fluid and cuttings in the borehole, plus the pressure necessary to overcome the frictional pressure losses associated with the upward flow in the borehole.

Liquid-Volume Fraction. To evaluate the change in the liquid volume fraction with elevation within the borehole, the compressibility of the foam must be considered. Therefore

$$f_a + f_w + f_s = 1 \ , \tag{7.1}$$

where

f_a = relative volume fraction of air,
f_w = relative volume fraction of water,
f_s = relative volume fraction of solids.

The foam density, $\rho(x)$, is a function of distance from the surface, x, and is given by

$$\rho(x) = f_a\rho_a + f_w\rho_w + f_s\rho_s \ ,$$

where

$\rho(x)$ = foam density at distance x,
ρ_a = air density,

ρ_w = water density,
ρ_s = solid density.

ρ_w and ρ_s will remain constant, whereas ρ_a is a function of the pressure and the distance x from the surface.

The total mass flow rate of foam in the well is given by

$$M = A + W + S \ ,$$

where

M = total mass flow rate,
A = air mass flow rate,
W = water mass flow rate,
S = solid mass flow rate.

These all remain constant for steady flow.

The total volume flux of foam at a distance x from the surface is

$$Q(x) = \frac{A}{\rho_a} + \frac{W}{\rho_w} + \frac{S}{\rho_s} \tag{7.2}$$

where

$Q(x)$ = total flux volume at distance x.

From Equations 7.1 and 7.2,

$$\rho(x) = \frac{M}{Q(x)} \ . \tag{7.3}$$

If frictional pressure losses are ignored, the pressure within the well $P(x)$ is related to the foam density $\rho(x)$ by

$$\frac{\partial P}{\partial x} = g\rho(x) \ , \tag{7.4}$$

where

g = gravitational acceleration.

However, the pressure within the well defines the air density, according to

$$P(x) = \rho_a gRT \tag{7.5}$$

where

R = universal gas constant (for air, $R = 53.3$) [ft lb/ lb/°R],
T = foam absolute temperature in °R (°F + 491), assumed to be constant within the well.

From Equations 7.2, 7.3, 7.4, and 7.5 it is possible to write

$$\left(\frac{1}{\rho_a} + \frac{1 - f_{ao}}{f_{ao}\rho_{ao}} \right) \frac{d\rho_a}{dx} = \frac{M\partial x}{ART} \, ,$$

and integrating from the surface $x = 0$, where the air density is ρ_{ao} to some depth x gives

$$f_{ao} \log D\left(\frac{\rho_a}{\rho_{ao}} \right) + (1 - f_{ao})\left(\frac{\rho_a}{\rho_{ao}} - 1 \right)$$
$$= f_{ao}\left(\frac{Mx}{ART} \right) \, , \quad (7.6)$$

where

ρ_{ao} = air density at atmospheric pressure,
f_{ao} = volume fraction of air in the foam at atmospheric pressure.

This equation defines the density of air ρ_a at depth x in terms of its density at the surface ρ_{ao}, the relative volume fraction of air at the surface f_{ao}, and the ratio of the total mass flux of foam M to the mass flux of air A. From Equation 7.5 the well pressure and air density at any point in the well are proportional so that Equation 7.6 can also be used to find the pressure in the well as a function of depth.

Graphs of the ratio of well pressure to atmospheric pressure, $P(x)/P_0$, as a function of depth for various air volume fractions at the surface, f_{ao}, are shown in Figure 7.5.

Note that from Equation 7.1 the liquid volume fraction f_w is given by

$$f_w = 1 - f_a - f_s \, . \quad (7.7)$$

Equation 7.1 and mass conservation of water, gas, and solid species enable the computation of the change in the relative air and water volume fractions with pressure. It is found that

$$\frac{f_a}{f_{ao}} = \left[f_{ao} + (1 - f_{ao})\left(\frac{P}{P_0} \right) \right]^{-1}$$

and

$$\frac{f_w}{f_{wo}} = \left[(1 - f_{ao}) + f_{ao}\left(\frac{P_0}{P} \right) \right]^{-1} .$$

Furthermore, since f_s is generally very small, the last relationship can be rewritten as

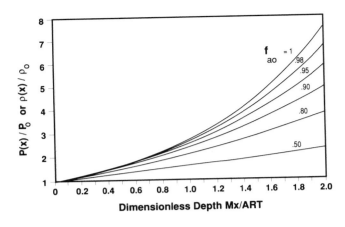

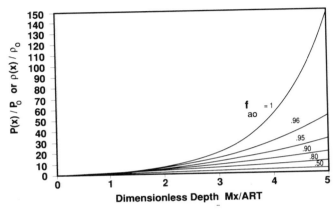

Figure 7.5. Ratio of well pressure to atmospheric pressure as a function of depth.

$$\frac{f_w}{f_{wo}} = \left[f_{wo} + 1 - f_{wo}\left(\frac{P_0}{P} \right) \right]^{-1} . \quad (7.8)$$

These relationships, together with Equation 7.7, enable the liquid-volume fraction to be specified as a function of depth, given the liquid volume fraction f_{wo} at the surface at pressure P_0.

Mass conservation applied to the air volume enables the velocity of flow in the well at depth x to be related to the surface borehole velocity, v_0. The relationship is

$$\frac{v}{v_0} = f_{ao} \frac{P_0}{P} + (1 - f_{ao}) \, . \quad (7.9)$$

Thus the velocity of flow in the well can be defined through use of Equation 7.7. This relationship is graphed in Figure 7.6, which shows the reduction in borehole flow velocity with depth in an air-foam system given a liquid-volume fraction f_{wo} at the surface.

In all of the preceding calculations the liquid-volume fraction can be interpreted as the combined volume fraction

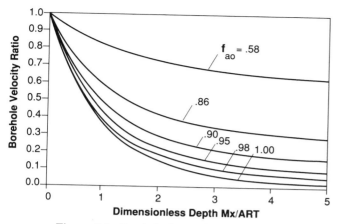

Figure 7.6. Borehole flow velocity vs. depth.

TABLE 7.2 Change in LVF, Pressure, Density, and Velocity with Depth, for Foam at 70°F at the Surface and Drilling Rate of 10 ft/min in a 16-in. Borehole

Depth	LVF	Pressure (lb/in.²)	Foam Density (lb/ft³)	Foam Velocity (ft/sec)
Surface	0.020	14.7	1.60	102
500 ft	0.027	20.2	2.20	73
1000 ft	0.039	29.4	3.20	53
2000 ft	0.082	64.7	7.04	22

of both liquids and solids, since both are essentially incompressible at normal borehole pressures. In addition the volume fraction of solids is small if there are no clays or polymers added to the drilling fluid.

The following example serves to illustrate how the depth of the well will influence borehole fluid velocity and the liquid-volume fraction of the foam. Consider a 16-in. diameter well that is being drilled at a rate of 10 ft/min. This represents a solid mass flow up the well of about 0.64 lbs mass/sec, based on a specific gravity of 2.65 for the borehole material. If a borehole velocity of approximately 100 ft/min is to be maintained at the well surface and the water-surfactant flow rate represents a 2% LVF at the surface, then the mass flow rates given in Table 7.1 can be computed.

From data in Table 7.1 and the well diameter, it is possible to find the foam velocity at the surface. Ignoring the cross-sectional area of the drill string, it is found that the velocity is 1.7 ft/sec, or 102 ft/min. Computation of the pressure, liquid-volume fraction, velocity, and density at various depths can be accomplished from Equations 7.7, 7.8, and 7.9, or from the appropriate graphs. The results are given in Table 7.2 This example illustrates that for a drilling rate of 10 ft/min and the water-surfactant and air flows proposed, the foam velocity is probably adequate to depths of about 1000 ft. At depths below 1000 ft the foam velocity may be too low to lift drill cuttings, unless it is stiffened by polymeric

additives. At such depths it is possible that the air flow rate may have to be increased, and water flow rate increased accordingly, to maintain a 2% LVF at the surface.

Compressor requirements for the system can be calculated by methods outlined previously for air drilling but will be greater at start-up if the borehole is at any time filled with water (Appendix I). Restarting a foam system requires a higher pressure than that for air drilling with a dry hole.

7.3 REVERSE CIRCULATION ROTARY

In reverse-circulation rotary drilling systems, the drilling fluid with cuttings returns inside the drill string and is discharged into a settling tank or pit. Downward flow is in the annulus between the drill string and borehole. The system components, as shown in Figure 7.7, are similar to those of the direct rotary except for circulation. A centrifugal pump may be used for drilling shallow holes (up to 400 ft), whereas a compressor is used to airlift the fluid for deeper holes.

When an airlift is used, the air is introduced into the drill string at an appropriate depth by a small-diameter pipe that is external to or sometimes concentric with the drill pipe. The expanding air reduces the density of the fluid in the drill string. This causes the drilling fluid and cuttings to rise in the drill pipe. Liquid flowing down the borehole annulus passes through the bit and enters the drill pipe as flow commences. The return velocity of fluid and drill cuttings is not the same as for air, since the air tends to form large bubbles or slugs that rise up the drill pipe through the liquid, bypassing the fluid next to the drill pipe. Increasing the flow of air beyond a certain rate does not pump any additional fluid because a greater proportion of air simply fills the drill pipe.

The compressor capacity and pressure required for circulation are defined by the depth of the well. Airlift pumping does not work until the borehole is more than 30 to 40 ft deep, since the necessary air expansion will not occur unless the air can drop in pressure by at least 1 atm during the lift.

TABLE 7.1 Foam Properties

Fraction	Mass Flow (lb/sec)	Density (lb/ft³)	Volume Flow (ft³/sec)	Volume Fraction
Air	0.178	0.0765	2.23	0.98
Solid	0.64	165.4	0.00387	0.0016
Water	3.00	62.4	0.048	0.02
Foam	3.818	1.60	2.382	1.00

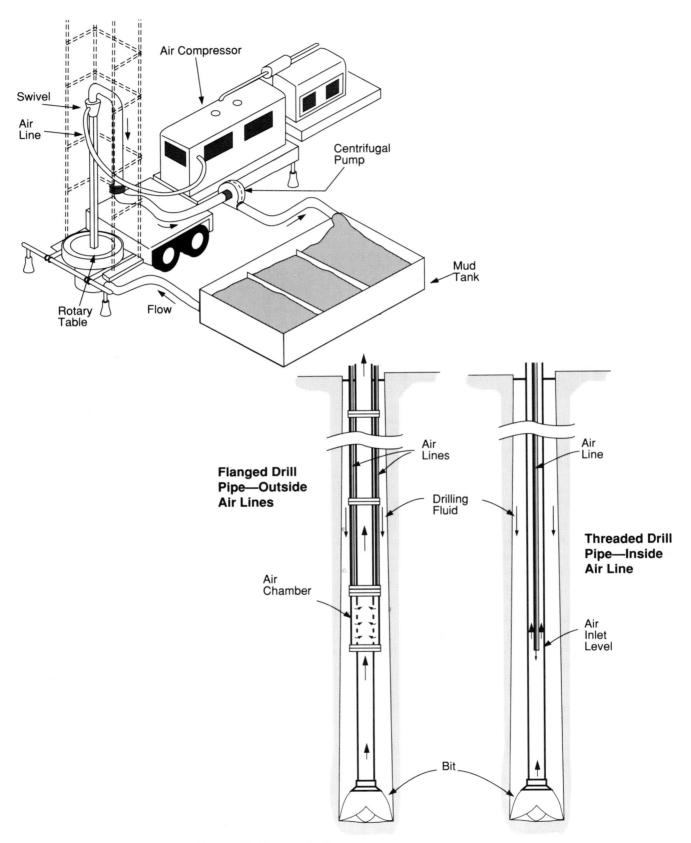

Figure 7.7. Reverse-circulation rotary system components.

The airlift pump has other uses such as dredging and, most recently, ocean mining (5). Its application to pumping shear-thinning or thixotropic suspensions such as drilling fluids has been investigated in some detail by Heywood et al. (6). Their studies were directed specifically at comparing the efficiency of the airlift pump with mechanical pumps in transport of flocculated kaolin suspensions but are appropriate for the analysis of pumping of drilling fluids. Maximum water-lifting efficiency was found to occur at an air flow velocity of about 2.5 ft/sec, or 150 ft/min, but overall efficiency is still less than that of a mechanical pump at the same water flow rate. However, for thixotropic mixtures of water, air, and solids, the airlift pump does show some efficiency advantage, but efficiency drops rapidly for air flow velocities in excess of 200 ft/min.

Airlift Pumping Systems. The results of these studies provide a basis for design of airlift pumping systems (also discussed in Appendix I). They indicate that velocities in the range of 75 to 200 ft/min result in reasonably efficient pumping.

An estimate of the rate at which drilling fluid and cuttings can be raised by a specific air flow is given by the energy made available by air expansion in the drill pipe. This is expressed by

$$W_a = Q_a P_0 \ln\left(\frac{P_d}{P_0}\right) ,$$

where

W_a = rate at which energy is made available by the expanding air $[L^2 M^{-1} T^{-3}]$,

Q_a = air flow rate $[L^3 T^{-1}]$,

P_0 = atmospheric pressure $[M L^{-1} T^{-2}]$,

P_d = downhole pressure = $\rho_w g H$, with H being the total submergence of the airlift pipe $[M L^{-1} T^{-2}]$.

The rate at which work is done by lifting the drilling fluid is

$$W_w = Q_w \rho_w g h ,$$

where

W_w = rate at which work is performed by lifting the drilling fluid $[L^2 M^{-1} T^{-3}]$,

Q_w = rate of flow of drilling fluid $[L^3 T^{-1}]$,

ρ_w = drilling fluid density $[M L^{-3}]$,

h = total hydraulic head loss through the system $[L]$,

g = gravitational acceleration $[L T^{-2}]$.

Note that h includes frictional head loss in flow down the borehole annulus, frictional head losses in flow up the drill pipe, and the net hydraulic lift involved. It is possible to write $h = \phi H$, where ϕ is the fraction of H, the depth of injection. With this formulation, and denoting the pumping efficiency by η, the required volume flux of air is given by

$$Q_a = \frac{Q_w \rho_w g \phi H}{\eta P_0} \ln\left(\frac{\rho_w g H}{P_0}\right) . \qquad (7.10)$$

Ratios of the quantity of air required for a given fluid flow rate are shown in Figure 7.8.

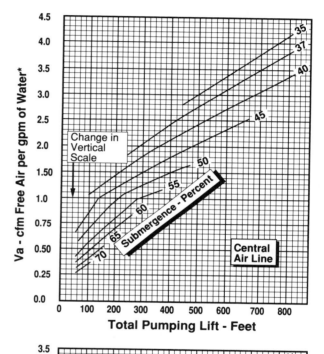

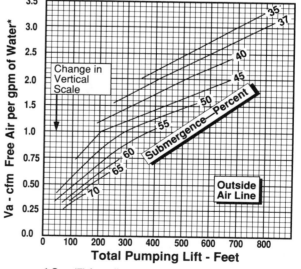

* See (7) for adjustments in Va for water quantity.

Figure 7.8. Air required for fluid flow rate (courtesy of Ingersoll-Rand Corporation).

In general, efficiency is dependent upon the diameter and depth of the well, as well as the actual air flow rate. Field tests show η varying from a high of 0.3 to a much lower value as the air flow velocities begin to exceed 150 to 200 ft/min. Assuming a maximum pumping efficiency η of 0.20 gives a conservative estimate for the air flow required.

Equation 7.10 in a slightly different form has been used as an empirical basis for determining the free air requirement per gallon of water pumped (7). The form quoted by Loomis is

$$V = \frac{L}{C \, \log_{10}[(S + 34)/S]} \, ,$$

where

L = total pumping lift [ft],
S = pumping submergence [ft],
C = empirical constant.

The constant C incorporates efficiency, water density, and atmospheric pressure, as given in the more general result in Equation 7.10. Values of C given by Loomis are presented in Figure 7.9. Note that the value of C depends upon whether the air line is inside or outside the drill string.

Advantages and Disadvantages of the Reverse-Circulation Rotary System

Reverse-circulation rotary drilling has a number of advantages under some drilling conditions. These include:

1. Lower capital cost than equivalent-capacity direct-rotary equipment.

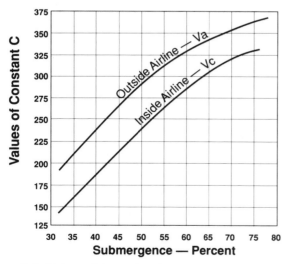

Figure 7.9. Values of C as a function of submergence (courtesy of Ingersoll-Rand Corporation).

2. Most economical system for drilling large-diameter boreholes in soft, unconsolidated alluvial formations.
3. Formation sampling more accurate than with direct rotary.
4. High-return velocity lowers drilling-fluid viscosity requirements.
5. Lower noise levels with insulated compressors.
6. Lower transportation costs than equivalent-capacity direct rotary.
7. Simpler, less costly circulating system.
8. Lower bit costs than with direct rotary.
9. Lower development pumping time where water without additives is used as drilling fluid.

Disadvantages include:

1. Drilling efficiency declines rapidly below 800 to 1000 ft.
2. Large water supply requirements (up to 1000 gpm in highly permeable formations).
3. The system is not suitable for drilling large boulders, consolidated rock formations, and karstic formations. When drilling long sections of clay and shale, drilling fluid additives must be used.
4. *SP* and resistivity logs are not reliable when water without additives is used as the drilling fluid.
5. Maintaining borehole alignment is more difficult than with direct rotary because of the relationship of drill collar diameter and weight to the large-diameter borehole. This is examined in detail in Appendix E.
6. Boreholes smaller than 18 in. cannot be drilled due to the eroding effect of the higher-velocity fluid down the annulus.
7. Difficult to use where the static water level is less than 15 ft.

7.4 CABLE TOOL METHOD

Cable tool drilling methods have been employed continuously in the drilling of water wells for more than 4000 years. In this system all drilling tools are operated in the borehole on a wire line. The surface equipment with basic components is shown in Figure 7.10. These include hoists (for handling the drilling tools, bailer, and casing), mast, power plant, and means of imparting reciprocating motion to the drilling tools.

The drill string consists of a drill bit, drill stem, drilling jars (optional), sinker bar (optional and used only with drilling jars), rope socket, and drilling line. An alternative string used primarily for drilling unconsolidated alluvium consists of a mud scow, drilling jars, rope socket, and drilling line.

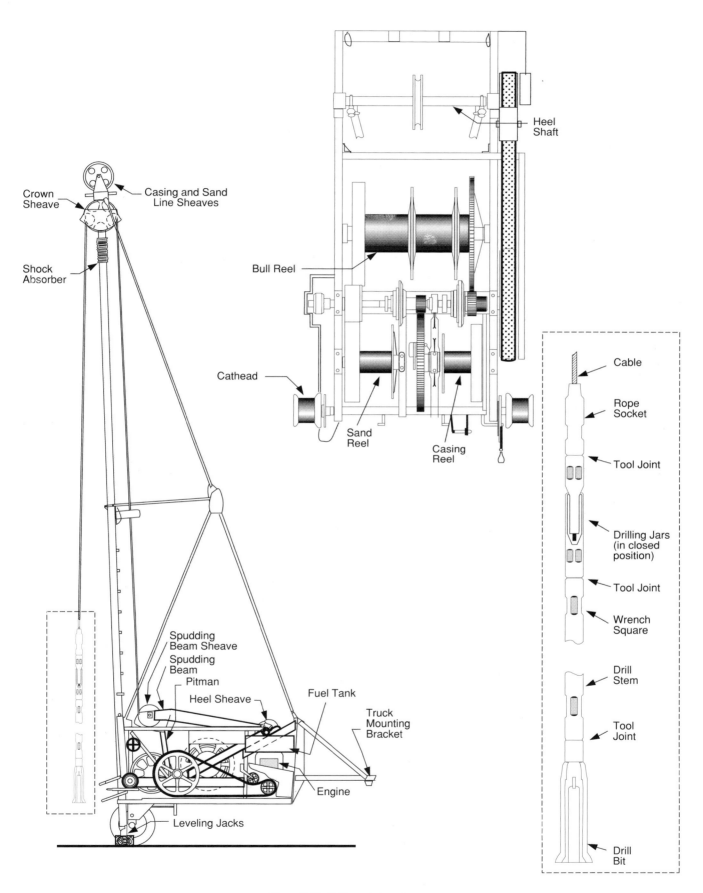

Figure 7.10. Cable tool rig and components.

Crown Sheave

Casing and Sand Line Sheaves

Shock Absorber

Bull Reel

Cathead

Sand Reel

Casing Reel

Heel Shaft

Cable

Rope Socket

Tool Joint

Drilling Jars (in closed position)

Tool Joint

Wrench Square

Drill Stem

Tool Joint

Drill Bit

Spudding Beam Sheave

Spudding Beam

Pitman

Heel Sheave

Fuel Tank

Truck Mounting Bracket

Engine

Leveling Jacks

Repeatedly raising and dropping the heavy drilling tools onto the bottom of the borehole causes the impacted formation material to fracture or loosen and mix with the fluid in the borehole. Drill cuttings size is a function of formation type and hardness. They range from fine-grained rock, ground to small particles by the bit, to alluvial gravels removed intact by a mud scow. Since no fluid is circulated in the borehole, cuttings must be removed by bailing at relatively short drilling intervals (4 to 10 ft). The drilling process is thus intermittent, alternating with cuttings removal.

The cable tool bit is a shaped steel bar, generally 4 to 8 ft long with a tool joint on top and a bottom design that varies with the formation drilled. Sharper bits are used in hard-rock drilling, whereas soft-formation bits are relatively blunt. Bits require frequent sharpening (dressing) by forging or electric-arc welding. Cable tool bits are made by various manufacturers from high carbon and molybdenum-silicon alloy steels in a number of patterns. Larger-diameter bits (12 in. or more) can be cast from SAE 4140 steel. When drilling hard abrasive formations, durable tungsten-carbide inserts can be added to maintain bit diameter.

The drill stem, a cylindrical steel bar generally 10 to 20 ft long, is connected directly to the bit. Its diameter and length depend on the hole size and amount of weight desired. This tool stabilizes and adds weight to the drill string.

Jars are heavy steel links that telescope within each other and connect the stem to the rope socket. If the bit sticks in a soft formation, jars produce a sharp upward blow on the tools to release them. Drilling jars normally have strokes of 9 to 18 in. and are often omitted in hard-rock drilling.

Tool joints used in both cable tool and rotary drill strings are tapered, coarse-threaded, right-hand connections machined in accordance with API standards. The thread design allows fast makeup and long thread contact, with necessary tightness obtained from metal-to-metal fit at the flat shoulders of the joint. Proper joint tightness is essential to prevent the severe vibrations during drilling from either unscrewing the tools or further tightening them (making disassembly difficult).

The rope socket may be a swivel or KAK ratchet type. Either type facilitates the bit rotation caused by varying tension in the drilling line during each stroke.

Drilling Lines

Most cable tool rigs have three lines or cables used for various purposes. These are the sand line, calf (casing) line, and drilling line.

The sand line, spooled on the sand reel, passes over a sheave at the top of the derrick and normally attaches to a bailer. The bailer stands to one side of the derrick floor while drilling is under way. The drill string must periodically be removed from the hole and the bailer run to remove accumulated cuttings. Sand line load and service requirements are light compared to drilling, allowing use of smaller-diameter wire rope. Usually $\frac{7}{16}$ to $\frac{5}{8}$-in., 6×7 construction, improved-plow steel cables are used.

The casing line, spooled on the casing reel, is used to run casing into the well. Line loads are controlled within desired limits by varying the number of lines strung between the block and derrick. Casing lines are normally $\frac{3}{4}$ to 1-in. diameter, 6×25 construction, improved-plow steel cables.

The drilling line, spooled on the bull reel, passes around the heel sheave, under the spudding-beam sheave, and over the main-line sheave at the top of the derrick. The spudding beam, driven by the pitman and crank, imparts reciprocating motion to the drilling line. This line is subjected to extremely severe service. It supports not only its own weight but also that of the tools. It is also subjected to the jarring, fluctuating loads of drilling, as well as severe abrasion against the casing and borehole.

The most common type of cable used for this service is a 6×19, fiber-core, plow or improved-plow steel cable. Diameters of $\frac{3}{4}$ to $1\frac{1}{8}$ in. are normally used with larger-diameter boreholes, although larger tapered strings may be found on deep wells. Tapered strings consist of sections of different line diameters, with the largest at the top and the smallest at the bottom. This practice allows lower beam loads and places the largest diameter line at the top where the tensile load is greatest.

Bailers and Sand Pumps

Bailers for removal of cuttings consist of a pipe equipped with a bail at the top and a valve at the bottom. The bottom valve, generally a disc or flapper type, is opened and closed as the bailer is alternately raised and lowered. This motion causes the cuttings to wash into and be trapped within the bailer until emptied at the surface. It is sometimes necessary to use a sand pump to recover coarse cuttings in the absence of drilling fluid having adequate gel properties. This device is similar to a bailer except that it incorporates a piston which, when pulled up inside the bailer above the valve, creates sufficient vacuum to draw the cuttings and fluid into the tube.

Mast

The mast must afford the vertical clearance and capacity necessary for conducting such operations as handling tools and running casing. Mast loads vary with the depth and diameter of the hole and casing. Some, used in drilling deep, high-capacity wells, are designed to handle up to 250,000 lb.

Dynamics of Cable Tool Drilling

Successful cable tool drilling depends upon imparting a reciprocating motion that maximizes the bottom-hole impact of the drilling tools. The bit, stem, jars, rope socket, and cable all have mass that throughout the drilling cycle is supported by the elastic drilling line. The objective is to deliver maximum kinetic energy to the bottom of the hole.

Although the process appears straightforward, experience has shown that efficient drilling depends upon raising and dropping the tool at just the appropriate frequency for the drill string weight, cable size, and depth of hole. It is somewhat surprising that there is presently no definitive method for determining the most effective frequency, other than the driller's judgment of tool action evaluated by the "feel" of the line.

There have been attempts to find mathematical principles that define the drilling operation. The most notable of these are by Sprengling and Stephenson and by Bonham (8, 9). In their analyses, the basic concept of the drill string behaving as a vibrating spring-mass system was recognized. When a heavy mass is suspended on a spring, there is a natural frequency of vibration for the system that is defined by the weight of the mass and the elasticity of the spring. This frequency is given by the formula

$$f_w = \frac{30}{\pi} \sqrt{\frac{gk}{W}} ,$$

where

f_w = oscillation frequency [cycle/min],
g = gravitational acceleration [ft/sec^2],
W = weight of the tool [lb],
k = spring constant of the cable [lb/ft].

The spring constant k is defined by the formula

$$k = \frac{EA}{L} ,$$

where

E = the elastic modulus of the cable,
A = cross-sectional area of the cable,
L = length of cable.

This analysis does not include the dynamic motion of the heavy cable directly, but Timoshenko suggested it could be accounted for by adding one-third the weight of the cable to the weight of the tools (10). Adding this cable weight clearly reduces the natural frequency of oscillation.

This analysis is based on the assumption that if the system is driven at the natural frequency of oscillation of the drilling tools and cable, maximum energy will be transmitted to the tools through the resonance of the motion. The idea is fun-damentally sound, but as will be discussed, the analysis ignores the fact that cable dynamics and mass play an important role after the tool impacts the formation.

Bonham's analysis of cable tool drilling was based on an energy argument that relates the kinetic energy of the drilling string to the strain energy stored in the stretched cable. He believed that the energy available for useful work would be a maximum at some critical tool weight and that this weight would be given by

$$W = \sqrt{\frac{2GAE}{LK^2}} ,$$

where

G = net energy input to the formation (taken by Bonham as $1325d$ ft-lb, with d the hole diameter in inches),
K = ratio of cable tension to tool weight (K was selected as having a value of 2, since this would impart an upward acceleration of g to the tool when the cable reaches the end of the stroke).

The difficulty with this analysis is that there is an assumption that action of the spudding beam is directly transmitted to the tool. The fact that the beam motion is transferred to the tool by elastic waves in the heavy cable is ignored, and the effect of the cable mass on the frequency of system oscillation is not properly included. Nor is the fact considered that oscillation is interrupted by the tool impacting the formation.

The first of these two factors can be accounted for quite readily. It can be shown that a heavy cable oscillating vertically under its own weight has a natural frequency of oscillation (cycles per minute) given by

$$f_c = 15 \sqrt{\frac{gk}{w}} ,$$

where

w = total cable weight.

In order to include the effect of both the cable and drilling tools correctly, it is necessary to solve equations jointly for the elastic waves in the cable and the cable spring and tool mass system. The natural frequency of oscillation is then no longer given by a simple formula but by the solution to the equation

$$\sqrt{\frac{w}{W}}\left(\frac{f_r}{f_w}\right) + \arctan\left[\sqrt{\frac{W}{w}}\left(\frac{f_r}{f_w}\right)\right] = \frac{\pi}{2} ,$$

where

f_r = resonant frequency of oscillation of the combined system, including the cable mass,

W/w = the ratio of the weight of drilling string to weight of cable,

f_w = resonant frequency of the cable and drilling string, ignoring the cable mass.

In the special case where the cable and drilling string are equal in weight, the natural frequency is reduced to $0.83f_w$. From the results of the analysis it is possible to establish the ratio of energy available at the bit in each stroke to the energy input by the beam motion. It is found that

$$\frac{\text{Energy at tool}}{\text{Energy at beam}} = \frac{\sin 2\theta}{\sin 2[\sqrt{w/W}(f_r/f_w) + \theta]},$$

where

$$\theta = \arctan\left[\sqrt{\frac{W}{w}}\left(\frac{f_r}{f_w}\right)\right].$$

This analysis ignores any frictional dampening of the drilling string motion, such as results from submergence in drilling fluid. Inclusion of the frictional dampening is simple in principle but requires evaluation of complex transcendental functions.

It is apparent from the analysis that resonant frequencies are not simple multiples of the fundamental frequency, as suggested by Sprengling and Stephenson. Furthermore, since the dampening induced by the fluid friction is not included, the energy ratio given here is only a qualitative estimate. Fluid friction will not significantly affect the resonant frequency, except to lower it slightly. It will, however, have a marked effect on the amplification of the drilling string motion. This will always exceed the amplitude of the beam motion, provided that beam frequency is less than the resonant frequency of the system. The principle is analogous to the

operation of a child's swing, wherein a small push, if correctly timed, can keep a high swing amplitude. Swing amplitude is ultimately limited by the friction in the system. In the same way drilling string motion is limited by drilling fluid friction and the energy expended in fracturing the formation.

This simple resonant frequency analysis also ignores another important point. If any energy is to be transmitted to the formation being drilled, the bit must have some velocity when it strikes the formation. The motion of the drilling string is therefore not a simple harmonic one, since the harmonic motion will be interrupted at the instant of impact.

In practice, the bit strikes the formation before the cable has completely stopped. The cable's inertia causes its motion to continue. When it finally stops, its kinetic energy is stored in the strain of the slack cable. At this point the cable, which was originally accelerated by the weight of both the cable and drill string, is stretched by its own inertia. The cable then retracts, both from the upward motion of the beam and the elastic energy stored in the stretch. It accelerates vertically, pulling the drilling string with it as its length is reduced. Cable mass is therefore an important part of the system. The beam frequency must be intimately tied to the frequency of oscillation of the waves in the cable. If it is moved at the wrong frequency, it can work completely counter to the motion of the cable and result in little energy being transmitted to the bit.

Although this analysis does not properly account for bit impact, it provides a reasonable estimate of the beam frequency that will be appropriate for a given drilling situation. To see how this works, consider the following example in which the drilling system includes a 1-in. cable with elastic modulus $E = 16 \times 10^6$ psi, cross-sectional area 0.405 in.^{2}, and weight of 1.6 lb/ft. Table 7.3 gives the resonant frequency in cycles per minute for a given drilling depth and drill string weight for this cable. f_r is the resonant frequency, f_w is the resonant frequency ignoring the cable weight, and f_c is the resonant frequency ignoring the tool weight. The resonance of the system is fundamental to the low energy consumption and high efficiency of the cable tool drilling system.

TABLE 7.3 Resonant Frequencies of Cable Tool Drilling String as a Function of Cable Length and Tool Weight (in cpm for 1.6 lb/ft Cable)

Cable Length (ft)	Weight of Tools (lb)								
	0	1000		2000		5000		10,000	
	f_c	f_r	f_w	f_r	f_w	f_r	f_w	f_r	f_w
500	343	173	195	129	138	85	87	61	62
1000	171	110	138	86	98	59	62	43	44
2000	86	66	98	55	69	40	44	29	31
5000	34	31	62	28	44	22	28	17	20
10,000	17	16	44	15	31	13	20	11	14

Note: f_r is resonant frequency including cable weight; f_w is resonant frequency excluding cable weight.

Casing Installation

The amount of open borehole that can be drilled with cable tools depends upon borehole stability, the effect of line whip and the ability to keep cuttings in suspension. The possibility of being able to drill a limited amount of open borehole makes it prudent to begin construction with a larger-diameter casing than is required to house the pump. Ultimately, the borehole may have to be cased to allow further drilling. Casing programs can be carried out in several ways.

When installed in an open borehole, casing diameters are 2 to 4 in. less than the borehole diameter. If borehole instability occurs at greater depths, additional strings can be installed with a suitable lap (10 to 20 ft) inside the upper string. Each diameter reduction is normally 2 to 4 in.

Since there are practical limitations to the number of diameter reductions, techniques have been developed to facilitate completion with fewer strings. With formations that tend to drill oversize, casing can be lowered or driven from the surface as drilling proceeds. Where formations drill to gauge, the borehole can be underreamed to a diameter large enough to clear the shoe. In deeper wells, friction between the casing and borehole wall will tend to build to a level exceeding the weight of the casing string. In this event casing can be advanced by driving. Casing installation can be materially enhanced by maintaining a bentonite slurry in the annulus (if the casing is installed from the surface). This reduces wall friction and also seals the casing.

When cable tool equipment is used to drill wells in unconsolidated alluvial formations, the amount of open hole that can be drilled is limited to a few feet. In this situation the casing must continually be advanced to support the walls of the borehole. The conventional drilling string is used when drilling large boulders, or above the water table where alignment is critical. In well-defined unconsolidated alluvial formations below the water table, a system using a mud scow for drilling and hydraulic jacks for casing installation has proved to be more efficient. A mud scow resembles a bailer but is of much heavier construction, with the bottom reinforced by a heavy ring protected by hard facing. As the scow strikes the bottom of the borehole, it crushes the formation. The loosened material is lifted by the suction created when the scow is raised with the flapper valve closed. When it drops, the flapper valve opens and traps the suspended cuttings. More than half the length of the scow barrel can be filled with cuttings before it becomes necessary to empty it.

Use of hydraulic jacks requires an anchor hole, typically 15 ft long, 7 ft wide, and from 12 to 15 ft deep, as shown in Figure 7.11. Anchors fabricated from heavy steel channels are placed at the bottom of the hole. Thick planks are used to brace and cover the anchors when the hole is back-filled. Hydraulic jacks are attached to the anchors with heavy eye bolts. Either two or four casing jacks are set, depending upon the expected difficulty of moving casing. Links at the top of the jack stems fasten over a heavy steel cap on top of the two-ply well casing (see Chapter 11), or if longer lengths of casing are used, to an inverted heavy-duty spider-and-slip arrangement.

A 3000-psi positive-displacement water pump operates the jacks. Pull-downs of several hundred tons achieved by this system can force casing to depths greater than 1500 ft. The possibility of joint separation is remote, since the casing remains in compression (see Appendix G).

The mud scow and jacking system is very efficient in moving casing under very difficult drilling conditions and through loose formations below static water level. Frequently, it is possible to advance casing to much greater depths without reduction than with drilling and driving, since the vibration of driving often builds up wall friction. Casing is installed as drilling progresses and high penetration rates are usually achieved. Since no drilling fluid additives are used and samples are removed directly from the bottom of the borehole, this method produces very reliable formation and water samples.

Reduction in Casing Diameter in Unconsolidated Formation

Occasionally, casing cannot be jacked or driven to the programmed depth in unconsolidated formation, and a reduction in diameter is needed. This may be done in two ways. If final depth is known, a reduced diameter string is lowered into the well, with its length sufficient to reach that depth plus a 10-ft lap inside the upper casing. Construction continues by drilling through this casing, using an inside driver to drive it into position. A heavy drive shoe and casing reinforcement protects the top of the reducing string.

If final well depth is uncertain, the reducing string extends to the surface and is either jacked or driven as drilling proceeds. After completion, the string is cut off 10 ft above the bottom of original casing and removed. A tapered cone set on top of the lower casing serves as a tool guide.

Advantages and Disadvantages of the Cable Tool Method

The cable tool method continues to be employed in the drilling of water wells because of its greatest efficiency in a number of geologic conditions, particularly where large-diameter wells are to be drilled in hard formations or under conditions of loss of circulation. Advantages of this method include:

1. Lower capital cost of equipment.
2. Lower daily operating cost, including lower fuel consumption and lower maintenance and personnel expense.

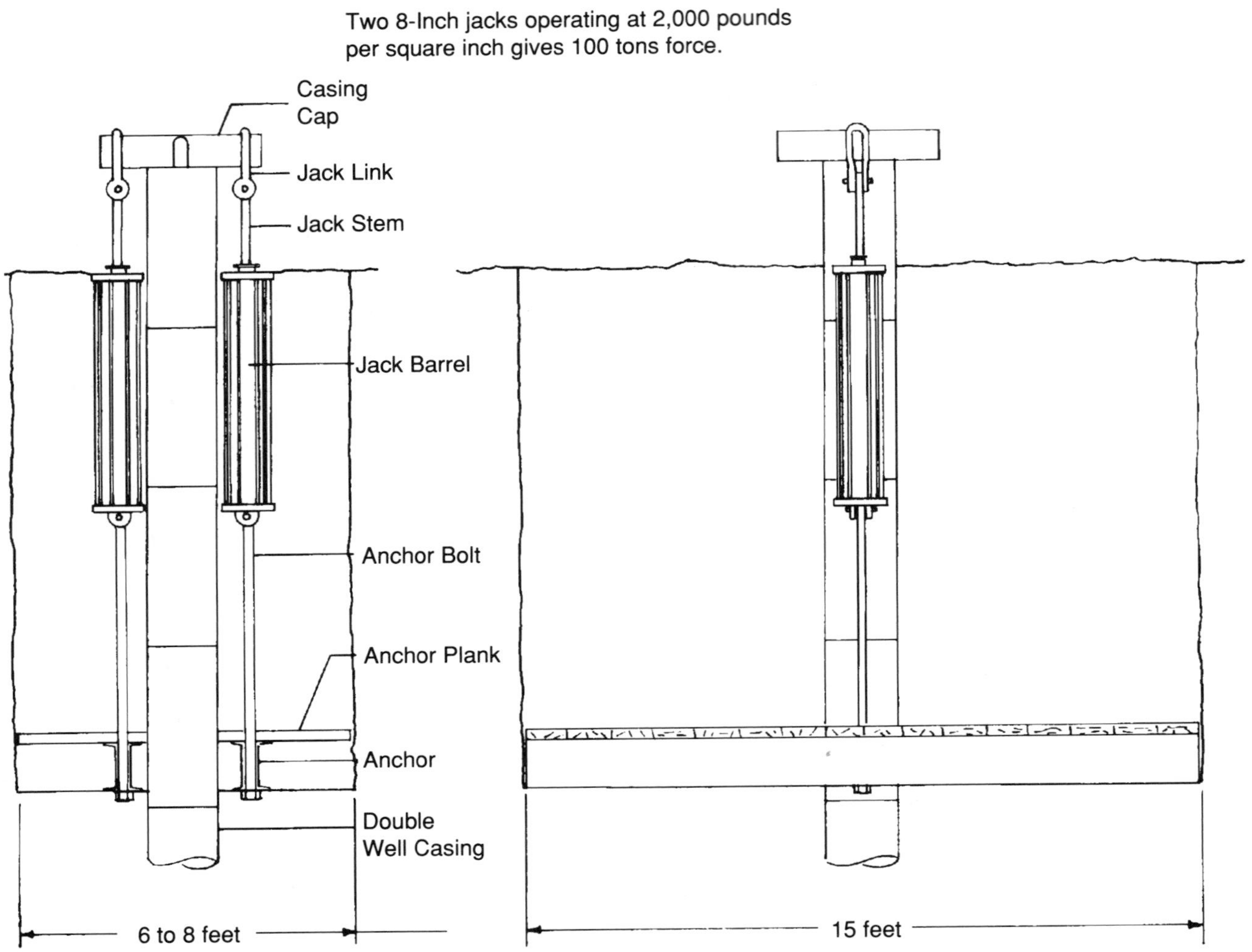

Two 8-Inch jacks operating at 2,000 pounds
per square inch gives 100 tons force.

Figure 7.11. Typical jacks and anchors arrangement.

3. Makeup water requirements when drilling above static water level that may be as low as 50 to 100 gal/hr.

4. Lower transportation costs.

5. Use of casing rather than drilling fluid to stabilize formations.

6. Simpler well design and generally more durable and maintenance-free well operation.

7. More reliable formation and water sampling.

8. Better ability to seal off undesirable zones.

9. Less chance of aquifer contamination while drilling.

10. Easier to operate in remote areas, because of operating simplicity, size of equipment, and low requirements for personnel and logistical support.

11. More efficient drilling operations in karstic formations, where loss of circulation is a severe problem.

Some disadvantages include:

1. Inability to drill open borehole when gravel envelope completion is required, except in shallow wells.

2. Possibly more costly, heavier wall, and larger-diameter casings.

3. Declining drilling efficiency with greater depth.

4. Relatively low penetration rates in soft, fine-grained alluvial formations.

7.5   VARIATION OF THE BASIC DRILLING SYSTEMS AND SPECIAL DRILLING TOOLS

Down-hole Hammer Drill

The down-hole hammer, illustrated in Figure 7.12, is a pneumatic hammer and drill bit combination designed prin-

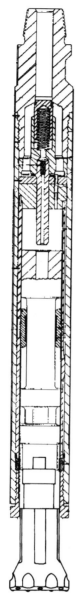

Figure 7.12. Downhole hammer.

cipally for hard-rock drilling. The tool is operated at the end of a standard rotary drill string. Drill collars are required only to maintain borehole alignment because relatively little weight is carried on the bit. The string is rotated at speeds from 10 to 30 rpm in order to maintain alignment and expose the formation to the uneven bit cutting surface.

The air-operated hammer delivers percussion blows to an alloy steel bit fitted with tungsten carbide inserts that fracture the rock formation by impact. Carbide inserts can be sharpened or replaced when they are worn.

This system achieves very high rates of penetration in consolidated hard rock and is the method of choice for

drilling up to 8-in. low-capacity domestic wells in granite and metamorphic rock. Since cuttings are removed continuously, the bit is always striking a clean surface. Requirements for air and admixtures are the same as for direct rotary drilling using air as the circulating fluid, except that 100 to 200 psi minimum pressure is required to actuate the hammer.

Boreholes drilled by the down-hole hammer are generally limited to 8 in., but tools are available in diameters up to 16 in. Large quantities of air are required with these larger diameters unless foam is used.

Dual-Tube Reverse Circulation Rotary System

The dual tube is a closed reverse-circulation system that is particularly useful in acquiring accurate formation and water samples, drilling through loss of circulation zones, and drilling large-diameter boreholes with comparatively low circulation volumes. Applications include test holes and water wells, slim-hole mineral-exploration drilling, and large-diameter shaft drilling.

The dual-tube drill pipe consists of two concentric tubes. The outer is similar to a conventional drill pipe. The inner is made of tubing with a slip-fit pin and box designed to seal the end connections. It is constructed in such a manner that both ends are fixed between shoulders within the outer tube.

Air-, water-, or freshwater-base drilling fluid, as described in Chapter 8, can be used. Drilling fluid is pumped into the annulus between the inner and outer tube, as shown in Figure 7.13. A standard tricone bit is attached to a sub connected to the inner tube. The drilling fluid passes through ports in the sub to the borehole-outer tube annulus. The fluid then flows through passages in the bit into the inner tube, carrying cuttings from the bit face. When a down-hole hammer is used, air flows down the hammer through the bit and up around the hammer, carrying the drill cuttings through a port in a crossover sub to the inner tube.

In a variation of the dual-tube drilling system, air may be injected for circulation by airlift pumping. Usually, the air is added through the swivel, and it passes down the dual-tube annulus through the orificed ports of the jet sub and into the inner tube. The air rises through the drilling fluid, creating a mixture of air, fluid, and cuttings. The heavier column of fluid in the borehole flows through the drill bit ports into the inner tube, picking up cuttings from the bit face, as illustrated in Figure 7.14.

Air volume requirements are based upon the inside diameter of the inner tube, the borehole diameter, the optimum drilling rate, the efficient velocity range, and the best operating range of lift and submergence. The starting submergence (the distance from the static water level to the air inlet) controls the maximum depth of drilling. With a compressor

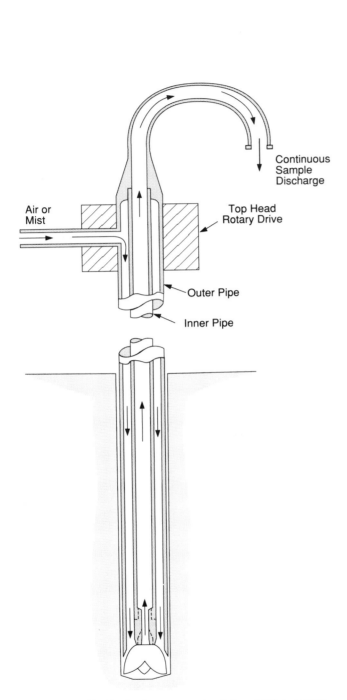

Figure 7.13. Dual-tube reverse-circulation drilling system.

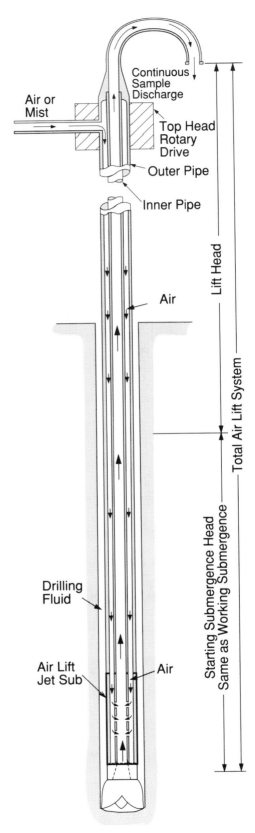

Figure 7.14. Dual-tube air-lift drilling system.

pressure output of 125 to 200 psi and 50% to 70% submergence, the total depth is limited to approximately 500 ft. Compressor boosters are available that can extend depth capability up to 1500 ft.

Samples are taken from air systems with a centrifugal separator. With liquid-base systems, samples are collected by standard methods.

The dual-tube closed-reverse-circulation system has a number of advantages, particularly for ground water exploration.

1. Reliable formation and water samples can be taken from an open borehole. Water samples are accurate because the fluid does not flow down the borehole annulus to the bit.

2. Approximations of aquifer yield can be made under favorable conditions.

3. With high penetration rates and elimination of sealing or packer requirements for water sampling, the cost of drilling and sampling is relatively low.

4. Loss of circulation problems experienced with conventional hydraulic rotary systems are materially reduced.

5. Since the bit diameter is only about 2 in. larger than the drill string, the string itself helps maintain borehole alignment.

Top-Drive Rotary Using Drill-Through Casing Driver

A direct-air rotary system with the capability of simultaneous driving of casing is very efficient for drilling small-diameter wells and exploratory holes in a variety of formations. Drill pipe and casing lengths must match and be assembled and installed as a unit. A drive shoe is fabricated to the bottom of the casing. The casing, seated in the driver anvil, is driven by an air-activated piston, as illustrated in Figure 7.15.

The casing is advanced more-or-less coincident with drilling. The size of the open hole is maintained according to the rate of penetration and cuttings removal. When sampling is required, the casing is driven 0.5 to 1.5 ft ahead of the borehole while circulation continues.

This system is particularly efficient in drilling smaller-diameter wells through boulders or coarse, highly stratified alluvial deposits. Simultaneous driving of casing ensures borehole stability, eliminates loss of circulation problems, ensures accurate formation and water sampling, and eliminates any need for freshwater-base drilling fluids. To complete the well, a screen is normally installed with pump housing (riser) casing. The driven casing is either pulled or jarred back by the casing driver. Alternately, the driven casing may be perforated in place.

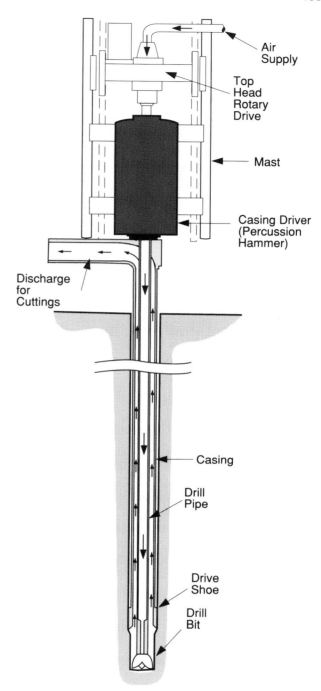

Figure 7.15. Top-drive rotary with casing driver (courtesy of Aardvark Corporation).

REFERENCES

1. Lyons, W. C. 1984. *Air and Gas Drilling Manual.* Gulf Publishing, Houston, TX.

2. Angel, R. R. 1958. *Volume Requirement for Air and Gas Drilling,* Gulf Publishing, Houston, TX.

3. Scott, J. O. 1957. "How to Figure How Much Air to Put Down the Hole." *Oil and Gas J.,* December 16, 1957.

4. Driscoll, F. G. 1986. *Groundwater and Wells.* Johnson Division, St. Paul, MN.

5. Griot, M. 1979. "Le Systeme de pompage air-lift et la remontee des modules polymetalliques marins." *La Houille Blance*, no. 6/7.

6. Heywood, N. I., R. A. Michalowicz, and M. E. Charles. 1981. "A Preliminary Experimental Investigation into the Air-Lift Pumping of Shear-Thinning Suspensions." *Can. J. Chem. Engr.* 59 (February).

7. Loomis, A. W. (ed.). 1982. *Compressed Air and Gas Data.* 3rd ed. Ingersoll-Rand Company, Washington, NJ.

8. Sprengling, K., and E. A. Stephenson. 1940. "Cable Tool Drilling, API Drilling and Production Practices," pp. 64–72.

9. Bonhan, C. F. 1955. "Engineering Analysis of Cable Tool Drilling." *The Petroleum Engineer* (December), pp. B-93–B-100.

10. Timoshenko, S. 1941. *Strength of Materials: Advanced Theory and Problems.* Part II. D. Van Nostrand, New York.

READING LIST

Brantly, J. E. 1948. *Rotary Drilling Handbook.* Palmer Publications, New York.

Campbell, Michael D., and J. H. Lehr. 1974. *Water Well Technology.* McGraw-Hill, New York.

Craft, Benjamin C., W. R. Holden, and E. D. Graves, Jr. 1962. *Well Design: Drilling and Production.* Prentice-Hall, Englewood Cliffs, NJ.

Gatlin, Carl. 1964. *Petroleum Engineering: Drilling and Well Completions.* Prentice-Hall, Englewood Cliffs, NJ.

McCray, Arthur W., and F. W. Cole. 1960. *Oil Well Drilling Technology.* University of Oklahoma Press, Norman.

CHAPTER 8

Drilling Fluid

8.1 INTRODUCTION

Drilling fluid can be simply defined as that combination of fluid and solids required in certain drilling processes to facilitate the production and removal of cuttings from a borehole. Commonly, and especially in the field, liquid drilling fluids are referred to as mud. Although in one sense the properties and use of drilling fluid can be considered construction techniques and as such are not within the scope of this book, they are also relevant to well design, logging, completion, and development. For these reasons this important subject is considered here.

A unique technology has been developed for use of drilling fluid in water well construction. This technology restricts drilling fluid additives to those compatible with potable water, which eliminates many used with oil well or other applications. Of great significance is the fact that though the cost of drilling fluid is very small in comparison to the cost of a well installation, a well-designed mud program is vital to the successful completion of an efficient well. This chapter will review the general functions and properties of drilling fluids and their application to water well construction.

8.2 FUNCTIONS OF DRILLING FLUIDS

Drilling fluid should be designed to promote efficient drilling and completion operations and to maximize developed well efficiency. Achieving these objectives necessitates controlling its properties. The varying properties needed for good performance in different formations necessitates a variety of compositions.

Bit Cooling and Lubrication of Rotary Drill Strings

Since the work performed by an uncooled rotary bit will quickly raise its temperature to damaging levels, bit cooling is an important drilling-fluid function. "Bit balling" and high temperature failure can be the result of improper bit design and/or insufficient circulation. Cooling effectiveness of the drilling fluid decreases as the suspended solids content increases.

There is also considerable frictional resistance to the rotation of the drill pipe in the borehole. If no mud were present, the drill pipe would be severely abraded. Use of a circulating liquid reduces the coefficient of friction and dissipates any rotational heat generated. The presence of a slick, low resistance film on the wall of the borehole also reduces rotary torque while drilling, and the friction load when the drill string is pulled.

Bit and Bottom-Hole Cleaning

Effective bit cleaning depends upon a number of factors, including bit design and drilling fluid velocity and properties. Cuttings not removed promptly from the bottom of the hole are reground, resulting in reduced rate of penetration, undesirable drilling-fluid properties, and shortened bit life (1).

Cuttings Transport

The large quantity of material removed in the drilling of a large-diameter borehole is frequently not appreciated. For example, a 30-in.-diameter hole drilled to a depth of 1500 ft requires removal of about 7400 ft^3 of material weighing over 400 tons. The drilling fluid must convey this material to the surface.

The cuttings transport rate in a rising drilling fluid depends on the difference between its upward velocity and the cuttings settling rate. The settling rate is dependent upon particle size and shape of the cuttings, the density difference between these particles and the drilling fluid, and the viscosity and thixotropic properties of the drilling fluid. Thixotropy is defined as a reversible isothermal transformation of a colloidal sol to a gel.[1] Such gels tend to liquefy when stirred and to solidify when standing. This is characterized by an increasing gel strength with time, at a decreasing rate until a maximum value is reached.

Gel development (or thixotropy) and flow properties are closely related in most water-based muds because both depend on the presence of clay particles of colloidal dimensions.

---

1. The jellylike substance of the gel consists of a colloidal suspension that is semisolid in consistency. The sol is a colloidal dispersion composed of a liquid solvent and a solid solute (a dissolved substance).

The force necessary to disrupt the gel structure is called "gel strength." Colloidal particles have both positive and negative charges on their surfaces and edges, and they immobilize part of the water in which they are suspended. While the drilling fluid is in motion, these factors affect viscosity. When stops moving, the attractive forces between the particles create a structure that requires the application of a measurable force to initiate movement. After a period of rest, a thixotropic mud will not flow unless the applied stress is greater than the strength of the gel structure. Gel strength increases when circulation stops, tending to hold the particles in suspension.

If formation particles are reground to a smaller size the settling rate is decreased, and drilling-fluid density and viscosity are increased. However, regrinding cuttings increases costs by slowing the penetration rate. A successful drilling fluid program maintains proper balance between three conflicting requirements: minimum weight, viscosity, and solids content.

Cuttings removal is limited by mud pump (or air compressor) capacity and is directly affected by the dimensions of the annulus between drill string and borehole. Drilling-fluid hydraulics is discussed in Chapter 7.

Cuttings Separation

Cuttings are the particles produced by a drill bit penetrating the formation. The particle size depends upon drill bit design, the drilling program, and formation characteristics. It is measured in microns (0.001 mm, or 0.0000394 in.). As reference points, 200 mesh screen retains particles larger than 74 microns; 325 mesh larger than 44 microns; with particles smaller than 2 microns arbitrarily classified as colloidal (by API). Sand particles are usually 77 microns or larger.

When cuttings are recirculated, their presence adversely affects drilling-fluid properties so that, insofar as possible, they must be removed at the surface. The desired properties for ease of cuttings separation are low viscosity and low gel strength, both contrary to the requirement for transporting cuttings in the borehole annulus.

Structural integrity in a thixotropic drilling fluid impedes the separation of drilled solids. When quiescent, particles will not fall unless the stress created by the difference in density between the particles and the mud is greater than the gel strength. The high shear rates prevailing in mechanical separators (shale shakers and hydrocyclones) promotes the release of solids by reducing structural[2] viscosity (2).

Deposition of Low-Permeability Filter Cake

Ideally, the drilling fluid will deposit a thin, low-permeability filter cake that seals pores and other openings in the formations

2. Structural viscosity represents the resistance to shear caused by the tendency for the particles to build a structure.

penetrated by the bit. If the drilling fluid contains suspended particles of critical size, they will stick at the bottlenecks in the flow channels and form a bridge just inside the surface pores, as shown in Figure 8.1. Once a primary bridge is established, successively smaller particles, down to the fine colloids, are trapped, and thereafter only filtrate (the liquid phase) invades the formation (1). To begin the bridging process, the particle diameter must be about one-third the pore diameter. The first stage formation is called the "mud spurt".

In building a good filter cake, the quantity and quality of solids is important. Solids with flat platelets are highly desirable. Since they are not ubiquitous in nature they must be added in most drilling systems.

With deposition of the filter cake, the liquid phase of the drilling fluid leaves the borehole at a rate dependent on filter cake permeability and the pressure differential across it. Without a filter cake, poorly consolidated formations such as alluviums and glacial drift, will take water from the well at a rate dependent on formation permeability and the pressure differential between borehole and formation. Both liquids and solids enter fractures induced by excessive mud pressure, preexisting open fractures, or large openings with structural strength (such as large pores or solution channels). This condition, known as "loss of circulation," will exist with any opening as long as the difference between the mud pressure and the formation pore pressure exceeds the tensile strength of the formation plus the compressive stresses surrounding the borehole, unless the drilling fluid contains particles large enough to bridge the opening.

Formation Pressure Control

Since formation pressure in water wells is usually equal to the hydrostatic pressure, typical drilling-fluid properties are

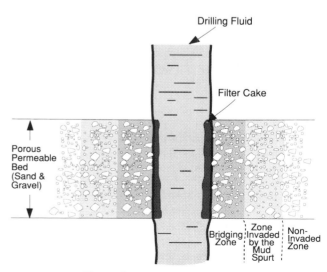

Figure 8.1. Filter cake formation.

usually adequate to prevent caving. If artesian aquifers are encountered, they have higher pressures, depending upon the elevation of their source relative to the surface elevation of the well. Even in this case typical drilling fluids provide ample pressure as long as the borehole is kept filled.

Most difficulties with formation-pressure control in water well construction occur when artesian zones are encountered at a depth of less than 150 ft. The problem is accentuated if the static water level is high, making it difficult to maintain sufficient drilling-fluid pressure to control formation heaving. Mud density may be increased, but often the only solution to this problem is to set and cement the surface and/or the conductor casing through the unstable zone before continuing to drill.

Borehole Stability

Maintenance of borehole stability is important to drilling time, cost, and successful completion. Borehole enlargement may occur in poorly consolidated formations, such as loose sand and gravel, or from swelling or hydration, which leads to disintegration of clays by water invasion. Formation of filter cake on unconsolidated sand results in some stabilization, which increases with lower cake permeability and the hydrostatic head against it.

A distinct filter cake does not form on shale (or clay) zones with hydration. The rate at which water is adsorbed by the shale is influenced by its mineral content, degree of compaction, amount and salinity of water in the shale, drilling fluid salinity, and resistance offered to water invasion at the shale surface. "Shale wetting" can change shale back to clay and is a major cause of borehole instability.

Rate of Penetration

The properties of drilling fluid that affect the penetration rate are density, filtration properties, viscosity, solids content, and size distribution. The most critical property is density. The penetration rate decreases with increased density due to increased pressure differential between the drilling-fluid column pressure and formation pore pressure. This slows down the release of cuttings (chip hold-down effect).

Penetration rate also decreases with high viscosities and high solids content (especially particles less than 30 microns in size). These characteristics decrease the amount of hydraulic energy delivered through the bit nozzles to the bottom of the borehole. The chips should be generated in a regime of turbulent flow. This is a function of velocity and bit nozzle or port design. The rate is also affected by drilling fluid composition to the extent that the bottom-hole hydration rates and bit balling tendencies of some clays are affected.

The generalizations that air drills faster than water and water drills faster than mud are confirmed by years of experience. The low density of air permits rapid equalization of pressure between drilling fluid and formation fluid, allowing

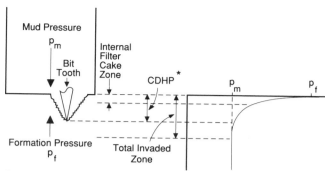

* Chip Hold Down Pressure (CDHP) is approximately equal to $p_m - p_f$, where p_m = mud column pressure and p_f = formation pressure.

Figure 8.2. Bottom-hole filter cake, and pressure distribution ahead of the bit (after Grey and Darley, 1980).

prompt bottom-hole release of chips generated by the bit. Since water has a higher density than air, chip release is slowed. The still higher density of mud, together with fine solids in close contact with the rock fragments, tends to form a filter cake at the bottom of the hole, as illustrated in Figure 8.2. This cake slows pressure equalization and impairs chip removal. It has been found that the drilling rate slows drastically when mud weight exceeds 9.5 lb/gal.

Well Completion and Productivity

Maintenance of proper drilling-fluid properties is important to well completion in several ways. Tight hole, caving, or an excessively oversize borehole may interfere with installation and cementing of casing, and similarly hinder gravel envelope placement. The presence of a thick filter cake is particularly detrimental in placing envelope material. The filter cake should be thin, usually less than 2/32 in. However, unless a filter cake is rapidly formed, solids will enter the aquifer until a constriction is encountered, causing restrictions to water flow at some distance from the well bore.

The filter cake must be removed during well completion and development. The filter cake properties affecting ease of removal are consistency and thickness. A thick, sticky cake seriously hampers pack placement and well completion.

Environment

The drilling fluid used in water wells should not contain any substance that, in the concentrations used, is toxic or injurious or produces objectionable odors, colors, or tastes. The properties of many ingredients commonly used in oil well drilling muds are objectionable in drilling fluids used in potable water wells. For example, oil well muds are commonly thinned with lignosulfonates and lignite. These color the mud and filtrate brown or black, which could require prolonged cleanup if they enter an aquifer. Also organic matter may decompose to produce gases that may be detected by a test for the presence of coliform bacteria.

Construction Cost

A cost/benefit analysis should be conducted in a drilling-fluid program. Design and control significantly affect the success of the entire well construction operation. Sometimes water used without additives can result in the greatest drilling cost. Drilling fluid selection should consider not only performance characteristics related to drilling the anticipated formations and required production, but also personnel and equipment limitations, climatic conditions, material availability, and waste disposal.

8.3   CLASSIFICATION OF DRILLING FLUIDS

For convenience, drilling fluids used in water well construction are classified on the basis of the principle component as:

1. Natural muds (no treatment).
2. Freshwater-base drilling fluids.
3. Air-base drilling fluids.

In addition to the principal component, any composition may contain one of the other fluids together with various dissolved and suspended substances. The primary classifications are divided into subsystems according to composition, characteristics, or application. In general, these systems are not rigidly defined, and often one will be modified with another. For example, air may be injected into a freshwater-base drilling fluid to reduce hydrostatic head in a zone of loss of circulation.

Natural Muds (No Treatment)

The term "clear water drilling" is often used when no materials are added to the water and the cuttings supply the solids in suspension. Solids may make up more than 10% by volume of the suspension, and recirculation of mud weighing as much as 11 lb/gal may occur with clear water drilling.

The importance of maintaining a minimum content of solids has been discussed. For most drilling, mud weight should be kept within a range of 9.0 to 9.3 lb/gal, or lower if possible. The content of solids is especially important while drilling through the aquifer(s) because the incorporation of silt increases the thickness of the filter cake deposited and affects well completion. The effect of solids content on mud weight is shown in Table 8.1.

If a very large settling area is available and the solids concentration is not too high, a flocculant can be used to promote separation of the cuttings. An acrylamide polymer,

TABLE 8.1   Effect of Solids Content on Drilling-Fluid Weight[a]

Solids (%)	Mud Weight (lb/gal)
0	8.33
1	8.47
2	8.60
3	8.74
4	8.88
5	9.02
6	9.15
7	9.29
8	9.43
9	9.57
10	9.70
11	9.84
12	9.98
13	10.12
14	10.25
15	10.39
16	10.53
17	10.67
18	10.80
19	10.94
20	11.08

Source: Courtesy of Baroid Corporation.
[a] Assumed solids specific gravity = 2.65.

thoroughly premixed in water, is added in amounts of about 0.02 lb/100 gal of water, and the flocculated solids are allowed to settle. However, a large settling area is rarely available, making removal of solids by hydrocyclones and shakers the only practical method.

Freshwater-Base Drilling Fluids

A freshwater-base drilling fluid is a mixture of solids, liquids, and chemicals, with fresh potable water the continuous phase. The active solids that react with the water phase usually consist of hydratable clays. Chemicals are added to restrict the activity of these solids, facilitating maintenance of desirable mud properties.

Air-Base Drilling Fluids

Air-base drilling fluids may appropriately be called "reduced pressure" muds because the objective is to drill faster and avoid loss of circulation. As was pointed out earlier, the differential between the borehole pressure and the formation pressure significantly affects drilling progress. One function of drilling fluid, control of pressures, is not supplied by air-base fluids, and their use is restricted to situations where this is not required. Air-base drilling fluids are discussed at the end of this chapter.

8.4 RHEOLOGICAL AND WALL-BUILDING PROPERTIES OF DRILLING FLUIDS

Rheology is the science dealing with the deformation and flow of matter. With regard to drilling fluid, this science is concerned with the relationships of flow pressure and rate to flow characteristics. An analysis of drilling fluid mechanics is presented in Chapter 7. This section will be confined to a description of basic rheological properties. These and wall-building properties are determined largely by the colloidal fraction of the drilling fluid.

Viscosity

Viscosity is the resistance offered by a mud to flow under an applied pressure. More precisely, viscosity may be defined as the frictional drag of one platelet of drilling fluid sliding over another platelet, divided by the relative velocity of the platelets. It is calculated by dividing the shear stress by the shear rate, or the pressure by the velocity. When the shear stress is expressed in dynes per square centimeter and the shear rate in reciprocal seconds, the unit of viscosity is the poise. The unit used for drilling fluids viscosity is the centipoise (cp).

A primary objective in drilling fluid design is to maintain hole stability. If the hole enlarges, higher viscosity and gel strength will be required, resulting in decreased penetration rates. The conflict between these rheological requirements will be minimized by using a shear-thinning mud, which sets to a gel sufficient to suspend cuttings when circulation is stopped but which breaks up quickly to a thin mud when disturbed. Such a drilling fluid will have a high yield point/plastic viscosity ratio.[3] As a rule, it is desirable to maintain the lowest possible plastic viscosity by mechanical removal of drilled solids at the surface, keeping the yield point no higher than required to provide adequate carrying capacity. The yield point is controlled by adding or withholding thinners when drilling in colloidal clays, and by adding bentonite when drilling in other formations.

Flow Properties. Viscosity is expressed in terms of flow models. If a plot of shear stress against shear rate gives a straight line passing through the origin, the fluid is called "Newtonian." Water, glycerine, oil, and salt solutions, are Newtonian fluids. Suspensions containing particles larger than molecules do not show the straight line relationship and are called "non-Newtonian." Drilling fluids are non-Newtonian, and their behavior is commonly expressed in terms of the Bingham plastic or power-law flow model, although their consistency curves may not coincide with either model. Consistency curves are plots of flow pressure

versus flow rate, or of shear stress versus shear rate, as illustrated in Figure 8.3. The equations applicable to these models are:

Newtonian

$$\text{Shear stress} = (\text{constant}) \times (\text{shear rate}).$$

Bingham Plastic

$$\text{Shear stress} = (\text{shear rate}) \times (\text{yield point})$$
$$+ \text{ plastic viscosity.}$$

Power Law

$$\text{Shear stress} = (\text{constant}) \times (\text{shear rate})^n,$$

where the constant is the consistency index, which corresponds to the viscosity of a Newtonian fluid; n is the flow behavior index, which indicates the degree of departure from Newtonian behavior.

Drilling fluids do not show a constant ratio of shear stress to shear rate. Their viscosity depends on the conditions at the time of measurement and can be expressed as effective viscosity at a given shear rate.

The component factors in the viscosity of drilling fluid are:

- The viscosity of the liquid phase.
- The size, shape, and number of suspended particles.
- The forces (electrostatic and gravitational) existing between the particles.
- The forces between the particles and the liquid.

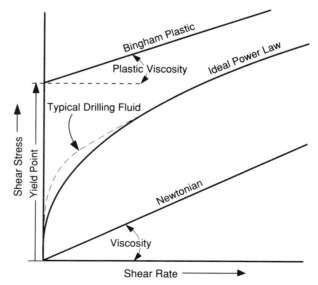

Figure 8.3. Representation of flow models. Source: Gray and Darley, 1980 (Reprinted with permission).

3. Plastic viscosity is defined as shear stress in excess of the yield stress that will induce unit rate of shear.

Shear Stress–Shear Rate Relationships. Field measurement of shear stress–shear rate relationships requires use of a concentric cylinder viscometer, which is not ordinarily used except by mud engineers. However, analysis of these relationships provides insight on modification of mud properties to meet specific drilling requirements.

The Bingham plastic model of flow behavior has been applied for controlling flow properties of drilling fluids. Considering its terms (yield point, plastic viscosity, and shear rate), the yield point is regarded as a measure of the electrostatic forces between the particles while the drilling fluid is in motion. These forces involve both attraction and repulsion. The Bingham yield point is the shear stress required to initiate laminar flow. The yield point is reduced by the addition of chemical thinners that diminish the attractive forces between the particles.

Plastic viscosity can be related to mechanical friction among the solids, friction between the solids and the liquid, and resistance of the liquid to flow. Plastic viscosity is lowered by adding water or by mechanically removing solids. These concepts indicate, for example, that chemical thinners should not be used to lower plastic viscosity when the yield point is already low. The Bingham plastic model better fits the flow behavior of drilling fluids having high solids content than those of either low solids content or those containing polymers.

Shear rate is important in determining flow properties, and it influences such factors as the pressure required to restart circulation after a period of quiescence and the carrying capacity of the drilling fluid. However, significant changes in shear rate occur in the course of circulation, and some time is required for flow conditions to stabilize. Because of this, shear rate is seldom measured in the field, and it has no correlation with factors that are commonly monitored.

Filtration and Wall-Building Characteristics

Filtration properties are a measure of the ability of the solid components of the mud to form a thin low-permeability filter cake. They are intimately related to hole stability, freedom of movement of the drill string, and satisfactory completion of the well. Filter cake formation was described earlier in this chapter.

While circulating, part of the cake is continually formed and washed away. The amount of water passing through the cake is controlled by its permeability. When the erosion of the filter cake by the mud stream becomes equal to the rate of cake deposition, the thickness of the cake and the rate of filtration become constant. When circulation is stopped, filtration continues at a decreasing rate as cake thickness increases.

Aquifer damage from the drilling fluid may result from the entry of liquid or solids or both. During the mud spurt, solid particles enter the openings in the formation (see Figure 8.1). After the filter cake has formed, these particles may be moved and rearranged by the invading filtrate. Unconsolidated sands and fractured rocks may be severely damaged by fine solids unless bridging material of suitable size is present, although sufficient bridging particles are almost always provided by the cuttings. This again emphasizes the necessity of quickly forming a thin and impermeable filter cake.

When shales and clays imbibe freshwater, their cohesion is destroyed. Imbibition of water is much slower, and disintegration is retarded with use of low-filtration muds. This stabilizing effect is most evident when certain organic polymers are present in the system.

Although a filter cake is not formed on shale, field experience has established that borehole stability is improved by lowering the filtration rate. This observation is supported by laboratory studies of shale disintegration in freshwater as compared with low-filtration muds.

8.5 GENERAL CONSTITUENTS OF FRESHWATER-BASE DRILLING FLUIDS

The physical and rheological properties of drilling fluids depend on the composition of their interactive constituents. Water is always involved in these reactions, surfaces play an important part, and clays are a significant ingredient.

Water

Water is the most important constituent in drilling-fluid technology. Among the unusual properties of water compared to other liquids are:

- Highest surface tension.
- Highest dielectric constant.
- Highest heat of fusion.
- Highest heat of vaporization.
- Superior ability to dissolve a variety of substances.

The ionization capacity of water is low, but it dissolves many substances that ionize (acids, bases, and salts), and water molecules cluster around both cations and anions. The special molecular structure attributed to water accounts for its unique properties. The formula H_2O does not suggest the nonlinear, unsymmetrical structure of the molecule. Since the two hydrogen nucleii and the midpoint of the oxygen atom form an angle of about 105° the molecule is strongly polar (like a magnet). This polarity leads to the formation of structures through "hydrogen bonding."

The bonding of water to a surface, or an ion, is called "hydration," with the extent of hydration dependent on the atomic structure of the surface involved. Ionic hydration reduces the attraction between oppositely charged ions. Hydration, ionization, and surface forces control such measured mud properties as viscosity, gelation, and filtration.

Colloids

Drilling fluids frequently are referred to as colloidal systems, although the greater part of the suspended matter has larger than colloidal dimensions. Principles involving reactions at surfaces (surface chemistry) largely determine the behavior of drilling fluids. These surface reactions are solid–liquid, liquid–liquid, liquid–gas, and solid–gas.

Colloids are particles intermediate in size between molecules and suspended particles visible in a powerful optical microscope. This size range has been characterized as being so small that the particles are affected by the movement of the molecules (Brownian movement) and so large they will reflect a beam of light (observed as the Tyndall effect).

A suspension contains solids, but a solution contains ions. There is an arbitrary distinction between a colloidal suspension and a solution, based on settling rate. The settling rate depends on particle shape and the medium in which settling takes place. Forces acting between the particles and between the particles and the fluid in which they are suspended affect the results.

The term "colloid" was coined by Thomas Graham in 1861 from the Greek word for "glue" and was meant to distinguish the behavior of a substance like gelatin from that of a crystalline material like quartz, when suspended in water. The term "hydrophilic" was applied to "water-liking" substances, such as gelatin, distinguished from "hydrophobic" or "water-fearing" inorganic colloidal suspensions, such as clay. The "hydrophilic colloids," such as gums and starches, are usually called "macromolecular colloids" or "polyelectrolyte solutions." The term "association colloids" is applied to soap solutions and similar compounds in which large units of associated molecules ("micelles") are in equilibrium with dissolved molecules.

Particles may be brought to colloidal size from larger sizes by dispersion or fine grinding, or from molecular sizes by aggregation or polymerization. Grinding a particle of a given volume increases the surface area but does not change the volume. As the size of a particle is reduced by grinding, an electrostatic charge is created as chemical bonds are broken. Individual particles may have both positive and negative charges on exposed surfaces. The charge on the particle is that which predominates. When suspended in water, the particle will move toward an electrode of opposite charge. It will be neutralized by ions, or other colloidal particles of opposite sign, and numerous particles clump together or flocculate. Stability is lost when there is no charge.

Polymers, such as starch and polyacrylates, are built up from simpler units, called monomers which are colloidal in size. Not only the size but also the way the units are assembled affects the properties of the polymer. The stability of organic polymers is largely dependent on their ability to retain water. Thus a starch suspension can tolerate a far higher salt concentration than can a clay suspension because the starch is more highly hydrated.

Surface-active agents, commonly called surfactants, are colloidal substances that are adsorbed on surfaces and at interfaces. Adsorption is a type of adhesion that results in a difference between the concentration of a component at the surface of a solid or liquid phase and the concentration in the interior of the adjacent phase. Surfactants are used in drilling as emulsifiers, foamers, defoamers, detergents, and wetting agents, and to diminish hydration of clay surfaces. They are classed according to their dissociation into "anionics" having a large organic anion and an inorganic cation (e.g., sodium alkyl aryl sulfonates); "cationics" having a large organic cation and an inorganic anion (e.g., lauryl pyridinium chloride); and "nonionics," which are long-chain polymers that do not dissociate (e.g., polyoxyethylated nonyl phenol).

Clay

The technology of freshwater-base drilling fluid is largely involved with the chemistry of clays. When drilling with water, formation clays become suspended. Clay present in the producing formations may be chemically altered or physically moved by the invading filtrate.

The term clay is variously defined as follows:

1. As a rock, a fine-grained earthy substance having plastic properties, or if consolidated, shale.
2. As a composition, mainly composed of clay minerals that determine its properties.

The division between colloids and silt shown in Figure 8.4 is arbitrary and indefinite because colloidal activity depends on the specific surface, which varies with particle shape, and on surface potential, which varies with atomic structure. The upper limit of the particle size of clays is defined by geologists as 2 microns, so that virtually all clay particles fall within the colloidal size range. As they occur in nature, clays consist of a heterogeneous mixture of finely divided minerals, such as quartz, feldspars, calcite, pyrites, etc., but the most colloidally active components are one or more species of clay minerals.

Clays are formed by the mechanical and chemical disintegration of igneous rocks, and the term does not imply a specific composition of matter. Clay minerals are hydrous aluminum silicates usually containing alkalies (sodium, potassium), alkaline earths (calcium, magnesium), and iron in significant quantities. In addition to these clay minerals, quartz, feldspar, calcite, and other minerals contained in

Equivalent Spherical Radius of Particles, Microns

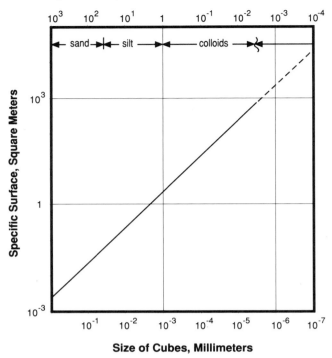

Figure 8.4. Specific surface of cubes. Source: Grey and Darley, 1980 (reprinted with permission).

the original rock are found in clays. Two broad classes of clay mineral structures are:

1. Sheet, or micalike thin platelets (e.g., montmorillonite, illite, kaolinite).
2. Bundles of laths (e.g., attapulgite, sepiolite).

The micalike group is identified by X-ray analysis on the basis of the number of sheets of atoms that form a single platelet, as two-layer unit cells or three-layer unit cells. Each layer is composed principally of alumina (Al_2O_3) or silica (SiO_2), although the substitution of other atoms for aluminum or silicon is common. Such substitutions profoundly affect the properties of the resulting clay minerals.

The variable compositions included in the term "clay" means that the behavior of clays can only qualitatively be interpreted in terms of the dominant clay mineral. The clay "bentonite" owes its water-retentive properties to the clay mineral montmorillonite, identifiable in X-ray analysis by its crystalline structure. However, complete evaluation of any bentonite sample requires additional tests.

Several terms used in discussing clays are defined as follows:

Face	The broad flat clay surface
Edge	The thin border of the clay platelet
Aggregate	Clay platelets stacked in a face-to-face orientation
Dispersion	Separation of clay platelets from a face-to-face orientation
Flocculation	Association of clay platelets in an edge-to-face orientation
Deflocculation	Edge-to-face dissociation
Cation	Positively charged ion
Anion	Negatively charged ion

Montmorillonite shows a typical three-layer atomic structure with the unit layers stacked together face-to-face to form what is known as the crystal lattice, as illustrated in Figure 8.5. Thousands of these units make up a single crystal, and the area of the face is large compared to the edge. Water molecules can be adsorbed between the oxygen layers, resulting in swelling of the crystal, partly because of high repulsive potential in the surface of the layers and partly because bonding between each three-layer segment is weak. Montmorillonite also contains iron and magnesium. These atoms, Fe^{2+} and Mg^{2+}, by replacing aluminum, Al^{3+}, leave an excess negative charge on the face of the crystal, which is offset by adsorbed cations, mainly calcium, Ca^{2+}, and sodium, Na^+. These loosely held cations largely control the behavior of bentonite in water (2).

Illite is another three-layer mineral. Illite does not have the crystal expansion of montmorillonite and does not take up water to the same degree. The difference is attributed to the potassium ion (K^+) that holds the silica sheets. Although illite does not swell in the same manner as montmorillonite, it does hydrate by holding water molecules on its surface.

Kaolinite is an example of the two-layer group of clay minerals, being composed of one silica sheet and one alumina sheet. Although Kaolin clays are important industrially in ceramic manufacture, they are objectionable in drilling fluids because they increase solids content without lowering filtration or raising viscosity except in high concentrations.

Mixed-layer clay minerals are common in shales because of their origin and deposition. Clay mineral of this type may consist of interlaminations of swelling and nonswelling units.

Needlelike clay minerals, such as attapulgite and sepiolite, are rarely used in water well drilling. Their primary use is as a suspending agent in saltwater-base drilling fluids.

Reference was made to the replacement of metal ions in the clay crystal and the significance of the exchangeable cations in clay behavior. This process of base exchange may be described as the exchange of ions in solution for others in the clay solid. The extent of the possible substitution is called the "base exchange capacity," and with the identity of the cations involved, it largely determines the clay's usefulness in the drilling fluid.

The base exchange, or cation exchange capacity (CEC), of a clay is expressed as milliequivalents per 100 g of clay. Clays can be roughly grouped on this basis as montmorillonitic

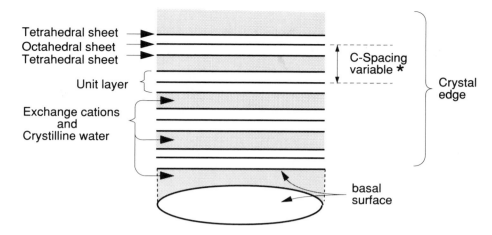

*C-Spacing is the distance between a plane in one layer and the corresponding plane in the next layer.

Figure 8.5. Diagrammatic representation of montmorillonite. Source: Grey and Darley, 1980 (reprinted with permission).

70 to 150 me/100 g, illitic 10 to 40 me/100 g, and kaolonitic 3 to 10 me/100 g. When two ions of different valences are present, the one with the higher valence is generally adsorbed preferentially. The order of preference usually is $H^+ > Ba^{2+} > Sr^{2+} > Ca^{2+} > Cs^+ > Rb^+ > K^+ > Na^+ > Lc^+$. Furthermore particle size is a factor affecting replacement, and sodium bentonite (sodium is the dominant mineral) is composed of smaller particles (larger surface area) and is therefore more reactive.

The base exchange of calcium for sodium occurs when sodium bentonite becomes contaminated by anhydrite ($CaSO_4$) or cement (CaO). The immediate effect is the clumping together of the clay particles (flocculation), as the negative charges on the faces of the clay platelets are diminished by the Ca^{2+} ions and the platelets associate in an edge-to-face arrangement. Water is entrapped in the resulting structure, and the mud thickens. As more calcium ions displace the sodium ions, negative charges are further reduced, and the particles aggregate in a face-to-face arrangement with the release of water from the structure. The solids begin to settle while water rises to the surface.

By addition of bicarbonate of soda ($NaHCO_3$) or soda ash (Na_2CO_3), the base exchange reaction is reversed, the particles are deflocculated, and the suspending properties are restored. The base exchange reaction of sodium for calcium is aided by the formation of insoluble calcium carbonate ($CaCO_3$), thereby lowering the Ca^{2+} ion concentration. Similarly, other deflocculating agents, such as the polyphosphates, remove the calcium ions and allow the base exchange reaction to proceed.

Properties of freshwater-base drilling fluids are interrelated with base exchange reactions. Finely ground dry clay consists of aggregates or packets of flakes. When the clay is placed in water the packets are separated by a film of water. Sub-

sequent behavior depends on the clay mineral composition, the exchangeable bases, and the electrolytes in the water. In the case of calcium bentonite (calcium is the dominant mineral), the packets are wetted by water but do not distintegrate like sodium bentonite. The Ca^{2+} ion restricts the formation of a stable double layer, and there is little dispersion and swelling. In the case of sodium bentonite placed in freshwater, as the particles begin to take up water they appear to swell. The volume of solids may increase tenfold. The flakes are surrounded by a film of water, and water may enter the clay mineral structure, promoting breakdown into smaller units. When sodium bentonite is added to water-containing calcium salts (hard water), base exchange of calcium for sodium occurs and the clay behaves like a calcium bentonite.

Addition of sodium chloride (NaCl) to a sodium bentonite suspension results in flocculation because the accumulation of sodium ions (Na^+) decreases the thickness of the electrical double layer and reduces the hydration of the clay particles. When sodium bentonite is added to salty water, the clay does not hydrate. The development of an electrical double layer is prevented by the highly conductive solution containing Na^+ and Cl^- ions. Sodium bentonite should not be used in drilling through saltwater zones in concentrations greater than about 10,000 ppm NaCl.

8.6 DRILLING-FLUID ADDITIVES

Many substances are added to drilling fluid to improve its effectiveness. Usually the addition is made to overcome a specific deficiency by altering some identifiable property. For example, bentonite is added to raise viscosity and thereby improve cuttings removal. Bentonite also reduces the filtration rate and cake thickness. If the filtration rate is to be reduced without increasing viscosity, a polyphosphate is added with the bentonite. Clearly, it is difficult to classify additives exclusively on the basis of function.

Specifications have been adopted by the American Petroleum Institute for the additives barite and bentonite. Products meeting these requirements are authorized to bear the API monogram, but the standards do not guarantee satisfactory performance under all conditions; nor is lack of the monogram necessarily proof of inferior quality.

Bentonite

The name "bentonite" was applied to an unusual clay of the Cretaceous Age found in the Fort Benton shale in Wyoming. This clay, first sold in the late 1800s as a component of hand soap, swelled in water and had exceptional sealing properties. Later studies showed that it consisted mainly of the clay mineral montmorillonite with sodium as the major exchangeable cation (as has been noted in the discussion of clays). Similar clay deposits are found in South Dakota and Montana. These "western" bentonites had superior mud-making qualities when compared to the "southern" bentonites common throughout the southern United States which had calcium as the dominant exchangeable cation.

For many years the quality of bentonite has been expressed in terms of "yield," which is defined as the number of barrels (1 bbl = 42 U.S. gal) of mud having an apparent viscosity of 15 cp that could be made from 1 ton of clay mixed into freshwater. On this basis western (sodium) bentonite has a yield of 90 to 100, and southern (calcium) bentonite a yield of 45 to 65. Treatment of some southern bentonite with soda ash (base exchange of sodium for calcium) raises their yield to 85–95.

The addition of a small amount of an acrylic polymer to western bentonite (usually with some soda ash) greatly increases the yield, leading to "extra-high-yield," "peptized," or "beneficiated" bentonites. Less than half the amount of these treated bentonites is required to produce the same viscosity as natural western bentonite.

Similar treatment of southern bentonite serves to raise its yield to that of untreated western bentonite. Other properties of the treated southern bentonite are inferior.

Bentonite serves several purposes in freshwater-base drilling fluids:

1. Improves hole cleaning.
2. Promotes hole stability in poorly consolidated formations.
3. Seals porous formations, reducing water loss by forming a thin impermeable filter cake.

Table 8.2 shows the proportions used for preparation and treatment of typical freshwater-base drilling fluid.

TABLE 8.2 Approximate Amount of Wyoming API-Grade Bentonite Added to Freshwater or to Freshwater-Base Drilling Fluid[a]

	lb/100 gal	lb/bbl
Added to Freshwater		
Under normal drilling conditions	30–50	13–22
To stabilize caving formations	60–80	25–35
To stop circulation loss	70–95	30–40
Added to Freshwater-Base Drilling Fluid		
Under normal drilling conditions	10–25	4–10
To stabilize caving formations	20–45	9–18
To stop circulation loss	25–50	10–20

Source: Courtesy of Baroid Corporation.
[a]Method of addition: mix slowly through a jet mixer or sift slowly into the vortex of a high-speed stirrer.

In gel foam systems bentonite strengthens the film, supplies stabilizing properties, and provides the constituents to form an excellent filter cake. Bentonite may also be used alone or with cement in sealing around casing. It is also used for plugging abandoned wells and boreholes.

8.7 ORGANIC POLYMERS

The term polymer is applied to substances made of many similar or repeating units (monomers) that consist mainly of compounds of carbon. When dispersed in water, these substances may appear to dissolve, forming a clear solution. The particles, however, are far greater in size than water molecules. They are characterized as hydrophilic colloids. As polymers rapidly take water into their molecular structure, the particle swells until it is completely surrounded by an immobilized layer of water held by hydrogen bonding.

Organic polymers are especially useful in water well drilling fluids because they furnish many essential properties with a minimum of material. They are effective thickeners, reduce filtration, and promote formation of thin filter cakes. They stabilize clays, allowing the recovery of larger cuttings and better separation at the surface, and act as "friction reducers," lowering pump pressure and reducing the drill-string torque. It should be remembered that polymers alone do not form a filter cake, since cake formation requires the presence of solids.

Breakup and dispersion of cuttings can be reduced by adding certain polymers to the water, such as a partially hydrolized polyacrylamide or a copolymer of vinyl acetate and maleic anhydride. All polymers that form viscous films on the surface of cuttings afford some stability; however, those having anionic activity are most effective. The adsorbed polymer is removed with the cuttings, and sufficient polymer must be added to maintain an adequate concentration.

Organic polymers are available for water well use that are nontoxic, colorless, and biodegradable and that do not decompose to produce objectionable odors or tastes. Not all properties are present to the same extent in all products, and sometimes some of the properties are objectionable. Often polymers are more effective as an additive than as a main component.

Organic polymers must be water dispersable and may come from natural products such as starches and guar gum from seed, may be chemically derived such as sodium carboxymethylcellulose from wood pulp, or may be synthesized from simple molecules such as partially hydrolyzed polyacrylamide. In general, the natural products are subject to decomposition by microorganisms, enzymes, acids, and heat. The modified products, such as the cellulosic polymers, are much more stable. The pure synthetic products are very stable because the carbon-to-carbon linkage in their structure is difficult to break.

Guar Gum.

Guar gum is obtained from the seeds of the guar plant—a hardy, annual, nitrogen-fixing legume. The endosperm is separated from the hull and germ of the seed by grinding and sifting, and then it is packaged as guar gum. Guar gum produces viscous solutions in either fresh or salty water at concentrations of 2 to 5 lb/100 gal. Guar gum flocculates drill cuttings in low concentrations while drilling with water, but large settling areas must be provided.

Guar gum, unless protected by a biocide, is attacked by soil microorganisms, creating objectionable odors and flavors. When it breaks down, the natural gum leaves a solid residue of about 10%. Commercial products based on starch and guar gum usually incorporate a preservative, but fermentation can occur after the biocide concentration is lowered by dilution. Acid produced as the material decomposes is regarded as evidence that the filter cake is breaking up, and methylene blue may be included with the gum as an indicator. In some instances the gas produced by the decomposing gum has confused the interpretation of the test for coliform bacteria.

Sodium Carboxymethylcellulose.

Sodium carboxymethylcellulose (CMC) is made by chemical modification of cellulose which forms the backbone of the polymer. Purified cellulose from wood pulp or cotton linters is reacted with sodium hydroxide and monochloracetic acid. The resulting product is water soluble CMC. Products varying in solubility, solution viscosity, and adsorptive characteristics are produced by differences in the degree of substitution in the cellulose structure and the length of the chain.

CMC is a versatile additive, providing filtration control over products with a wide range of viscosity. It is used in concentrations ranging from 0.5 to 5 lb/100 gal. Certain modifications (referred to as "polyanionic cellulasic polymers") are particularly effective in hole stabilization in clays. The stabilizing mechanism appears to be formation of a very viscous adsorbed film on the clay surface, which retards entry of water between the clay flakes.

Acrylic Polymers.

Because of their distinctive properties, acrylic polymers are finding wider application in drilling fluids. Since they are synthetic materials, they can be assembled in units that contribute to the desired qualities. They may be synthesized by reacting a unit (monomer) with itself to form a homopolymer or by combining different monomers to form a variety of copolymers.

Acrylic polymers are available both as solids and liquids and are sometimes used to supplement the carrying capacity of bentonite. The liquid may be either a solution or a stable emulsion (latex) that can be added to water to produce the desired improvement in drilling-fluid properties. Synthetic polymers are more expensive than natural or modified polymers, and the added cost must be justified by convenience or superior performance. Acrylic polymers do not ferment and are not toxic in the concentrations normally used. Some are not effective in hard water, requiring addition of soda ash.

Acrylic polymers may be used to replace bentonite in well completion since exposure to sodium hypochlorite rapidly shrinks the polymer, promoting completion. Concentrations used vary widely depending upon the specific application, although 0.7 gal of sodium hypochlorite per 100 gal of volume is typical. Calcium hypochlorite (HTH) is also used, although it leaves an undesirable leatherlike and insoluble precipitate.

Complex Phosphates

Complex phosphates (polyphosphates) are the most effective agents for reducing flow resistance and gel development of drilling fluids containing bentonite. The sodium polyphosphates deflocculate and disperse bentonite that has been flocculated by the calcium ion, overcoming the deleterious effects of cement and anhydrite. As a thinner they improve the filtration properties of the drilling fluid. In well development the sodium polyphosphates are used to remove the filter cake from the walls of producing formations.

Three products are primarily used: sodium acid pyrophosphate, sodium tetraphosphate, and sodium hexametaphosphate. Sodium acid pyrophosphate (SAPP) ($Na_2H_2P_2O_7$) in solution has a pH of about 4.2 and consequently is the most effective in overcoming cement contamination, especially when used after treatment with sodium bicarbonate. Sodium tetraphosphate ($Na_6P_4O_{13}$) is more effective for treating anhydrite because its pH in solution is about 7.5. Sodium hexametaphosphate is a glass, with a pH in solution of about 7.

These polyphosphates are colorless, odorless, and nontoxic in the concentrations used. They are effective in low concentrations, from 0.2 to 1 lb/100 gal of mud, although more concentrated solutions may be added directly through the drill pipe to combat bit balling. For well development, a solution containing 10 to 20 lb/100 gal can be used for jetting and surging to remove filter cake.

Polyphosphates are effective dispersing agents. However, their use when drilling thick clay sections may increase clay solids in the drilling fluid and cause formation sloughing.

Surfactants

Surfactants are used in water well drilling as foamers, detergents, and friction reducers. Most of the products sold for these applications are proprietary, and their composition and concentration are not revealed. Frequently, their primary active substance is a sulfonate or sulfate made from hydrocarbons, fatty glycerols, or animal and vegetable oils, although mixtures are common.

Addition of surfactants to the drilling fluid may aid in bit cleaning by promoting release of cuttings. For reasons not fully understood, field experience has shown that some surfactants improve penetration rate and increase bit life.

Barite

Pressure control is rarely a problem in water well construction, and the hydrostatic pressure in the borehole is usually adequate. In fact, excessive pressure is very undesirable. If drilling fluid density greater than 10 lb/gal is needed, barite should be added. Commercial barite is about 4.2 times as heavy as water and nearly 1.7 times as heavy as clay. Even though clay or sand could be used to make the mud heavier, the increased volume of solids is objectionable.

Commercial barite is not pure barium sulfate ($BaSO_4$) but contains various iron minerals, quartz, or calcite and may contain celestite or gypsum. Gypsum and celestite are objectionable in freshwater-base muds. However, API grade barite does not have detectable impurities.

When penetrating an aquifer, a mud weighted with ground limestone may be preferable to barite, since limestone can be dissolved in acid, making filter cake removal easier than with the virtually insoluble barite.

Salts and Alkalies

Soda ash or washing soda (sodium carbonate Na_2CO_3) is used to:

1. Remove hardness from water by precipitating calcium and magnesium salts.
2. Raise the pH of acidic water for improving dispersion of bentonite and organic polymers.
3. Treat cement-contaminated mud.

Soda ash is a white, hygroscopic powder readily soluble in water. A 70% pure grade with a pH of 8.5 to 9.0 is used in the field. Soda ash is not classed as a hazardous chemical, although it may be irritating to sensitive people.

Baking soda (sodium bicarbonate, $NaHCO_3$) is used to counteract cement contamination, with about 0.9 lb required for each pound of cement being treated. It is not a hazardous chemical, and a solution of 7 lb/100 gal of water has a pH of about 8.4.

Caustic soda is a white hygroscopic powder and is a hazardous corrosive substance. It should not be used in water well construction.

Hydrated lime (calcium hydroxide $Ca(OH)_2$) is useful as an anticorrosion agent, particularly if aerated mud is used. Its use in water well construction is extremely rare. It is not considered hazardous, although it is an irritant to the skin, respiratory tract, and particularly the eyes. A saturated solution of about 2 lb/100 gal of water has a pH of more than 12.

8.8 SELECTION OF DRILLING FLUID

General

Drilling-fluid selection is based on analysis of the total drilling program. The factors to be considered are cost, the influence of the drilling fluid on drilling and completion time, program complexity, and transportation costs of materials and waste. The capabilities and limitations of the drilling equipment, including pumps, mixing equipment, and solids-removal apparatus, may require program modification.

Consideration of surface factors affecting drilling-fluid selection must be combined with knowledge about the subsurface. Information can be obtained from drillers' records on bit life, lithology and penetration rates, and geophysical borehole logs. Attention must be given to the likelihood of problems such as unstable borehole and loss of circulation. Drilling-fluid selection must be focused on securing maximum well efficiency and production.

Unconsolidated Alluvium. Direct-rotary drilling fluids are usually produced as a mixture of potable water and western bentonite. Additives are used to counter adverse situations, usually caused by drilling formations that modify fluid properties. Included are barite (to increase density), thinners (to decrease viscosity), polymers (to control rheology and filtration rate), and soda ash for water softening.

Although reverse-rotary drilling is typically initiated by using water, the drilling fluid becomes a natural mud. This mud is seldom treated, and mechanical solids removal is rarely done. Therefore its properties will vary from well to well, depending upon the character of the formations penetrated and upon the volume, number, and surface area of settling pits used.

When drilling sands, gravels, and clays to relatively shallow depths, use of natural mud is usually satisfactory. Complications occur when less favorable formations, such as hydrating clays, shales, or loss-of-circulation zones, are found. Under these conditions the drilling-fluid specification should be similar to that for direct rotary. Use of additives with reverse circulation drilling may be required and in many cases is cost effective.

Karstic Formations. The cavernous nature of many karstic formations makes them difficult to drill with a system requiring circulating fluid, since the usual procedures for control of loss of circulation do not work. Depending upon borehole diameter and depth, and the volume of water present, some form of air-base drilling is often used. In extreme cases the hole is cased off or drilled blind (without any fluid or cuttings returning to the surface).

Sandstone. Sandstone that is not water bearing or will not be screened is often drilled with conventional freshwater-base drilling fluid. Certain very fragile water-producing sandstones containing fractures or zones of incomplete cementing must be drilled with careful control of mud properties. These aquifers can be permanently damaged if improper techniques are employed. Drilling these formations with systems that do not require mud circulation is common local practice in some areas.

Igneous or Metamorphic Rock. Igneous rock formations may be drilled either with conventional freshwater-base or air-base fluids, depending upon the hole diameter and depth, rock character, and production capacity.

Cable Tool

In a sense cable tool drilling always incorporates a drilling fluid since the reciprocating action of the tools thoroughly mixes the cuttings with the water in the borehole. Ideally, cuttings remain suspended until bailed. Most unconsolidated alluvium contains a variety of materials that produce a natural mud capable of suspending the cuttings. However, if the formation consists of clean sands and gravels, the fluid will not have the gel strength to hold the larger cuttings. Drilling progress is speeded if either a satisfactory local clay or bentonite is added. Some drillers simply add bentonite in granular or chip form, or polymers in freezer bags, to water in the hole.

Although the reciprocating action of the tools tends to hold the cuttings in suspension, large and heavy particles will settle rapidly if the density, viscosity, and gel strength of the drilling fluid are not satisfactory. If settling is too fast, bailing will not satisfactorily clean the borehole, and the cuttings must be redrilled until they are small enough to remain suspended.

Karstic Formations. Since karstic material seldom makes a suitable drilling fluid, a pre-mixed solution (usually made from bentonite) is placed with a bailer. Similar situations are encountered in fractured or cavernous basalt. Fluid cannot be retained in the borehole and cuttings cannot be bailed when the drilling occurs above the standing water level in some karstic formations. Water must be added continuously. If underwater, drilling is usually continued until the cavern or fracture is passed and cuttings are noted in the borehole.

Sandstone, Igneous, and Metamorphic Rock. Since sandstone, igneous, and metamorphic rock formations also seldom make a satisfactory natural mud, other materials, usually bentonite, must be added. Since the mud height is seldom more than the length of the cable tool drilling string, there is minimum head differential. This reduces the buildup of filter cake and fluid loss into sandstone, a distinct advantage when drilling fragile sandstone aquifers.

8.9 DRILLING-FLUID PROBLEMS

A number of borehole problems can occur during drilling. They usually effect a change in some combination of drilling procedures and mud properties. These problems include:

- Solids buildup.
- Borehole instability and water loss.
- Hole cleaning.
- Stuck drilling tools.
- Loss of circulation.

Modification of drilling procedures in response to borehole problems is complicated, and depends on the type and capacity of equipment used and on the experience of the operators. However, remedial changes made to drilling-fluid properties usually follow recognized standard practices. Table 8.3 shows the additives used to influence various properties of drilling fluids.

Solids Buildup

Perhaps the most common problem with the circulation of drilling fluid is solids buildup and excessive fluid density resulting from failure to separate solids at the surface. Use

TABLE 8.3 Additives to Influence Drilling-Fluid Properties

Property	Additive
Increase density	Barite
Improve viscosity	Bentonite
	Polymer(s)
Thin mud	Polyphosphates
Filtration control	Bentonite
	Polymers
Reduce hardness	Soda ash

of a shale shaker and hydrocyclones and/or lowering of viscosity with water dilution and chemical treatment is the usual solution for direct-rotary drilling.

For reverse drilling, increasing pit size or the number of pits lowers fluid velocity and lengthens settling time, allowing the solids to drop out. Use of hydrocyclones is also advantageous.

Borehole Instability and Water Loss

Borehole instability often occurs in unconsolidated formations and in certain types of clays and shales. This condition is noted by an increase in the drilling torque, rough drilling, loss of circulation, and loss of hole during round trips. Remedial measures usually consist of slowing the penetration rate, maintaining borehole fluid level at all times, and increasing viscosity by adding bentonite and an acrylic polymer. The bentonite should be premixed before polymers are added.

High gel strength and a greatly reduced pumping rate may be used to hold loose gravel in place while slowly drilling through the gravel interval. Greater gel strength is achieved by use of increased amounts of the highest-yield sodium-based bentonite and, if necessary, the addition of some polymer.

Another common problem is the swelling and/or sloughing of clays and shales, due to a filtrate invasion (high water loss) that causes hydrous disintegration. This happens frequently with reverse drilling when the drilling fluid does not contain filtration control additives to retard fluid loss.

High-filtration rates in permeable zones result in excessive invasion and filter cake deposition. This is a particularly severe problem in fine-grained aquifers such as fine sand or sandstones. In this situation undersize borehole opposite permeable zones is frequently noted when pulling out of or reentering the well with drilling tools. Clay particles deposited in the aquifer can be difficult to remove during development. Corrective action normally consists of the use of bentonite- and/or water-loss-inhibiting additives.

Hole instability should always be avoided. Drilling practices such as rapidly raising or lowering the drill string, which produces major pressure surges, are particularly troublesome.

Hole Cleaning

Soft, easily drilled clay can cause problems in hole cleaning. So-called "gumbo clays" may form mud rings, causing drilling-string drag and fluctuations in the mud pump pressure. Longer exposure causes sloughing or caving, increased fill during trips, bridges or tight spots in the borehole, and progressive enlargement. If the drilling fluid fails to clean the borehole, a "pill" of prehydrated bentonite or polymer, including some shredded paper, can be used periodically to remove larger cuttings or cavings.

Detergents and friction reducers can minimize bit balling and mud rings, whereas cellulosic and acrylic polymers inhibit swelling and dispersion of clays. Clays should be drilled as quickly as possible, at a speed consistent with hole cleaning since problems increase with exposure time.

Higher mud density increases the fluid buoyancy effect, reducing the apparent weight of the drill string and retarding cuttings settling in the annulus. However, these incidental benefits are overshadowed by a marked slowing in penetration rate caused by excessive pressure differential. Therefore a low-solids drilling fluid is essential. Solids that do not contribute beneficial properties to the drilling fluid are also objectionable because they interfere with the transfer of heat from the bit to the circulating liquid.

Stuck Drilling Tools

The thickness of the filter cake deposited on permeable formations when circulation is stopped increases with the solids content of the mud. This increases the risk of a stuck drilling string. Friction is reduced by maintaining a thin, slick filter cake through removal of sand and silt by hydrocyclones. A biodegradable vegetable oil derivative added to the drilling fluid will reduce drill pipe torque.

The accumulation of filter cake during static filtration may lead to severe problems. The severity of these problems is increased if the mud has high solids content and rapid filtration rate. Sometimes the filter cake may fill the hole when not circulating, creating a bridge that must be drilled out when reentering the borehole.

The drill pipe may become stuck when circulation is stopped to make a connection. If the pipe is in direct contact with the filter cake on porous, permeable formation, it may be held firmly in place by a pressure differential and become wall stuck. This mechanism is illustrated in Figure 8.6.

The filter cake continues to form on the exposed hole surface, increasing the resistance to pipe movement. When the pipe is stuck in this manner, there is no interference with circulation, and immediate steps should be taken to reduce the pressure differential by pumping water or air, or a slug of an environmentally acceptable vegetable oil to the spot where sticking has occurred. Oil is preferred because of its lower density and its wetting effect on steel (which tends to reduce frictional drag).

Loss of Circulation

Loss of circulation is a common problem that may result in decreased well production, due to invasion of the aquifer(s) by drill cuttings and/or lost circulation materials). In areas where loss of circulation is anticipated, use of a reduced-pressure drilling system may be desirable.

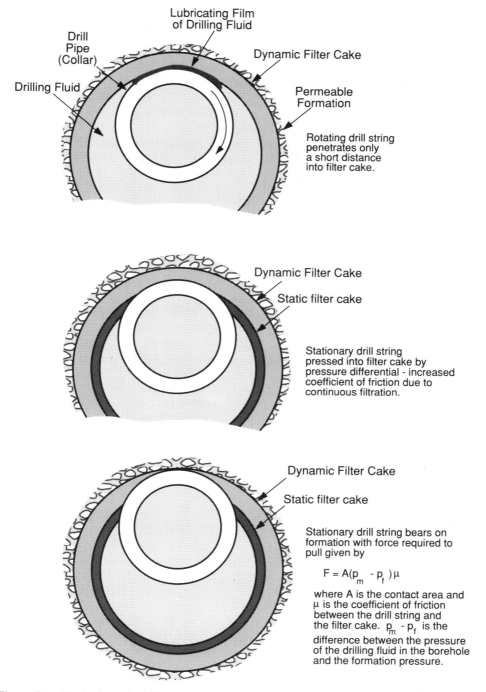

Figure 8.6. Mechanism of sticking drill pipe (collar) in filter cake opposite permeable formation. Source: Grey and Darley, 1980 (reprinted with permission).

Loss of circulation is countered by using lost-circulation materials that are compatible with the formation into which the loss is occurring. Materials that decompose to produce objectionable odors or tastes must be avoided. These include cottonseed hulls, bran, and some other plant products. Larger particles such as wood fiber, mica, and nut hulls are acceptable.

These products, however, should not be introduced into aquifers that will be screened.

If seepage losses occur when drilling a stable borehole with water, adding a small amount of bentonite for viscosity and support followed by bridging material will usually prevent an increase in the rate of loss and accompanying borehole problems. Finely ground nut hulls or mica in amounts of 5 to 15 lb/100 gals of water (and additional bentonite to increase drilling fluid viscosity) or bentonite mixed in 5 to 10 lb/100 gal are effective.

In difficult cases the loss of circulation may be overcome by forming a stiff plug (bentonite,[4] cement) within the porous rock. In some situations continued drilling by air, foam, or cable tools may be necessary. Many loss of circulation problems can be resolved by a change in the casing program, either by setting deeper surface or conductor casing or by installing and cementing intermediate casing.

When loss is anticipated, frequent measurements of mud properties should be made and treatment applied to maintain:

1. Minimum density—to control hydrostatic head and solids content.
2. Minimum viscosity—to reduce flow resistance in the annulus.
3. Minimum gel strength.
4. Minimum filtration—to minimize permeability on the borehole wall.

When partial loss of circulation occurs while drilling with freshwater-base drilling fluid, the drill pipe is usually pulled while the mud properties are reviewed. The hole may be filled with water to see if the reduced density fluid will mitigate loss. If this is partially successful, the density of the drilling fluid is reduced by adding water, although a large quantity may be needed to make a significant reduction. Use of an aerated, reduced-pressure mud system may be helpful if equipment is available. Otherwise, lost-circulation materials must be used.

Pressure surges caused by rapidly running the drill pipe into the borehole are often great enough to cause a loss of circulation. Even though the mud will sometimes return to the hole when pressure drops, the formation is weakened and further loss is likely to occur, frequently at a lower pressure.

8.10 PROCEDURES FOR EVALUATING DRILLING-FLUID PROPERTIES

Drilling fluid effectiveness depends on:

1. Physical properties such as density, viscosity, thixotropy, filtration, and lubricity.

---

4. A stiff bentonite plug is made from unyielded (not fully hydrated) bentonite. Benseal™, EZ-Mud Slurry®, or Holeplug™ are very effective for this purpose. These are products of Baroid Corporation, and similar products are available from other manufacturers.

---

2. Chemical interaction and interparticle forces within the suspension and between the drilling fluid and formations drilled.
3. Size, shape, number, and abrasiveness of the suspended solids.

Certain measurements are useful in controlling fluid properties. Methods for conducting them have been published by the American Petroleum Institute Committee on Standardization of Drilling Fluid Materials as RP 13B, *Standard Procedure for Testing Drilling Fluids*. The procedures given here are those commonly used in the water well drilling industry with direct hydraulic rotary drilling. Density, viscosity, water loss, and sand content are the primary properties measured. The remaining tests are not required in every case.

Density

Density is the most important and easily measured property relating to the use of a liquid fluid for drilling. Weight measurement provides an estimate of the total suspended solids in freshwater-base drilling fluids not containing barite. Density is defined as weight per unit volume and can be expressed in any convenient unit, such as pounds per gallon, pounds per cubic foot, grams per cubic centimeter, or kilograms per cubic meter. The weight can be compared with that of an equal volume of water and reported as specific gravity.

Density can also be expressed as a pressure gradient per unit of depth. For example,

$$\text{Pressure gradient in psi/ft} = \frac{\text{lb/ft}^3}{144}$$

$$= \frac{\text{lb/gal}}{19.24}$$

$$= 0.433(SG),$$

where SG is specific gravity.

The mud balance is the most convenient instrument for measuring mud weight. In use, the balance cup is filled to capacity with the fluid to be weighed, and the balance and cup are thoroughly cleaned. After the knife-edge is placed in the notched fulcrum on the stand, the rider is moved along the balance arm until the bubble in the glass level is centered. Weight is read from the rider marker, and is recorded to the nearest 0.1 lb/gal. One lb/gal $\times$ 0.052 equals psi/ft. Psi/ft $\times$ depth equals hydraulic head in the borehole.

Viscosity

The Marsh funnel provides an estimate of viscosity through measurement of the time to the nearest second required for one quart to flow from the funnel. Flow time of freshwater is 26 ($\pm$ 0.5) sec at 70° ($\pm$ 5)F, which can be used for instrument calibration. The Marsh funnel viscosity of drilling fluids is affected by their density and gel development. It is important to note that this flow time cannot be used for correlating viscosities with other types of viscosity measurement devices.

The device consists of a funnel and a one-liter cup that has a gradation showing one quart (946 cc). API standards call for one quart of liquid to be run from the funnel filled to the 1500 cc mark. A 14-mesh bronze screen filters out larger particles from the drilling fluid. In order to reduce errors due to thixotropy, the sample is usually taken from a part of the stream where flow is relatively turbulent.

Gel Strength

The thixotropic properties of drilling fluids may be measured on a concentric cylinder viscometer. It is agitated to reduce gel strength as near to zero as possible, and then the force necessary to start spindle movement is measured. If allowed to remain quiescent for lengthening periods of time, increasing force with time is necessary to start spindle rotation.

Filtration and Wall-Building Tests

The filtration (water loss) and wall-building (cake) test is commonly known as the API Low Temperature, Low Pressure Filtration Test. It consists of measuring the amount of liquid forced from a filter press containing a mud sample under specified conditions of pressure and time, usually 100 psi and 30 min, and measuring the thickness of the residual solids film deposited on the filter paper.

The test is normally made at ambient temperature in a filter press. The standard filter has an area of 7 in.[2], although a one-half size filter press is also available. Pressure is supplied by a carbon dioxide cartridge that fits into the regulator assembly. A sheet of hardened filter paper is placed on a wire screen. After assembling the cell on the base cap, mud is poured into it to within ½ in. from the top. The unit is placed on the frame, and the top cap is held firmly in place by a T-screw. Pressure of 100 psi is applied and the filtrate is collected in the graduated cylinder. The API standard filtration time is 30 min. However, if as much as 10 cm^3 of filtrate have been collected in 7½ min, the test may be ended, and the volume of filtrate doubled to arrive at the 30-min value.

At the end of the test period, the volume of filtrate is recorded, and the thickness of the cake is measured in ¹⁄₃₂ of an inch. The measured cake thickness is doubled if the filtration time was 7½ min. The texture of the cake is examined, particularly with regard to stickiness, flexibility, and hardness. Too often attention is given only to the water loss (volume of the filtrate). Permeability of the filter cake is the fundamental parameter that controls filtration and cake thickness.

Solids Content

Although the distinctive characteristics of a drilling fluid are largely a product of the types of solids in suspension, only the total volume fraction is usually measured. The procedure consists of distilling the water from a measured volume, collecting and measuring the liquid, and calculating the solids as the difference.

The solids content of unweighted, freshwater-base drilling fluids can be estimated from the measured density, with use of Table 8.4. Clay, sandstone, shale, dolomite, limestone, feldspar, granite, marl, quartz, siltstone, and slate cuttings range in specific gravity from about 2.2 to 2.8. Consequently no significant error is made in assuming a specific gravity of 2.65 for estimating solids content from density, unless the water is salty or barite has been added.

Sand Content

Abrasive solids cause excessive wear, especially of pump and bit parts. Abrasiveness depends on particle hardness and shape and is not limited to the so-called "sand fraction," which is defined as the particles retained on a 200-mesh screen. Severe abrasion can result from "silt fraction" particles smaller than 325 mesh. The sand content test, a measurement of material retained on a 200-mesh screen, is a valuable field test with direct hydraulic rotary drilling. Samples are usually taken at the pump suction.

Test equipment consists of a 200-mesh screen, a funnel, and a graduated glass tube. Measurements are made by pouring mud into the glass tube to the indicated mark, with water added to the upper mark. The mouth of the tube is covered and the mixture shaken vigorously. The mixture is poured onto the previously wetted screen and any adhering drilling fluid is washed from the sand. The funnel is placed upside down on the screen, the assembly inverted, and the funnel tip inserted in the glass tube. Water is sprayed on the screen, washing the sand into the tube. After the sand has settled, the percent by volume is read from the graduations on the tube.

pH

The hydrogen ion concentration is a significant factor in the behavior of drilling fluid. Most drilling fluids with clay have optimum properties when the pH is about 8.5 to 9.5. Performance is poor at pH below 7. Acidic waters reduce the effectiveness of organic polymers and organic thinners. If necessary, the pH of the water used for preparing drilling fluids can be raised by the addition of soda ash before

TABLE 8.4 Densities and Solids Content of Freshwater-Base Drilling Fluids

Bentonite:Water (lb/100 gal)	Density			Hydrostatic Head (psi/100 ft depth)	Clay (% weight)	Clay (% volume)	Total Volume (gal)
	(lb/gal)	(lb/ft³)	(g/cc)				
5.0	8.4	62.5	1.0	43.4	0.6	0.2	100.2
10.0	8.4	62.8	1.0	43.6	1.2	0.5	100.5
15.0	8.4	63.0	1.0	43.9	1.8	0.7	100.7
20.0	8.4	63.2	1.0	43.9	2.3	1.0	101.0
25.0	8.5	63.4	1.0	44.1	2.9	1.2	101.2
30.0	8.5	63.6	1.0	44.2	3.5	1.4	101.4
35.0	8.5	63.9	1.0	44.4	4.0	1.7	101.7
40.0	8.6	64.1	1.0	44.5	4.6	1.9	101.9
45.0	8.6	64.3	1.0	44.7	5.1	2.1	102.2
50.0	8.6	64.5	1.0	44.8	5.7	2.3	102.4
55.0	8.7	64.7	1.0	44.9	6.2	2.6	102.6
60.0	8.7	64.9	1.0	45.1	6.7	2.8	102.9
65.0	8.7	65.1	1.0	45.1	7.2	3.0	103.1
70.0	8.7	65.4	1.0	45.4	7.8	3.3	103.4
75.0	8.8	65.6	1.1	45.5	8.3	3.5	103.6
80.0	8.8	65.8	1.1	45.7	8.8	3.7	103.8
85.0	8.8	66.0	1.1	45.8	9.3	3.9	104.1
90.0	8.8	66.2	1.1	46.0	9.8	4.1	104.3
95.0	8.9	66.4	1.1	46.1	10.2	4.4	104.6
100.0	9.0	66.6	1.1	46.3	10.7	4.6	104.8
110.0	9.0	67.0	1.1	46.6	11.7	5.0	105.3
120.0	9.0	67.4	1.1	46.8	12.6	5.4	105.8
130.0	9.1	67.8	1.1	47.1	13.5	5.9	106.2
140.0	9.1	68.2	1.1	47.4	14.4	6.3	106.7
150.0	9.2	68.6	1.1	47.6	15.3	6.7	107.2
175.0	9.3	69.6	1.1	48.3	17.4	7.8	108.4
200.0	9.4	70.5	1.1	49.0	19.4	8.8	109.6
250.0	9.7	72.3	1.2	50.2	23.1	10.7	112.0
300.0	9.9	74.0	1.2	51.5	26.5	12.6	114.4
350.0	10.1	75.8	1.2	52.6	29.6	14.4	116.8
400.0	10.3	77.4	1.2	53.7	32.4	16.1	119.2
450.0	10.6	78.9	1.3	54.8	35.1	17.8	121.6
500.0	10.7	80.4	1.3	55.8	37.5	19.4	124.0

bentonite or polymers are added. Cement raises pH, in addition to the objectionable effect of adding calcium ions.

Paper strips impregnated with dyes that develop characteristic colors are used for measuring pH. An indicator paper strip is placed on the surface of the drilling fluid. As soon as the color has stabilized (10 to 15 sec), it is compared with pH color standards, and the pH is estimated to the closest matching color.

Chemistry

The chemical content of the water used in the drilling fluid affects its performance. Since satisfactory makeup water is usually available where water wells are drilled, chemical treatment is seldom required.

Hardness is the most common water-quality problem, since salts of calcium and magnesium impair the swelling and filtration properties of bentonite. The effectiveness of some organic polymers and foaming agents is also reduced by calcium ions. A simple test for their presence is made by adding a few drops of ammonium oxalate solution to the water. If a white precipitate appears, revealing the presence of calcium enough soda ash should be added to the water to remove the hardness before using it to make drilling fluid. The calcium ion concentration should be less than 100 ppm.

Drilling into anhydrite ($CaSO_4$) or gypsum ($CaSO_4 \cdot 2H_2O$) causes immediate thickening of bentonite mud. Although calcium ions in the filtrate will increase slowly because some are held by the clay, sulfate ions will increase at once and can be identified by the formation of a precipitate when tested with acidified barium chloride solution.

If the salinity of the makeup water is questionable, the water should be tested for chlorides, commonly by titrating a sample with standard silver nitrate solution, using potassium chromate solution as an indicator. The results are expressed as parts per million chloride ion, although actually measured as milligrams per liter. The distinction is not significant in concentrations below 5000 ppm. Frequently, the concentration is expressed as sodium chloride, which is obtained by multiplying the chloride ion concentration by 1.65.

A "fluffy" thickening of the drilling fluid may reveal an abrupt increase in the chloride ion concentration in the filtrate. This indicates influx of saltwater or penetration of salt, a cause for immediate concern if a potable water well is being constructed.

8.11 FIELD TESTS

Field tests by the drilling crew are usually only made to measure density, viscosity, filter cake thickness, 30-min water loss, and sand content. Other tests are customarily performed by a mud engineer.

A typical well construction specification will require field tests to be performed hourly. Allowable drilling-fluid properties will vary, depending on the formations penetrated and the stage of the well construction. A standard often used for drilling of alluvial formations requires the following:

- Weight, a maximum of 75 lb/ft^3 (10.0 lb/gal).
- Funnel viscosity, a maximum of 38 sec/qt.
- 30-min water loss, a maximum of 15 cm^3.
- Filter cake, a maximum of $3/32$ in.
- Sand content, a maximum of 3%.

These properties are modified during the graveling and swabbing operations. Because of the large quantities of filter cake and formation fine materials dislodged, constant attention is required to maintain the following standards:

- Weight, a maximum of 68 lb/ft^3 (9.1 lb/gal).
- Funnel viscosity, a maximum of 30 sec/qt.
- Sand content, a maximum of 1%.

8.12 DRILLING SYSTEMS USING AIR

Air drilling is a term applied to any reduced-pressure circulating system that uses air, with or without additives, as the circulating fluid. These systems are segregated into the following:

Dry air	Compressed air circulated at velocities adequate to lift the cuttings from the borehole
Mist	The addition of surfactants to the air along with a small amount of water
Foam	Unlike mist, the addition of larger amounts of surfactants and water to the air of which water is the continuous phase and air (since it is contained in bubbles within the water) is the discontinuous phase
Gel (stiff) foam	A low-velocity, low-air-volume system in which a slurry of bentonite and polymer is added to water and surfactant to produce a foam having the consistency of thick shaving cream
Stable foam	A completely premixed air and liquid (surfactant and polymer) dispersion in which the liquid is the continuous phase and the air the discontinuous phase
Aerated mud	A conventional mud system in which air is introduced to decrease the hydrostatic head and reduce the tendency toward loss of circulation. Compressed air, introduced at the pump discharge, must be thoroughly removed from the fluid at the surface before it reaches the pump

Capabilities and Limitations of Air-Drilling Systems

Air is the ultimate low-density drilling media. Compared with liquid-drilling fluid, the low pressure against the bottom of the hole permits instant removal of the material crushed by the bit. This results in high penetration rates and long bit life. High air volume cools the bit.

Other advantages include:

1. Improved borehole plumbness and alignment because of lowered requirements for weight on bit.
2. Tight hole problems likely to be reduced.
3. Differential-pressure drill string sticking eliminated.
4. Loss of circulation problems likely to be reduced.
5. Less drilling water required.
6. Formation fluid sampling often performed on a continuous basis.
7. Significant reduction in time for recovery of formation samples.
8. Damage to the permeability of producing zones likely to be substantially reduced.
9. Relative ease of use in cold, dry climates.

Air drilling, however, does not intrinsically accommodate the drilling-fluid functions of carrying formation water out of the hole and offsetting formation pressures. This leads to the conclusion that for water well drilling, air systems are most suitable for drilling hard, generally low yield, consolidated formations, including igneous, metamorphic, and some sedimentary rock.

Some disadvantages of air-drilling systems are:

1. Operating costs may be higher where compressor volume and pressure requirements are high, or if continuous injection of gel or stable foam is required.
2. The weight of the drill string is increased because buoyancy is reduced.
3. The penetration of water zones may lead to formation of mud rings and sticking of the drill string.
4. High-velocity air returns can lead to borehole erosion in loosely consolidated formations.

Air-drilling equipment and accessories, including volumetric and pressure requirements for direct air drilling, are covered in Chapter 7.

Dry Air. Dry air is simple and efficient, if certain conditions are met. These are:

- There must be steady flow conditions through the annulus.
- The formation cuttings and air must move at approximately the same velocity up the annulus. Relative motion between the air and formation particles results in friction losses that reduce the energy available to circulate up the annulus.
- The minimum kinetic energy at the bottom of the annulus should be equivalent to standard air moving at 3000 ft/min.

Dry air drilling may not work in extremely porous dry rock since air forced into the rock can return into the borehole as an implosion on release of pressure, resulting in borehole enlargement and instability in that zone.

With dry air drilling, the chips that are removed range in size from fine to coarse. As these particles start up the annulus, the larger sizes are ground and pulverized by the drill string, and by impacting each other and the borehole walls. This reduces the cuttings to the dustlike particles seen at the surface. The time required to remove these particles should not materially exceed 1 min per 1000 ft of borehole depth. Excessive annular velocities can cause problems by eroding the borehole walls in weakly consolidated formations. This will result in inadequate velocity to clear the cuttings in the enlarged zones.

Air drilling is usually started with dry air. Formation water requires modification of the system. As water is encountered, a mixture of water and cuttings can adhere to the borehole wall, causing cleaning problems. Disappearance of dust at the surface indicates wet down-hole conditions or loss of circulation. If an increase in air pressure is accompanied by a corresponding decrease in air volume output at the surface, wet-hole conditions are probable.

The degree of wetness ranges from damp to water flows. Dampness can be dried up by injecting a slug of water and surfactant. When formation water cannot be handled, mist or foam drilling may prove satisfactory.

Mist. When water saturated formations are drilled, adding small quantities of water to the air forms an air-mist system that assists in preventing mud ring formation. Sufficient water must be added to saturate the air. When this takes place, the formation water can be removed as particles. However, some shales and clays become unstable in the presence of water, requiring caution in the use of mist drilling when these formations are encountered.

With most mist drilling a small amount of surfactant foaming agent is added with water to the air stream. The surfactant prevents water and cuttings from balling up in the annulus. Mist systems will require about 30% to 40% more air than dry air. Standpipe pressures will be greater, 200 to 400 psi as compared to 100 to 300 psi for dry air. These additional requirements are due to the weight of the water in the annulus.

The concentration of surfactant should be adjusted according to the volume of formation water in the borehole. If insufficient surfactant is used, there will be considerable heading.[5] If too much is used, the well will head with slugs of heavy foam and no liquid discharge. The lowest concentration of surfactant that gives a near-constant discharge with steady standpipe pressure is the amount desired. Pressure surges caused by heading are detrimental to the hole. For this reason slug drilling, which consists of alternating columns of water and air up the annulus, should be avoided as a continuous operation. Too little air volume, or penetration rates exceeding the capacity of the equipment to clean the borehole, are the primary causes of slug drilling.

Foam. Foaming requires injection of sufficient surfactant into the air stream to lower the surface tension of the water. Suitable surfactants include anionic soaps, alkyl polyoxyethylene nonionic compounds, and cationic amine derivatives. Ideal foams consist of small, tightly joined cells about the consistency of shaving cream. Since foams tend to collapse in a short time, longevity must be considered when selecting a foamer, and a 90-min life is usually the minimum feasible. Foam is difficult to confine at the surface. It should rapidly break down after leaving the well annulus to avoid surface

---

5. Heading is the process of undesirably high pressure buildup under a restraining load (usually water), until the pressure is relieved by the liquid moving as a slug.

buildup. This is difficult to reconcile with the necessary foam stability for drilling. Defoamers are available but are expensive, and often not environmentally acceptable.

Mist and foam systems have two phases, air and liquid. Distribution of the two phases depends on the relative amounts of each present. This ratio is usually expressed as either the volume fraction of the gas (foam quality) or the liquid-volume fraction (LVF). In the foam quality range from 0 to about 0.54, the foam consists of independent bubbles dispersed in the gas; in the range from 0.54 to 0.96, the system is analogous to an emulsion with gas as the internal phase and the liquid as the external phase. Above 0.96, the system consists of ultra microscopic droplets of water dispersed in air and is termed a mist. This is illustrated in Figure 8.7.

The cuttings-carrying capacity of foam depends on its rheological properties and the square of the annular velocity. The rheological properties depend mainly on the viscosity of the air and liquid, and on the quality of the foam. Foam drilling is particularly applicable to large-diameter borehole drilling since dry air drilling would entail enormous volume flow rates to maintain the velocity necessary to raise the drill cuttings.

Foam drilling materially reduces air flow requirements compared to dry air drilling, and water demands are much lower than for a complete liquid drilling fluid system. The water-injection rate varies with water production from the well. Large volumes of infiltrating water can be lifted in the foam.

Since flow velocities in the borehole are reduced, the erosion of poorly consolidated formations, which might occur with high-velocity air, is reduced. The reduction in required up-hole velocities may be quite dramatic, to as low as 150 ft/min. The surfactant concentration level to accomplish this varies between 0.2% to 3.0% of the water injected into the air.

Gel Foam. Gel foam is a freshwater-base drilling-fluid (bentonite and polymers added) surfactant and air combination that has excellent hole-cleaning and wall-building characteristics in freshwater formations. As compared with dry air and mist, gel foam is a low-velocity, low-air-volume system that extends the range of air-drilling applications. This technique has been applied successfully in areas of severe loss of circulation where low density is required.

The addition of polymers and bentonite to the foam causes the formation of filter cake on the borehole wall, requiring removal by development. Typical bentonite concentrations vary from 25 lb to 50 lb per 100 gal of injection water. Surfactants and additives are generally mixed in a tank near the drilling rig and injected into the airstream at a rate equivalent to 2% of the standard cubic feet per minute of air flow.

Guidelines call for 70 to 90 ft per minute of annular velocities, with air to liquid volume ratios between 100:1 and 300:1. Excess air causes the air to channel through the foam column, whereas insufficient air increases back pressure on the hole bottom. This compresses the air and changes the liquids-to-air-volume ratio affecting the nature of the foam developed at the bit.

Stable Foam. Stable foam as a circulating medium has distinct advantages over air mist and gel foam. It is a completely premixed air and liquid dispersion in which the liquid is the continuous phase and air is the discontinuous phase. For this mixture to remain homogeneous, the proportions of liquid (a water, surfactant, and polymer solution) and air must be kept within narrow limits (between 2% and 5% by volume).

A foam-generating mixer, developed by Chevron Corp., is used to mix the air and liquid phases at the surface. The compositions of the foaming agent and the polymer are selected to meet down-hole conditions.

For the stable foam to have a solids-carrying capability of 10 times that of the liquid, the liquid volume proportion of the foam must be between 2% and 5%. At this state the liquid phase represents a molecularly thin film between air bubbles. As the liquid-volume fraction of the foam increases, the solids-carrying capacity declines abruptly so that, at a 10% liquid-volume fraction, the capacity is about five times that of the liquid. At higher liquid fractions the foam tends to break down into frothy liquid. Stable foam develops air pockets and slugs with less than 2% liquid by volume.

If stable foam is to be used effectively as a circulating fluid, it is necessary that its quality remain consistent from

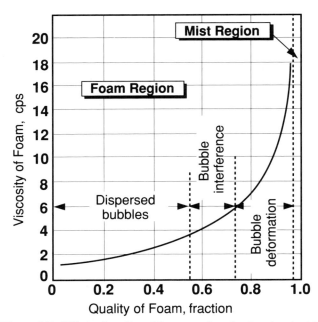

Figure 8.7. Effect of quality of foam on viscosity (reprinted with permission from the *Oil and Gas Journal*, September 6, 1971).

TABLE 8.5 Foam Mixtures with a Typical Biodegradable Agent

	Quik-Foam™ Added to Injection Water (per 100 gal)
Dry air drilling as dust suppressant	0.5–1 pint
Mist drilling in sticky clay	1–2 quart
Foam drilling, all purposes	0.5–2 gal[a]
Gel foam drilling	0.5–2 gal
Slug drilling to clean annulus	1 pint in drill pipe followed by 3–5 gal water

Source: Courtesy of Baroid Corporation.
[a] After mixing Quik-Gel™ and/or Quik-Trol™.

TABLE 8.6 Foam Mixtures with a Typical Organic Polymer

	(lb/100 gal)
Quik-Trol™ Added to Freshwater	
Stabilize water-sensitive formations	3–5
Stop drill pipe vibration, reduce torque, lower circulating pressure	0.5–1
Added to Mud (Made with Freshwater)	
Improve performance (better hole cleaning, thinner filter cake, increased hole stability) Extend and beneficiate bentonite	0.5–1.5
Added to Injection Liquid in Air/Foam Drilling	
Improve foam performance and hole condition	0.5–1.5

Source: Courtesy of Baroid Corporation.

TABLE 8.8 Injection Rates for Standard Mixtures

Overburden (unconsolidated)	7–10 gpm
Rock (broken)	5–7 gpm
Rock (solid)	3–5 gpm
Dust control/water	0.25–2 gpm

Source: Courtesy of Baroid Corporation.

bit to surface. Its high solids-carrying capability is lost if, due to pressure, the discharge at the bit is water or if, due to release of pressure up the annulus, it turns to mist.

Aerated Mud. This system has been rarely used in drilling water wells and is principally applied to loss of circulation conditions. With aerated mud, freshwater-base drilling fluid is the continuous phase. Control of gel strength must be such that breakout of air occurs not within the annulus but in the pits. If breakout occurs in the annulus, the slugging of the mud will erode the hole. Failure to separate on the surface will cause pumping problems.

Basic Additives. Principal air-drilling additives include surfactants, organic polymers, and high-yield bentonites. The following mixtures are typical and will vary, depending upon the characteristics of the product used.

Quik-Foam™, a biodegradable liquid foaming agent, is a mixture of anionic surfactants. This liquid is added to water, depending upon application in the concentrations shown in Table 8.5. The water mixtures are injected into the air stream at rates from 10 to 20 gal/hr. Gel foam mixtures are injected at rates from 7 to 10 gal/ft^3 of hole drilled.

Quik-Trol™ is a nonfermenting organic polymer that forms a clear colloidal solution in fresh or salty water. This liquid is added to water in concentrations outlined in Table 8.6.

Quik-Gel™ is a high-yield bentonite (twice the viscosity produced by an equal weight of standard API bentonite) mixed with water to a viscosity of 36 to 40 sec/ft and combined with Quik-Foam™ to form gel foam.

TABLE 8.7 Standard Mixtures[a]

Injection Slurry	Water (gal)	Quik-Gel™ (lb)	Quik-Trol™ (lb)	Quik-Foam™ (% volume)	
Quik-Foam™	100	—	—	0.02–3.0	Increase as required to compensate for down-hole water dilution, etc.
Trol-Foam™	100	—	0.5–1	0.1–2.0	Mix well
Mud mist	100	25	—	0.3–1.0	Mix well viscosity, 32–40 sec/qt
Gel foam	100	12–15	1	0.3–1.0	Mix well viscosity, 32–40 sec/qt

[a] Mix only in this sequence.
Source: Courtesy of Baroid Corporation.

Tables 8.7 and 8.8 give standard mixtures, and injection rates for different borehole conditions. Quik-Foam™, Quik-Trol™, and Quik-Gel™ are products of Baroid Corporation, Inc. Similar products are available from other manufacturers.

REFERENCES

1. Rogers, W. 1963. *Composition and Properties of Oil Well Drilling Fluids*. Gulf Publishing, Houston, TX.

2. Gray, G., and H. Darley. 1980. *Composition and Properties of Oil Well Drilling Fluids*. Gulf Publishing, Houston, TX.

3. Angel, R. R. 1957. "Volume Requirements for Air or Gas Drilling." *Pet. Trans., AIME* 216.

4. Lyons, W. 1984. *Oil and Gas Drilling Manual*. Gulf Publishing, Houston, TX.

CHAPTER 9

Stresses on Well Casing and Screen

9.1 INTRODUCTION

The stresses imposed on the well casing and screen during installation and well operation are often overlooked by the well designer, contractor, and operator. Many wells have experienced structural failure from causes impossible to diagnose. While some stresses, such as the collapsing force on well casing from unbalanced hydrostatic pressures, are easily and accurately calculated; others, such as those caused by formation slumping, are not.

This chapter identifies the loads and stresses of concern in water wells and discusses how they may be quantified. The physical characteristics of commonly used casing and screen materials are considered and compared. Specific recommendations for avoiding structural failure are presented. Proper use of this information will prevent easily foreseen well failures and some others whose predictability is difficult.

Stresses on well casings and screens can be categorized by the direction of the forces imposed. Axial forces, if tensile, act to pull the casing and screen apart but tend to buckle or collapse it, if compressive. Radial forces directed toward the well axis act to collapse the casing or screen but create bursting loads if directed outward. Bending forces result from uneven distribution of radial forces along the well and greatly reduce the casing and screen collapse and tensile strengths. Figure 9.1 illustrates these forces.

9.2 TENSILE STRESSES ON WELL CASING AND SCREEN

In a typical casing and screen installation, the weight of the casing is $\rho g L A$, where

ρ = density of casing $[ML^{-3}]$,
g = acceleration due to gravity $[LT^{-2}]$,
L = length of casing $[L]$,
A = average cross-sectional area of casing $[L^2]$.

At the bottom of the casing there is a buoyant upward force due to the fluid pressure, equal to $\rho_f g h A$, where

ρ_f = density of fluid $[ML^{-3}]$,
h = depth of fluid $[L]$.

For the geometry shown in Figure 9.2, the total force (downward) at the top of the casing is

$$\mathbf{F} = gA(\rho L - \rho_f h) \qquad (9.1)$$

When $L = h$, Equation 9.1 reduces to $(\rho - \rho_f)gAL$, the buoyant weight of the casing.

For a casing of uniform cross section and density, the maximum tensile stress occurs at the top, and is given by

$$\frac{\mathbf{F}}{A} = \boldsymbol{\sigma}_m = g(\rho L - \rho_f h)$$

where $\boldsymbol{\sigma}_m$ is the tensile stress, or the force per unit area directed along the casing axis. The tensile stress at any distance x from the top of the casing is

$$\boldsymbol{\sigma}(x) = \boldsymbol{\sigma}_m - \rho g x = g[\rho(L - x) - \rho_f h] . \qquad (9.2)$$

When $x = L$ (the bottom of the casing), $\boldsymbol{\sigma}(L) = -\rho_f g h$ (compression) and $\boldsymbol{\sigma}(x) = 0$ when $x = L - (\rho_f/\rho)h$. Note that the stress is independent of the cross-sectional area, a result that is true only if the casing is of uniform cross section. In the general case, where the cross-sectional area $A(x)$ is a function of distance along the casing

$$\boldsymbol{\sigma}(x) = \frac{\int_x^L \rho g A(x)\,dx - W(x)}{A(x)} , \qquad (9.3)$$

where $W(x)$ is the weight of the fluid displaced by the casing between L and x.

Safety Factors

It is good engineering practice to calculate the expected tensile load at the top of the casing while ignoring buoyant forces; in this way the tensile stress at any level is measured by the weight of the casing below that level divided by the cross-sectional area bearing the load. Neglecting the buoyant force provides two safety factors.

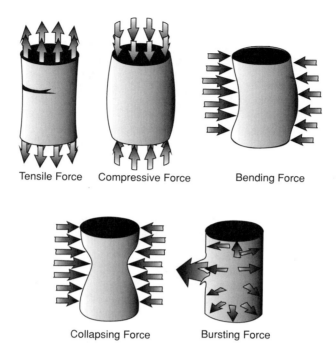

Figure 9.1. Forces acting on well casing.

The first is in the design tensile stress, $\boldsymbol{\sigma}_d$, which for a uniform casing is

$$\boldsymbol{\sigma}_d = \rho g (L - x) \quad \text{(tension)} . \qquad (9.4)$$

Equation 9.4 is derived from Equation 9.2, with the buoyant force neglected. The second safety factor results from the increased collapse resistance at the bottom of the casing which results from axial compression.

Tensile strength safety factors range from 1.6 to 2.0 or more in the oil industry. This range takes into account bending, acceleration, and other dynamic forces during installation, and unknown forces that may be encountered.

In the water well industry, safety factors are also used, but they tend to be on the higher side (2.0 or more). The reasons for this difference are that casing in the oil industry is usually made from high tensile steel in diameters smaller than typical for water wells, and because loads on the casing can change due to gradual settlement over 20 years or more. A high safety factor is usually a wise choice. Although in practically all circumstances steel casing is sufficiently strong to withstand normal tensile loads, this should be verified by calculation.

Yield Point

When tensile stresses are measured, the yield point is the critical physical characteristic of the casing material. This value varies from material to material and also within a material classification (e.g., steel). The minimum yield points and tensile strength for commonly used casing and screen materials are shown in Table 9.1. A more complete listing of materials and their physical strengths is shown in Appendix H.

Example

An example illustrates these principles. A 2000-ft deep well calls for nominal 12-in.-diameter casing. Is the material strength sufficient to support the expected tensile load?

Answer: Assume carbon steel well casing with a yield point of 35,000 psi; a safety factor of 2.0 is applied so that the maximum allowable tensile stress is 17,500 psi. From equation 9.4

$$\boldsymbol{\sigma}_d = \frac{\rho g (L - x)}{144} = \frac{(490)(2000 - 0)}{144} = 6800 \text{ psi} ,$$

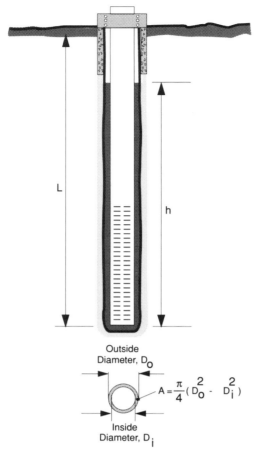

$$A = \frac{\pi}{4}(D_o^2 - D_i^2)$$

Outside Diameter, D_o

Inside Diameter, D_i

Figure 9.2. Well casing suspended in an open hole.

$$\sigma(x) = \frac{\int_0^x w(X)\,dX}{A(x)},$$

where w is the weight per foot, A is the cross-sectional area in square inches, and X is measured in feet from the bottom. If 2000 ft of 12-in. diameter, ¼-in. wall well casing (33.4 lb/ft) is used, then the total weight is 66,800 lbs. If the casing at the top is undersize in thickness by 12.5% (maximum allowable by ASTM and API specifications), then the stress at this point is

$$\frac{66,800 \text{ lb}}{\pi D t (0.875)} = 7776 \text{ psi},$$

still well below the yield point. Therefore the casing is safe if it is within the standards for thickness. The stress is independent of the diameter and thickness as long as the casing is uniform over its length.

The plastic pipe for water wells is generally manufactured in accordance with ASTM F 480. These plastics typically have moduli of elasticity of $3 - 4 \times 10^5$ psi, tensile strengths of $5 - 10,000$ psi, and compressive yield strengths of approximately 10,000 psi. It must be recognized that the elastic properties of these materials differ substantially from steel, which typically has an elastic modulus of 3×10^7 psi, a tensile strength in excess of 55,000 psi, and a compressive yield strength of 30,000 psi or more. Care is also necessary in distinguishing between the various subclasses of a given plastic material.

which is well below that maximum allowable stress of 17,500 psi.

Note that Equation 9.4 assumes a constant cross-sectional area. A reduced cross-sectional area will proportionally increase the stress. Thus the tensile stress at a point x on the casing is given by

Tensile Strength of Well Screen

In most circumstances screens are not subjected to high tensile loads. The tensile stress at a given level in the screen is calculated by dividing the weight below the level by the cross-sectional area supporting the weight (Equation 9.3). The maximum tensile stress is usually found at a point of minimum cross-sectional area near the top of the screen.

TABLE 9.1 Minimum Yield and Tensile Strengths

Description	Yield Strength	Tensile Strength
Carbon steel casing manufactured to ASTM and API Grade B requirements	35,000 psi	60,000 psi
Casing manufactured from: High-strength low-alloy steel to ASTM A 714 Grade I requirements	50,000 psi	70,000 psi
Type 304 & 316 stainless steel casing	30,000 psi	75,000 psi

Note: Most steel pipe standards allow as much as 12.5% negative deviation on the wall thickness, although the overall weight tolerance is ±5%. Tolerances should be taken into account when strength calculations are made.

For the wire wrap screen the area is taken as the cross-sectional area of the longitudinal rods that are the weight-carrying elements. For other types, it is also taken as the smallest load-bearing cross-sectional area, considering that the screen openings reduce this area. The maximum tensile stress should be compared with the yield strength of the weight-carrying material.

As with casing, safety factors should be applied to the computed screen strength. For tensile strength, a factor of 2.0 is recommended in order to account for additional loads from bending or other sources. Again buoyant forces should be ignored.

9.3 RADIAL FORCES ON WELL CASING AND SCREEN

Unbalanced radial hydrostatic forces occur as a result of fluid pressure differences between the outside and inside of the casing and screen. Excessive force will lead to collapse. Other collapse forces can be caused by formation pressures such as slumping, gravel movement, or a combination of these and hydrostatic forces.

The maximum external hydrostatic pressure on a casing (dry interior) at a given depth h is

$$\mathbf{P} = \rho_f g h \,, \qquad (9.5)$$

where

$\mathbf{P}$ = hydrostatic pressure [$ML^{-1}T^{-2}$],
ρ_f = fluid density [ML^{-3}],

g = acceleration due to gravity [LT^{-2}],
h = depth [L].

[For freshwater, $\mathbf{P}$ (psi) = 0.433 h (ft).]

It is important to note that ρ_f may differ significantly from the value for freshwater, as in saltwater production wells. Equation 9.5 gives the full radial pressure on the casing (or screen) if the interior is dry. Below the water level the radial pressure is the difference between the external and internal pressures.

When the borehole is filled with drilling fluid, the hydrostatic pressure can be calculated with the formula $\mathbf{P}$ (psi) = 0.052 $\rho_m h$ (ft), where ρ_m is the drilling-fluid density in pounds mass per gallon and h the depth in feet.

Although the present discussion deals with hydrostatic forces, it is important to note that other forces, including dynamic forces, are sometimes the critical design elements. These will be discussed in several sections throughout this chapter. Because static loads are so easily calculated compared to dynamic loads, the latter are often overlooked, but they can cause failures in materials otherwise quite adequately designed.

Overburden Pressures

The overburden pressures at depth in a well is a function of two phenomena: the pressure of the ground water, and the pressure from the weight of the overburden material and its fluid content. In most cases the weight of the formation is supported by the formation itself. However, in a few circumstances, some or all of the formation load may be transmitted to the fluid content.

Figure 9.3 gives a schematic drawing of a generalized cross section of an aquifer. It shows confined and unconfined

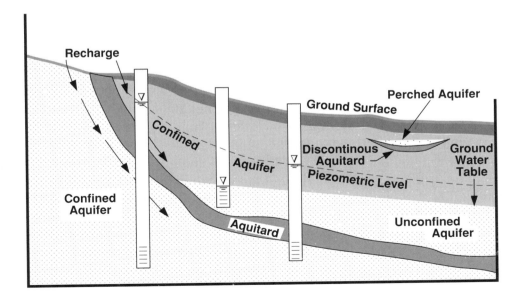

Figure 9.3. Hydrostatic pressures in aquifers.

aquifers and perched water conditions. For an unconfined aquifer, the piezometric level coincides with the top of the saturated zone. For confined aquifers, the piezometric level generally differs considerably from the upper confining surface; in the case of flowing artesian wells, it extends above the ground surface. The head in deep aquifers may be considerably higher than found in upper aquifers.

In some areas overburden pressures may be found in deep, unconsolidated confined aquifers. As a consequence of compaction, repacking, or other sedimentary processes, the rocks at a given depth may not support all the weight of the overburden, and some or all of this weight will be supported by the interstitial fluid. Soil movement within the formation may also temporarily expose the casing and screen to overburden pressures. It may result from slumping, caving, or quick conditions. Under these circumstances the pressure exerted on the casing and/or screen will exceed the hydrostatic pressure, rising as high as the full overburden pressure or even exceeding it momentarily.

Quick conditions usually arise when an upward movement of the ground water effectively suspends the formation so that the formation particles are in loose contact and cannot support the weight of the formation above (i.e., the effective pressure is reduced to zero). Transient quick conditions may arise when a loosely compacted sand is disturbed; during the resettlement the sand effectively behaves as a liquid whose density is defined by the combined mass of liquid and solid per unit volume.

In the petroleum industry a relationship used commonly to estimate the overburden pressure is

$$P = 2.3\rho_w gh ,$$

or equivalently,

$$P \text{ (psi)} = 1.0 \, h \text{ (ft)} , \qquad (9.6)$$

where P is the pressure at depth h and ρ_w is the density of freshwater. The factor of 2.3 is the average specific gravity of rock. Equation 9.6 can be used to estimate the maximum overburden pressure at depth, but it must be remembered that this equation is valid under static conditions. In areas where sudden subsidence or slumping of material may occur, or where quick conditions are prevalent, higher pressures may result from dynamic forces. If experience in a region indicates slumping is a possibility, it would be prudent to apply a larger than usual safety factor for casing and screen collapse strength.

Example

What is the expected pressure in a well at a depth of 500 ft, with a water table 40 ft below the surface?

Answer: The hydrostatic pressure due to the water is calculated from the surface of the water table from Equation 9.5

$$P = 0.433 \, (500 - 40) = 200 \text{ psi} .$$

The maximum overburden hydrostatic pressure expected (e.g., due to settling of unconsolidated sediments) would be (Equation 4.6)

$$P = 1.0 \, (500 \text{ ft}) = 500 \text{ psi} .$$

Thus the maximum hydrostatic pressure due to water (i.e., the maximum external pressure if the casing is dry inside) is 200 psi. If the full overburden pressure is a possibility, then the maximum external pressure could be as high as 500 psi, or even higher if there is sudden slumping.

9.4 COLLAPSE RESISTANCE OF CASING

When the external pressure on casing exceeds the internal pressure by a sufficient amount, the casing will collapse. The unbalanced hydraulic pressure is the same throughout the length of the casing below the pumping level. However, collapse will normally occur at or below this level at a point of maximum eccentricity. The casing will start to deform, first assuming an elliptical shape, and then will fold in on itself, as depicted in Figure 9.4. Because the casing is de-

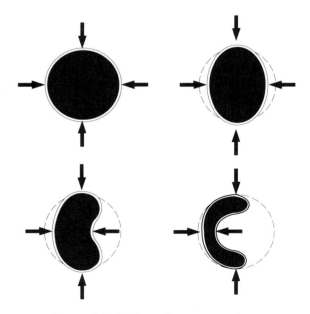

Figure 9.4. Casing collapse progression.

formed above and below the point where collapse starts, it is weakened, and the collapsing region continuously extends. The failure thus propagates along the casing, and can run from a few tens of feet to a hundred feet or more. Although for steel casing the main determinants of resistance to collapse are the diameter and wall thickness, eccentricity (out-of-roundness) is important.

There are several ways in which a circular cylinder can collapse; these are controlled by the diameter to thickness ratio, D/t. With a thin cylinder for which D/t is large (greater than 45 for 35,000 psi yield strength steel), collapse will take place elastically and a buckling type instability will be produced. The exact value of D/t above which elastic collapse takes place is determined by the yield strength of the material. When a cylinder is so thick that failure occurs because the material yield strength is exceeded before it buckles, it is said to collapse by yield failure.

Between yield failure and elastic collapse is a range of D/t for which a cylinder will collapse plastically because the loads are such that plastic deformation takes place before an elastic collapse is possible. This type of collapse would be rare in water well casing, except perhaps for exceptionally thick thermoplastic casing.

In the following section several methods for calculating collapse strength are presented. In almost all cases encountered in water well construction, the elastic collapse formulas are appropriate. For completeness, several other formulas from the American Petroleum Institute are presented.

Timoshenko's Elastic Formula with Eccentricity

Timoshenko's method uses the theoretical elastic collapse strength with adjustment for the eccentricity of the casing. The adjustment is always necessary because a small eccentricity can greatly reduce the collapse resistance of a cylindrical shell.

The theoretical collapse pressure for a perfect cylinder (i.e., with no eccentricity and provided the length is more than eight times the diameter) is

$$\mathbf{P}_{cr} = \frac{2E}{(1 - \nu^2)(D/t - 1)^3} , \qquad (9.7)$$

where

P_{cr} = critical collapse pressure of a perfect cylinder [psi],
E = Young's modulus (3×10^7 psi) for steel,
ν = Poisson's ratio (0.28 for steel),
D = outside diameter of the casing [in.],
t = wall thickness of the casing [in.].

The collapse pressure $\mathbf{P}_d$, as derived by Timoshenko for a cylinder with eccentricity $e = D_M/D_m - 1$ (where D_M is the major diameter and D_m is the minor diameter), is the solution of the quadratic equation

$$\mathbf{P}_d^2 - \left\{ \frac{2Y_p}{D/t - 1} + \left[1 + 3\left(\frac{D}{t} - 1\right)e \right] \mathbf{P}_{cr} \right\} \mathbf{P}_d$$
$$+ \frac{2Y_p \mathbf{P}_{cr}}{(D/t - 1)} = 0 , \quad (9.8)$$

where

$\mathbf{P}_d$ = design collapse pressure [psi],
$\mathbf{P}_{cr}$ = critical collapse pressure of a perfect cylinder (Equation 9.7) [psi],
Y_p = yield strength (35,000 psi for ASTM A 139 Grade B casing, 30,000 psi for Types 304 and 316 stainless steel),
e = eccentricity = $D_M/D_m - 1$ (frequently assumed as 0.01, but may be as high as 0.015 for some standards).

In Equation 9.8 the value of $\mathbf{P}_{cr}$ is calculated from Equation 9.7. Values of $\mathbf{P}_d$, calculated for carbon steel casing with $S = 35,000$ psi and $e = 0.01$ (1%), are tabulated in Appendix H.

9.5 API COLLAPSE FORMULAS

The American Petroleum Institute has conducted extensive tests on casing collapse and recommends the following formulas for use with API casing.[1] *API Bulletin 5C-3* provides a complete description of their derivation and application. Note that these formulas are valid only for steel. In most cases water well casing collapses in the elastic range. The Timoshenko formula with eccentricity, Equation 9.8, usually but not always gives more conservative values than the API elastic formula (Equation 9.12).

1. Reproduced by permission of the American Petroleum Institute from API Bulletin 5C-3, 5th edition, effective July, 1989. Copies of this publication in most recent edition may be obtained from API Publications and Distribution Section at 1220 L Street N.W., Washington, D.C., 20005.

Constants

In the API formulas, the following constants are used:

$$A = 2.8762 + 1.0679 \times 10^{-6} Y_p$$
$$+ 2.1301 \times 10^{-11} Y_p^2$$
$$- 5.3132 \times 10^{-17} Y_p^3 ,$$

$$B = 0.026233 + 5.0609 \times 10^{-7} Y_p ,$$

$$C = -465.93 + 0.030867 Y_p$$
$$- 1.0483 \times 10^{-8} Y_p^2$$
$$+ 3.6989 \times 10^{-14} Y_p^3 ,$$

$$F = \frac{46.95 \times 10^6 \left[\dfrac{3B/A}{2 + B/A} \right]^3}{Y_p \left[\dfrac{3B/A}{2 + (B/A)} - (B/A) \right] \left[1 - \dfrac{3B/A}{2 + (B/A)} \right]^2} ,$$

$$G = \frac{FB}{A} ,$$

where

Y_p = yield strength [psi].

Yield Strength Collapse Formula

The yield strength collapse formula is based on the minimum pressure that generates the yield stress on the inside wall of the casing. This formula will seldom be used for water wells. It is theoretical and calculated by means of the Lamé equation. For

$$\frac{D}{t} \leq \left(\frac{D}{t} \right)_{yp} = \frac{[(A-2)^2 + 8(B + C/Y_p)]^{1/2} + (A-2)}{2(B + C/Y_p)} ,$$

$$P_c = \left[2 Y_p \frac{D/t - 1}{(D/t)^2} \right] ,$$

(9.9)

where

$(D/t)_{yp}$ = cutoff value for yield strength collapse (see Table 9.2).

Plastic Collapse Formula

The plastic collapse formula is based on a regression analysis of experimental data of casing collapse tests. The pressures and casing materials were such that the casing collapsed with stresses in the plastic range of the material. This formula will also seldom be needed in the water well industry. For

$$\left(\frac{D}{t} \right)_{yp} < \frac{D}{t} \leq \left(\frac{D}{t} \right)_{pt} = Y_p \frac{(A - F)}{C + Y_p (B - G)} ,$$

$$P_c = Y_p \left(\frac{A}{D/t} - B \right) - C ,$$

(9.10)

where

$(D/t)_{pt}$ = cutoff value for plastic collapse (see Table 9.2).

Transition Collapse Formula

Equation 9.10 usually intersects the elastic collapse curve, but in some cases it falls below it. To correct this anomaly, a transition curve is provided by the API. For

$$\left(\frac{D}{t}\right)_{pt} < \frac{D}{t} < \left(\frac{D}{t}\right)_{te} = \frac{2 + B/A}{3B/A} ,$$

$$\mathbf{P}_c = \mathbf{Y}_p \left(\frac{F}{D/t} - G\right) , \qquad (9.11)$$

where

$(D/t)_{te}$ = cutoff value for the transition formula (see Table 9.2).

Elastic Collapse Formula

The elastic collapse formula, which is similar in form to the Timoshenko formula, is based on experimental collapse data. For

$$\frac{D}{t} > \left(\frac{D}{t}\right)_{te} , \qquad (9.12)$$

$$\mathbf{P}_c = \frac{46.95 \times 10^6}{(D/t)(D/t - 1)^2} .$$

The elastic collapse formula, Equation 9.12, is quasi empirical and is equal to $0.7125(D/t-1)/(D/t)$ times Equation 9.7 (the Timoshenko formula for a perfect cylinder). This equation takes into account tolerances based on API standard casing. It is best to compare values obtained from calculating Equations 9.8 and 9.12, and to use the smaller.

Use of the API Formulas

Equations 9.9 through 9.12 are empirically derived formulas for API standard casing. The minimum yield strength for this casing (grade H-40) is 40,000 psi. Thus the constants A through F are extrapolated from these data when the yield strength of the casing under consideration is less than 40,000 psi. Differences in tolerances between other standard materials and API casing should also be taken into account if these differences are large. These formulas are not valid for plastic casing. Table 9.2 gives the values of A, B, C, F, and G as well as the transition values of D/t for a yield strength ($\mathbf{Y}_p$) of 35,000 psi steel.

For most water well casing the value of D/t is larger than 45, and the elastic collapse pressure formula (Equation 9.12) is valid. The yield point and plastic collapse formulas (Equations 9.9 and 9.10) are very rarely needed for water wells.

TABLE 9.2 Constants for API Formulas, $\mathbf{Y}_p = 35,000$ psi Steel

$A = 2.9374$	$(D/t)_{yp} = 17.21$
$B = 0.04395$	$(D/t)_{pt} = 26.48$
$C = 603.16$	$(D/t)_{te} = 44.84$
$F = 2.1804$	
$G = 0.0326$	

Note: Refer to Equations 9.9 through 9.12.

The Timoshenko formula, which includes eccentricity effects, is usually more conservative than the API elastic collapse formulas but not always, such as when D/t is large and e is relatively small. The value from Equation 9.8 should be checked against the appropriate API formula, both to verify that the arithmetic is correct and that a conservative value is used.

Figure 9.5 shows a comparison of collapse strengths according to the formulas discussed here.

9.6 COLLAPSE STRENGTH OF DOUBLE WELL CASING

Two-ply casing (discussed in Chapter 11) collapse values are calculated from an empirical formula (*API Bulletin 5C-2*, 1940).[2] The design collapse pressure for two-ply well casing is given by

$$\mathbf{P}_{dw} = \frac{0.65(2E)}{(1 - \nu^2)(D/T)(D/T - 1)^2} , \qquad (9.13)$$

where

D = nominal diameter [in.],

and

$$T = \sqrt{t_1^2 + t_2^2} ,$$

where

t_1 = inside casing thickness [in.],

t_2 = outside casing thickness [in.].

2. Reproduced by permission of the American Petroleum Institute from API Bulletin 5C-2 effective March, 1940. Copies of this publication in most recent edition may be obtained from API Publications and Distribution Section at 1220 L Street N.W., Washington, D.C., 20005.

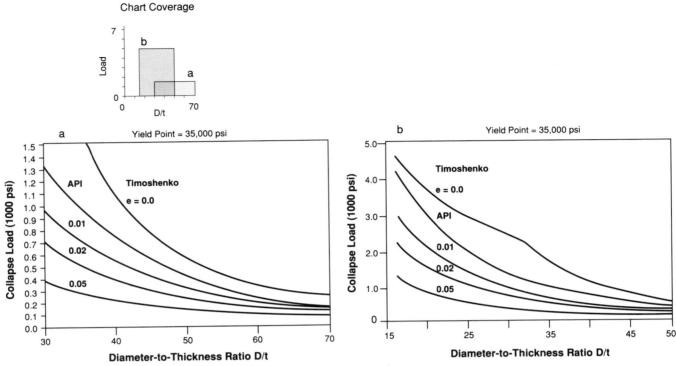

Figure 9.5. Comparison of collapse formulas.

The factor 0.65 is an empirical value derived from collapse measurements. Values of $\mathbf{P}_{dw}$ calculated for standard double well casing sizes are tabulated in Appendix G.

9.7 ADDED STRENGTH DUE TO CEMENTING

Cementing may provide added resistance to collapse. Evans and Harriman measured the collapse pressure of casing with a 1-in. sheath of cement (1). Their results are interesting in that they found little increase in collapse resistance for high-strength steel casing (API N80) but up to a 23% increase in collapse strength for lower-strength casing (API H40). Moreover they concluded that the existence of either radial or longitudinal voids in the cement resulted in no improvement in strength.

Good bonding between the cement and casing is required in order to realize any added strength. As discussed later, the heat generated by setting cement can result in substantial temperature increases. For steel casing one would normally expect a significant enlargement of the casing diameter during cementing due to thermal expansion; later, when the cement has cooled, the casing will shrink. There is therefore no guarantee of maintaining a good bond. Although there may

be some strength added due to cementing, it is not good practice to depend upon it.

9.8 SAFETY FACTORS

Safety factors for collapse pressure should be large. For water wells, it is prudent to assume the casing has at most 85% of the calculated collapse strength; many engineers use 50%. Although casing may meet rigorous production standards, rough handling in transit or on the job may result in out-of-roundness. Equation 9.8 shows that small eccentricities greatly reduce collapse resistance, so it may be prudent to take a large safety factor even though nominal eccentricities have been accounted for. A large safety factor may be especially warranted if the design loads do not include possible formation pressure.

Example

In this example, four potential collapse pressures on 500 ft of 14-in. OD casing are analyzed. Assuming that the casing is cemented, the maximum pressure will be

$$\mathbf{P} = \mathbf{P}_0 - \mathbf{P}_1 = \rho_c g L - \rho_f g L \ ,$$

where ρ_c is the cement density and ρ_f is the fluid density in the pipe. Assume freshwater in the casing. Then

$$\mathbf{P} = (\rho_c g - \rho_f g)L = (118 - 62.4)\,\frac{\text{lb}}{\text{ft}^3}\left(\frac{500\text{ ft}}{144\text{ in.}^2/\text{ft}^2}\right) = 195\text{ psi}\;.$$

If the casing is empty the maximum external water pressure subjecting the casing to collapse is

$$\mathbf{P} = \rho_f g L = \left(\frac{62.4\text{ lb}}{\text{ft}^3}\right)\left(\frac{500\text{ ft}}{114\text{ in.}^2/\text{ft}^2}\right) = 216\text{ psi}\;.$$

If drilling fluid is in the annulus and the well is bailed dry, one should calculate the collapse pressure due to drilling-fluid alone. Assuming a 74.7 lb/ft^3 mud, Equation 9.5 yields

$$\mathbf{P} = 0.5184\,(500\text{ ft}) = 260\text{ psi}\;.$$

This pressure would apply if the annulus were filled with drilling fluid and the casing empty—an extreme case—but it is important to keep in mind the maximum loads possible.

In the unlikely event that the formation is unconsolidated and there is a possibility of the casing taking the full overburden pressure, Equation 9.6 yields

$$\mathbf{P} = 1.0(500\text{ ft}) = 500\text{ psi}\;.$$

Casing Strength

AWWA Standard A 100-84 specifies a minimum ¼-in. thickness for 14-in. nominal OD casing at 500 ft. For $D = 14$ in., $t = ¼$ in., Equation 9.7 yields 391 psi for the perfectly round cylinder collapse pressure. ($E = 30 \times 10^6$ psi, $\nu = 0.3$.) Assuming that $Y_p = 35{,}000$ psi and eccentricity is 0.01, Equation 9.8, the Timoshenko formula with eccentricity yields 241 psi as the critical collapse pressure. For this size, $D/t = 56$, so the API elastic formula applies (Equation 9.12). This formula yields $\mathbf{P} = 277$ psi. The more conservative Timoshenko formula should be used. Thus the ¼-in. thickness is adequate for the cement and external water pressures, slightly thin with full length mud column pressure, and very inadequate if formation pressure is expected.

Designing for the formation pressure, $t = ⅝$₁₆ in. and $t = ⅜$ in. are tried. Equation 9.8 yields 419 psi and 636 psi, respectively. For $t = ⅜$ in., $D/t = 37.33$ which is less than $(D/t)_{te}$, so the API formula for transition (Equation 9.11) should be used for verification (see Table 9.2). This formula yields $\mathbf{P} = 864$ psi, so the ⅜-in. thick casing is adequate for the full formation pressure of 500 psi.

9.9 THERMOPLASTIC COLLAPSE FORMULA

The API formulas are empirical, reflect collapse measurements made on API-approved casing, and do not apply to thermoplastic material. Kurt has made numerical calculations to find the collapse pressure of plastic casing under various circumstances, including effects of pressure gradients and soil resistance (2). When soil resistance and pressure gradients are not considered (the conservative approach and good design practice), the following collapse formula best fits numerical results:

$$\mathbf{P}_c = \frac{0.75(2E)}{(1 - \nu^2)(D/t - 1)^3}\;, \qquad (9.14)$$

where

$\mathbf{P}_c$ = collapse pressure [psi],
D = outside diameter [in.],
t = thickness [in.].

Note the similarity between Equations 9.14 and 9.12; the latter is based on measured data. Kurt compared Equation 9.14 with ASTM Standard F 480,

$$\mathbf{P}_c = \frac{2E}{(1 - \nu^2)(D/t)(D/t - 1)^2}\;, \qquad (9.15)$$

and found that for the standard dimension ratios (SDR) of thermoplastic casing, Equation 9.14 yields collapse pressures about 25% less than those calculated with the equation used in ASTM Standard F 480-76 (Equation 9.15).

Johnson, Kurt, and Dunham measured collapse pressures of short lengths of thermoplastic casing (length to diameter ratios from 8 to 10) and found good agreement with Equation 9.15 when the minimum thickness (rather than the nominal thickness) of the sample casing was used in the formula (3). This is the accepted practice. However, their sample was relatively small, and typically had a 0.5% variation in inside diameter. A large sample may show more variation in eccentricity and thickness than allowed by ASTM standards, making use of Equation 9.14 rather than Equation 9.15 appropriate.

A second and very important consideration in the selection of thermoplastic casing is the temperature to which the casing will be exposed. Johnson et al. found that the pressure required to collapse thermoplastic casing decreased approximately 0.5 psi per degree Fahrenheit above 70°F (6.2 kPa with each degree Celsius increase above 21°C) (3, 4). This can result in a considerable loss in collapse resistance, especially during cementing operations.

Johnson et al. measured the temperature rise during the cementing of thermoplastic casing (4). Their results indicated that under "normal" conditions with a 1½-in.-thick annulus (3-in. oversize borehole), temperature increases on the order of 17°F to 26°F could be expected. When the annulus thickness was 4 in., the temperature rise was as much as 67°F. In extreme cases, as would be found in a large annulus formed by slumping, the rise in measured temperature was as high as 180°F. Maximum temperature was found to occur eight to ten hours after the pour. ASTM Type I portland cement was used in the experiments.

Temperature rises during cementing can seriously endanger thermoplastic casing, especially in formations subject to slumping. The expected rise of 26°F in normal operations, or possibly even greater temperature rises, and the resulting decreased collapse resistance of 0.5 psi/°F, should always be a design consideration when cemented thermoplastic casing is used.

Effect of Packing

In practice, the resistance of soil to movement may increase the collapse resistance of any casing, since the collapse of a cylinder requires that the diameter increase in at least one direction. As shown by Kurt, soil resistance can greatly increase the collapse resistance of a circular cylinder (and is taken into account in buried pipelines where construction practices allow control over the bedding and backfilling) (2). It is claimed in ASTM Standard F 480-76 that thermoplastic "casing with gravel or concrete packing can operate with almost unlimited head differential." As discussed in the section on cementing, cementing potentially increases the collapse resistance of steel casing by less than 25%. Greater increases in resistance would be expected for plastic casing if bonding is good. However, the added strength due to packing is certainly not "almost unlimited," and good practice in well design requires that any added strength not be considered without sound technical reasons for doing so. Formation properties are never completely known, and many other factors, such as slumping, compaction of the gravel envelope, and creation of voids, are involved. The possibility of increased strength due to soil resistance should not be considered in design calculations.

9.10 COLLAPSE STRENGTH OF WELL SCREEN

Collapse strength is perhaps the most important strength criterion for well screens. The strength of a screen is determined by its design and diameter. Because of the complex shapes of most well screens, the collapse strength is difficult to calculate, and the manufacturers are often the only source of information.

Wire Wrap Screen

The pressure tending to collapse a well screen that is not plugged and not subject to formation loads is simply the pressure drop across the screen. This pressure is applied to the fraction of the screen consisting of the non-open area. However, pressure on sand or gravel plugging some of the open area will transmit additional collapsing forces. If there are formation loads, then the collapsing load will be equal to the formation pressure applied to the screen area. Thus, for a wire wrap screen, the pressure force tending to collapse the screen may vary from the screen pressure drop applied only to the wire to the formation pressure applied to the entire screen surface area.

Calculation of the collapse resistance of a wire wrap screen can be made assuming that very little collapse resistance is provided by the longitudinal rods and that all of the collapse load is carried by the wire wrapped around and welded to the rods. Following the work of Timoshenko, the pressure at which a single wire ring will collapse is given by

$$\mathbf{P}_{ww} = \frac{24EI}{(w + s)D^3} , \qquad (9.16)$$

where

$\mathbf{P}_{ww}$ = collapse pressure of continuous wire wrap screen [psi],

I = moment of inertia of the wire cross section [in.]^4,

w = width of the wire on the external face [in.],

s = slot width of the screen [in.],

D = mean diameter of the screen [in.].

In this equation it is assumed that the ring will collapse at the pressure $\mathbf{P}_{ww}$ applied over both the ring and the adjacent slot width. This will clearly be true if the load on the screen arises from the formation pressure. However, if the only pressure force is the pressure drop across the unplugged screen, then the actual collapse resistance is $\mathbf{P}_{ww}$ $(w + s)/w$. For normal screen sizes, $(w + s)/w$ is between 1.1 and 1.5, so using Equation 9.16 introduces a factor of safety between 1.0 and 1.5, depending upon how much the screen is plugged. These principles apply to all types of screen.

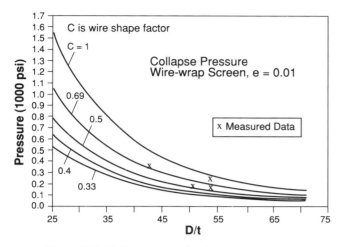

Figure 9.6. Collapse strength of wire wrap screen.

The moment of inertia I depends on the shape of the wire, but it generally is between $wt^3/12$ (a rectangle) and $wt^3/36$ (a triangle), where t is the thickness of the wire. Within the limits of the assumption that the longitudinal rods add little to the collapse strength (which is valid for screens long compared to their diameter, the usual case), Equation 9.16 provides a conservative estimate of the collapse strength of wire wrap well screen. Note that this value will always be less than that for blank casing of the same diameter as the screen and thickness of the screen wire (where $s = 0$, and $I = wt^3/12$). If keystone-shaped wire is used, the screen is less than one-third as strong as blank casing.

Figure 9.6 shows collapse strengths of wire wrap screen for various wire shapes, derived from Equation 9.16. In this figure the ratio of the collapse pressure to Young's modulus is plotted against the screen thickness to diameter ratio for several values of $C = I/[I_0(1 + s/w)]$, where I_0 is the moment of inertia of a rectangle of width w and height t, and I is the moment of inertia of the wire used. (For triangular wire when $s = 0$, C is equal to ⅓.) For a trapezoid of major base w, minor base b and height t,

$$C = \frac{1/3[1 + 4b/w + b^2/w^2]}{(1 + b/w)(1 + s/w)} .$$

If b/w is 0.1 and $s = 0$, $C = 0.43$. These values may be slightly conservative because the small added strength of the longitudinal rods is neglected.

Also shown on Figure 9.6 are measured collapse data for wire wrap screen. The wire used was not quite triangular but shaped more like a trapezoid, with I/I_0 about 0.57. Note that a very small deviation in the shape of the wire from a perfect triangle to a trapezoid or rectangle increases the strength of wire wrap screen significantly.

Pipe Base Well Screen

A wire wrap screen is sometimes formed on or fitted over a pipe base for strength and to protect the somewhat vulnerable wires from tools inserted in the well during well development or rehabilitation. Collapse strength is dependent upon the pipe diameter and wall thickness and the size, type, and number of perforations in the pipe. It is best to check with the manufacturer for collapse strength data.

Shutter Screen

The shutter or louvered well screen is manufactured from casing by perforating openings in such a way that no material is removed. As the name implies, the slots form horizontal louvers. In effect, these louvers represent small arches around the circumference of the screen. The slots are dense enough in the vertical direction that the arches are able to add to the collapse strength despite the openings in the horizontal direction. Thus shutter screen is stronger in collapse strength than the casing from which it is made. It can be up to 60% stronger in configurations providing a higher number of openings per unit surface area.

Hydrodynamic collapse test data indicate that a 50% increase in strength over the theoretical value for blank casing is common if the standard eccentricity of 0.01 (1%) is used. These data also indicate that shutter screen is about 20% stronger than the theoretical value for blank casing, using the measured eccentricity of the test specimen. Because the screen shape is complicated, a collapse pressure relationship based on theoretical considerations would be difficult to formulate. Manufacturers' test data should be consulted for actual values (typical values, obtained from hydrostatic tests and three-edge bearing tests, are shown in Figure 9.7). To be conservative, a collapse strength for shutter screen can be assumed to be the same as for equal thickness blank casing.

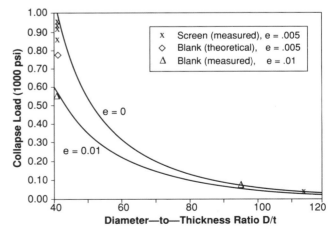

Figure 9.7. Collapse strength of shutter screen and blank casing.

Bridge Slot

Although no material is removed in the process of manufacturing the bridge slot screen, the slots are numerous and vertically oriented. This results in a structure considerably weaker than casing of equivalent wall thickness and diameter. The tensile strength is also reduced (because of the staggered nature of the slots), although the exact reduction is difficult to calculate.

Perforated Pipe Screens

Perforated pipe screens are generally manufactured from well casing. Openings include vertical machined slots, oxyacetylene-torch-cut slots, and punched slots. In all cases the strength of the screen is dependent upon the size, orientation, and density of slots, the manner in which they are made, and the residual stresses left by the perforation process (thermal stresses, jagged corners, etc.). Because the holes are usually relatively large, a perforated pipe screen has a lower collapse resistance than the material from which it is manufactured. Horizontally oriented slotted screens have a higher collapse strength than vertically oriented slotted screens.

Screen Selection

Since each screen type is available in a variety of configurations, thicknesses, and so forth, the designer, after examination of the operating conditions, should select a screen built to adequate tensile and collapse strength requirements. For the wire wrap screen the major selection is the wire's size. For vertical slot configurations, it is the number of slots per unit length. For shutter screen, safety and consistency are ensured if the screen is manufactured from the same thickness material as the casing. Selection of casing and screen is discussed in Chapter 11.

9.11 COMPRESSIVE LOADS

Buckling

Although the casing and screen usually remain under tensile stress due to their own weight, subsidence of the formation or poor construction practices may put them into compression. Because the string is effectively a long slender column, it is extremely vulnerable to compressive loads, and it is important to examine their nature.

Buckling Due to Weight

The vulnerability of casing to buckling is easily seen with a simple example. Consider a round pipe with mean diameter

D, wall thickness t, and length L, subjected to an axial compressive load $\mathbf{F}$. The pipe will buckle if

$$\mathbf{F} > \left(\frac{\pi D}{2}\right)^3 \frac{Et}{L^2} .$$

A 12-in. nominal diameter steel casing, ¼-in. thick and 200 ft long, will buckle with a load greater than 10,000 lb. This load corresponds to 300 ft of the same casing (nominal weight 33.4 lb/ft) so that a 500-ft length of 12-in. casing allowed to rest unsupported in an open hole will subject the lower 200 ft to a buckling load. Clearly, only very short, large-diameter strings can support their own weight.

It is for this reason that long lengths of casing should not be allowed to rest on the bottom of an open hole; without lateral support, buckling becomes a real danger. Because of its reduced cross-sectional area, the well screen has even less resistance to buckling. A screen resting on the bottom of an open hole while supporting the entire weight of the casing above is particularly vulnerable.

Dropping the casing is also very dangerous. If it is dropped only 1 ft, its speed at impact will be over 8 ft/sec. Depending on the type of formation encountered (e.g., hard rock), the impact load can be enormous. The dynamic loading caused by the impact of the casing on the bottom, even for very short drops, can far exceed the compressive load of its own weight, and failure is very likely to result. Designing the casing and screen to withstand buckling loads under these circumstances would be nearly impossible; proper handling is the only way to insure against failure.

Jacking Casing

In the southwestern United States, the casing placed during drilling is often installed by jacking. Two hydraulic jacks are anchored in a large pit surrounding the borehole, and positive-displacement water pressure pumps operate the jacks at pressures up to 3000 psi. The casing can be installed to depths greater than 1500 ft by this system. When larger diameters are used, four jacks may be employed, which will typically apply a downward force of up to 300 tons. Often a mud scow is used as the drilling tool, removing material ahead of the casing shoe. The stresses on the casing during jacking are axial (due to the jacks) and shear (due to the wall friction). As long as the force from the jacks divided by the cross-sectional area of the casing remains below the yield point of the casing material, the process is usually safe. Joint separations are rare because the casing is almost always in compression.

There is, however, a slight possibility of casing failure due to buckling. This is unusual since under normal circumstances the casing fits tightly into the hole and thus is supported peripherally along its length. The aboveground portion of the casing is, of course, unsupported, and if the casing cannot be moved easily, it is here that failure will occur.

Double well casing is generally used when jacking, but the load applied at the top is applied to a single section, which has half the thickness of the lower, two-ply sections. The compressive stress at the top is then

$$\sigma_c = \frac{F}{A} ,$$

where **F** is the applied load and *A* is the cross-sectional area of a single ply of the casing. The applied load should always be kept below the yield strength of the casing.

Driving Casing

The history of water well drilling with cable tools is filled with unfortunate examples of casing that failed under the stress of installation by driving. Driving is accomplished through the impact of the weight of the drilling tools on the casing. A drive head, which acts as an anvil, is positioned on top of the casing. Clamps are attached to the tool string to serve as a hammer face. Blows may be struck singly by raising and dropping the tool string, or repeatedly by using the spudding-beam motion.

Resistance due to friction at the casing shoe suggests that driving forces are compressive, and thus they tend to hold the casing together. However, this is not always the case, and an analysis of what actually occurs to a casing string being driven may help to avoid operating conditions that cause failure.

Measurements of the forces involved in driving casing have been made by Knauss and Liechti (5). Using high-speed photography and strain gauges, they were able to measure the stresses and the motion of casing being driven into the ground. Figure 9.8 shows a typical measurement of the movement of the top of the casing with time. The well casing was approximately 250 ft long.

After the casing is struck, a compressive wave travels down the casing, is reflected, and returns to surface in 30 msec. Smaller preceding waves may be from ground movement (possibly shaking the camera) and the bouncing of the drive clamp. The formation at the bottom breaks as the compressive wave returns a second time and allows the casing to move down.

Figure 9.9 shows the typical motion of casing measured by Knauss and Liechti under different circumstances. In this case the casing moved in several discreet jumps after impact. Knauss and Liechti explained that this behavior is to be expected if the bottom end of the casing is stress free, such as when it encounters a formation that provides little resistance.

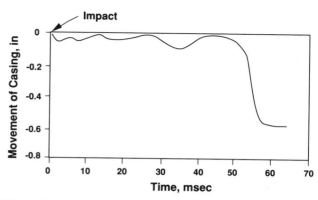

Figure 9.8. Movement of the top of driven casing after initial impact with resistance at the casing shoe.

Initially, the casing is at rest, supported by friction between the borehole wall and the casing. As the drive clamps strike, a compressive wave starts down the casing and travels at a speed of approximately 17,000 ft/sec for steel. The bottom of the casing is absent of stress and free to move, thus the reflected wave is negative (tension), as opposed to a compressive reflection when the casing bottom is restrained. The tension wave returns to the surface, where the casing is also free, and reflects as a compressive wave. However, in order to maintain the stress-free condition at the end of the casing, the velocity doubles. This doubling of velocity is seen in Figure 9.9 where the second hop is almost exactly twice the initial hop. The motion slowly dampens because of friction between the casing and the borehole.

Stress measurements by Knauss and Liechti revealed maximum compressive stresses of approximately 25,000 to 70,000 psi during driving, although average stresses (averaged over the length of the compressible wave) were on the order of 10,000 psi. When the casing is restrained at the shoe, the maximum stresses during reflection may, for short periods

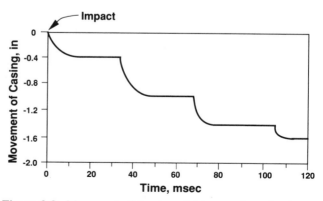

Figure 9.9. Movement of the top of driven casing after impact with little or no resistance at the casing shoe.

of time, exceed the yield strength of the casing, since for a restrained end, the stress doubles and is of the same sign.

When the casing is not restrained a more serious problem develops because there are now significant stress reversals imposed on the casing. Knauss and Liechti point out that these stress reversals combined with the typical number of blows required to drive the casing can result in fatigue failure. Welded joints, particularly those near the bottom, are usually the first to fail. Butt-welded casing is especially vulnerable; double well casing to a much lesser degree. It is for this reason that strict quality control of welded joints is essential when casing is driven during well construction.

9.12 BURSTING STRENGTH

In addition to the fundamental tensile, collapse, and compressive loads that apply to almost every water well, there are two other forces that can be considered in a general sense and may be important in some installations. They are bursting and bending.

The burst strength of well casing is given by

$$\mathbf{P}_b = \frac{2\mathbf{Y}_p}{D/t} ,$$

where

$\mathbf{P}_b$ = burst strength [psi],
D = diameter of casing [in.],
t = thickness of casing wall [in.].

It would be extremely unusual for water well casing to fail by bursting because there are seldom situations in which the internal pressure exceeds the external pressure, except during cementing operations by the "Bradenhead" method, discussed in Chapter 12. During this operation, cement is pumped under pressure through tubing inside the fluid-filled casing and forced up the annulus to the surface. Near the top of the casing the internal pressure exceeds the external pressure by the sum of the buoyant weight of the cement and the pressure needed to overcome the viscous forces involved in moving it. The net cement-pressure difference may be larger if the casing is not completely filled with fluid. However, in the experience of most drilling contractors, internal pressures will not exceed 150 psi during this operation, well within a safe range.

9.13 BENDING STRESSES

Bending stresses sufficient to cause misalignment generally occur either from crooked boreholes or where caving or other formation or ground movement imposes a lateral force. Such stresses are difficult to assess in advance but have resulted in joint or material failures. Bending stresses can be calculated from

$$\mathbf{S} = \frac{ED}{2R} ,$$

where

$\mathbf{S}$ = bending stress [psi],
D = casing diameter [in.],
R = radius of curvature of the bend [in.].

If the deflection of the well from the vertical is X feet in a vertical depth of Y feet, then R can be calculated from:

$$R = \frac{X^2 + Y^2}{2X} .$$

Thus, for 16-in. diameter steel casing, a deflection of 1 ft in 100 ft of depth yields $R \simeq 5000$ ft and $\mathbf{S} = (16/12 \text{ ft})$ $(30 \times 10^6 \text{ psi})/2(5000 \text{ ft}) = 4000$ psi. In this example, the stress is well below the yield point of steel, and in general, bending stresses due to crooked holes are not an important problem. The fact that these stresses may exist is one reason for applying safety factors in design calculations.

Screen-Bending Loads

A screen can also be subject to bending loads. Usually bending is compensated for by using a large safety factor in the column tensile strength. If estimates of the bending strength of the well screen need to be made, one must consider the fact that the strength depends upon the screen configuration.

9.14 COMBINED LOADS

To this point only loads acting separately have been considered. Frequently, two or more of the loads discussed can act together and in combination may induce stresses exceeding the yield strength of the material, whereas taken singly, they would not. For example, casing normally has both radial loads (from hydrostatic pressures) and axial loads acting simultaneously. For very deep water wells, the lowest section of the well casing or screen may be in compression, and this will increase the collapse strength of the section, as discussed in *API Bulletin 5C-3*.

Other stresses acting in concert that may be important include axial tension and bending. The increased axial stresses imposed on casing because of crooked holes may be significant when combined with tensile stresses. In general, these added stresses are taken into account by applying safety factors.

Torsional loads are usually unimportant. However, the wire wrap screen has almost no torsional resistance; it should always be handled carefully.

9.15 EXTRAORDINARY LOADS

Specific abnormal loads on the casing and screen must be considered. Under normal circumstances they are unlikely to be experienced. Poor construction practices or abnormal conditions are occasionally encountered, and a review of some known abnormal loads is useful.

Possible Loads during Construction

There are several avoidable extraordinary practices that can cause casing failure. The usual method of cementing with a tremie is illustrated in Chapter 12. If the tremie is slowly withdrawn during the operation, the outside pressure remains close to the hydrostatic value during the cementing operation. Since cement has about twice the density of water, the unbalanced pressure will be

$$\mathbf{P} = (\rho_c - \rho_f)gh \ ,$$

where ρ_c is the density of the cement and ρ_f is the density of the fluid in the casing.

The casing may be easily collapsed if the tremie is not withdrawn as cementing proceeds. When the tremie end is near the surface of the cement, there is little resistance (back pressure) to the cement in the tremie. However, if the tremie is submerged in the cement, the cement leaving the tremie must, with each stroke of the pump, push up all the cement in the annulus above in order to make room for the added cement volume. The pressure at the tremie end then equals the sum of the hydrostatic pressure of the cement, the pressure from the viscous resistance of the moving cement, and the pressure needed to accelerate the cement upward. Since a positive displacement reciprocating pump is used, the collapse strength of the casing easily may be exceeded without warning. It must be emphasized that hydrostatic pressures are, as the name implies, valid only in static equilibrium, and potentially important dynamic forces often are not considered.

Installation by Floating

A dangerous but unfortunately common practice is the "floating" of casing into an open borehole. This operation is usually undertaken when the total weight of the string approaches or exceeds the safe mast capacity. The casing is plugged at some level with a plate that can be ruptured after the string is installed. The force of the water pressure on the plate reduces the load carried by the mast (Figure 9.10). It can be calculated from

$$\mathbf{W} = g[\rho LA + \rho_m dS - \rho_f (l - h)S] \ ,$$

where

$\mathbf{W}$ = load on the mast [lb],
ρ = density of the casing [slug/ft^3],
ρ_m = density of fluid in the casing [slug/ft^3],
ρ_f = density of the fluid in the borehole [slug/ft^3],
A = cross-sectional area of the casing wall [ft^2],
S = total cross section of the casing [ft^2],
l = distance from the surface to the plug [ft],

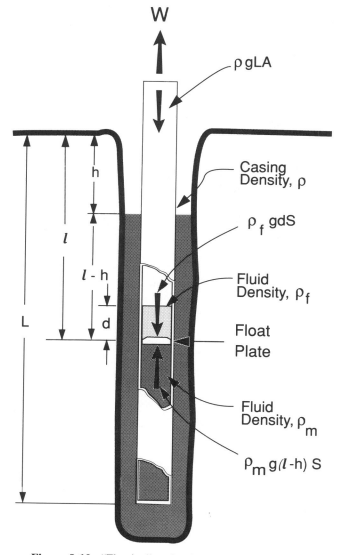

Figure 9.10. "Floating" casing into an open borehole.

h = distance from the surface to the water level [ft],

d = depth of fluid in the casing above the plug [ft].

This can be rewritten as

$$\frac{W}{S} = g\left[\frac{\rho LA}{S} + (\rho_m - \rho_f)d - \rho_f(L - h - d)\right],$$

which states that the load on the mast per unit cross-sectional area of well casing is equal to the weight of the casing plus the small pressure differential resulting from the difference in density in fluids inside and outside the casing minus the maximum collapsing pressure on the casing. Thus, if the load on the mast is held constant as the casing is lowered, the collapse stress must increase and it can reach the failure point. If the collapse stress is held constant as the casing is lowered, then the depth of liquid inside the casing must increase as will the load on the mast.

As the operation proceeds, the collapse stress $g\rho_f(L - H - d)$ should be computed and compared with the safe limit as previously defined as Equations 9.7 through 9.13. The mast load, W, must also be maintained within safe limits.

The value of W should be kept less than two-thirds of the safe mast capacity for the rig. Water or drilling fluid should be poured into the casing from time to time as it is being floated down the borehole in order to keep the upper portion of the casing in tension. If a compressive load is allowed over the length of the casing, a failure by axial buckling or collapse may occur, although for most casing sizes collapse will occur first. The compressive stress above the plate is proportional to S/A. Since S/A is in the order of ten for common casing diameters, if care is not taken, large compressive stresses can develop quickly with depth of installation. It should be kept in mind that the success of the operation depends upon mast capacity and the casing collapse strength, both of which may be near failure.

Bailing the Well

Bailing has caused casing failures when performed with insufficient forethought. The water level can be easily and quickly bailed down, subjecting the casing to the full external hydrostatic pressure. This is particularly true when the casing shoe is in an impervious formation.

Gravel Installation and Well Development

Pack installation in a gravel envelope well must be performed with care. Tensile failures of casing have occurred when gravel bridges in the annulus before reaching the bottom. The weight of the gravel can be transmitted to the casing by friction.

Well development can subject the well casing and screen to a variety of stresses. Swabbing leads to the movement and consolidation of the gravel, and its subsidence may force the casing or screen out of round. Downward gravel movement while swabbing may produce enormous tensile loads on the casing. Some failures that have occurred indicate loads exceeding 50 tons.

Use of a single rather than a double swab during drilling fluid conditioning prior to gravel installation and development can also collapse casing. As can be seen in Figure 9.11, the circulating fluid can reenter the well instead of rising to the surface through the annulus behind the blank casing. A pressure differential develops due to the density difference between the drilling fluid and the circulating fluid. This may be sufficient to collapse the casing if the well is sufficiently deep. A second swab located in the blank casing causes the circulating fluid to return to the surface through the annulus.

Insufficient well development can lead to collapse. If the screen or gravel envelope is plugged, bailing or pump start-up can excessively lower the water level in the well. Development pumping can lead to similar results if the screen becomes plugged and water levels are not carefully monitored.

Postconstruction Loads

Excessive collapse pressures can result from too much or too rapid drawdown, an abnormally high water table (caused by surface flooding or water injected into upper aquifers), and formation pressures caused by movement due to subsidence or earthquake.

During pump start-up the casing can be temporarily subjected to pressures in excess of the design pressure. If the pump is started rapidly and brought immediately to a high flow rate, water in the well will be pumped out while the water level in the surrounding formation will be declining at a much lower rate (Figure 9.12). This will result in a collapse pressure on the casing equal to the difference between the external head and the pumping level. In one case, rapid pump start-up took place when water table levels were seasonally high. The casing collapsed both above and below a screened section due to hydrostatic pressure. The shutter screen did not collapse because its strength was greater than that of the casing and the pressure differential at the screen was limited to the screen head loss. A conservative design should ignore the eventual formation of a cone of depression and assume that the full potential pressure exists on the casing.

Water from heavy rain or snow melt (or, in some cases, from the recharge of a perched aquifer) can also cause casing collapse if it flows into the well-borehole annulus. Relief holes in the casing can help prevent this. This problem is particularly insidious because it can be seasonal, and a well may operate for a long period of time, only to collapse

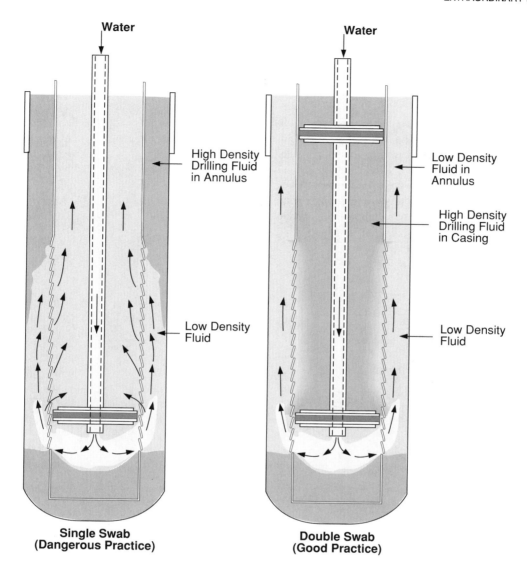

**Single Swab
(Dangerous Practice)**

**Double Swab
(Good Practice)**

Figure 9.11 Circulation of fluid prior to gravel installation.

suddenly. Similarly, collapse can occur during flooding if the well is not properly sealed at the surface, or during development or test pumping if water is allowed to flow into the well-borehole annulus.

Dropped pumps and tools are not uncommon occurrences and may result in damage to some types of screens. The wire wrap screen is particularly vulnerable in this regard. The internal screen configuration as well as overall strength should be considered in screen selection.

Earthquake Loads

Earthquake-induced ground movement has resulted in well damage or failure. It is not feasible to propose well design capable of resisting such large formation displacements.

Therefore wells should not be located near active faults or unstable ground when continuous supply is critically important. Further information is available in a paper by Nazarian (6).

Subsidence

In many regions water depletion results in compression of the formations and ground surface subsidence. In some instances subsidence may be very large, up to 30 ft or more. When it occurs, the casing and screen may be put into compression, resulting in one or more breaks. These breaks are usually a telescoping type of rupture.

Because of the high modulus of elasticity of steel well casing, relatively small strains can easily cause axial compressive loads that exceed its strength. This can be illustrated with the relationship between stress and strain:

$$\sigma = E \frac{\Delta L}{L} ,$$

where E is Young's modulus and L is the length of the casing. For a 1000-ft casing string, a 1-ft change in length will result in a stress near the yield strength of typical steel casing material.

Because axial loads can be so large under subsidence conditions, it is futile to try to prevent failure by strengthening the casing. A better solution is use of a compression section

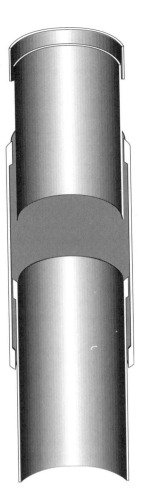

Figure 9.13. Typical compression section.

(Figure 9.13) that allows the casing to telescope. Compression sections are often installed at the bottom of the pump housing casing, although best results are obtained by examining the history of the area or by photographic or video examination of failed wells, and then positioning them at depths of known failures. Two or three sections incorporated in a well are not uncommon in areas of known subsidence.

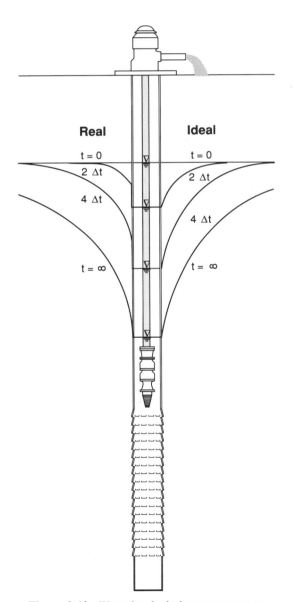

Figure 9.12. Water levels during pump start-up.

REFERENCES

1. Evans, G. W., and D. W. Harriman. 1972. "Laboratory Tests on Collapse Resistance of Cemented Casing." Society of Petroleum Engineers of AIME. 47th Ann. Fall Mtg., Paper No. SPE4088, 6 pp.

2. Kurt, C. F. 1979. "Collapse Pressure of Thermoplastic Water Well Casings." *Ground Water* 17, 6, pp. 550–555.

3. Johnson, R. C., C. E. Kurt, and G. F. Dunham. 1980b. "Experimental Determination of Thermoplastic Casing Collapse Pressures." *Ground Water* 18, 4, pp. 346–350.

4. Johnson, R. C., C. E. Kurt, and G. F. Dunham. 1980a. "Well Grouting and Casing Temperature Increases." *Ground Water* 18, 1, pp. 7–13.

5. Knauss, W. G., and K. M. Liechti. 1978. "Dynamic Fatigue of Impact Driven Water Well Casing." Report to Rosoce Moss Company. California Institute of Technology, Pasadena, CA.

6. Nazarian, N. H. 1973. "Water Well Design for Earthquake-Induced Motions." *J. Power Div. ASCE* 99, P02. Proc. Paper 10176, pp. 377–394.

READING LIST

American Petroleum Institute. 1985. "Bulletin on Formulas and Calculations for Casing, Tubing, Drill Pipe, and Line Pipe Properties." API BUL 5C-3, February 1, 1985. Washington, DC, 42 pp.

American Petroleum Institute. 1985. "Specification for Line Pipe." API Spec 5L, May 31, 1985. Washington, DC, 94 pp.

American Society for Testing and Materials. 1984. "Standard Specification for Pipe, Steel, Black and Hot-Dipped, Zinc-Coated, Welded and Seamless." ASTM A 53-84c. Philadelphia, PA, 25 pp.

American Society for Testing and Materials. 1988. "Standard Specification for Thermoplastic Water Well Casing Pipe and Couplings Made in Standard Dimension Ratio." ANSI/ASTM F 480-88. Philadelphia, PA, 14 pp.

American Society for Testing and Materials. 1985. "Standard Specification for Welded Large Diameter Austenitic Steel Pipe for Corrosive or High Temperature Service." ASTM A 409-85. Philadelphia, PA, 10 pp.

American Water Works Association. 1984. "AWWA Standard for Water Wells." AWWA A100-84. Denver, CO, 75 pp.

American Water Works Association. 1980. "AWWA Standard for Steel Water Pipe 6 Inches and Larger." AWWA C200-80. Denver, CO, 15 pp.

American Water Works Association. 1982. "AWWA Standard for Field Welding of Steel Water Pipes." AWWA C206-82. Denver, CO, 7 pp.

American Water Works Association. 1982. "Ground Water." AWWA Manual M21. New York, 130 pp.

Gatlin, C. 1960. "Oil Well Cementing and Casing Practices." In *Petroleum Engineering, Drilling and Well Completion*. Prentice-Hall, Englewood Cliffs, NJ, ch. 14.

Ritchie, E. A., et al. 1981. "Water Well Standards: State of California." State of California, Resources Agency, Dept. of Water Resources Bulletin 74-81, 92 pp.

Saye, F. J., and T. W. G. Richardson. 1954. "Field Testing of Casing String Design Factors." *API Drilling and Production Practices*, pp. 23–38.

Schuh, F. J. 1968. "Failures in the Bottom Joints of Surface Intermediate Casing Strings." *J. Pet. Tech.* 20, 1, pp. 93–101.

Timoshenko, S. 1941. *Strength of Materials. Advanced Theory and Problems*. Part II. D. Van Nostrand, New York, 510 pp.

Uren, G. C. 1956. *Petroleum Production Engineering, Oil Field Development*. McGraw-Hill, New York, 792 pp.

CHAPTER 10

Corrosion and Incrustation

10.1 INTRODUCTION

Corrosion and incrustation are two major factors that affect the design, material selection, and operation of water wells. Although corrosion is not the usual cause of well failure (aquifer contamination, pressure differential caused by drawdown in the water-bearing formation, or subsidence of the overlying soils are more common causes), metallic casings and screens will fail eventually from corrosion. Similarly, although few wells fail from incrustation, many experience a decrease in specific capacity as a result of this phenomenon.

This chapter is devoted to understanding chemical equilibria, water-quality parameters, and their interaction, which result in corrosion and incrustation of casing and screen material. By understanding the mechanisms involved, the design engineer can select materials appropriate to the design conditions and desired well life.

For water wells, steel is the traditional casing and screen material because of its availability, low cost, workability, and physical properties. These attributes are partially offset by steel's susceptibility to corrosion in most ground waters. The corrosion of steel in water wells occurs in four distinct zones; the external surface of the casing and screen in contact with the formation or gravel envelope, the internal surface above the static level, the internal surface between the static level and the pumping level (splash zone), and the internal surface below the pumping level. Each zone experiences corrosion unique to its exposure to the chemical and bacteriological constituents of the surrounding formation, the atmosphere, and the ground water. As a result of alternate wetting and drying cycles and high dissolved oxygen levels, corrosion is usually most severe in the splash zone, followed by the zone below the pumping level.

As steel oxidizes (loses electrons to an oxidant, such as dissolved oxygen), structural deterioration occurs that may eventually lead to failure of well casings and screens. Enlarged screen openings enable material from the filter zone and aquifer to enter the well, and in extreme cases, this leads to structural failure. The extent of damage caused by corrosion has been well documented by prior investigators (1–6). In each instance the investigators were able to point to specific water-quality parameters that were incompatible with the installed well material.

The term incrustation is used with wells to describe any undesirable substance deposited on well components, including corrosion products, precipitated minerals, incrusted sand and silt from the geologic formation, and bacterial growths. Whereas corrosion enlarges screen openings and reduces structural strength, incrustation reduces the size of screen apertures and coats exposed surfaces. The net effects of incrustation are to create blockages in the screen apertures and casing, reduce production, and increase the head loss through critical flow channels. Studies have shown that well capacity can be reduced from 80% to 100% by incrustation (7, 8).

This chapter begins with a preliminary discussion of chemical equilibria, which form the foundation for both the corrosion and incrustation reactions. This discussion is necessarily brief, and the reader is referred to chemistry texts, such as *Aquatic Chemistry* by Stumm and Morgan (9), for more details.

The discussion of chemical equilibria is followed by sections on theory of corrosion, corrosion of casings and screens, and incrustation.

10.2 CHEMICAL EQUILIBRIA

In order to compare product concentrations with reactant concentrations for a typical reaction such as

$$a\text{A} + b\text{B} \rightarrow c\text{C} + d\text{D} , \qquad (10.1)$$

where a, b, c, and d are the stoichiometric coefficients of the reactants A and B, and of the products C and D, respectively; each of the species is first expressed in terms of its activity,

$$\alpha_i = m_i \gamma_i = \text{activity of } i\text{th species} , \qquad (10.2)$$

where

$m_i =$ molarity of ith species,
$\gamma_i =$ activity coefficient of ith species.

The molarity m_i is determined by chemical analysis of the water sample, and the activity coefficient γ_i is computed from the Debye–Huckel equation (9). The activity of each species α_i is further defined in terms of the chemical potential differences between the observed μ_i and standard μ_i° states by the expression

$$\mu_i - \mu_i^\circ = RT \ln \alpha_i , \qquad (10.3)$$

where

$$
\begin{aligned}
R &= \text{universal gas constant,} \\
T &= \text{absolute temperature [°K],} \\
\mu_i, \mu_i^\circ &= \text{chemical potentials of the observed and standard} \\
&\quad \text{[25°C, 1 atm pressure] states, respectively.}
\end{aligned}
$$

For the entire reaction expressed in Equation 10.1, the differences in chemical potentials may be summed to yield

$$\Delta\mu - \Delta\mu^\circ = RT \ln \frac{(\alpha_C)^c (\alpha_D)^d}{(\alpha_A)^a (\alpha_B)^b} . \qquad (10.4)$$

The product

$$\frac{(\alpha_C)^c (\alpha_D)^d}{(\alpha_A)^a (\alpha_B)^b}$$

is defined as the activity product (AP), or for a solution at equilibrium, as the equilibrium constant (K).

If the solution is at equilibrium,

$$\Delta\mu = 0 ,$$

and hence

$$-\Delta\mu^\circ = RT \ln K .$$

If the solution is not at equilibrium, then

$$\Delta\mu = RT \ln \left(\frac{AP}{K}\right) . \qquad (10.5)$$

The natural logarithm of the AP/K ratio times the gas constant R and the absolute temperature T gives a direct measure of the sum of the chemical potentials of species involved in the reaction and hence gives a direct measure of the potential existing for the reaction under consideration (Equation 10.1).

If the solution is supersaturated (i.e., large chemical potentials are available to drive the reaction), the activity product (AP) is greater than the equilibrium constant (K). If the solution is at equilibrium, $AP = K$; and if the solution is undersaturated, AP is less than K.

For example, the change in chemical potential associated with the precipitation of calcite, $CaCO_3$, can be determined readily if the molarities of Ca^{2+} and CO_3^{2-} are known, together with the reaction temperature. At 25°C,

$$
\Delta\mu = (8.314 \text{ Joule/°K}
$$
$$
\cdot \text{ mole) } (298°\text{K}) \ln\left[\frac{\alpha CaCO_3/(\alpha Ca)(\alpha CO_3)]}{10^{-8.34}}\right] ,
$$

where $K = 10^{-8.34}$ and $\alpha CaCO_3 = 1$.

Equation 10.5 gives a direct measure of the potential existing for a specific reaction to occur. Only the total chemical potential for a reaction can be determined, and hence the choice is entirely arbitrary as to which species are considered present in supersaturated or unsaturated concentrations. In simple precipitation reactions, the excess (or deficiency) of the driving chemical potential is generally assigned to one species, such as the metallic cation, an anion, or the hydrogen ion.

For electron transfer reactions (such as corrosion), the chemical potential is a function not only of species concentration, temperature, and pressure but also of the electrochemical properties of the species, since an electron acceptor must be available to receive the electrons produced in the oxidation reaction. In an electron transfer reaction, the electrical potential (Ehm) is measured relative to a standard hydrogen half cell. The standard expression for the calculated electrical potential (Ehc) is given by the following equation:

$$E\text{hc} = E^\circ + \frac{RT}{nF} \ln \frac{(\alpha_C)^c (\alpha_D)^d}{(\alpha_A)^a (\alpha_B)^b} , \qquad (10.6)$$

where

$$
\begin{aligned}
E^\circ &= \text{standard electrode potential (referred to hydrogen} \\
&\quad \text{half cell),} \\
n &= \text{number of electrons involved in reduction-oxidation} \\
&\quad \text{(redox)} \\
F &= 96,494 \text{ coulombs [Faraday].}
\end{aligned}
$$

The Ehc is the electrode potential the solution should have if the identified oxidation-reduction (redox) reaction is controlling the observed Ehm. Thus the calculated (equilibrium) electrode potential, Ehc, can be related to the measured Ehm by the expression

$$\Delta\mu = nF (E\text{hc} - E\text{hm}) . \qquad (10.7)$$

In this manner, the redox potential is a measure of the relative activities of oxidized and reduced ions.

In operational terms, the solution Eh can be measured in the field relative to a standard hydrogen half-cell potential.

This measured Eh (Ehm) may then be correlated with a calculated Eh (Ehc) for a particular half-cell reaction at a particular temperature. Although measurement of Eh would appear to give a straightforward reading of the electrical potential involved in a redox reaction, difficulties arise in a natural system in which Eh is related to multiple redox reactions. In other words, Eh is a general solution property not specifically related to a single species but related to all reactions involving electron transfer. The difficulty arises in determining the particular reactions that control solution Eh. An identified redox reaction may be too slow to control solution Ehm and also too slow to affect solution composition over a short period of time. Other, more rapid reactions in the solution may control system Eh.

Despite these difficulties departure from internal equilibrium can be treated for electron transfer reactions in the same manner as departure from equilibrium in mass action (precipitation) reactions. In one investigation (8), the difference in chemical potential was arbitrarily assigned to the metal cation so as to provide an initial estimate of the predicted precipitants. In this manner Barnes was able to make predictions as to the likely mass action and electron transfer reactions that would occur, based on bulk solution composition. These predictions could then be compared with actual composition of deposits on well casings and screens.

Some of the likely chemical reactions postulated by Barnes for the formation of reaction products on well casings and screens are as follows:

Hydrolysis Reactions

$$Fe^{2+} + CO_3^{2-} \rightarrow FeCO_3 \quad (siderite)$$

$$Ca^{2+} + CO_3^{2-} \rightarrow CaCO_3 \quad (calcite)$$

$$Cu^{2+} + 2H_2O \rightarrow Cu(OH)_2 + 2H^+$$

$$Fe^{2+} + H_2S \quad \rightarrow 2H^+ + FeS \quad (mackinawite)$$

Redox Reactions

$$Fe \quad \rightarrow Fe^{2+} + 2e^-$$

$$3Fe^{2+} + 4H_2O \rightarrow 8H^+ + 2e^- + Fe_3O_4 \quad (magnetite)$$

$$Fe^{2+} + 3H_2O \quad \rightarrow Fe(OH)_3 + 3H^+ + e^-$$

$$H_2S + 4H_2O \quad \rightarrow SO_4^{2-} + 10H^+ + 8e^-$$

$$3Mn^{2+} + 4H_2O \rightarrow 8H^+ + 2e^- + Mn_3O_4 \quad (hausmannite)$$

10.3 THEORY OF CORROSION

Chemical Reactions

Metals used in well casings and screens are thermodynamically unstable in the presence of water, particularly when the water contains an electron acceptor, such as dissolved oxygen, chlorine, or sulfate. For the corrosion reaction to occur, a change of oxidation state must occur in the metal, with the production of electrons as shown in the following reaction:

$$Fe \rightarrow Fe^{2+} + 2e^- . \quad (10.8)$$

Since free electrons cannot accumulate because of the repulsive nature of like electrostatic charges, this oxidative half reaction must always be coupled with a reductive half reaction, which consumes electrons. A reductive reaction that occurs in a typical oxygenated water supply is

$$O_2 + 2H_2O + 4e^- \rightarrow 4OH^- . \quad (10.9)$$

The oxidative half reaction (Equation 10.8) and the reductive half reaction (Equation 10.9) do not need to take place at the same location, since electricity, as represented by electrons e^-, flows freely through metals.

The location of the metal oxidation is termed the "anode" and the location of the oxygen reduction is the "cathode" (Figure 10.1). The chemical reaction between the coupled half reactions is termed a "galvanic cell" when the reaction proceeds spontaneously and "electrolysis" when the current flow originates from an external source. After the initial electrochemical reaction, the chemical forms of the reacting species are determined by their solubilities, as related to the chemical equilibria among the constituents in the water, and their rate of dissolution, as described by chemical kinetics.

Under acidic situations (pH less than 7.0) a second reductive reaction that may occur is the reduction of available hydrogen ions:

$$2H^+ + 2e^- \rightarrow 2H \rightarrow H_2 . \quad (10.10)$$

For the corrosion of iron in water, equilibrium is reached very quickly because of the low solubility of hydrogen in water (Equation 10.10). In order for the corrosion reaction to proceed further, this equilibrium must be upset by the removal of either the atomic hydrogen (H) or the molecular hydrogen (H₂), a process termed "depolarization."

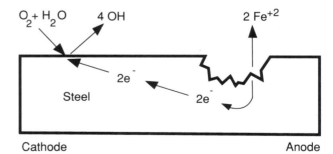

Figure 10.1. Corrosion of steel in typical well water.

With respect to the anodic reaction (Equation 10.8), an equilibrium condition is reached eventually between the metal ion concentration (Fe^{2+}) in the water and the availability of a species to accept the additional electrons generated. At equilibrium the atomic hydrogen forms a thin layer over the cathode site, preventing additional hydrogen ions (H^+) from reaching the metal surface and accepting the electrons. The formation of this atomic hydrogen layer is called "polarization."

Under aerobic conditions this hydrogen layer can be removed by dissolved oxygen or another oxidant, such as dissolved chlorine. The reaction with dissolved oxygen is

$$O_2 + 4H^+ + 4e^- \rightarrow 2H_2O . \qquad (10.11)$$

In the presence of another oxidant such as chlorine, this depolarization takes the form of Equation 10.12, in which hypochlorous acid reacts with the hydrogen ions and available electrons to form the chloride ion and water:

$$HOCl + H^+ + 2e^- \rightarrow Cl^- + H_2O . \qquad (10.12)$$

As a result of these depolarization reactions at the cathode, the basic corrosion reactions shown in Equations 10.8 and 10.9 can proceed. At the anode two reactions can take place to accomplish the removal of ferrous ions, the choice of reaction depending upon the carbonate concentration. In low carbonate solutions ferrous ions initially react with hydroxide ions, forming insoluble ferrous hydroxide (Equation 10.13). Under oxidizing conditions ferrous hydroxide can be further oxidized to ferric hydroxide or even the more insoluble hematite (Equation 10.14). In high carbonate solutions ferrous ions react with bicarbonate to form siderite (Equation 10.15) or, under oxidizing conditions, hematite (Equation 10.16):

$$Fe^{2+} + 2H_2O \rightarrow 2H^+ + Fe(OH)_2 \quad \text{(ferrous hydroxide)}$$
$$(10.13)$$

$$4Fe(OH)_2 + O_2 \rightarrow 4H_2O + 2Fe_2O_3 \quad \text{(hematite)}$$
$$(10.14)$$

$$Fe^{2+} + HCO_3^- \rightarrow H^+ + FeCO_3 \quad \text{(siderite)} \quad (10.15)$$

$$4Fe(HCO_3)_2 + O_2 \rightarrow 4H_2O + 8CO_2 + 2Fe_2O_3 \quad \text{(hematite)}$$
$$(10.16)$$

Corrosion Cells

In the preceding subsection the anode and cathode reactions were discussed separately in order to distinguish the chemical reactions that occur at each site. However, for corrosion to occur not only must the chemical reactions at each site proceed, but electrical current must flow between the two sites, hence the definition of corrosion as an electrochemical reaction in which a metal is oxidized (anode reaction) and an oxidizer is reduced (cathode reaction). Neither of these reactions will occur without the other, and means must be provided by which the electrons can flow from anode to cathode.

Furthermore, an electrolytic solution must exist in order to (1) transport reacting species, such as dissolved oxygen, to the corrosion site, (2) transport corrosion products, such as ferrous ions, away from the corrosion site, and (3) provide charge-neutralizing anions, such as chloride and sulfate, to balance the cations produced during corrosion. In general, the corrosion rate increases with increasing concentrations of dissolved ions; for example, seawater is more corrosive than freshwater. However, equal importance must be attached to the actual ionic species present, since some species such as chloride and sulfate enhance corrosion, whereas others such as calcium and bicarbonate effectively coat the reactive surfaces. Hence very soft, low alkalinity waters are generally very corrosive; but hard, alkaline waters are usually protective of metal surfaces. Three measures of ionic species—hardness, alkalinity, and total dissolved solids (conductivity)—are often used together to estimate the corrosivity of a water supply.

In order to simplify our approach to corrosion of well casings and screens, the formation of corrosion cells is presented in terms of oxidation-reduction potentials, localized effects (heterogeneities), water qualities, physical properties, and biological activities. Although undoubtedly more than one mechanism is operative in any corrosion cell, for ease of discussion they have been broken down into these five groupings.

Oxidation-Reduction Potentials (Galvanic Cells). One of the causes of corrosion is the difference in standard oxidation-reduction potentials between different metals, or between different sites of the same metal. When two dissimilar metals are placed in an electrolyte, a galvanic cell results in which the metal with the more positive standard potential will be reduced, and the metal with the more negative standard potential will be oxidized. Metals with the more positive standard potential are referred to as "noble," whereas metals with more negative standard potential are referred to as "active."

Table 10.1 presents selected standard reduction-oxidation potentials for some common metal half reactions. The magnitude of the potential for a particular coupled pair of half reactions is found by adding the standard potential of the reduction reaction to the standard potential of the oxidation reaction, with its sign reversed, since the reaction proceeds in reverse from the direction shown in Table 10.1. The potentials shown in Table 10.1 were derived under standard conditions (1 atm, 25°C) relative to a "standard hydrogen electrode." Whether or not the coupled reactions can proceed spontaneously under actual conditions (and not under the

TABLE 10.1 Electrochemical Series: Selected Standard Oxidation–Reduction Potentials

Reaction	Potential
$Au = Au^{3+} + 3e^-$	+1.498
$O_2 + 4H^+ + 4e^- = 2H_2O$	+1.229
$Fe^{3+} + 1e^- = Fe^{2+}$	+0.771
$4OH^- = O_2 + 2H_2O + 4e^-$	+0.401
$Cu = Cu^{2+} + 2e^-$	+0.337
$Sn^{4+} + 2e^- = Sn^{2+}$	+0.150
$2H^+ + 2e^- = H_2$	0.000
$Pb = Pb^{2+} + 2e^-$	−0.126
$Sn = Sn^{2+} + 2e^-$	−0.136
$Ni = Ni^{2+} + 2e^-$	−0.250
$Fe = Fe^{2+} + 2e^-$	−0.440
$Zn = Zn^{2+} + 2e^-$	−0.763
$Al = Al^{3+} + 3e^-$	−1.662

standard conditions with unit molar concentrations of dissolved species) is determined by a host of physical and chemical factors, such as temperature and water quality. The standard redox potentials form what is called the "electrochemical series," based on thermodynamically reversible reactions, which are not practically obtainable.

A more realistic portrayal of the electrochemical potentials that exist between dissimilar metals in an electrolyte is termed the "galvanic series," as presented in Table 10.2. In the galvanic series the least noble metal is oxidized when connected to a more noble metal, which serves as the site for the reduction of dissolved oxygen in the electrolyte (Equation 10.9). In general, the galvanic series parallels the standard redox potentials with some minor variations.

The relative positions of some metals on the galvanic series can be changed by environmental factors. For example, although zinc is lower than iron in Table 10.2, the potentials are reversed above 140°F. Thus in galvanized iron pipe systems at temperatures greater than 140°F, the zinc layer no longer serves as a sacrificial anode, but instead becomes the cathode, causing rapid pitting failure of the underlying iron at holidays in zinc coating. Similarly, severe attack of the more noble aluminum occurs in the vicinity of the less noble magnesium, a phenomenon caused by hydroxyl ions originating from the reduction of oxygen attacking the aluminum metal (10).

Localized Effects (Heterogeneities). Galvanic corrosion such as induced by the coupling of dissimilar metals may also occur within the same metal. In fact severe corrosion has been found to occur at heterogeneities introduced into the same piece of metal. For example, accelerated corrosion of mild steel specimens exposed to highly corrosive well water was found at points of cold working, such as bridge pieces in bridge-slotted specimens and the edges of holes drilled in casing specimens (11). Similarly, torch-cut slots in mild steel casing were shown to corrode at two to three times the rate of machined slots. Subsequent photo micro-

graphs indicated a martensitic structure in the vicinity of the torch-cut slots as opposed to the undisturbed structure of the steel in the machined slots (12). The formation of a martensitic crystalline structure in steel has also been found to occur when electrical resistance welding is used to form the weld. The structural differences that exist between the steel formed in the weld and the steel that comprises the remainder of the pipe can be alleviated by annealing, but if this operation is not carried out, the weld seam serves as the site of the most severe corrosion (13).

Water-Quality Effects. Given the thermodynamic basis for corrosion described earlier (oxidation-reduction potentials), the rate of corrosion is influenced by both the chemical and physical properties of the water. The principal chemical factors discussed here are dissolved gases and dissolved salts; the principal physical factors are water temperature and water velocity. These properties have been observed to affect the nature of the corrosive attack on metals, and the resulting longevity of casings and screens.

Of all the chemical characteristics of water that influence rate of corrosion, dissolved gases are probably the most important. Of the three dissolved gases commonly found in wells—oxygen, carbon dioxide, and hydrogen sulfide—the most important in terms of its effect on casings and screens is dissolved oxygen.

TABLE 10.2 Galvanic Series

Platinum	
Gold	(most noble)
Silver	
Stainless steel (Type 316, passive)	
Stainless steel (Type 304, passive)	
Inconel (75% Ni, 15% Cr, passive)	
Nickel (passive)	
Monel (65% Ni, 30% Cu)	
Copper-nickel alloys	
Bronzes	
Copper	
Brasses	
Tin	
Lead	
Cast iron	
Steel or iron	
Aluminum	
Zinc	
Magnesium	
Magnesium alloys	(least noble)

Note: Metals grouped as shown above can be safely used together without harmful galvanic corrosion. Passive means metals have developed a thin protective oxide coating.

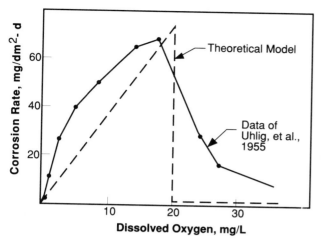

Figure 10.2. Influence of dissolved oxygen concentration on rate of iron corrosion (from Uhlig, H., D. Triadis and M. Stern, "Effect of Oxygen, Chlorides and Calcium on Corrosion Inhibition," *Journal of Electrochemical Society*, 102, p. 61, 1955). Reprinted by permission of the publisher, The Electrochemical Society, Inc.

The rate at which oxygen reaches the metal surface controls the rate of corrosion and accelerates the action of carbon dioxide and hydrogen sulfide. The importance of dissolved oxygen (DO) as an electron acceptor has been shown in Equations 10.9 and 10.11. Increasing the level of DO in the bulk solution increases the rate at which oxygen is transported to the corroding metal surface. As a result the rate of corrosion for most metals increases with an increase in the level of DO up to about 20 to 25 mg/L (Figure 10.2). At higher concentrations the effect may slowly reverse because of passivation of the metal by oxygen, but this is not a practical concern in pumping ground waters at moderate flow velocities. The solubility of DO is proportional to the partial pressure of oxygen in the air above the surface of the water. Increases in atmospheric pressure will increase oxygen solubility, whereas increases in temperature will reduce oxygen solubility (Figure 10.3).

Differences in oxygen concentrations along the casing and screen metal surfaces may occur for a variety of reasons, including grains of sand, microbial colonies, weld seams, and heterogeneities in the surface of the metal itself. A small area may be shielded from the dissolved oxygen in water by a deposit, thereby resulting in a differential oxygen concentration between two locations. In the pH range of most natural waters, electrons produced by corrosion of iron are typically consumed by the reduction of dissolved oxygen rather than of hydrogen ions. Under these conditions the overall reaction is described by the following redox equations:

Reaction	$E°(v)$
$Fe = Fe^{2+} + 2e^-$	$-(-0.440)$
$\frac{1}{2} O_2 + H_2O + 2e^- = 2OH^-$	$(0.401)/2$
$\frac{1}{2} O_2 + H_2O + Fe = Fe^{2+} + 2OH^- + 0.640$	

$$(10.17)$$

Note that the standard potential for the reduction of ferrous ion (Table 10.1) has been corrected by the factor of -1 because the direction of the reaction is reversed to show oxidation of iron metal; the standard potential for the reduction of oxygen must be corrected by a factor of $\frac{1}{2}$ to balance the number of electrons transferred from the oxidation reaction. The chemical reactions involved are precisely the same as those that occur in a galvanic cell and hence the potential of any oxygen concentration cell will be exactly the same as in a galvanic cell. The oxygen concentration cell is responsible for the majority of well casing and screen corrosion.

Once started, the oxygen concentration cell becomes self-perpetuating. A crust of metal oxide (tubercle) forms over a pit that further lowers the oxygen concentration under the tubercle. The tubercle, a dark and relatively hard and brittle structure, consists of magnetite (Fe_3O_4) or goethite ($FeOOH$), or mixed mineral phases of magnetite and goethite. To the extent that the tubercle contains magnetite, it becomes a relatively good electrical conductor that may act as a cathode to the underlying pit in the base metal (the anode). An operating corrosion cell introduces ferrous ions into the interior of the tubercle and releases hydroxide ions into the bulk solution. Maintenance of electroneutrality within a turbercle requires that anions penetrate into the interior by passing through the enveloping layer. The outer layer exhibits some

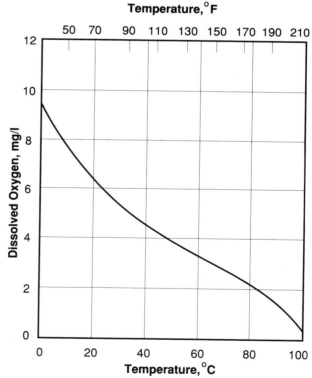

Figure 10.3. Solubility of dissolved oxygen in water at 1 atm pressure as a function of water temperature (from Coburn, S.K., "Corrosion in Fresh Water"). Reprinted with permission from Metals Handbook, Vol. 1, 9th Ed., American Society for Metals, Metals Park, OH, 1978, p. 734.

selectivity for the various anions present in the bulk solution, favoring the penetration of chlorides. Often the interior of the tubercle contains a solution of ferrous chloride and sulfite ions in concentrations greater than those of the bulk solution, and a slightly acidic pH. Eventually, the tubercles become impenetrable even to chloride ions and, since the solution inside must maintain electrical neutrality, the corrosion reaction at this location ceases (14).

The presence of different concentrations of dissolved oxygen at different sites on the same metal surface leads to "differential oxygen corrosion." The Nernst equation for corrosion of iron with oxygen as the electron acceptor,

$$E = E° + \frac{2.303 RT}{nF} \log \frac{(Fe^{2+})^2}{[O_2 \, (aq)] \, (H^+)^4} \, , \quad (10.18)$$

where

E = potential [v],
R = universal gas constant,
T = absolute temperature [°K],
n = number of electrons transferred,
F = 96,494 coulombs [Faraday],

can be used to show that if two sites on a steel surface have the same pH and ferrous ion concentration, but the dissolved oxygen concentration is 5 mg/L at one site and 0.1 mg/L at the other, the potential difference between the two sites will be 25 mv (15). A schematic of this situation, as it might occur in well casings or screens, is shown in Figure 10.4. Common areas of differential oxygen corrosion are between two metal surfaces, under organic deposits, or in crevices. Since oxygen is not produced by either the anodic or cathodic reaction, corrosion will proceed in such a way that the higher

oxygen concentration is reduced. The part of the metal in contact with the higher oxygen concentration is the cathode; and the part in contact with the lower oxygen concentration is the anode and will corrode (15).

By itself, carbon dioxide does not participate directly in the corrosion reaction. Rather, it reacts with water to form carbonic acid, which in turn weakly dissociates to yield a lower solution pH. The lower pH has two effects on the corrosion reaction; first, the hydrogen ion can act as an electron acceptor, as per Equation 10.10 (acid corrosion); and second, the lower pH increases the solubility of calcium carbonate, thus hindering its ability to form a protective scale over the metal surface.

The reaction that leads to the production of carbonic acid,

$$CO_2 + H_2O \rightarrow H_2CO_3 \rightarrow H^+ + HCO_3^- \, ,$$

can be reversed by a reduction in pressure on the solution. This occurs, for example, when naturally carbonated water is brought to atmospheric pressure. Some authors have attributed the formation of calcium carbonate at the well screen to the release of CO_2, caused by the differential pressure between the aquifer and the interior of the screen (16). However, the partial pressure of carbon dioxide in well water seldom exceeds a small fraction of an atmosphere. Hence bubbles of carbon dioxide cannot form, since the total pressure in the flow path will always exceed the carbon dioxide partial pressure. Rossum (10) has calculated that screen velocities on the order of several tens of feet per second, with screen slots shaped as perfect venturi throats, would be required to reduce atmospheric pressure plus the head associated with submergence to a level sufficient to allow carbon dioxide to escape. Thus the incrustation often formed at well screens likely results from the alkaline mi-

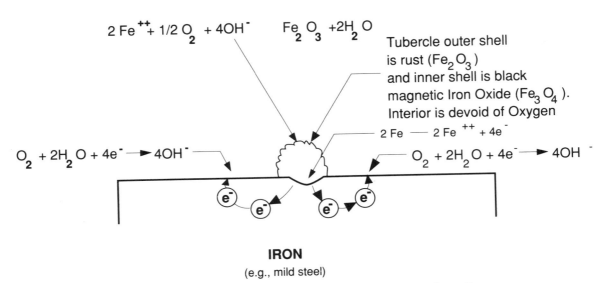

Figure 10.4. Formation of tubercle at an oxygen concentration cell.

croenvironment created at the cathode of a corrosion cell, as opposed to the release of carbon dioxide from solution.

In solution, hydrogen sulfide also occurs as a weak acid, but depression of the pH is not the reason it causes corrosion. Rather, the sulfide ion reacts with the steel surface to form black, insoluble, iron sulfide deposits:

$$H_2S + Fe^{2+} \rightarrow FeS + 2H^+ . \qquad (10.19)$$

In addition to its direct consumption of iron by this reaction, hydrogen sulfide produces intergranular networks of low-strength metallic sulfides. Furthermore the liberated atomic hydrogen (Equation 10.10), tends to penetrate and embrittle the steel beneath the sulfide deposits. Both types of internal attack are especially pronounced in hard alloy steels and in cold worked areas (17). Clarke (18) attributed the breaking away of bridge sections from steel bridge-slotted screens to hydrogen embrittlement at stress points.

Waters containing hydrogen sulfide are anaerobic; otherwise, any oxygen available would rapidly oxidize the sulfide to sulfate. In addition waters containing hydrogen sulfide are free of dissolved iron and manganese, since these metallic sulfides are relatively insoluble and would precipitate from solution, leaving only the excess sulfide (19). Although waters containing hydrogen sulfide are not corrosive in the usual sense of the word (i.e., through the use of oxygen as an electron acceptor), the attack of steel surfaces by hydrogen sulfide in ground waters can nevertheless be rapid. Since a considerable electrical potential exists between iron sulfide and steel, sulfide deposits invariably accentuate galvanic attack, as characterized by localized pitting. Severe sulfide pitting has been reported for a steel pump plunger during only 35 days service in water containing 1 mg/L of hydrogen sulfide (20).

In the corrosion process the electrolyte solution conducts ions between the anode and cathode. The positive ions (Fe^{2+}) generated at the anode migrate to the cathode, and the negative ions (OH^-) generated at the cathode migrate to the anode, in response to concentration gradients and to maintain an electrically neutral solution. In general, increasing the concentrations of ions present in the electrolyte increases its conductivity, resulting in more localized corrosion cells. The increased conductivity permits cathodic areas to become larger and enables the anode and cathode reaction products to combine in the water instead of at the corroding surface.

Aside from the effect on conductivity, some anions such as chloride and sulfate accelerate corrosion, other anions such as nitrate are neutral toward corrosion, and some anions such as bicarbonate and carbonate retard corrosion. Each operates by a different mechanism. One offsetting beneficial effect of increased salts dissolved in water is the decreased solubility of oxygen.

The corrosion rate of mild steel invariably increases with increasing concentrations of chloride and sulfate, independent of the available cations. The rates have been found to be less in buffered solutions, resulting in an expression relating the total molar concentrations of chloride and sulfate to the bicarbonate concentration. Known as "Larson's ratio" after its originator, the ratio should be held to less than five in order to minimize corrosion in steel pipes (21). The chief ionic culprit for well casings and screens is the chloride anion, which is particularly noted for its penetration of the passivity in many ferrous alloys and is one of the main causes of pitting in stainless steels (22).

The principal benefits of the bicarbonate anion are its ability to buffer pH changes within the bulk solution and to react with calcium in the vicinity of the corrosion cathode, thereby producing calcium carbonate scale. At the cathode active corrosion maintains a pH of 9.5 near the steel surface, almost irrespective of the pH of the solution (23). At this pH the bicarbonate anion dissociates, yielding the carbonate ion needed for precipitation with calcium.

Calcium, the principal hardness ion, serves to inhibit corrosion by forming carbonate scales on metal surfaces. A minimum of about 60 mg/L hardness (as $CaCO_3$) must be present along with a minimum alkalinity of 50 to 100 mg/L (as $CaCO_3$) at normal temperatures (0 to 70°C) in order for an eggshell scale to begin to form. The best predictor of whether a scale will form or not is the calculated Langelier Index, which serves as a measure of the saturation of calcium carbonate. For scale formation, three conditions must be met: (1) the water must have a Langelier Index equal to or greater than 0.0, (2) it must contain a significant bicarbonate ion concentration, and (3) it must be flowing over the metal surface.

The Langelier Index (LI) is calculated precisely by the following equations:

$$LI = pH - pH_s , \qquad (10.20)$$

where

$$
\begin{aligned}
pH_s &= pK_2 - pK_{so} - \log(Ca^{2+}) - \log(alk) \\
&\quad + \log(1 + 2K_2/(H^+)) - \log(f_m),
\end{aligned}
$$

pK_2 = dissociate constant for bicarbonate,
pK_{so} = dissolution constant for calcium carbonate,
Ca^{2+} = calcium ion concentration,
alk = alkalinity,
H^+ = hydrogen ion concentration,
f_m = ionic strength correction factor.

A satisfactory approximation to the expression for pH_s is as follows:

$$pH_s = A + B - \log(Ca^{2+}) - \log(alk) ,$$

TABLE 10.3 Constant "A" as Function of Water Temperature

Water Temperature (°C)	A
0	2.60
4	2.50
8	2.40
12	2.30
16	2.20
20	2.10
25	2.00
30	1.90
40	1.70
50	1.55
60	1.40
70	1.25
80	1.15

where values of A, as a function of water temperature, and B, as a function of total filterable residue, are found in Tables 10.3 and 10.4, respectively.

If the calculated LI is negative, the water is under-saturated with calcium carbonate and an eggshell layer is not expected to form. The expected thickness of the calcium carbonate layer will increase at higher concentrations of calcium and alkalinity. However, too great a supersaturation of calcium carbonate may lead to objectionable deposits of scale, especially at temperatures above the temperature at which water is saturated with calcium carbonate.

The effect of water quality on corrosion can be summarized in a general way based on relationships between mineral content, pH, and dissolved oxygen content. The first case is low mineral content with dissolved oxygen present. In this case increasing pH decreases the corrosion rate. Pitting can occur, especially in crevices that experience oxygen depletion, making them susceptible to local pitting attack. Differences in oxygen concentration, such as occur on metal surfaces under deposits at the water level, can cause severe corrosion.

The second case involves low minerals and little or no dissolved oxygen. In this case the principal cause of corrosion is a hydrogen sulfide attack often mediated by sulfate-reducing bacteria under anaerobic conditions.

TABLE 10.4 Constant "B" as Function of Total Filtrable Residue

Total Filtrable Residue (mg/L)	B
0	9.70
100	9.77
200	9.83
400	9.86
800	9.89
1000	9.90

The third case involves the presence of noncarbonate minerals, such as sulfates or chlorides, in the presence of dissolved oxygen. The chloride and sulfate ions, known as aggressive ions, increase the corrosion rate at all pH levels as their concentrations increase in solution.

The fourth case involves the presence of carbonate minerals along with dissolved oxygen. The carbonate minerals, measured as the alkalinity in drinking water supplies, inhibit corrosion at increasing levels, opposing the attack of chloride and sulfate ions. This inhibition works best at pH in the range 6.5 to 7.0 and at bicarbonate concentrations 5 to 10 times greater than the molar sum of chloride and sulfate concentrations.

The fifth case consists of water supplies with dissolved minerals, but without dissolved oxygen. In this case corrosion is generally low, except for that resulting from the activities of sulfate-reducing bacteria.

The sixth and final case involves calcium salts with dissolved oxygen present. In general, waters in this category will be corrosive unless some event occurs to upset the natural equilibrium of the calcium and carbonate species. For example, a water high in calcium but low in alkalinity might blend at the well screen (when the well is shut down) with another water low in calcium but high in alkalinity, resulting in a solution supersaturated in calcium carbonate. A minimum alkalinity of 50 to 100 mg/L and a minimum calcium concentration of 60 mg/L must be present at normal temperatures (0 to 70°C) for even a small degree of precipitation of the protective calcium carbonate scale. Corrosive action at the cathode itself may be adequate to result in calcium carbonate precipitation, since the pH at the corroding metal surface approaches 9.5.

Physical Properties. In addition to water chemistry, two physical properties, velocity and temperature, have a marked effect on corrosion. The effects of velocity on corrosion rate are dependent upon a number of properties of the solution and the metal. In the usual case in which metal corrosion is limited by the rate of oxygen transfer to the surface, increases in velocity will improve mass transfer and increase corrosion rates. At some point the corrosion rate becomes independent of velocity since the presence of oxygen is no longer the limiting parameter. At a much higher "critical velocity" erosion-corrosion may occur, dramatically increasing the overall corrosion rate.

In general, entrance velocities through well screens are less than 1 foot per second (fps), resulting in fairly low corrosion rates, since the reaction is oxygen limited. For entrance velocities between 1 and 5 fps, the oxygen level near the surface is continually renewed, resulting in moderate corrosion rates. However, if the calcium carbonate level is near saturation, scale may form on the cathode to limit oxygen transfer, thereby resulting in lowered corrosion rates.

At velocities greater than 4 fps for copper pipe and 10 fps for steel pipe, erosion-corrosion removes dissolved ions and even solid corrosion products from the metal surface. This results in pipe loss characterized by grooves or gullies on the inside of the pipe, especially near points of turbulence.

High velocity results in increased corrosion rates for low alkalinity, low hardness waters. This is caused by the reduction in thickness of the boundary layer and hence the more rapid diffusion of dissolved oxygen to the metal surface. This effect is reduced in waters with a positive Langelier Index. When protective films such as calcite or siderite form, velocities up to 5 fps promote denser and more protective coatings. Excessive velocities (greater than 5 fps) still result in a combination of corrosion and erosion by sweeping away protective coatings, but these velocities are unlikely to be encountered in a properly designed well screen.

The importance of hydraulic conditions (turbulence) and length of exposure must be considered when evaluating the effects of fluid velocity on corrosion (24). For example, during the initial period of exposure, increased velocity (and turbulence caused by surface roughness) will generally increase the corrosion rate. As corrosion products or protective coatings form on the surface, reactions leading to continued corrosion take place beneath the products, resulting again in diffusion-controlled corrosion (controlled by the rate of diffusion of dissolved oxygen to the metal surface).

This eventually leads to a situation where corrosion proceeds almost uniformly, as long as no changes occur in velocity or turbulence. Thus tests conducted over a few days or weeks are not likely to provide useful data on the effects of velocity or dissolved oxygen content on the long-term corrosion rate or passivation of various metals utilized in well screens.

As with many chemical reactions the corrosion reaction will increase with increasing temperature, according to the Arrhenius equation:

$$K = Ae^{-E_a/RT} , \qquad (10.21)$$

where

K = corrosion rate constant,
A = empirical constant,
E_a = activation energy,
R = universal gas constant,
T = temperature [°K].

Though the actual impact will vary depending on the chemical composition of the water and the physical properties of the metal, Butler and Ison (25) have found that the corrosion rate for iron increases 75% for every 10°C rise in temperature. This increase may be partially offset by the lowered solubility of certain important gases, such as oxygen, at elevated temperatures. In addition calcium carbonate is less soluble at higher temperatures, and thus scale may form in hot water pipes and heaters, reducing corrosion. The effects of temperature are generally not significant in the design of well casings and screens.

Biologically Mediated Corrosion.
Microorganisms can play a significant role in fostering corrosion, but their effects are often difficult to quantify. Certain species of bacteria enhance corrosion by catalyzing reactions that lead to the oxidation of ferrous metals. Other biological growths result in chemical or physical conditions that favor the formation of oxygen concentration cells, or disrupt the formation of protective surface films. In these cases the action is physical rather than biological. All of these may affect the rate of metallic corrosion at specific sites in wells.

The ability of some bacteria to obtain energy by utilizing local gradients in the redox potential suggests several possible mechanisms for biologically mediated corrosion. Three groups of bacteria frequently implicated are the iron bacteria (*Crenothrix polyspore*), sulfate-reducing bacteria (*Desulfovibrio desulfuricans*), and the sulfur bacteria (*Beggiatoa alba*). Other groups that may be active are methane producers and nitrate reducers, but sulfate-reducing and iron bacteria are most often encountered in wells. These particular groups are ubiquitous in many ground water environments and seem to disperse between wells in operating well fields (26).

Iron bacteria can occur in one of three physical forms; the encapsulated coccoid form (*Siderocapsa*), the stalked iron-fixing form (*Gallionella*), and a filamentous form (*Crenothrix, Sphaerotilus, Clonothrix,* and *Leptothrix*) (27). All are aerobic organisms that in the presence of organic matter catalyze the oxidation of ferrous ions to ferric ions or to insoluble hydroxides. This serves to remove ferrous ions, which are the immediate reaction products of the oxidation of iron at the anode. The following equation illustrates a typical reaction:

$$2Fe(OH)_2 + H_2O + \tfrac{1}{2}O_2 \xrightarrow[\text{catalyst}]{\text{bacterial}} 2Fe(OH)_3 + \text{energy} .$$

$$(10.22)$$

How much this reaction accelerates corrosion is not known. A much greater impact on the rate of corrosion may result from the formation of oxygen concentration cells. Insoluble hydroxide compounds, such as goethite ($FeOOH$), result in the mineralization of iron bacteria colonies and the formation of incrustations (28). These incrustations in turn diminish the diffusion of oxygen to the metal surface, resulting in differential oxygen concentration cells. Iron bacteria can also adversely affect the aesthetic quality of well water.

Sulfate-reducing bacteria such as *Desulfovibrio desulfuricans* obtain their energy by reducing sulfate to sulfide in an anaerobic environment such as might occur in deep wells. Organic matter is also necessary for the bacteria to proliferate. The resulting formation of odorous hydrogen sulfide can render a water unsuitable for domestic use unless it is treated (see Chapter 18). The presence of hydrogen sulfide causing severe corrosion is puzzling because sulfide can only exist in waters devoid of oxygen, and corrosion of steel in such water is expected to be minimal. A reasonable explanation is that hydrogen sulfide is a product of the corrosion reaction:

$$4Fe + 6H_2O + SO_4^{2-} \xrightarrow[\text{catalyst}]{\text{bacterial}} 4Fe^{2+} + H_2S$$
$$+ 10OH^- + \text{energy} . \quad (10.23)$$

This reaction occurs in the absence of oxygen and in the presence of sulfate-reducing bacteria. The sulfate ion (SO_4^{2-}) is a potent oxidizing agent, but without bacteria, its reduction to sulfide is too slow to significantly affect corrosion rates. Bacteria such as *Desulfovibrio desulfuricans* catalyze this reaction to a point where oxidation of ferrous metals occurs rapidly.

The reduction of the sulfate ion is followed by the formation of insoluble ferrous sulfide or hydroxide corrosion products as shown in the following equations:

$$Fe^{2+} + H_2S = FeS + 2H^+ , \quad (10.24)$$
$$Fe^{2+} + 2OH^- = Fe(OH)_2 . \quad (10.25)$$

The uptake of the ferrous species, which is the reaction product of the corrosion of iron (Equation 10.23), may further enhance the kinetics in the forward direction.

A rough correlation has been proposed between the redox potential and sulfate-reducing bacterial activity in those situations where bacteria are known to be present. The redox potential of natural soils and water is useful for establishing whether or not the environment is favorable for microbiological corrosion activity by sulfate-reducing bacteria. Table 10.5 indicates the proposed correlation between redox potential and bacterial activity.

In addition to their presence in wells under low oxygen or anaerobic conditions, sulfate-reducing bacteria can survive in an aerobic environment. This can occur in a mature biofilm community where anaerobic conditions exist beneath the aerobic layer. The anaerobic bacteria are known to feed successfully on the by-products of the aerobic community, which include acetate, carbon dioxide, and propionate (30).

TABLE 10.5   Redox Potential versus Sulfate-Reducing Bacterial Activity

Eh Range	Activity
Below 100 mv	Severe
100–200 mv	Moderate
200–400 mv	Slight
Above 400 mv	None

Source: Starkey and Wight (29).

Under aerobic conditions sulfur bacteria (*Beggiatoa alba*) carry out a two-step oxidation of hydrogen sulfide in order to obtain energy for metabolic processes. Representative equations for this oxidation of hydrogen sulfide to elemental sulfur and eventually to sulfuric acid are

$$2H_2S + O_2 \xrightarrow[\text{catalyst}]{\text{bacterial}} 2S + 2H_2O , \quad (10.26)$$

$$2S + 3O_2 + 2H_2O \xrightarrow[\text{catalyst}]{\text{bacterial}} 2H_2SO_4 + \text{energy} .$$
$$(10.27)$$

Metallic corrosion is enhanced by the formation of sulfuric acid, which is very corrosive. However, experience suggests that this type of corrosion rarely occurs in active wells. As with other biological growths, colonization of metal surfaces by the sulfate-reducing iron or sulfur bacteria may interfere physically with protective scale formation or lead to the establishment of oxygen concentration cells.

10.4   CORROSION OF CASINGS AND SCREENS

Well casings and screens are subject to different types and degrees of corrosive attack, depending upon their physical location within the well. In freshwater wells the four general areas of interest are external exposure, atmospheric exposure (casing above the static level), splash exposure (casing between static and pumping level), and submerged exposure (casing and screen below the pumping level). In most wells the greatest corrosion occurs in the "splash zone" because the casing is exposed to a humid atmosphere and to alternate wetting and drying cycles that accelerate the attack on steel. In some wells the most severe corrosion occurs in the "submerged zone" (especially the well screen), where differential oxygen levels, adverse water quality, the relatively large exposed steel surface area, and/or the presence of bacteria accelerate the corrosive attack.

External Zone

External corrosion of carbon steel casings is generally confined to the upper 5 to 25 ft of casing. The principal causes are either an oxygen differential cell, such as contact between an open porous formation and a tight clayey formation, or galvanic corrosion caused by contact with a more noble metal. Oxygen differential corrosion is alleviated by use of a cemented casing, whereas galvanic corrosion is mitigated by installation of a dielectric coupling between the discharge head and the discharge line.

Cathodic protection has been used to control corrosion on the exterior surfaces of well casings, especially in the oil and gas industry. The installation usually consists of a rectifier to convert ac power to dc, a positive terminal connected to an anode bed located distant from the well casing, and a negative terminal attached to the well. If power is not available at the well to create impressed current protection, magnesium anodes can be coupled to the casing, thereby providing a protective effect by their oxidation potential. In either instance, the design of an effective cathodic protection system depends on soil resistivity and casing-to-soil potential as measured by a copper/copper sulfate half cell. Cathodic protection using sacrificial anodes of magnesium, aluminum, or other materials less noble than the casing material is more effective in low-resistivity soils. In high-resistivity soils the impressed current protection is more effective.

Although cathodic protection for the exterior surface of casing has been utilized, it has not found application to interior surfaces. Several problems occur, including finding a suitable anode for installation between the casing and the column pipe, between the column pipe and the line shaft, and inside the screen. The cost of these anodes, together with the care required for proper installation, is a drawback, as is the fact that the well casing itself now becomes the cathode of a corrosion reaction. The change in its role from anode to cathode means that additional deposition and incrustation may occur as a result of operating the cathodic protection system (31, 4, 18). As a result of these limitations, cathodic protection has found beneficial usage in protecting exterior well surfaces and storage tanks, but not the interior surfaces of casings and screens which in water wells are the areas of greater attack.

Atmospheric Zone

Within the "atmospheric zone" the principal factors leading to corrosion are the condensation of vapor on the well casing and the presence of oxygen and small amounts of carbon dioxide which readily diffuse through the thin layer of condensed water. Although this low-salt, highly oxygenated condensed water is corrosive, the resulting corrosion is uniform over the casing surface, and ordinary carbon steel casings last many years in this zone.

Early experiments established that the addition of small amounts of copper (up to 0.2%) to carbon steel enhanced its atmospheric corrosion resistance. In general, atmospheric corrosion of a copper-bearing steel is about one-half of a residual-copper carbon steel (32). The improved atmospheric corrosion resistance of copper-bearing steel is a result of the formation of a dense, adherent oxide on the surface resulting from alternating wetting and drying cycles (33, 34). Although the advantages of using copper-bearing steel in the atmospheric zone of a well are well documented, no advantage occurs from using it for submerged or buried structures.

If greater corrosion resistance is required, high-strength, low-alloy (HSLA) steels having an atmospheric corrosion resistance four to six times greater than carbon steel may be used (35). Extensive testing of HSLA steels has been conducted over a period of 15 years in the three principal environments of concern: rural, industrial, and marine (36). The corrosion rate for carbon steel essentially leveled off after five years of service, copper-bearing steel after three years of service, and HSLA steel after two years of service. The data indicate that copper concentrations from 0.05% to 0.2% cut the corrosion rate in half and that the addition of low concentrations of other alloys (in particular, chromium, molybdenum, manganese, phosphorus, and nickel) reduce the corrosion rate by another 50% (37).

Alloying (stainless steel) and protective coatings are two reasonably economical and effective methods of protecting steel from atmospheric corrosion. For well casings, alloying is recommended if conditions warrant it, since abrasion of a protective coating—incurred during handling, installation, or use—can produce random bare spots, or "holidays," that corrode at a greater rate than if the casing were completely unprotected. Furthermore coatings on field connections are impossible to apply properly. Any holidays or crevices in the coating quickly become anodic to large cathodic surface areas, resulting in rapid pitting corrosion.

An exception to the ineffectiveness of coatings is galvanizing, which protects holidays by acting as the anode in a galvanic cell, corroding preferentially to protect the underlying steel exposed at pores or breaks in the coating. The suitability of galvanizing is highly dependent upon water-quality conditions.

Although corrosion in the atmospheric zone must be considered, in general, either the splash zone or submerged zone will experience more severe corrosion.

Splash Zone

The splash zone between the static and pumping levels is often the site of relatively severe corrosion, resulting primarily from differential oxygen concentrations. Other factors that tend to speed corrosion in the splash zone are the alternating wet and dry cycles, and elevated TDS levels caused by evaporation of the ground water remaining on the well casing.

The key to understanding the corrosion mechanism involved in the splash zone is to examine the sequence of water flow as the well is pumped. When pumping begins, the ground water gradually declines to the pumping level resulting in a wetted casing surface. Oxygen from air in the casing rapidly diffuses through the thin water layer to the metal. This creates an oxygen differential cell between the highly oxygenated surface just above the pumping level and the relatively low oxygen surface immediately below the pumping level. When pumping ceases, the rising water level again wets the casing surface and gradually lowers the oxygen level to that of the ground water. This alternating wet and dry cycle, with high and low oxygen levels, corrodes the steel surface in contact with the low dissolved-oxygen ground water.

A similar situation occurs with cascading water, caused by the flow of water (1) through a screened interval above the pumping level, (2) a hole in the casing, or (3) a hole in the pump column. Oxygen in the entrained air causes accelerated corrosion in the splash zone.

As with corrosion in the atmospheric zone, the first line of defense against rapid corrosion in the splash zone is material specification. A distinct advantage is obtained by adding low concentrations of copper to mild steel for protection in either the atmospheric or the splash zone. Improvements in the corrosion resistance of the casing in the splash zone can also be achieved by using high-strength, low-alloy steels. The improvement is about the same as in the atmospheric zone.

The highest resistance to corrosion in any potable water environment is provided by stainless steel. Type 304 stainless steel (18% chromium, 8% nickel) has been found especially effective, except in chloride concentrations greater than 500 mg/L where it has a tendency to pit. It also has a tendency to pit in high dissolved-oxygen environments. In cases involving high chloride or other halide ion concentrations, Type 316 stainless steel (18% chromium, 12% nickel, and 2.5% molybdenum) is recommended. For the type of corrosion that usually occurs in the splash zone, Type 304 stainless steel probably represents over design and unnecessary extra cost, since alloys containing lower percentages of expensive chromium and nickel do just as well in similar environments (38). Unfortunately, the limited availability of such products restricts their use.

Fiberglass has been successfully used in small-diameter, shallow wells under very corrosive conditions (39). The chief disadvantages of fiberglass are its physical characteristics—poor tensile and compressive strengths and handling problems.

Submerged Zone

The submerged zone, consisting of casing and screen below the pumping level, is an area of the well likely to experience corrosion for a variety of reasons, including low dissolved-oxygen levels, high CO_2 or H_2S levels, and conditions favorable to the growth of sulfate-reducing organisms. One of these effects may have a synergistic effect on another; for example, the low oxygen levels at greater depths cause the portion of the screen and casing there to act as an anode and thereby to produce ferrous ions in solution (Equation 10.8). These ferrous ions are in turn available for reaction with hydrogen sulfide to produce iron sulfide, an insoluble corrosion product (Equation 10.19). The iron sulfide in turn forms tubercles that create even more anaerobic conditions. In addition to combining with soluble iron, hydrogen sulfide can form a substrate for sulfur bacteria under more aerobic conditions (Equation 10.26) (19).

Submerged Casing. Casing in the submerged zone does not experience as great a corrosive attack as the screen or perforated area because less metal surface is exposed and a smooth homogeneous surface is provided. If corrosion in the submerged casing is expected to be severe, stainless steel should be specified. However, in most situations alloying is not necessary. In lieu of alloying, the casing wall thickness can be increased as heavier wall thickness extends life considerably. As a rule of thumb doubling the wall thickness should extend casing life by more than four times, provided the corrosion is fairly uniform. Increasing wall thickness is also effective when pitting is the primary corrosion mechanism.

The pit depth, P, in ferrous metals buried in soil, may be related to time of exposure, t, as follows:

$$P = Kt^n , \tag{10.28}$$

where K depends on the local environment and n takes values of $\frac{1}{6}$ to $\frac{2}{3}$ (40, 41). Rossum (40) states that this relationship can be adapted to well waters, where n is $\frac{1}{6}$ in cases where a protective film forms and $\frac{1}{3}$ where no film forms. Thus, if the wall thickness is doubled, the time required for a pit to penetrate the new thickness will be 64 times greater than the time required to penetrate the original thickness, provided a protective film forms, and 8 times greater if no protective film forms. This principle also applies to casings exposed to atmospheric and splash zone conditions.

A method suggested for controlling corrosion in the submerged zone is the addition of corrosion inhibitors, which have been successfully applied by the petroleum industry especially in recharge wells. The inhibitors act by either coating the casing, and thus protecting it, or modifying the pH or other characteristics of the water to make it less aggressive. For recharge wells, inhibitors are added directly to the injected water either continually or periodically. This is not possible for a producing water well because the chemicals are rapidly removed during pumping and cannot be replaced. Tubes have been used to discharge hypochloride solutions below the pump (19), but they have a high likelihood of plugging if lime is utilized as the corrosion inhibitor. In addition to the plugging problem, adding even the cheapest neutralizing chemical (lime) could be prohibitively expensive based on the need for pH change in the ground water. Expenses include not only the chemical itself but also maintenance and operator time. Other chemicals, such as sodium silicate or polyphosphates, are even more expensive and generally not as effective as lime. In general, use of cathodic protection or film-forming inhibitors has been either ineffective or too expensive for practical applications.

Submerged Screen. Well screens are more susceptible to corrosion than casings because of their large exposed surfaces. If experience in a locale indicates that screen corrosion will be a problem, the material of choice is Type 304 stainless steel because of its availability, cost, and very low corrosion rate in potable water.

From a theoretical viewpoint one would expect that an alloy material, such as stainless steel screens, should not be directly welded to low carbon steel casing in order to avoid dissimilar metal contact. The common opinion is that galvanic action between the two dissimilar metals will cause the "less noble" carbon steel to deteriorate rapidly. In fact, although galvanic action does take place initially, the carbon steel rusts and polarizes rapidly, effectively inhibiting further deterioration (42). In this case, the key design parameter is that the area of the cathode (stainless steel) should be less than the area of the anode (carbon steel). Alternatively, a special joint can be installed with a nonconducting seal of plastic, rubber, or some other material placed in between the two metals. Or, if threaded connections are used, the male threads of the stainless steel screen should be threaded into the female threads of the carbon steel. In this case the heavy wall thickness of the carbon steel is adequate to protect the structure from corrosive failure.

Bacterial Corrosion

Well screens serve as excellent places for bacterial growth because of their large surface areas; the continuing supply of reactive agents, such as sulfate, hydrogen sulfide, or ferrous ions; and the redox potential gradient which often exists where anaerobic ground waters are initially exposed to an aerobic environment. Even in an anaerobic aquifer, an aerobic environment may be created by cascading water. Similar conditions may be created by mixing waters from aerobic and anaerobic aquifers. If soluble iron is present, severe iron bacteria problems may occur in the vicinity of the well screen and nearby casing. This is a major reason for not screening aquifers that separately contain iron and oxygen. These two elements cannot exist in the same aquifer since the oxygen will oxidize the iron to ferric oxide.

Initially, the species of bacteria that predominate in a well will depend on ground water quality. Aerobic conditions favor the growth of iron or sulfur bacteria. Anaerobic conditions favor the sulfate-reducing bacteria. Eventually, however, anaerobic bacteria can be found throughout as corrosion and bacterial action create anaerobic zones beneath deposits. Because iron and sulfur bacteria are ubiquitous, maintenance of sterile conditions during well construction is not a means for avoiding these problems. Rather, care should be taken in drilling and casing and screen installation so as not to introduce gross organic contamination into the aquifers. Should either form of bacteria become established, suitable procedures for disinfection and acid cleaning of the well must be used to maintain water quality and yield. These procedures are described in Chapter 17.

As discussed earlier, the principal aerobic bacteria are the iron bacteria which occur in three physical forms: encapsulated coccoid form (*Siderocapsa*), the stalked iron fixing bacteria (*Gallionella*), and a filamentous group (*Crenothrix, Sphaerotilus, Clonothrix,* and *Leptothrix*) (27). The *Galli-onella* and *Leptothrix* forms are free-floating organisms and hence do not tend to clog well screens. They are nuisance bacteria that may result in discoloration and an unpleasant taste to water. The other filamentous bacteria attach themselves to surfaces and form a wet tenacious slime. Iron bacteria flourish in waters containing as little as 0.02 mg/L of iron but are most prolific in waters that provide a continued supply of ferrous ions at a concentration of 1 mg/L or greater.

In an anaerobic environment the principal species associated with corrosion are the sulfate-reducing bacteria. Under favorable environmental conditions sulfate-reducing bacteria can proliferate around a well, combining with available hydrogen to form hydrogen sulfide. The hydrogen sulfide in turn reacts with the iron to produce insoluble iron sulfides. In addition to their presence in wells under low oxygen or anaerobic conditions, sulfate-reducing bacteria can survive in an aerobic environment and can grow prolifically in a mature biofilm community beneath the aerobic layer. These anaerobic bacteria are known to feed successfully on the by-products of the aerobic community, which includes acetate, carbon dioxide, and propionate (30).

Mitigating factors fall into two categories: those occurring before and during well construction, and those those occurring after bacterial problems have been identified. Field experiences of water well contractors and regulatory personnel may identify an aquifer that harbors significant natural populations of iron-precipitating or sulfate-reducing bacteria (43). Then the well might be sited in another location in order to avoid the known problem. In general, shallow wells and springs are vulnerable to bacterial contamination and thus iron precipitation problems, namely, those associated with high oxygen and nutrient levels. Very deep, iron-containing waters may deposit iron sulfides if the sulfates are successfully reduced by sulfate-reducing bacteria.

Saline Water Corrosion

Wells are sometimes constructed to either extract or inject saline water. For such wells, the most severe corrosion agent is seawater. Seawater is the most corrosive of saline waters, acting more aggressively toward steel than water with either higher or lower salt concentrations (44). A commonly cited corrosion rate for carbon steel immersed in seawater is 5 mils per year for up to eight years. Then the corrosion rate decreases to a steady but slower rate. In addition to salinity, the principal factors affecting corrosion rate in seawater are dissolved oxygen content and water temperature. Alloying small amounts of either copper or chromium appears to have no effect on the corrosion rate. Stainless steel Types 316 or 304 offer good protection, even though Type 304 is subject to pitting in the chloride environment. The presence of oxygen in seawater is necessary to passivate stainless steel. Brackish waters, such as might be found in estuaries or industrial discharges, exhibit corrosive properties similar to seawater.

10.5 INCRUSTATION

Incrustation of well screens and casings exposed to ground water can occur as a result of two types of chemical reactions: mass action (precipitation) or electron transfer (redox). Both reactions can and often do occur simultaneously at the same location. Simple precipitation reactions are primarily functions of solute concentrations, temperature, and pressure. Within the well itself, the temperature and pressure are not likely to vary widely, hence precipitation reactions will primarily be controlled by solute concentrations. For precipitation to occur, the products of the reaction must have a lower chemical potential than the reactants (8). Since the reactants have been present for a long time in the ground water, equilibrium between them has probably already been attained. Hence, when precipitation takes place, a change in chemical potential has been caused by environmental conditions in the immediate vicinity of the well casing or screen. Such conditions include the rapid mixing of ground waters from two different aquifers, corrosion, biological activity, and changes in dissolved oxygen content.

The fact that the products of a reaction, such as the formation of calcite,

$$Ca^{2+} + HCO_3^- \rightarrow CaCO_3 + H^+ , \quad (10.29)$$

have a lower chemical potential than the reactants themselves does not mean that the reaction necessarily proceeds to equilibrium. What can be said is that the free-energy change is favorable for the reaction and subsequent observation will verify whether overall conditions were suitable for the reaction to proceed.

Some reactions, such as corrosion, are definitely electron-transfer reactions, but they also produce hydroxide species that raise the microenvironment's pH, often leading to the precipitation of calcium carbonate or other hydroxide species (i.e., mass-action reactions). Corrosion-induced deposition is a common phenomenon; it often results in products not predicted by bulk solution water quality analyses, since the corrosion reaction can raise the pH at selected cathodic microenvironments to approximately 9.5.

Likely Incrustation Species

Deposition caused by the mixing of two unsaturated waters from different aquifers is a relatively infrequent occurrence. Occasionally, mixing of a low-calcium, high-alkalinity water with a high-calcium, low-alkalinity water may create a supersaturated mixture that produces a thick calcite incrustation. When this occurs in an operating well, only the upper screen interval becomes incrusted, but when the well is shut down, small differences in hydrostatic head can result in flows from one aquifer to another, plugging the gravel envelope or formations.[1]

1. Jack Rossum, personal communication, 1985.

Incrustation has been reported in fiberglass well screens in both supersaturated (3) and unsaturated (6) water sources. Although supersaturated aquifer waters are potentially incrusting, the triggering mechanism undoubtedly occurs at the material surface, possibly because of a corrosion reaction in steel casings or screens or because of bacterial activity in either metallic or nonmetallic screens. For steel casings and screens in an aerobic environment, iron bacteria form insoluble iron hydroxide forms (Equation 10.22), and in an anaerobic environment sulfate-reducing bacteria form insoluble iron sulfide (Equation 10.19). Even in the absence of steel casings and screens, sulfate-reducing bacteria or sulfur bacteria can react with sulfate ions or hydrogen sulfide, respectively, to modify the microenvironment's pH, thereby precipitating previously unsaturated reactants.

Deposits are generally found in three locations: (1) in the slots of the well screen, (2) on the aquifer side of the screen, and (3) on the inside wall of the screen. An analysis of deposits from one well in West Pakistan by optical, X-ray diffraction, and chemical tests indicated that although the proportions of specific deposits vary with location, the types were the same (8). For example, the material filling the slots immediately adjacent to the carbon steel screen was principally siderite ($FeCO_3$), followed by spinel (Mn_3O_4) and copper hydroxide. The deposits filling the interior portion of the slot were principally copper hydroxide, followed by mackinawite (FeS) and quartz. Deposits on the aquifer side of the screen were principally siderite, followed by copper hydroxide and spinel (Mn_3O_4); deposits on the inside wall of the screen were principally copper hydroxide, followed by mackinawite and siderite. The authors concluded that the iron-bearing deposits were related to the corrosion of the well casing and hence probably would not have formed on a nonmetallic casing material. The initial deposition of corrosion-induced species undoubtedly has a catalytic affect on further precipitation, by modifying both the surface of and the microenvironment in contact with the well casing.

Predicting Incrustation Species

Barnes (8) developed a technique for analyzing the incrusting species and comparing these data with the precipitants predicted from water-quality data. A computer model was utilized to determine the states of reaction (supersaturation, equilibrium, and unsaturation) with respect to possible oxides, hydroxides, carbonates, sulfides, and sulfates of the metal cations, utilizing the known hydrolysis and redox reactions outlined earlier. With this computer program the predicted supersaturations (or unsaturations) of a wide variety of product species were calculated based on available water-quality data. These data included major cations, major anions, trace metals, pH, Eh, and alkalinity. In this procedure the actual precipitants were compared with predicted reaction products, thereby giving an estimate of the effect of bulk solution water-quality properties on well incrustation.

Several limitations to this type of analysis are immediately apparent. First, the reactions are based on achieving equilibrium within the bulk phase, a factor that does not take into account the speed with which the reaction may occur. Certain of the reactions are kinetically very slow, which means that the reacting species may be pumped from the well before adequate time has elapsed to form the reaction product. A second problem develops from the time dependence of the concentrations of the various reacting species. As the well is pumped, water quality may vary, thereby producing corrosion or precipitation during a time when water quality is not sampled. Third, sampling the pumped water from the well only gives bulk phase concentrations of each species and hence does not sample the microenvironment at the casing and screen surface where corrosion and deposition are occurring. Also within this microenvironment, active microbiological communities may be modifying solution properties, especially dissolved-oxygen content and hydrogen ion concentration, the two principal factors controlling the solubilities of many species. Finally, the catalytic effect of corrosion has been clearly demonstrated; especially important is the fact that in the vicinity of an active corrosion cathode, the solution pH may rise to approximately 9.5, a value likely to precipitate many carbonate and hydroxide species.

Despite these difficulties valuable comparisons are available from analyses of this sort. Using the water-quality parameters of the bulk solution, one can calculate a degree of supersaturation, or unsaturation, for the various species present. This provides a measure of the degree of departure from equilibrium existing in the bulk phase solution. Details may be found in Barnes (8).

Correlation with Corrosion

In addition to incrustation studies, Barnes (8) attempted to correlate the formation of possible protective phases with the measured average corrosion rates of mild steel for nine wells found in Egypt and Pakistan. Each of the well waters examined was unsaturated with respect to iron, and hence one would expect gradual corrosive attack on the steel screen and casing. Once in solution, the potential was available for the precipitation of this iron in other iron-bearing species (corrosion products).

The intent of their investigation was to determine if three species—calcium, iron, and copper—were likely to form solid phases that would incrust the well surface, thereby reducing the average corrosion rate. The results indicated that only two of the solids tested—calcite ($CaCO_3$) and limonite ($Fe(OH)_3$)—yielded a consistent relationship between their formation and the corrosion rate. In both cases, when either the calcium concentration or the iron concentration was great enough to yield supersaturated (incrusting) solutions,

lower corrosion rates were observed than when the calcium or iron concentrations were unsaturated. Since the formation of a thin calcium carbonate scale has long been associated with reduced corrosion of galvanized steel pipe in water systems, these results indicate that slightly supersaturated calcium carbonate is a more likely corrosion inhibitor than supersaturated ferric hydroxide. Furthermore ferric hydroxide is a common component in tubercular corrosion products, indicating that it is not an effective corrosion inhibitor. The other phases considered—including siderite ($FeCO_3$), magnetite (Fe_3O_4), hausmannite (Mn_3O_4), and copper hydroxide—gave little if any corrosion protection to carbon steel.

Despite the difficulties associated with prediction of probable incrustation, especially the inability to collect reaction products immediately at their point of formation and water-quality data in the microenvironment adjacent to the corroding surface, a strong correlation does exist between corrosion rates in deep water wells and supersaturation with calcite and limonite. Refinement of data-gathering capability should enhance the capability to predict the behavior of natural ground water systems in causing corrosion and incrustation. The authors (8) reported that only three solids—calcium carbonate, ferric carbonate, and ferric hydroxide—of 29 possible reactions exhibited near-equilibrium behavior in the natural water systems observed. Obviously, more study of the kinetics of the reactions are needed before reliable predictions can be made of the probability of incrustation from a particular water source. Details may be found in Barnes (8).

In those cases involving steel casing and screens, corrosion of the steel has a catalytic effect on subsequent incrustation, especially by increasing pH at cathodic surfaces which results in precipitation of calcium carbonate. At the present time modeling capabilities are inadequate to predict what effects water-quality changes brought about by corrosion will have on subsequent incrustation. In those situations where corrosion does not participate in the reaction, such as with stainless steel and plastic materials, current modeling does provide a limited predicting capability of the likely supersaturated species. By separating corrosion problems from incrustation problems, a more effective selection of well materials can be made. If, for example, corrosion is expected to be relatively minor but precipitation of supersaturated species is expected, well construction materials should be chosen to resist the effects associated with periodic cleaning of the well screen; conversely, if the water-quality conditions are known to be corrosive to carbon steel, a more corrosion resistant construction material should be selected. By controlling corrosion, one similarly controls corrosion-catalyzed incrustation.

REFERENCES

1. Clarke, F. E. 1962. "Evaluation and Control of Water Well Corrosion Problems in Kharga and Dakhla Oases, Western Desert, Egypt, UAR." U.S. Geological Survey. July.

2. Clarke, F. E. 1968. "Significance of Corrosion in Water Well Development." Water Supply Technology.

3. Clarke, F. E. 1964. "Selection of Metal Components for Long-Term Development of Egypt's Corrosive Ground Waters." U.S. Geological Survey. July.

4. Sudrabin, L. P. 1962. "Preliminary Study of Corrosion of Deep Well Casings and Screens, Kharga and Dakhla Oases, Western Desert, Egypt." U.S. Geological Survey. March.

5. Mogg, J. 1971. "What Experience Teaches Us about Corrosion." *Johnson Driller's J.* (March–April) 1973.

6. Mogg, J. 1973. "Corrosion and Incrustation, Guide Lines for Water Wells." *Water Well J.* (March).

7. Clarke, F. E., and I. Barnes. 1964. "Preliminary Evaluation of Corrosion and Encrustation Mechanisms in Tube Wells of the Indus Plains, West Pakistan." U.S. Geological Survey. July.

8. Barnes, I., and F. E. Clarke. 1969. "Chemical Properties of Ground Water and Their Corrosion and Incrustation Effects on Wells." U.S. Geological Survey Professional Paper 498-D. Washington, DC.

9. Stumm, W., and J. Morgan. 1981. *Aquatic Chemistry.* J. Wiley and Sons, New York.

10. Rossum, J. 1980. "Fundamentals of Metallic Corrosion in Fresh Water." Report Prepared for Roscoe Moss Company, Los Angeles, CA.

11. Clarke, F. E. 1963. "Appraisal of Corrosion Characteristics of Western Desert Well Waters, Egypt." U.S. Geological Survey. June.

12. Oilwell Research, Inc. 1950. "Electrolytic Corrosion Comparison of Mill Cut and Torch Cut Liners." Long Beach, CA.

13. Trussell, R. R., and I. Wagner. 1985. "Corrosion of Galvanized Pipe." *Internal Corrosion of Water Distribution Systems.* AWWA Research Foundation, Denver, CO.

14. Sontheimer, H., W. Kolle, and A. Kuch. 1985. "Uniform Corrosion and Scale Formation." *Internal Corrosion of Water Distribution Systems.* AWWA Research Foundation, Denver, CO.

15. Snoeyink, V., and A. Kuch. 1985. "Principles of Metallic Corrosion in Water Distribution Systems." *Internal Corrosion of Water Distribution Systems.* AWWA Research Foundation. Denver, CO.

16. Edward E. Johnson, Inc. 1955. *The Corrosion and Incrustation of Well Screens.* St. Paul, MN.

17. Walch, M. 1985. "Bugs and Hydrogen Embrittlement." *Science News* 128 (July).

18. Clark, S., and E. Longhurst. 1962. "The Corrosion Behavior of Metals and Protective Coatings in Tropical Atmospheric Exposure Tests." *Metallic Corrosion—First Internatinal Congress on Corrosion.* Butterworth's, London.

19. Pomeroy, R. D., and H. H. Bailey. 1981. "Iron Bacteria and Sulfide Problems in Wells." *Opflow,* 7, 12 (December).

20. Amstutz, R. 1959. "Corrosion Problems in Water Flooding." *Corrosion* 14.

21. Larson, T. 1970. "Corrosion of Domestic Waters." Illinois State Water Survey Bulletin 56. Urbana, IL.

22. Okamoto, G., and T. Shibata. 1978. "Passivity and Breakdown of Passivity of Stainless Steel." (In Frankenthal and Kruger (eds.)), *Passivity of Metals.* Corrosion Monograph Series. Electrochemical Society, Inc.

23. Coburn, S. K. 1978. "Corrosion in Fresh Water." Corrosion Characteristics of Carbon and Alloy Steels, American Society of Metals. Metals Handbook. 9th Ed.

24. Romeo, A. J., R. J. Skunde, and R. Eliassen. 1958. "Effects of Mechanics of Flow on Corrosion." *J. Sanitary Eng. Div.,* Proceedings ASCE. July.

25. Butler, G., and H. Ison. 1966. *Corrosion and Its Prevention in Waters.* Reinhold, New York.

26. Van Beek, C., and D. Kooij. 1982. "Sulfate-Reducing Bacteria in Groundwater from Clogging and Non-Clogging Wells in the Netherlands River Region." *Groundwater* (June).

27. Stott, G. 1973. "The 'Tenacious' Iron Bacteria." *Johnson Driller's J.* (August).

28. Kolle, W., and H. Rosch. 1980. "Untersuchungen an rohmetzenkustierungen unter mineralogischin Gesichtspunkten." *Vom Wasser* 55, 159.

29. Starkey, R., and K. Wight. 1945. "Anaerobic Corrosion of Iron in Soil." American Gas Association Report.

30. Raloff, J. 1985. "The Bugs of Rust." *Science News* 128 (July).

31. Ahrens, T. 1966. "Corrosion in Water Wells." *Water Well J.* (April).

32. Buck, D. M. 1913. *Industrial and Engineering Chemistry.* Vol. 5, p. 447.

33. Swan. J. 1982. "Relative Corrosion Resistance of Certain Steels in Soils and Waters." *Materials Performance* (June).

34. American Society for Metals. 1980. "The Selection of Mild Steel for Corrosion Service." *Metals Handbook.* Vol. 1, 8th ed.

35. ASM Committee on Mild Steel Corrosion. 1954. "The Selection of Mild Steel for Corrosion Service." *Proc. ASTM* 54.

36. Larrabee, C. P. 1958. "Corrosion-Resistant Experimental Steels for Marine Applications." *Corrosion.* Vol. 14.

37. Coburn, S. K. 1975. "Atmospheric Corrosion." *Metals Handbook,* 9th Ed. American Society of Metals.

38. Moss, R. 1966. "Evaluation of Materials for Water Well Casings and Screens." *NACE Proc.* (October).

39. Clarke, F. E. 1967. "Corrosion in Water Well Development." *Groundwater Age* (November).

40. Rossum, J. 1969. "Prediction of Pitting Rates in Ferrous Metals from Soil Parameters." *JAWWA* 61, 305.

41. Romanoff, M. 1957. "Underground Corrosion." National Bureau of Standards Circular 579. U.S. Government Printing Office.

42. Roscoe Moss Company. 1982. *A Guide to Water Well Casing and Screen Selection.* Los Angeles, CA.

43. Smith, S. 1985. "Iron Bacteria Problems and Wells." *Groundwater Age* (May).

44. Coburn, S. K. 1975. "Corrosion in Seawater." *Metals Handbook,* 9th Ed. American Society of Metals.

CHAPTER 11

Selection of Casing and Screen

11.1 INTRODUCTION

Water well casing and screen comprise the finished well structure and are essential to the well's usefulness and longevity. Their selection is a critical component of the well-design process.

Since underground conditions vary, the selection process requires investigation and analysis. This chapter discusses the functions of casings and screens in general use today, and how they are choosen and applied. Also covered are the important topics of end connections and casing and screen accessories.

Other important information relevant to the casing and screen "string" design is found in Chapters 6, 7, 9, 10, 12, and 13.

11.2 WELL CASING

The casing serves several essential purposes in water wells. Together with the well screen it supports the borehole, protecting the well structure and pump. It also protects the ground water from contamination originating at the surface, and in some cases plays an important role in the well construction process.

There are four types of water well casings. They are classified according to function: surface casing, conductor casing, intermediate casing, and pump housing casing. Each category has some common requirements:

1. A smooth interior to permit the installation and operation of drilling tools, development tools, and pumps.
2. Physical properties adequate to withstand the stresses of installation and other forces that may be applied during well completion, development, and operation (see Chapter 9).
3. Casing material selected to ensure satisfactory life in the well environment.

Surface Casing

Occasionally, a 10- to 20-ft length of large-diameter steel casing is used to stabilize the ground around the borehole and drilling equipment, because of unstable surface formations or high water levels.

Conductor Casing

A high-capacity well design usually specifies that conductor casing be installed and cemented to a minimum of 50 ft, or to the first impervious formation, in order to prevent well contamination from the surface. Conductor casing also stabilizes the upper borehole while drilling.

In gravel envelope wells where the pump housing casing is not sealed, the annular space between the conductor casing and pump housing casing is used as a reservoir for gravel. The conductor casing also serves as a foundation on which to suspend the well casing and screen to avoid buckling and to prevent its movement if the gravel and formation should settle.

Intermediate Casing

Intermediate strings of casing may be required to facilitate completion in deep boreholes where difficult drilling conditions are encountered (hydrating shales, loss of circulation zones, etc.). A multiple-diameter casing program may be required for deep cable tool wells where the casing cannot be advanced to total depth in a single string by driving or jacking.

Pump Housing Casing

Pump bowl diameter, depth of setting, and diameter of column pipe are established by maximum anticipated production, lift, and total head. Pump housing casing diameter and length are based on the pump design needed to meet these requirements (see Chapters 6 and 16).

11.3 SELECTION OF CASING MATERIAL

Installation in Open Borehole

Since installation of casing in an open borehole involves lower stresses than incurred in driving, more latitude is available for material selection. Steel is by far the most common casing material in water wells. It is easily formed into tubes and, when exposed to the atmosphere, soils, or water, builds up a protective oxide coating that ensures adequate life under mildly corrosive conditions. Life can be extended through modifications in chemistry (see Chapter 10).

Steel possesses the high yield and tensile strengths required for water wells. Chapter 9 discusses the physical requirements of well casing and screen. Appendix H, Table H.1 lists the physical properties of steel casing. Elasticity and resiliency are other important characteristics inherent in steel. Weldability facilitates field installation.

In some wells nonferrous materials have been successfully used as casing. They include concrete, plastic, and fiberglass. Concrete casing is used in some shallow installations, but its weight, difficulty in handling, and special connecting-joint requirements have rendered it impractical for general use.

Plastic well casing has found widespread use in shallow domestic wells up to 8 in. in diameter. The standard specification for thermoplastic water well casing and couplings made in standard dimension ratios (SDR) is ASTM F 480. The standard dimension ratio (SDR) refers to the ratio of diameter to wall thickness.

Three materials used in the manufacture of thermoplastic well casing are covered in this standard. The most commonly used is rigid poly(vinyl chloride) (PVC) manufactured to ASTM specification D 1784, with a cell classification of 12454C or 14333 C or D). Less frequently used are ABS (acrylonitrile-butadiene-styrene) and SR (styrene-rubber). ABS is covered by ASTM specification D 1788, with a cell classification of 533 or 434. SR is covered by ASTM specification D 1892, with a cell classification of 4434 A.

Collapse resistance of thermoplastic well casing remains constant for all diameters with a constant SDR and material. Dimensions and collapse pressure for PVC, ABS, and SR casings manufactured from pipe size (IPS) standards are shown in Appendix H, Tables H.4 to H.6. Plastic pipe does not possess physical properties comparable to those of steel and is not as suitable for large-diameter or deep wells.

Fiberglass-reinforced plastic casing has been used in some areas where waters are known to be highly corrosive or encrusting. Connecting-joint limitations have restricted it to use in wells up to 800 ft in depth. Although mechanical joints designed for this end use have been adequate for installation purposes, they have been known to present difficulties in well rehabilitation. Fiberglass is relatively costly, particularly for the larger-diameter, heavier-wall tubes required for high-capacity wells.

Installation by Driving or Jacking

Only steel casing has the physical properties to withstand the stresses imposed by driving or jacking. Use of casings that present a smooth exterior is advantageous since wall friction is reduced. This is satisfied by the use of butt-welded casing or double well casing. The assembly of these casings is explained in Appendixes F and G. The system for casing installation by use of hydraulic jacks is covered in Chapter 7.

11.4 MANUFACTURING PROCESSES—STEEL CASING

Practically all steel pipe manufactured worldwide falls into two categories:

1. Transmission (line) pipe for the conveyance of water, oil, or gas.
2. Oil country tubular goods such as oil field casing, heavy wall tubing, and drill pipe.

It is useful to review the manufacturing processes of steel pipe and relate them to their primary purposes, as well as their use as water well casing.

Electric Resistance Welded

Most transmission pipe of diameters from 4 to 16 in. is produced by the electric resistance weld (ERW) process. First raw material in coil form (skelp) is unrolled and flattened. The skelp then moves through a series of forming rolls that stage by stage shape it into a cylinder. The seam is welded as it passes beneath sliding contact shoes. Resistance encountered by the current at the seam edge heats the metal to a plastic state. Simultaneously, pressure is applied forging the edges together. The pipe then travels through a series of finishing rolls. These stages reduce the diameter slightly, ensuring proper diameter and roundness. Since resistance-welded pipe is designed for transmission pipe use, it is manufactured in approximate 40-ft lengths for transportation and field assembly.

Seamless

Oil country tubular goods, such as high-pressure casing and drill pipe, require wall thicknesses and chemistries that are difficult to weld by ordinary procedures. These products and other special purpose pipe and tubing in diameters 20 in. and smaller are manufactured by the seamless process. A billet of steel is heated to a plastic state and pierced by a spear or lance. The hollow billet is then gradually elongated as it is shaped over mandrels and sized by rollers until pipe is produced. For water well applications, seamless pipe has no inherent advantage over welded pipe, but its cost is higher. Furthermore wall thickness uniformity may not be as consistent as the pipe or casing produced by other methods of manufacture.

Press Formed

A substantial portion of transmission pipe 18 in. in diameter and larger is manufactured by the press-forming process. This method is a three-stage operation:

1. In the first stage the edges of a flat steel plate are curved upward.
2. The plate is then pressed by dies into a U-shape.
3. In the final stage a third press closes the U-shaped plate to form a cylinder. The seam is then welded by the submerged arc electric weld (SAW) process. The SAW process requires the use of a bare-wire electrode and a granular flux. Contact of the electrode and the seam to be welded creates an electric arc. As a welding head traverses the seam, wire and flux are continuously added. The function of flux is to shield the molten weld from atmospheric contamination and add alloys to the weld deposit. This process allows the weld to be in a molten state long enough to purge impurities.

Fabricated

Generally used for production of large-diameter pipe for special end uses, the fabricated process is noncontinuous and multistage. Flat sheets or plates are first squared and sheared to the appropriate diameter requirements. In the second stage longitudinal edges are formed to the required curvature. The steel is then driven between rolls, bending it into a cylinder. Seam welding is performed by the SAW process. Longer lengths, when required, are manufactured by welding the necessary number of cylinders together.

Spiral Weld

A newer pipe manufacturing technology involves fabricating a spiral seam tube. The skelp is flattened and formed into a cylinder between rolls or a circular-shaped cage or shoe. The inside seam is welded at the first point of strip contact. Most spiral tubes today are welded from both the inside and outside, ensuring full weld penetration.

Spiral pipe mills are flexible, enabling the manufacture of pipe from many grades of steel and nonferrous weldable metals. Nonstandard diameters and thicknesses can be economically produced for water well installations where required.

11.5 CASING AND MATERIAL STANDARDS

Most well casings are manufactured to standards established by the American Society for Testing Materials (ASTM), the American Petroleum Institute (API), and the American Water Works Association (AWWA). All three agencies designate grades comparable in strength and quality. Standards applicable to water wells are shown in Appendix H, Table H.2.

Substandard Pipe

The line pipe produced to API and other standards must be subjected to various tests during and following manufacture. When the pipe fails one or more of these tests, it may be sold as a substandard product. Among the names applied to this product are "reject," "substandard," "limited service," and "structural grade" pipe. In many cases it is impossible at a glance to differentiate between "prime" and "reject" pipe.

A large proportion of rejected pipe has failed tests to determine weld soundness. In water well installations stresses may cause a seam to split, allowing formation material to enter the well.

Pipe may be rejected for reasons other than faulty welding. Variations in diameter or nonuniform wall thickness are frequently cited. Substandard pipe manufactured from laminated steel has substantially reduced collapsing strength.

Used pipe or casing should never be installed in a water well. Problems include out-of-roundness, varied wall thickness, and difficulties in assembly.

Diameter as Manufactured

Seamless, ERW, and press-formed pipe diameters are controlled externally during manufacture. Accordingly, nominal diameters are measured by the outside diameters (OD), and the inside diameter (ID) varies with the wall thickness of the pipe. Diameters up to 12 in. nominal (12¾ in. OD) are designed to have an ID at least equal to the nominal diameter. However, for diameters of 14 in. and above, the OD is the nominal diameter. For instance, 16-in. OD pipe 0.3125 in. thick would have an inside diameter of 15.375 in. Casing diameter is generally selected according to pump size. Therefore it is often possible to substitute an ID size for the next larger OD size; for example, 16-in. ID instead of 18-in. OD.

Material Selection

Steel casings may be manufactured from a variety of materials. The most common is carbon steel equivalent to Grade B transmission line pipe (35,000 psi minimum yield and 60,000 psi minimum tensile strength). Also available are copper-bearing steel, high-strength low-alloy steel and stainless steel in various grades. For stainless steels, Type 304 is the grade most frequently specified for water wells and Type 316 for saltwater production wells. Various grades of ferrous metals, their governing specifications, physical and chemical properties, and resistance to atmospheric corrosion are listed

TABLE 11.1   Recommended Minimum Thickness for Carbon Steel Well Casing

Depth (ft)	Nominal Casing Diameter (in.)									
	8	10	12	14	16	18	20	22	24	30
0–100	1/4	1/4	1/4	1/4	1/4	1/4	1/4	5/16	5/16	5/16
100–200	1/4	1/4	1/4	1/4	1/4	1/4	1/4	5/16	5/16	5/16
200–300	1/4	1/4	1/4	1/4	1/4	5/16	5/16	5/16	5/16	3/8
300–400	1/4	1/4	1/4	1/4	5/16	5/16	5/16	5/16	3/8	3/8
400–600	1/4	1/4	1/4	1/4	5/16	5/16	5/16	3/8	3/8	7/16
600–800	1/4	1/4	1/4	5/16	5/16	5/16	3/8	3/8	3/8	7/16
800–1000	1/4	1/4	1/4	5/16	5/16	5/16	3/8	3/8	7/16	1/2
1000–1500	1/4	5/16	5/16	5/16	3/8	3/8	3/8	3/8		
1500–2000	1/4	5/16	5/16	5/16	3/8	3/8	7/16	7/16		

Source: AWWA Standard A 100-84.

in Appendix H, Table H.3. The selection of material depends on a number of factors, especially cost and the underground environment. This is covered in Chapter 10.

Selection of Wall Thickness

Analysis of the relationship of casing wall thickness to physical properties is covered in Chapter 9. The relationship of casing wall thickness to longevity is discussed in Chapter 10. Recommended minimum wall thickness for selected casing diameter and total depth is given in Table 11.1 from AWWA Standard A 100-84.

11.6   END CONNECTIONS

No discussion of water well casings is complete without considering the connections through which they are joined in the field. The tensile strength of any column is limited by the strength of the connections between its components. Observation of well failures shows that many involve casing or screen rupture, collapse, or deformation. Frequently the problem originates in the connecting joints. Other considerations in addition to mechanical-strength requirements in connecting-joint design are smoothness of internal wall, minimization of external diameter, alignment, ease of installation, and cost.

The four major types of connections used for water well casing and screen, as shown in Figure 11.1, are:

1. Threaded and coupled.
2. Square or beveled machined ends.
3. Bell and spigot joints for lap welding.
4. Machined ends with welding collars for lap welding.

Threaded and Coupled

Threaded and coupled connections are commonly used in 4-in. and smaller-diameter wells where they provide relatively inexpensive, fast, and convenient connections. Strength requirements in such domestic low-capacity wells are not critical. The cost of threaded and coupled joints increases with diameter, and they are not generally available in diameters larger than 12 in.

Square or Beveled Machined Ends

Casing and screen joints prepared with square ends for welding are generally satisfactory up to 0.1875-in. wall thickness. With heavier wall thicknesses the ends should be beveled to facilitate weld penetration, leaving approximately 0.125-in. flat. Advantages of these connections (butt welding) are economy and smoothness of the external diameter, which reduces the tendency of gravel to bridge in gravel envelope wells and friction when driving. Disadvantages lie in greater assembly time and the difficulty of properly welding casing in the vertical position. A further problem occurs if removal and reassembly is required. The connection must be cut with a torch when the casing is withdrawn, requiring possible machining for reassembly, which is a time-consuming and expensive process.

Bell and Spigot

Use of bell and spigot joints overcomes a number of problems inherent in plain-ended or threaded and coupled joints. A down-hand fillet (lap) weld is used to connect the sections. Lap welds are easier to make than horizontal butt welds. These joints are also very economical. Their chief disadvantage involves more installation time to ensure proper alignment.

Welding Collars

The welding collar has proved very satisfactory for lap welding 6-in. and larger-diameter casing and screen. Width of the collar ranges from 2 to 6 in., with the casing end extending approximately midway through the length of the collar.

Alignment holes are provided in the collar to allow visual confirmation of joint butting.

A properly made welding collar connection is as strong or stronger than the casing. API threaded and coupled joint strength by comparison is less than 70% of the casing strength. The importance of having a field connection equal in strength to the casing or screen material cannot be overemphasized. Withdrawal of casing or screen sections requires only removal of the field weld at the top of the collar. Such sections are easily reinstalled because the original end faces of the tubes have not changed. Transportation and handling damage is reduced. The welding process for collared casing is discussed in Appendix F.

11.7 CASING ACCESSORIES

Stainless Steel to Carbon Steel Connections

Some concern has been expressed regarding the connection of stainless steel material to regular carbon steel in the field. It is commonly thought that galvanic action between the two dissimilar metals will cause the "less noble" carbon steel to deteriorate rapidly and fail. Although galvanic action does take place initially, the carbon steel rusts and polarizes rapidly, effectively inhibiting further deterioration. If stainless steel is welded directly to carbon steel, the carbon steel

section should be at least two times the thickness of the stainless section and its length at least three times its diameter. A connection that may be used to eliminate welding the two dissimilar metals is shown in Figure 11.2. Further information on this subject is presented in Chapter 10.

Landing Clamps

When casing and screen is installed in an open hole it should be suspended from the surface by a heavy-duty clamp. This clamp may be supported at the ground surface by beams or by the conductor casing. The conductor casing, in turn, must be supported by beams or cemented in place. The purpose of suspending the casing and screen is to ensure that it remains straight and centered in the borehole. It is also at risk as explained in Chapter 9. Figure 11.3 shows a typical range of clamp dimensions for various diameters and total casing and screen weights.

Float Plates

Float plates are installed in casing strings when the weight of the casing and screen approaches the lifting capacity of the installing rig. The plate, which must be manufactured from a frangible material, is installed between two joints of casing at a predetermined depth, where the collapsing strength

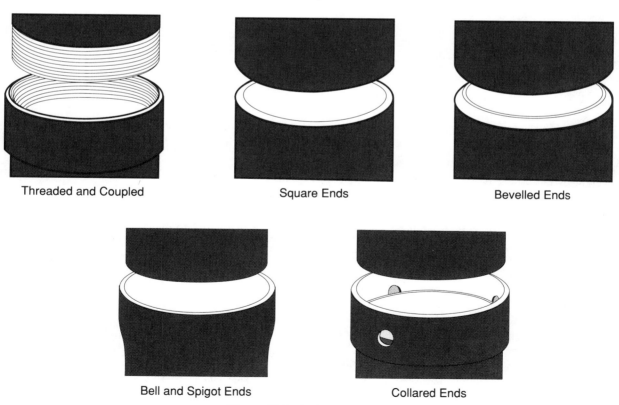

Threaded and Coupled   Square Ends   Bevelled Ends

Bell and Spigot Ends   Collared Ends

Figure 11.1. End connections.

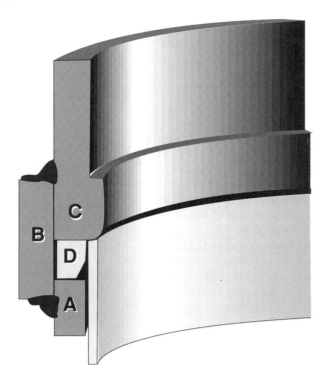

Slip carbon steel ring **A** with machined surface up over end of stainless steel joint.

Attach stainless steel ring **D** to top of stainless steel joint by welding with chamfer on bottom.

Assemble carbon steel rings **A**, **B**, and **C**. **B** is an open ring. Clamp **B** around **A** and **C** making sure that **A** and **C** are snug against **D**. Tack weld in place

Seam weld **B**.

Fillet weld **B** to **A** and **C**, respectively.

Figure 11.2. Stainless steel to carbon steel connector.

of the casing is not exceeded and the rig is not overloaded. The weight of the casing is reduced by the weight of the fluid displaced. Cast iron plates machined for a watertight fit between the casing joints have been found satisfactory. The use of welding collars simplifies the installation of float plates considerably.

To ensure that hydrostatic forces on the empty casing above the float plate do not exceed its collapsing strength, the casing may be partially filled with water during installation as the buoyancy increases. Once the casing and screen is installed, the pump housing casing is completely filled with water and the plate removed by striking with a bailer, drill pipe, or tubing. Float plates must be used with great caution, and under no circumstances are as safe a procedure as conventional installation with equipment having adequate mast capacity (see Chapter 9).

Reducing Sections

Occasionally, reductions are made in diameter between casing and/or screen sections. Regular bell pipe reducers have been used, but the reduction is rather abrupt. Reduction cones have been fabricated from pipe by torch-cutting segments from the pipe, bending the remaining segments inward, and welding the new seams together (orange peeling). But the resulting structure is frequently not uniform and weak. A better solution consists of a fabricated tapered cone with a short stub joint of each diameter casing welded at each end. The stub ends should be machined for greater alignment accuracy.

Although no supporting data exist, it has been suggested that the length of the cone of the reducing section should be at least 10 times the difference in diameter of the two ends for hydraulic efficiency and strength. A longer tapered section of cone mitigates gravel bridging at that point.

Shoe and Starter

When casing is installed by driving or jacking, the bottom must be reinforced with a casing drive shoe. Since drilling tools must pass through the casing, the borehole diameter may be smaller than the casing's outside diameter. Most of the forces involved in moving the casing are absorbed where the casing shoe cuts through the ring of formation not drilled by the bit.

Casing drive shoes may be fabricated from rings rolled from carbon or high-strength steel plate. The latter may be heat treated after machining for use in harder drilling conditions. Heat treating should produce a Rockwell C hardness of 30 to 32 with an SAE 1040 steel ring. Drive shoes are made in thicknesses up to 1½ in., and in lengths up to 14 in.

The shoe is welded onto a length of casing, which may be strengthened by a reinforcing ply welded on the casing above the shoe. This starter section is made 4- to 20-ft long, depending on the casing diameter and expected drilling conditions.

Installation of Well Casing

When installing casing and screen in an open borehole, a spider and slips are generally used temporarily to support the casing as new lengths are added. A nondestructive device, such as a casing grips and sling assembly or a casing elevator, connects to and lowers the string into the well. Use of any method of installation that requires holes to be cut into the casing is not advisable.

11.8 WELL SCREEN

No aspect of well design is subject to more hyperbole than selection of well screen. Screen geometry, aperture size,

Dimensions given are in inches

Casing O.D.	A	B	C	D	E	F	G	H	Allowable Wt. Tons
6-5/8	5/8	6	22-9/16	8	11/16	4	3	7-7/8	18
8-5/8	5/8	6	28-27/32	8	11/16	4	3	9-7/8	16
10-3/4	3/4	6	35-7/8	8	15/16	4	3	12-1/4	17
	3/4	8	35-7/8	8	15/16	4	5	12-1/4	31
12-3/4	3/4	6	42-5/32	8	15/16	4	3	14-1/4	16
	3/4	8	42-5/32	8	15/16	4	5	14-1/4	28
14	3/4	8	46-3/32	8	1-1/16	4	5	15-1/2	26
	1	8	46-25/32	8	1-1/16	4	5	16	35
14-1/2	3/4	8	46-25/32	8	1-1/16	4	5	16	26
	1	8	48-3/8	8	1-1/16	4	5	16-1/2	34
16	3/4	8	52-3/8	8	1-1/16	4	5	17-1/2	24
	1	8	53-1/16	8	1-1/16	4	5	18	32
	1	10	53-1/16	8	1-1/16	6	3-1/2	18	50
16-5/8	3/4	8	54-11/32	8	1-1/16	4	5	18-1/8	23
	1	8	55-1/32	8	1-1/16	4	5	18-5/8	31
	1	10	55-1/32	8	1-1/16	6	3-1/2	18-5/8	49
18	3/4	12	58-21/32	8	1-1/16	6	4-1/2	19-1/2	50
	1	12	59-11/32	8	1-1/16	6	4-1/2	20	65
18-5/8	3/4	10	60-5/8	8	1-1/16	6	3-1/2	20-1/8	34
	1	10,12	61-5/16	8	1-1/16	6	3-1/2,4-/12	20-5/8	45,65
20	3/4	10	65-1/8	8	1-1/16	6	3-1/2	21-1/2	32
	1	10,12	65-5/8	8	1-1/16	6	3-1/2,4-1/2	22	43,62
20-5/8	3/4	10	66-29/32	8	1-1/16	6	3-1/2	22-1/8	31
	1	10,12	67-19/32	8	1-1/16	6	3-1/2,4-1/2	22-5/8	42,60

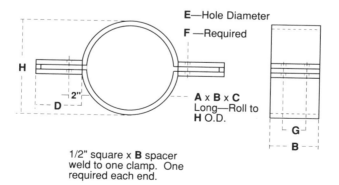

E—Hole Diameter
F—Required

A x **B** x **C**
Long—Roll to
H O.D.

1/2" square x **B** spacer
weld to one clamp. One
required each end.

Figure 11.3. Landing clamp dimensions.

diameter and length, material selection, and hydraulic properties have been the subjects of many studies and publications. Although some general principles have been established, aquifer characteristics in locations separated by only a few feet can vary considerably. Evaluation of relative performance on the basis of selected wells is rarely valid. Experience and common sense suggest that there is no single best type for all conditions. Selection and application of a screen to a specific well environment should be based on a thorough evaluation of all factors involved.

Well screen serves as the intake component of a well and supports and stabilizes the aquifer and filter zone and in the case of gravel envelope wells, it filters the pack. Although perforations made by down-hole tools in casing in place meets these needs, they will be referred to as down-hole perforations.

Technical Considerations

Screen placement and its selection are subject to the well environment and hydraulic principles, and in gravel envelope wells practical considerations related to well development. Discussed next are theoretical analyses of screen length and diameter, area of opening, and entrance velocity.

Screen Length

General Design Criteria. An important design factor directly affecting the production of a well is the amount of saturated aquifer thickness penetrated by the screen. This may be seen by examining the steady-state equation for aquifer drawdown. The drawdown in this example (denoted by s) is measured far enough away from the well screen so that no turbulent head-loss terms are present—usually less than 1 ft). Rewriting the Theim equation in terms of specific capacity,

$$\frac{Q}{s} = \frac{Kb}{528 \log(r_0/r_e)} , \qquad (11.1)$$

where

Q/s = specific capacity of the well [gpm/ft],

Q = well discharge [gpm],

s = aquifer drawdown [ft],

b = saturated aquifer thickness (assumed equal to screen length [ft],

K = hydraulic conductivity of the aquifer [gpd/ft^2]

r_e, r_0 = effective well radius and radius where drawdown is zero [ft].

Equation 11.1 can be rearranged to obtain

$$Q = \frac{Kbs}{528 \log(r_0/r_e)} . \qquad (11.2)$$

As can be seen in Equation 11.1, the specific capacity (Q/s) is directly proportional to the length of well screen.

Equation 11.2 shows the relationship between well discharge and other parameters. For example, consider two wells constructed in an identical fashion in the same hydrogeologic environment except that one well has twice the length of screen of the other. If the wells were pumped and limited to the same drawdowns, the well containing the longer screen would have twice the production of the other. This assumes that well losses and minor head losses, such as convergence, are neglected. Consequently for maximum production it is important to screen all productive zones throughout the depth of a well.

Gravel Envelope Wells

In gravel envelope wells there is an overriding practical consideration related to gravel installation and replenishment, and development and redevelopment, that dictates the interval screened. As discussed in Chapter 6, most gravel envelope wells should be screened continuously from the pump housing casing to within about 20 ft of total depth. This avoids the possibility of casing off productive formation, particularly in alluvium with thin interbedded aquifers. During installation gravel tends to bridge at the top of a screen section and multiple screened intervals compound this problem. Gravel bridges and voids are difficult to eliminate behind blank casing.

Effect of Vertical Flow in Gravel Envelope Wells

Some individuals still justify shorter lengths of screens with the belief that most (or all) of the production from nonscreened aquifers may be gained through vertical flow in a coarse gravel envelope. The fallacy of this belief is shown in the following example.

Consider the well shown in Figure 11.4 having a borehole diameter of 20 in., a well screen diameter of 12 in., and a gravel envelope hydraulic conductivity of 5000 gpd/ft^2. The amount of flow migrating vertically from the upper 100-ft aquifer to the screen section below 200 ft with a 50-ft drawdown imposed is calculated as

$$Q = \left(\frac{AK}{1440}\right)\left(\frac{dh}{dx}\right) , \qquad (11.3)$$

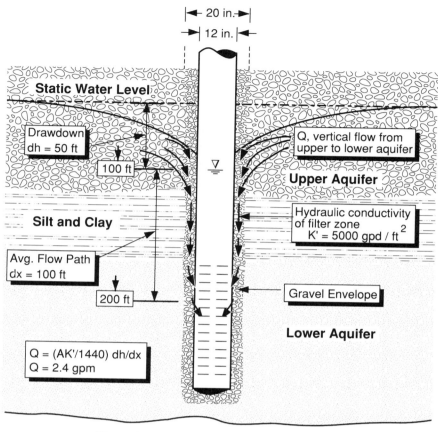

Figure 11.4. Vertical flow in a gravel envelope.

where

Q = amount of vertical flow through the gravel envelope [gpm],

A = $(20^2 - 12^2)/183.3$ (annular area between borehole and well screen) = 1.4 ft².

K' = hydraulic conductivity of gravel envelope = 5000 gpd/ft².

dh = drawdown in the well = 50 ft.

dx = length of vertical flow path through the gravel = 100 ft.

Solving Equation 11.3 for the example yields

$$Q = \frac{1.4(5000)(50/100)}{1440} = 2.4 \text{ gpm} . \quad (11.4)$$

Test for Minimum Screen Length Based on Critical Radius

In Chapter 5, there is an explanation of the phenomenon of the critical radius. The critical radius is defined as that distance measured from the center of the well to the point where the flow changes from predominantly turbulent to laminar. In most cases, the critical radius is less than the nominal well radius due to the usual combinations of filter material, screen length, and diameter and is not a major design concern. However, in some cases, a combination of high production rate, thin aquifers, and small screen diameter may cause turbulent-flow head losses to extend into the filter zone. In these cases, additional well depth and screened intervals may be considered. The following is an example.

Assume that 2000 gpm is produced by a well having a 12-in.-diameter screen with an average aquifer material size (50% passing) of 0.87 mm. A pack/aquifer ratio of 6:1 is selected, resulting in the filter pack having an average grain size of 5.2 mm (small pebble gravel). From Figure 11.5 the specific aquifer discharge for a critical radius of 6 in. is found to be 9 gpm/ft.

The minimum length of well screen (b) required to eliminate any turbulent filter-zone head losses would be

$$b = \frac{2000}{9} = 222 \text{ ft} .$$

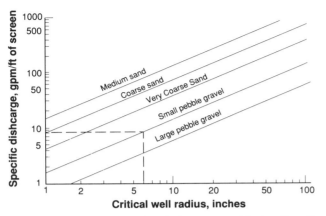

Figure 11.5. Specific aquifer discharge *vs.* critical well radius (reprinted from *Journal* American Water Works Association, Vol. 77, No. 9 (September 1985) by permission. Copyright © 1985, American Water Works Association).

For screen lengths shorter than 222 ft, the critical radius exceeds the nominal radius, creating near-well turbulent head losses. These additional head losses can mean higher pumping lifts and thus higher operating costs over the lifetime of the well. If only 150 ft of screen were installed, then the specific aquifer discharge (Q/s) would be (2000/150) = 13 gpm/ft. From Figure 11.5, the critical radius is found to be 10 in. In this case, turbulent flow would extend 4 in. into the filter zone. If the screen length cannot be changed, due to limited aquifer thickness the screen diameter may be increased to 20 in.

Effect of Screen Diameter on Specific Capacity

The effect of the screen's diameter on well production and specific capacity is not a critical design factor as the following example shows.[1] For a 12-in.-diameter well completed in an aquifer having a transmissivity of 100,000 gpd/ft and a storativity of 0.001, the specific capacity can be calculated after one day of pumping (Equation 5.6) as

$$\frac{Q}{s} = 100,000 / \left\{ 264 \log \left[\frac{0.3(100,000)(1)}{(0.5^2)(0.001)} \right] \right\}$$

$$\frac{Q}{s} = 47 \frac{\text{gpm}}{\text{ft}} \quad \text{(12-in.-diameter well)} .$$

If the well screen diameter is doubled, the specific capacity for the same time period will be:

$$\frac{Q}{s} = 51 \frac{\text{gpm}}{\text{ft}} \quad \text{(24-in.-diameter well)} .$$

Therefore doubling the diameter in this example increases the specific capacity by only 9%.

1. This example assumes no well losses (turbulent flow). An example of the effect of screen diameter on well losses is given in Chapter 5.

Screen Diameter as a Function of Discharge Rate

Regardless of hydraulic considerations screen-diameter selection is usually a direct function of pump housing casing diameter. This is explained in Chapter 6.

Naturally Developed Wells

In naturally developed wells, screens or down-hole perforations are only placed opposite desired aquifers. Usually 1 to 3 ft at the top and bottom of the aquifer(s) are not screened or perforated. The distance depends on aquifer(s) thickness and the gradation of the material.

11.9   SCREEN INLET AREA AND ENTRANCE VELOCITY

Concept of Effective Area of Opening

Water entering the screened section of a well only passes through part of the total screen open area since a portion of the apertures are blocked by particles. The ratio of the amount of area open to flow to the total open area of the screen, similar in concept to the effective porosity in the aquifer, is expressed as the "effective open area" (%) (1).

Entrance Velocity

There has been considerable discussion and controversy regarding the upper limit for screen entrance velocity corresponding to good well design. The transition from laminar to turbulent flow as water approaches and enters the screen is discussed in Chapter 5. Recent investigations have produced data that refute earlier conceptions of the empirical relationship between head losses through screens and entrance velocities (2). Results of these investigations show that for both gravel envelope and naturally developed wells, efficiencies do not significantly increase if percentage open areas of screens are above 3% to 5% (see Figure 11.6). Similar findings have been reported by Ahmad et al. (3).

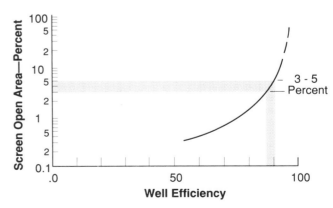

Figure 11.6. Efficiency vs. open area.

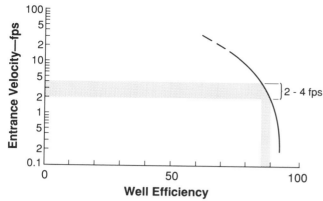

Figure 11.7. Efficiency vs. entrance velocity.

This 3% to 5% open area corresponds to an upper limit for entrance velocity of 2 to 4 ft/sec (much higher velocities were observed in the tests with no measurable movement of sand) (see Figure 11.7). In most field design installations, the entrance velocity is well below this upper limit as the following example shows.

Example

A well with 100 ft of 10-in.-diameter screen having 3% open area is pumped at 1000 gpm. The entrance velocity through the screen can be calculated from

$$V = \frac{Q}{235rbP} , \qquad (11.5)$$

where

V = screen entrance velocity [ft/sec],
Q = well discharge = 1000 gpm
r = radius of the screen = 5 in.
b = length of well screen = 100 ft.
P = percentage open area of the screen = 0.03.

From Equation 11.5, the entrance velocity is calculated as

$$V = \frac{1000}{235 \times 5 \times 100 \times 0.03} = 0.28 \ \frac{ft}{sec} .$$

The example shows that even if the discharge from the well were increased by a factor of five (a highly unlikely field situation), the calculated entrance velocity would still be less than the upper limit recommended by AWWA standard A 100-84 of 1.5 ft/sec.

Effect of Porosity on Screen Open Area

Where formation effective porosities are very low (1% to 2%), well losses are negligible for all ranges of discharge, and the percentage of open area is not a design factor.

Axial Velocity and Head Loss

An evaluation of pressure head losses within the well should include the frictional losses from axial flow within the screen section. Vertical flow velocities near the top of the screen can reach 5 to 6 ft/sec, making this consideration important for long screens. The computation of these losses is made using the Darcy–Weisbach friction formula:

$$h = \frac{fLV^2}{12gD} , \qquad (11.6)$$

where

h = head loss due to frictional resistance [ft],
f = equivalent friction factor,
L = screen length [ft],
D = screen internal diameter [in.],
V = mean axial velocity [fps],
g = acceleration due to gravity.

The calculation is complicated by the fact that the axial velocity is affected by the screen inflow. This inflow also increases the friction factor because the axial flow feels the screen inflow as additional roughness elements. However, there appears to be no relationship between screen geometry and axial flow head loss. Simon performed laboratory studies to evaluate the effect of this inflow and concluded that the friction factor can be increased by as much as 80% over that for blank casing (4).

A conservative estimate of the magnitude of the screen flow frictional loss can be obtained by assuming that the friction factor is twice that of the blank casing and that the maximum velocity of axial flow, which occurs at the top of the screen, applies over the entire length of the screen. With these assumptions there would be 0.015 ft of head loss per foot in a 12-in.-diameter screen flowing at 5 ft/sec. Taking an extremely conservative position and assuming that the screen inflow equates to extreme roughness elements, the friction factor could be as high as 0.07 (see Moody diagram, which relates friction factor f to magnitude of roughness elements and Reynolds number, Figure 11.8). Thus a very high estimate of head loss in such flow is 0.027 ft per foot of screen. A hundred feet of screen would under these drastic assumptions, have a total head loss of less than 3 ft. In general, screen head losses for axial flow can be conservatively assessed by assuming that each foot of screen is equivalent to 2 to 3 ft of blank casing.

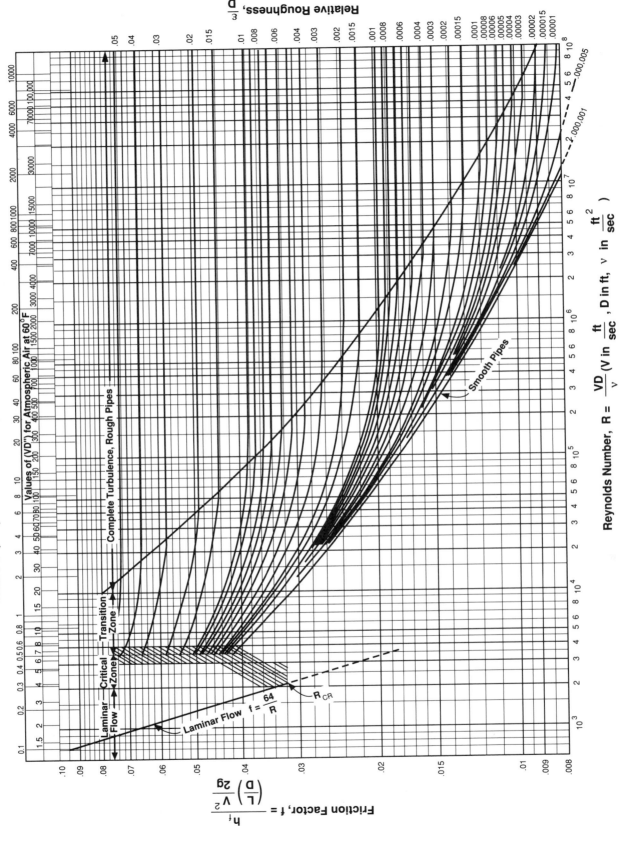

Figure 11.8. Moody diagram (courtesy of the Hydraulic Institute).

11.10 TYPES OF WELL SCREEN

Although the well screen is an important component of well design, it must be realized that it is not an answer to all well problems. In a gravel envelope well the pack filters the formation material, not the screen. The screen simply retains the gravel envelope. Screen selection must be based on theoretical and practical considerations. The best screen design cannot automatically correct incomplete pack installation or inadequate well development. Screen can only facilitate the construction of an efficient well and help ensure its satisfactory operation.

There are a number of design and performance criteria that must be weighed in well screen type selection. These include:

1. Application to gravel envelope or naturally developed wells.
2. Physical properties (discussed in Chapter 9).
3. Material selection and resistance to corrosion (discussed in Chapter 10).
4. Application to development methods.
5. Clogging and stabilization of the filter zone.
6. Aperture size.
7. Hydraulic properties.

A wide variety of screens have been developed for water well construction. Each type has been designed for application to a general set of underground conditions and well design.

Principal Application to Gravel Envelope Wells

Torch-Cut Slots. The torch-cut practice has fallen into disfavor and is rarely used today. However, where alternatives are not available, oxygen-acetylene torch-cut vertical slots can be made in a variety of patterns in casing in the field to produce a screen. Torch-cut slots are characterized by low area of opening, reduced corrosion resistance at the ragged edges, irregular openings, weakness (particularly in collapse strength), and a tendency to clog.

Vertical Machined Slots. Another type of vertically slotted screen is manufactured from casing into which openings are milled with axially oriented cutters (Figure 11.9). This product is designed for use in oil wells where fluid production rates are very low. Its disadvantages include clogging due to the parallel surfaces within the opening. Since slot clogging is directly related to wall thickness, thicker material encourages greater clogging. Some vertical slotted casings are machined with an undercut to reduce this tendency. A second drawback is the low open area, although this can be overcome at higher cost by increasing the number of slots. The collapse strength

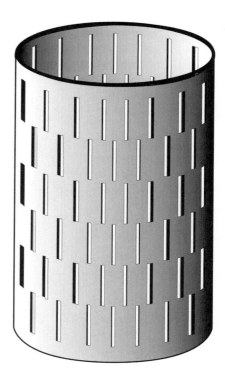

Figure 11.9. Vertical machine slotted screen.

of vertically slotted screens, however, is substantially reduced when the number of openings is increased.

Field experience has proved that development is generally slower in wells completed with vertical machined slot screens. Stabilization of particles is less effective with vertical slots than with equivalent horizontal slots. The aperture size must be selected to retain 90% to 100% of the gravel envelope.

Horizontal Machined Slots. Slots machined perpendicular to the axis of the casing are generally prepunched in flat sheets or plates prior to fabrication into tubes. The minimum slot width is generally limited to the wall thickness of the steel being punched. The horizontal machined slot has similar characteristics to the vertical machined slot, except that gravel envelope material will more easily stabilize around the slot opening.

Bridge Slot. Another well screen type is referred to as the "bridge slot" (Figure 11.10). This screen is manufactured on a press from flat sheets or plates. The slot opening is usually vertical and provides two orifices, longitudinally aligned to the axis. The perforated steel sheets, or plates, are then fabricated into tubes, and the seam is welded. Normally, 5-ft sections of bridge slot screen are welded together into longer lengths for field installation.

The bridge slot screen is usually installed in gravel envelope wells. Its chief advantage is a reasonably high area of opening at a relatively low cost. One important disadvantage is its

Figure 11.10. Bridge slot screen.

low collapsing strength due to the large number of vertically oriented slots. The manufacturing process is generally limited to 0.250-in. wall thickness. Since gravel control is more difficult with any vertical opening as compared with a horizontal opening, a smaller-aperture size relative to the pack material should be selected.

Pipe Base Screen. The pipe base screen is manufactured by fabricating a wire wrap screen to a perforated pipe (Figure 11.11). This type of screen is usually only installed in deep wells where high collapse strength is required. Two sets of openings are presented by the wrapped wire and the perforated pipe base. The total area of perforations is generally less than the area of the outer slots. The pipe base is usually carbon steel with stainless steel used for the wrapping wire.

Shutter Screen. One type of screen designed many years ago for use in gravel envelope wells has enjoyed considerable success. The shutter screen was originally manufactured by punch-forming downward-facing louver apertures into short lengths of pipe, then welding them together forming sections up to 20-ft long. A newer method has been developed that permits manufacture from tubes up to 50 ft in length. This

process incorporates the use of an internal mandrel that perforates the louver against external die blocks. The shutter screen is manufactured in a variety of patterns with various percentage area of openings (Figure 11.12).

One of the major physical advantages of the shutter screen is its high collapse strength which is greater than the collapse strength of blank casing of the same wall thickness (Chapter 9). This is due to the corrugating effect of the louver-shaped openings. Its resistance to collapse does not vary according to aperture size.

Another characteristic of the shutter screen is that tight-fitting swabs can be safely used to develop and redevelop wells. This is due to its high mechanical strength and full circular cylinder interior. For the same reason wells can be repaired and deepened more easily.

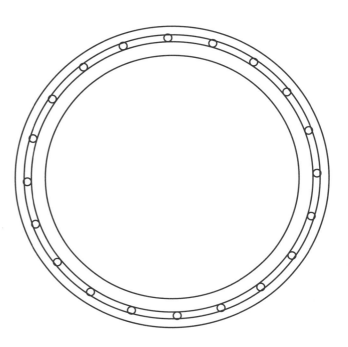

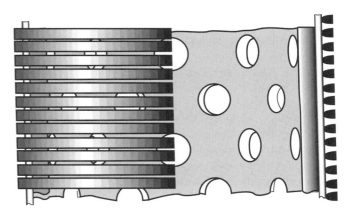

Figure 11.11. Pipe base screen

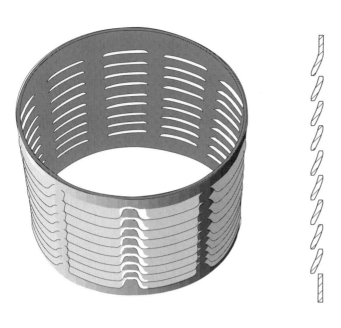

Figure 11.12. Shutter screen.

Use of down-the-hole cameras, model studies, and years of experience confirm that the louver-shaped aperture resists clogging as well as, or better than, any other type of screen. Its shape is the same for any wall thickness. Better gravel control during installation, development, and operation is characteristic of shutter screen because of the hood-shaped, downward-facing orifice. Tolerance of the filter pack range is enhanced. Many wells have been successfully completed with greater than 80% of the pack capable of passing through the apertures. This latitude provides protection against variations in the gravel envelope gradation due to segregation or other reasons.

Principal Applications to Naturally Developed Wells

Wire Wrap Screen. One type of well screen is manufactured by wrapping a wire around longitudinal rods. The wire is welded to the rods by resistance welding, which produces a cage-shaped cylindrical configuration. It is commonly known as the wire wrap or continuous slot screen and is usually available manufactured from Type 304 stainless steel, galvanized steel, and carbon steel (Figure 11.13).

The continuous slot design originated in the early part of this century to overcome the problems of ground water development from distinctive aquifers associated with the north central United States. These aquifers were generated from rock picked up, broken, and pulverized by advancing glaciers during the Ice Ages. Although glacial till is not well sorted, occasionally thin layers of fine-grained, uniform sands were washed from the original deposits. Such materials can be high-yielding aquifers. Prior to the advent of rotary well construction and the gravel envelope well design, it

was difficult to produce the full capacity of sand-free water from these aquifers. A well design incorporating wire wrap screen was successfully developed to meet these conditions.

The characteristics of the wire wrap screen are well suited for its original purpose. This design offers the highest percentage of open area of any screen. Consequently, with very small aperture sizes (0.010 in. to 0.035 in.) necessary to control fine sands from thin aquifers without a gravel envelope, sufficient area of opening is still available to minimize frictional head losses through the screen. However, under such circumstances stainless steel must be used since enlargement of openings results in sand pumping. The manufacturing process lends itself to close tolerances required for very fine aperture sizes, and the V-shaped slot configuration reduces clogging.

Careful consideration must be made of the use of wire wrap screen in situations or under conditions for which it was not originally designed. It is generally more expensive than other types and usually has lower collapsing strength than the casing it is installed with.

Furthermore the high area of opening, which may be an asset under certain conditions, acts adversely in the carbon steel version. The surface area exposed to corrosion is on the order of three times that of other types of screen. This results in faster loss of weight and strength. The screen geometry is such that the mass at any point varies considerably from relatively large at the exterior of the wrapping wire to small at the point of contact between the wire and longitudinal rods. Application of the pit depth–time relationship (Chapter 10) to this design suggests earlier structural weakening compared with other screen types.

Wire wrap screens are more difficult or impossible to repair or restore to their original shape and structural integrity if damaged. Extra caution must be exercised in using down-the-hole tools.

Well Casing Perforated in Place. With cable tool wells drilled in unconsolidated alluvium, casing is normally driven or jacked to the total depth and then perforated in place. This type of well, commonly drilled in the southwestern United States, is most successful where the aquifers are clearly defined and consist of materials that can be naturally developed. Extreme care and attention must be given to logging the aquifer materials. The perforating program must be designed to permit maximum flow while preventing the entrance of formation particles.

The mills knife and hydraulic louver-type perforators are the most frequently used down-hole perforating tools. The mills knife is a mechanically operated perforator that produces a vertical opening about 1/2 in. wide and 5 in. long. The hydraulic louver-type perforator produces a horizontal louver-shaped aperture facing downward. The aperture varies in length from 2 in. to 2 1/2 in. and in width from 1/8 in. to 5/16 in.

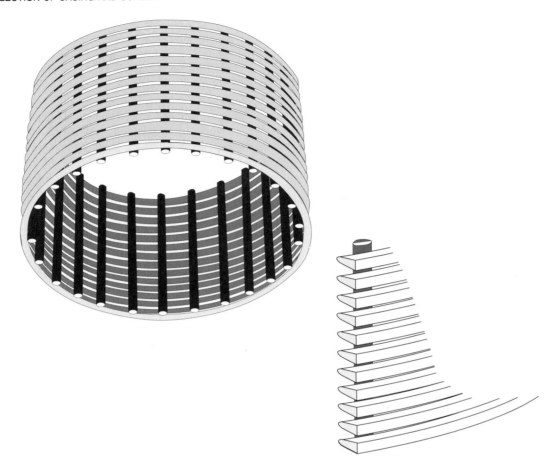

Figure 11.13. Wire wrap screen.

Generally the hydraulic louver perforator is the superior tool because:

1. Superior physical characteristics are imparted to the casing.
2. The hydraulically operated perforator controls the aperture size more precisely.
3. The louver-shaped opening achieves much greater stabilization of particles in relatively fine formations.

There are limitations, however, in the use of this perforator for screening fine materials. The smallest possible aperture is 1/8 in. to 5/32 in., and another well design must be used where the aquifer is composed of fine sand and silt without any well-graded coarser particles.

11.11 SELECTION OF MATERIAL

Screen materials and required corrosion resistance are covered in Chapter 10. The material and specifications available and used are generally the same as for casing. Normally, machine-slotted screens are produced from carbon steel transmission line pipe, bridge-slotted screens are manufactured from carbon steel and stainless steel; the shutter screen from carbon steel, copper-bearing steel, high-strength low-alloy steel, and stainless steel; the pipe base screen from stainless steel wire on carbon steel pipe; and the wire wrap screen from carbon steel, galvanized steel, and stainless steel.

The wall thickness of machine-slotted and shutter screen is usually the same as the casing when the material is carbon steel, copper-bearing steel, or high-strength low-alloy steel. The thickness of stainless steel may be reduced due to its durability.

11.12 SCREEN GEOMETRY

Effect on Development

Screens designed with a smooth interior can be developed with a larger variety of techniques than the wire wrap screen. This is particularly important with gravel envelope wells where swabbing is normally required during the preliminary development. The internal surface of the screen is not uniform due to the vertical rods. Swabbing efficiency is limited because water bypasses through the annulus between the rods instead of through the pack and formation. Development with air or high-velocity jetting is not as effective as other development methods, particularly with gravel envelope wells (see Appendix J).

Effect on Clogging and Stabilization of Particles

Machine-slotted openings have a much greater tendency to clog than louver- or V-shaped openings. For this reason the percentage of material passing should be limited to 10% or less, but this increases the tendency to clog in situations where incrustation is a problem. Pipe base screens have a greater tendency to clog over time because smaller particles tend to wedge between the wire and pipe, leaving only a small open area opposite each perforation in the pipe. Therefore the slot size selected should also allow only about 10% of the formation or pack material to pass.

Particles will tend to stabilize more readily on any horizontally oriented aperture. The downward-facing louver-shaped aperture of shutter screen is effective in this regard.

11.13 APERTURE SIZE

Gravel Envelope Wells

Aperture size for screen installed in gravel envelope wells is discussed in Chapter 13.

Naturally Developed Wells

Driscoll states that in a naturally developed well the screen slot size (and down-hole perforation size) is selected so that most of the finer formation materials near the borehole are brought into the screen and pumped from the well during development (5). This practice results in creating a zone of graded formation materials that extend 1 to 2 ft from the screen. The increased porosity and hydraulic conductivity of the graded materials reduces the drawdown near the well during pumping.

The following criteria have been developed for screen slot selection for naturally developed wells:

1. Where the uniformity coefficient of the formation is less than 6 and greater than 3, the aperture should retain 40% to 50% of the aquifer sample.

2. A more conservative slot selection may be advisable when there is some doubt about the reliability of the sample, the aquifer is thin and overlain by fine-grained loose material, and development time is at a premium.

3. If the water in the formation is corrosive, the aperture size should retain up to 60% of the material with a carbon steel screen. Generally, no allowance is required for stainless steel.

4. With coarse well-graded aquifer materials, larger openings (to 30% retained) can be selected. Well development time will correspondingly increase. Specific capacity can be increased, and the effects of incrustation are mitigated.

Screen Connections

The screen connections and the tools used for installation are the same as for casing.

11.14 SCREEN ACCESSORIES

Casing Guides

Casing guides are used to center the screen within the borehole. They should be of sufficient strength and surface area to provide support, yet not impede the installation of gravel. These are conflicting requirements, and some compromise is required. A simple guide, which has proved effective, is manufactured from $\frac{5}{16} \times 2$-in. steel, 30 in. long, bent to provide the proper centering distance. This distance or bend is usually the theoretical borehole radius minus the screen radius minus 1 in. The guides are attached to the screen by welding. Three or four guides are placed equidistantly around the screen at 40-ft intervals. Normally, they are not installed on the pump housing casing.

Bottom Plugs

A bull nose or plug is always attached to the bottom of casing and screen strings installed in gravel envelope wells. These may be fabricated by orange peeling a short joint of casing to a 1- to 3-ft taper, depending on diameter. Semi-elliptical tank ends provide a convenient fulfillment of this requirement.

REFERENCES

1. Williams, D. E., 1981. "The Well/Aquifer Model—Initial Test Results," Roscoe Moss Co., Los Angeles.

2. Williams, D. E. 1985. "Modern Techniques in Well Design. *J. AWWA*, 77, 9.

3. Ahmad, M. U., E. B. Williams, and L. Hamdan. 1983. "Commentaries on Experiments to Assess the Hydraulic Efficiency of Well Screens." *Ground Water* (May–June)

4. Simon, Z. 1987. "Solutions for Lateral Inflow in Perforated Conduit." *J. Hydraulic Eng.* 113, 9, pp. 1117–1132.

5. Driscoll, F. G. 1986. *Groundwater and Wells.* Johnson Division.

CHAPTER 12

Water Well Cementing

12.1 INTRODUCTION

In the past, cementing operations were generally confined to sealing the conductor casing installed in gravel envelope wells. However, regulatory agencies are increasingly concerned with ground water quality. Expanding problems of contamination require deeper multiple-zone completions, with cementing becoming more routine and the procedures used more complex. Oil-field equipment and techniques are being employed to meet these requirements.

12.2 PURPOSES OF WATER WELL CEMENTING

The primary reasons for cementing water wells are:

1. Protecting an aquifer, or aquifers, from contamination.
2. Protecting the well from surface contamination.
3. Preventing water movement from one aquifer to another.
4. Protecting casing from contact with formation water.
5. Reinforcing casing against external pressures and buckling.
6. Sealing selected zones.
7. Abandoning open boreholes and wells.

12.3 CEMENTING MATERIALS

Most cements used in industry today are referred to by the common name of "portland" (from Portland, England, where cement was first manufactured). Cement used in wells is manufactured from clay and limestone materials, composed chiefly of compounds of lime, silica, alumina, and iron oxide. The components, finely ground, are calcined to incipient fusion (in the general range of 2600°F to 3000°F). The calcined product, or clinker, is ground to a fine powder, along with a small amount of gypsum. Typical proportions of the materials used in the manufacture of portland-type cement for wells are given in Figure 12.1.

When these clinkered products hydrate with water, they combine to form four major crystalline phases. Table 12.1 shows these chemical formulas and standard designations.

Chemistry of Cement

The chemical reaction of dry cement upon the addition of water is a complex process. The tricalcium aluminate hydrates immediately and is first amorphous (or colloidal) and later crystalline. Calcium sulfoaluminate also crystallizes from the gypsum. The tricalcium silicate begins to react in about 24 hours, with the lime crystallizing and the less basic part forming a gel. As the other complex reactions proceed, the processes of hydration and hydrolysis are progressive and, in part, relatively slow.

The chemical composition of commercial cements varies to some degree, depending upon the raw materials used in the manufacturing and design of the cement. The fineness of the grind (percentage through 200 mesh) and the resultant specific surface obtained (square centimeters per gram) are very important in determining the physical properties and volume (Table 12.2) of set cement because the reaction between dry cement and water starts and proceeds at the surfaces of the cement particles.

Classification of Cement

Portland cements are usually manufactured to meet certain chemical and physical standards required by their application. In the United States there are several agencies that establish these standards. They include the ACI (American Concrete Institute), the AASHO (American Associates of State Highway Officials), the ASTM (American Society for Testing Materials), the API (American Petroleum Institute), and various departments of the federal government. Of these groups, the ASTM (construction and building applications) and the API (oil-well applications) provide the specifications most commonly applied in water well construction.

ASTM standard C 150 provides for five classes of portland cements (classified as Types I, II, III, IV, and V). Cements manufactured for use in oil wells are subject to a wide range of temperature and pressure conditions, and some differ considerably from the ASTM classes that are used at atmospheric conditions. For these reasons the API, under *Standards 10*, provides specifications covering eight classes of oil-well cements. These are identified as A, B, C, D, E, F, G, and H. API Classes A, B, and C correspond to ASTM

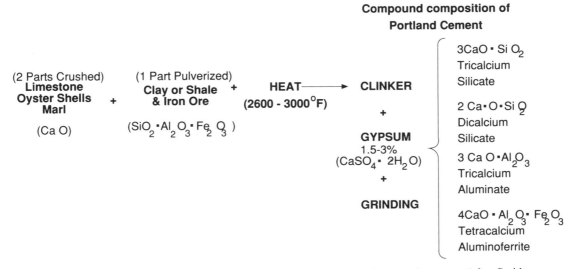

Figure 12.1. Proportions of materials used in the manufacture of portland cements (after Smith, 1987).

Types I, II, and III, whereas ASTM Types IV and V have no corresponding API class.

Comparative composition and properties of these two cement classifications are set forth in Table 12.3.

Cements Used for Water Wells Based on API Classifications

Only two API classes of cement are commonly used in water wells: the vast majority of all cementing uses Class A, and Class C is used occasionally when high early strength is needed.

Class A is intended for use to 6000 ft below ground surface when special properties are not required, and it is only available in one type (similar to ASTM C 150, Type I).

Class C is intended for use to 6000 ft below ground surface when conditions require high early strength. It is available in ordinary-, moderate- (similar to ASTM C 150, Type III), and high-sulfate-resistant types.

Properties of Cements Covered by API Standards

The physical and chemical requirements of all API classes of cement are defined in *API Standards 10*, and shown in Tables 12.4 and 12.5. Typical physical properties of the various API classes of cement are shown in Table 12.6.

Specialty Cements

A number of cementitious materials that are used very effectively for the cementing of wells do not fall under any API or ASTM system of classification. Although these materials may or may not be sold under any recognized specifications, their quality and uniformity are generally controlled by their supplier. These materials include:

1. Pozzolanic-portland cements.

2. Gypsum cements.

3. Expanding cements.

TABLE 12.1   Chemical Compounds Found in Set Portland Cement

			Influence on		
Chemical Compounds	Formula	Standard Designation	Setting Rate	Strength	Heat Liberated
Tricalcium silicate	$3CaO \cdot SiO_2$	C_3S	Fast	Strong	Moderate
Dicalcium silicate	$2CaO \cdot SiO_2$	C_2S	Slow	Strong	Cool
Tricalcium aluminate	$3CaO \cdot Al_2O_3$	C_3A	Fast	Weak	Torrid
Tetracalcium aluminoferrite	$4CaO \cdot Al_2O_3 \cdot Fe_2O_3$	C_4AF	Fast	Weak	Tropical

After Smith, 1976.

TABLE 12.2   Influence of Varying Surface Areas and Water Ratios on the Volume of Set Cement

Water Content (wt % of cement)	Volume of Slurry (ft³/Sack)	Free Water When Set (%)	Volume of Set Cement (ft³/Sack)
Specific Surface, 1890 cm²/g[a]			
40	1.069	0.00	1.069
50	1.220	0.74	1.211
60	1.370	2.34	1.338
70	1.521	4.75	1.449
Specific Surface, 1630 cm²/g[b]			
35	0.994	0.88	0.985
40	1.069	1.33	1.055
50	1.220	7.66	1.114
60	1.370	16.01	1.151
Specific Surface, 1206 cm²/g			
35	0.994	3.15	0.963
40	1.069	8.38	0.979
50	1.220	16.20	1.022
60	1.370	22.35	1.064

After Smith, 1976.

[a] Similar to API Class C cement.

[b] Similar to API Class A, B, and G cements.

Pozzolanic-Portland Cements.

Pozzolanic-Portland Cements. Pozzolans include any siliceous materials—either natural or artificial, processed or unprocessed—that develop cementitious qualities in the presence of lime and water. They can be divided into natural and artificial pozzolans. The natural pozzolans are, for the most part, materials of volcanic origin. The artificial pozzolans are mainly products obtained by the heat treatment of natural materials such as clays, shales, and certain siliceous rocks.

Fly ash, a combustion by-product from coal, is widely used in the oil industry and is the only pozzolan covered by both API and ASTM specifications. Fly ash has a specific gravity of 2.3 to 2.7, depending upon the source, compared to 3.1 to 3.2 for portland cements. This difference in specific gravity results in a pozzolan cement slurry of lighter weight than slurries of similar consistency made with portland cement. Table 12.7 compares typical chemical and physical properties of fly ash and cement.

When portland cement hydrates, calcium hydroxide is liberated. This chemical in itself contributes nothing to strength or watertightness and can be removed by leaching. When

TABLE 12.3   Typical Compositions and Properties of API and ASTM Classes of Portland Cements

ASTM Types	API Class	Compounds (%)				Wagner Fineness (cm²/g)
		C_3S	C_2S	C_3A	C_4AF	
I	A	53	24	8+	8	1500–1900
II	B	47	32	5−	12	1500–1900
III	C	58	16	8	8	2000–2800
—	G & H	50	30	5	12	1400–1700

Property	How Achieved
High early strength	By increasing the C_3S content and grinding finer
Better retardation	By controlling C_3S and C_3A content and grinding coarser
Low heat of hydration	By limiting the C_3S and C_3A content
Resistance to sulphate attack	By limiting the C_3A content

After Smith, 1987.

TABLE 12.4 Physical Requirements for API Cements

1	2	3	4	5	6	7	8	9	10	11	12
Well Cement Class				A	B	C	D	E	F	G	H
Water (% by weight of well cement)				46	46	56	38	38	38	44	38
Soundness (autoclave expansion), maximum (%)				0.80	0.80	0.80	0.80	0.80	0.80	0.80	0.80
Fineness[a] (specific surface), minimum, m²/kg				150	160	220	—	—	—	—	—
Free water content, maximum, mL				—	—	—	—	—	—	3.5[b]	3.5[b]

Compressive Strength Test, Eight Hour Curing Time

Schedule Number	Curing Temp, F	Curing Pressure psi	Minimum Compressive Strength, psi							
			A	B	C	D	E	F	G	H
—	100	Atmos.	250	200	300	—	—	—	300	300
—	140	Atmos.	—	—	—	—	—	—	1500	1500
6S	230	3000	—	—	—	500	—	—	—	—
8S	290	3000	—	—	—	—	500	—	—	—
9S	320	3000	—	—	—	—	—	500	—	—

Compressive Strength Test, Twelve Hour Curing Time

Schedule Number	Curing Temp, F	Curing Pressure psi	A	B	C	D	E	F	G	H
8S	290	3000	—	—	—	—	—	—	—	—

Compressive Strength Test, Twenty-four Hour Curing Time

Schedule Number	Curing Temp, F	Curing Pressure psi	Minimum Compressive Strength, psi							
			A	B	C	D	E	F	G	H
—	100	Atmos.	1800	1500	2000	—	—	—	—	—
4S	170	3000	—	—	—	1000	1000	—	—	—
6S	230	3000	—	—	—	2000	—	1000	—	—
8S	290	3000	—	—	—	—	2000	—	—	—
9S	320	3000	—	—	—	—	—	—	—	—
10S	350	3000	—	—	—	—	—	1000	—	—

Pressure Temperature Thickening Time Test

Specification Test Schedule Number	Maximum Consistency 15–30 Min Stirring Period, Bc[c]	Minimum Thickening Time (min)[d]							
		A	B	C	D	E	F	G	H
1	30	90	90	90	—	—	—	—	—
4	30	90	90	90	90	—	—	—	—
5	30	—	—	—	—	—	—	—	—
5	30	—	—	—	—	—	—	90 120 max.[e]	90 120 max.[e]
6	30	—	—	—	100	100	100	—	—
8	30	—	—	—	—	154	—	—	—
9	30	—	—	—	—	—	190	—	—

After Smith, 1987.

[a] Determined by Wagner turbidimeter apparatus described in ASTM C 115: *Fineness of Portland Cement by the Turbidimeter.*

[b] Based on 250 mL volume, percentage equivalent of 3.5 mL is 1.4%.

[c] Bearden units of slurry consistency (Bc).

[d] Thickening time requirements are based on 75 percentile values of the total cementing times observed in the casing survey, plus a 25 per cent safety factor.

[e] Maximum thickening time requirement for Schedule 5 is 120 minutes.

fly ash is present in the cement, it combines with the calcium hydroxide, contributing both to strength and impermeability.

Gypsum Cement.

Gypsum cements are normally available as a hemihydrate form of gypsum ($CaSO_4 \cdot \frac{1}{2} H_2O$). The unique properties of gypsum cement are its rapid setability, high early strength, and positive expansion (approximately 0.3%). Gypsum cements are blended in 8% to 10% concentration with API Class A cement to produce thixotropic properties. This combination is particularly useful in wells to minimize fallback after placement.

Because of the solubility of gypsum, it is usually considered a temporary plugging material unless placed downhole where there is no moving water. To avoid dissolution, gypsum cements are sometimes mixed with equal volumes of portland cements to form a permanent plugging material when fighting loss of circulation. Caution should be exercised in using these blends because they have very rapid setting properties and may set prematurely during placement.

TABLE 12.5 Chemical Requirements for API Cements

1	2	3	4	5	6	7
			Cement Class			
	A	B	C	D, E, F	G	H
Ordinary Type (O)						
Magnesium oxide (MgO), maximum, %	6.0	—	6.0	—	—	—
Sulfur trioxide (SO$_3$), maximum, %[a]	3.5	—	4.5	—	—	—
Loss on ignition, maximum, %	3.0	—	3.0	—	—	—
Insoluble residue, maximum, %	0.75	—	0.75	—	—	—
Tricalcium aluminate (3CaO · Al$_2$O$_3$), maximum, %[b]	—	—	15	—	—	—
Moderate Sulfate-Resistant Type (MSR)						
Magnesium oxide (MgO), maximum, %	—	6.0	6.0	6.0	6.0	6.0
Sulfur trioxide (SO$_3$), maximum, %	—	3.0	3.5	3.0	3.0	3.0
Loss on ignition, maximum, %	—	3.0	3.0	3.0	3.0	3.0
Insoluble residue, maximum, %	—	0.75	0.75	0.75	0.75	0.75
Tricalcium silicate (3CaO · SiO$_2$), { maximum, %[b]	—	—	—	—	58	58
{ minimum, %[b]	—	—	—	—	48	48
Tricalcium aluminate (3CaO · Al$_2$O$_3$), maximum, %[b]	—	8	8	8	8	8
Total alkali content expressed as sodium oxide (Na$_2$O) equivalent, maximum, %[c]	—	—	—	—	0.75	0.75
High Sulfate-Resistant Type (HSR)						
Magnesium oxide (MgO), maximum, %	—	6.0	6.0	6.0	6.0	6.0
Sulfur trioxide (SO$_3$), maximum, %	—	3.0	3.5	3.0	3.0	3.0
Loss on ignition, maximum, %	—	3.0	3.0	3.0	3.0	3.0
Insoluble residue, maximum, %	—	0.75	0.75	0.75	0.75	0.75
Tricalcium silicate (3CaO · SiO$_2$), { maximum, %[b]	—	—	—	—	65	65
{ minimum, %[b]	—	—	—	—	48	48
Tricalcium aluminate (3CaO · Al$_2$O$_3$), maximum, %[b]	—	3	3	3	3	3
Tetracalcium aluminoferrite (4CaO · Al$_2$O$_3$ · Fe$_2$O$_3$) plus twice the tricalcium aluminate (3CaO · Al$_2$O$_3$), maximum, %[b]	—	24	24	24	24	24
Total alkali content expressed as sodium oxide (Na$_2$O) equivalent, maximum, %[c]	—	—	—	—	0.75	0.75

After Smith, 1987.

Note: Methods covering the chemical analyses of hydraulic cements are described in ASTM C 114: *Standard Methods for Chemical Analysis of Hydraulic Cement.*

[a] When the tricalcium aluminate content (expressed as C$_3$A) of the Class A cement is 8% or less, the maximum SO$_3$ content shall be 3%.

[b] The expressing of chemical limitations by means of calculated assumed compounds does not necessarily mean that the oxides are actually or entirely present as such compounds. When the ratio of the percentages of Al$_2$O$_3$ to Fe$_2$O$_3$ is 0.64 or less, the C$_3$A content is zero. When the Al$_2$O$_3$ to Fe$_2$O$_3$ ratio is greater than 0.64, the compounds shall be calculated as follows:

$$C_3A = (2.65 \times \%Al_2O_3) - (1.69 \times \%Fe_2O_3) ,$$

$$C_4AF = 3.04 \times \%Fe_2O_3 ,$$

$$C_2S = (4.07 \times \%CaO) - (7.60 \times \%SiO_2) - (6.72 \times \%Al_2O_3) - (1.43 \times \%Fe_2O_3) - (2.85 \times \%SO_3) .$$

When the ratio of Al$_2$O$_3$ to Fe$_2$O$_3$ is less than 0.64, an iron-alumina-calcium solid solution [expressed as as (C$_4$AF + C$_2$F)] is formed and the compounds shall be calculated as follows:

$$(C_4AF + C_2F) = (2.10 \times \%Al_2O_3) + (1.70 \times \%Fe_2O_3)$$

and

$$C_3S = (4.07 \times \%CaO) - (7.60 \times \%SiO_2) - (4.48 \times \%Al_2O_3) - (2.86 \times \%Fe_2O_3) - (2.85 \times \%SO_3) .$$

[c] The sodium oxide equivalent (expressed as Na$_2$O equivalent) shall be calculated by the formula:

$$Na_2O \text{ equivalent} = (0.658 \times \%K_2O) + \%Na_2O$$

TABLE 12.6 Physical Properties of Various API Classes of Cement

Properties of API Classes of Cement	Class A	Class C	Classes G and H	Classes D and E
Specific gravity (average)	3.14	3.14	3.15	3.16
Surface area (range), cm²/g	1,500 to 1,900	2,000 to 2,800	1,400 to 1,700	1,200 to 1,600
Weight per sack, lbm	94	94	94	94
Bulk volume, ft³/sack	1	1	1	1
Absolute volume, gal/sack	3.6	3.6	3.58	3.57

Properties of Neat Slurries	Portland	High Early Strength	API Class G	API Class H	Retarded
Water, gal/sack (API)	5.19	6.32	4.97	4.29	4.29
Slurry weight, lbm/gal	15.6	14.8	15.8	16.5	16.5
Slurry volume, ft³/sack	1.18	1.33	1.14	1.05	1.05

Temperature (°F)	Pressure (psi)	Typical Compressive Strength (psi) at 24 Hours				
60	0	615	780	440	325	a
80	0	1,470	1,870	1,185	1,065	a
95	800	2,085	2,015	2,540	2,110	a
110	1,600	2,925	2,705	2,915	2,525	a
140	3,000	5,050	3,560	4,200	3,160	3,045
170	3,000	5,920	3,710	4,830	4,485	4,150
200	3,000			5,110	4,575	4,775

Temperature (°F)	Pressure (psi)	Typical Compressive Strength (psi) at 72 Hours				
60	0	2,870	2,535	—	—	a
80	0	4,130	3,935	—	—	a
95	800	4,670	4,105	—	—	a
110	1,600	5,840	4,780	—	—	a
140	3,000	6,550	4,960	—	7,125	4,000
170	3,000	6,210	4,460	5,685	7,310	5,425
200	3,000	a	a	7,360	9,900	5,920

Depth (ft)	Temperature (°F) Static	Temperature (°F) Circulating	High-Pressure Thickening Time (hours:min)				
2,000	110	91	4:00+	4:00+	3:00+	3:57	a
4,000	140	103	3:26	3:10	2:30	3:20	4:00+
6,000	170	113	2:25	2:06	2:10	1:57	4:00+
8,000	200	125	1:40[a]	1:37[a]	1:44	1:40	4:00+

After Smith, 1987.

[a] Not generally recommended at this temperature.

Expanding Cements. Under certain downhole conditions it is desirable to have a cement that will expand in the annulus. For such applications, the oil industry has evaluated various compositions that expand slightly when set. The reactions that cause this expansion are similar to the process described in the cementing literature as Ettringite. Ettringite is a crystal growth between sulfates and the tricalcium aluminate component in portland cement. These commercial expanding cements are portland types to which have been added a sulfoaluminate hydrate of calcium ($C_6AS_3 \cdot H_{32}$).

Commercial expanding cements are made in three types:

Type K[1] Contains the calcium sulfoaluminate component that is blended with a portland cement by licensed manufacturers. When slurried with water, the reaction created by hydration expansion is approximately 0.05% to 0.20%.

Type S Suggested by the Portland Cement Association. Consists of a high C_3A cement similar to API

1. Covered by U. S. Patent Klein 3,155,526.

TABLE 12.7 Typical Physical and Chemical Properties of Fly Ash as Compared with API Class A Cement

	API Fly Ash	API Class A Cement
Physical Properties		
Specific gravity	2.46	3.14
Weight equivalent in absolute volume to 1 sack (94 pounds) cement	74 lb	94 lb
Amount retained on 200-mesh sieve	5.27%	
Amount retained on 325-mesh sieve	11.74%	
Chemical Analysis		
Silicon dioxide	43.20	22.43
Iron and aluminum oxides	42.93	8.86
Calcium oxide	5.92	64.77
Magnesium oxide	1.03	1.14
Sulphur trioxide	1.70	1.67
Carbon dioxide	0.03	—
Loss on ignition	2.98	0.54
Undetermined	2.21	—

After Smith, 1987.

Class A, with approximately 10% to 15% gypsum. Expansion may vary within the ranges of Type K.

Type M   Obtained by adding small quantities of refractory cement to portland cement to produce expansive forces.

Expanding cement can also be formulated as follows:

1. API Class A with 5% to 10% of the hemihydrate forms of gypsum, which is the same mechanism as the Type S commercial expanding cement.

2. Pozzolan cements also create expansive forces in the hydration process because of a sulfoaluminate type of mechanism formed from the reaction of the alkali with Class A cement. Reactions are somewhat slower, and these cements require periods of curing, from 3 to 20 days, under moist conditions, before they reach their maximum expansion.

There is no current uniform testing procedure or specification within the API or ASTM standards for measuring the expansion forces in cement or concrete. Most laboratories use the expansive bar test with a molded $1 \times 1 \times 10$-in. cement specimen. Shortly after setting, the expansive force is measured for a base reference and again at various time intervals until the maximum expansion is reached. Hydraulic

bonding tests have also been used to evaluate the crystal growth of expanding cements.

12.4 CEMENT ADMIXTURES

Cement slurries can be tailored for specific well requirements with the use of additives. Most cement additives are free-flowing powders that can be dry blended with the cement before it is transported to the well. Most additives are also water soluble and can be dispersed in the mixing water at the job site when necessary.

Through the proper selection of additives, cement slurries can be designed having the following range of properties:

Density	9.5 to 25 lb/gal
Compressive strengths	200 to 20,000 psi
Setting times	Can be accelerated or retarded to produce a cement that will set within a few seconds or remain fluid for up to 36 hours
Cement filtration	Can be lowered to as little as 25 cc/30 min when measured through a 325-mesh screen at a differential pressure of 1000 psi.
Rheological properties	Can be varied over a wide range.
Cement slurry loss	Can be controlled by the use of granular, fibrous, or flakelike bridging agents or gelling agents.
Permeability	Can be controlled by densification and dispersion.
Expansion	Can be created in the set cement by the use of gypsum and/or sodium chloride in low percentages.
Heat of hydration	Liberated during the setting process from a chemical reaction between the cement and water. This heat can be controlled by the use of a filler-type material, such as sand, fly ash, or bentonite, in combination with water. These products dilute the cement content of the mixture and produce less reaction heat during the setting process.

Effects of cement additives are in Table 12.8. A discussion of some additives of the most interest in water well applications follows this table.

TABLE 12.8   Effects of Additives on the Physical Properties of Cement

	Bentonite	Pozzolan (Fly Ash)	Sand	Accelerator Calcium Chloride	Sodium Chloride	Retarder	Friction Reducer (Dispersant)	Low Water Loss Materials	Lost-Circulation Materials
Density									
Decreases	⊗	⊗							
Increases			⊗		×				
Water required									
Less									
More	⊗	×	×				×		×
Viscosity									
Decreases				×	×	⊗	⊗		
Increases	×	×	×						×
Thickening time									
Accelerated				⊗	⊗				
Retarded	×				×	⊗	×	×	
Early strength									
Decreases	×	×				⊗		×	×
Increases				⊗	⊗		×		
Final strength									
Decreases	⊗	×							
Increases						×	×		
Durability									
Decreases	×								×
Increases		⊗							
Water loss									
Decreases	⊗								
Increases						×	×	⊗	×

After Smith, 1976.

Note: × Denotes minor effects. ⊗ Denotes major effects and/or principal purpose for which used.

Cement Accelerators

Cement slurries to be used opposite formations may require acceleration to shorten thickening times and increase early strength, particularly at temperatures below 100°F. Compressive strength of 500 psi can be developed in as little as four hours with the use of accelerators, basic cements, and good mechanical practices. This strength is generally accepted as the minimum value for bonding and supporting casing.

Calcium Chloride. Calcium chloride, a very hygroscopic material when used as a grout accelerator, is available in flake and powder forms in the regular 77% calcium chloride grade, and in flake form in the anhydrous 96% grade. The anhydrous flake form is in more general use because it can absorb some moisture without becoming lumpy and is easier to maintain in storage. The amounts required for maximum acceleration normally range between 2% and 4%, based on the weight of cement. In some instances, 4% calcium chloride is used with cement mixtures requiring high water ratios, where large volumes of water dilute the concentration of the accelerator. The effects of calcium chloride are outlined in Table 12.9.

Sodium Chloride (Salt). Sodium chloride is an effective accelerator for neat cement at concentrations of 1.5% to 5% by weight of cement (Table 12.10). Concentrations of 2 to 2.5% give maximum acceleration, except when using higher water ratio slurries. High concentrations (14 to 16 lb salt per sack of cement) will, however, retard setting. Salt used in slurries may improve the bond to shaley and clayey formations by protecting these sections from sloughing and heaving during cementing.

Densified Cements

Cement slurries, prepared from any Class A cement with 0.75% to 1.0% dispersant mixed at 17.5 lb/gal (water ratio 3.4 gal per sack) can be accelerated by densifying the slurry. The addition of 15 to 20 lb of sand per sack of cement, mixed at 18 lb/gal with the same water ratio, will generate higher early strengths. In general, this slurry can provide relatively good strength within 8 hours at a bottom-hole static temperature above 80°F when designed for a 1½- to 2-hour pumping time.

Lightweight Additives

Cement slurries with the recommended amount of water will produce slurry weight in excess of 15 lb/gal. For formations that will not support a cement column of this density or where possible casing collapse is a consideration, additives are frequently used to reduce the weight of the slurry. Such additives, summarized in Table 12.11, also reduce cost, increase yield, and sometimes lower filter loss. There are three methods for reducing the weight of cement slurries:

1. Adding water.
2. Adding low-specific-gravity solids.
3. Combinations of 1 and 2.

Bentonite. The most commonly used weight-reducing material is bentonite because of its ability to tie up large volumes of water. Bentonite can be added to most classes or types of cement in concentrations from 2% to 16% by weight of the cement.

When dry-mixed with cement in amounts as high as 8% to 12%, bentonite has a water requirement of approximately 1.3 gal for each 2% bentonite added. When using 8% to 12% bentonite (gel) cement, dispersants are often used to reduce viscosity and to provide flexibility in the water ratio. The effects of bentonite on the composition and properties

TABLE 12.9   Effects of Calcium Chloride on the Thickening Time and Compressive Strength of API Class A Cement

	Thickening Time (hr:min)			
	API Cementing Tests for Simulated Well Depths (ft)			
	Casing Cementing		Squeeze Cementing	
Calcium Chloride (%)	1000	2000	1000	2000
0.0	4:40	4:12	3:30	3:29
2.0	1:55	1:43	1:30	1:20
4.0	0:50	0:52	0:48	0:53

		Compressive Strength (psi)				
		At Atmospheric Pressure and Temperature of			At API Curing Pressure and Temperature of	
					800 psi	1600 psi
Curing Time (hr)	Calcium Chloride (%)	40°F	60°F	80°F	95°F	110°F
6	0	NS	20	75	235	860
12	0	NS	70	405	1065	1525
24	0	30	940	1930	2710	3680
48	0	505	2110	3920	4820	5280
6	2	NS	460	850	1170	1700
12	2	65	785	1540	2360	2850
24	2	415	2290	3980	4450	5025
6	4	NS	755	1095	1225	1720
12	4	15	955	1675	2325	2600
24	4	400	2420	3980	4550	4540

After Smith, 1976.
Note: NS = not set. Water ratio = 5.2 gal/sack. Slurry weight = 15.6 lb/gal.

TABLE 12.10   Effects of Sodium Chloride on Thickening Time and Compressive Strength of API Class A Cement

Thickening Time (hr:min)

| | API Casing Cementing Tests for Simulated Well Depth (ft) | | |
Sodium Chloride (%)	1000	2000	4000
0.0	4:40	4:12	2:30
2.0	3:05	2:27	1:52
4.0	3:05	2:35	1:35

Compressive Strength (psi)

| Curing Time (hr) | Sodium Chloride (%) | At Atmospheric Pressure and Temperature of | | API Curing Pressure and Temperature of | |
		60°F	80°F	800 psi 95°F	1600 psi 110°F
12	0	70	405	1065	1525
24	0	940	1930	2710	3680
48	0	2110	3920	4820	5280
12	2	290	960	1590	2600
24	2	1230	2260	3200	3420
48	2	3540	3250	3900	4350
12	4	280	1145	1530	2575
24	4	1390	2330	3150	3400
48	4	3325	3500	3825	4125

After Smith, 1976.
Note: Water ratio = 5.2 gal/sack. Slurry weight = 15.6 lb/gal.

of Class A cement slurries are shown by the data in Table 12.12. Figure 12.2 shows the slurry weight, volume, and water requirement with cement-bentonite slurries.

In areas where bulk equipment is not available for the purpose of dry-blending bentonite in cement, it may be necessary to add it to the water (prehydrate) instead of the dry cement. One percent of prehydrated bentonite has the effect of 3.6% of the dry mixture. Prehydration is complete in about 24 hours when the wet mixture is allowed to set

TABLE 12.11   Summary of Lightweight Cement Additives

Type of Material	Usual Amount Used
Bentonite clay[a]	2 to 16%
Attapulgite clay	½ to 4%
Pozzolan, artificial (fly ash)	74 lb/sack of cement
Pozzolan-bentonite cement	5 to 20 lb/sack of cement

After Smith, 1976.
[a] High percentages of bentonite in cement cause a reduction of compressive strength and thickening time of both regular and retarded cements. Bentonite and water also lower the resistance of cement to chemical attack from formation water.

in a tank. But mixing it with a high-shearing-type mixer can shorten the setting time somewhat.

Pozzolans. Pozzolans, as discussed earlier, are also used as an additive to produce lightweight cement slurries. They are particularly useful in cementing long lengths of larger-diameter conductor or intermediate casing strings where casing collapse is a particular concern.

Cement Dispersants or Friction Reducers

Dispersants or friction reducers are commonly added to cement slurries to control filtration (Table 12.13) by dispersing and packing the cement particles. This is especially effective when the water–cement ratio is reduced.

Dispersing agents may also be added to cement slurries to improve their flow properties. Critical flow rates for turbulent flow are higher without a dispersant and lower with it. Dispersed slurries have a lower viscosity and can be pumped in turbulence at lower pressures, minimizing the horsepower required and lessening the chances of loss of circulation and premature dehydration. These agents function by lowering the yield point and gel strength of the slurry.

Polymers are the dispersants commonly added to cement slurries, whereas fluid-loss agents are used with bentonite cement slurries. The polymers are free-flowing powders and have no significant effect on the acceleration or retardation of settling in most slurries. Because of this property and the pronounced reduction of apparent slurry viscosity, they are well suited for use in wells. Although apparent viscosities[2] are greatly reduced, these polymers do not cause excessive free-water separation or settling of cement particles from the slurry.

Sodium chloride can act as a dispersant and can effectively reduce the apparent viscosity of cement-bentonite and cement-pozzolan slurries.

As the useful properties of dispersants are discovered, they are increasingly utilized in water well applications. Commonly used dispersants are listed in Table 12.14.

Gypsum Additives

About 4% to 10% gypsum is added to portland cement to produce:

1. Fast-setting cements for loss of circulation.
2. Gelling (thixotropic) properties.
3. Expansion properties in the set cement.

Gypsum added in concentrations of 3% to 6% will react with the tricalcium aluminate in the set cement to provide

2. The linear relationship between shear stress and shear rate in a Bingham plastic cement slurry.

TABLE 12.12   Effects of Bentonite on the Composition and Properties of Class A Cement Slurries

Bentonite (%)	Water Requirement		Viscosity (0–20 min Bc)	Slurry Weight		Slurry Volume (ft³/sack)
	(gal/sack)	(ft³/sack)		(lb/gal)	(lb/ft³)	
0	5.2	0.70	4–12	15.6	117	1.18
2	6.5	0.87	10–20	14.7	110	1.36
4	7.8	1.04	11–21	14.1	105	1.55
6	9.1	1.22	13–24	13.5	101	1.73
8	10.4	1.39	12–19	13.1	98	1.92

Pressure-Temperature Thickening Time (hr:min)

Bentonite (%)	API Casing Tests (ft)		API Squeeze Tests (ft)[a]
	2000	4000	2000
0	4:00+	4:04	3:58
2	4:00+	3:15	3:37
4	4:00+	3:04	3:08
6	4:00+	2:52	3:19
8	4:00+	2:58	3:05

Compressive Strengths (psi)

Bentonite (%)	12-Hour Temperature				24-Hour Temperature				72-Hour Temperature			
	60°F	80°F	95°F[b]	110°F[b]	60°F	80°F	95°F[b]	110°F[b]	60°F	80°F	95°F[b]	110°F[b]
0	NS	295	560	1040	190	950	1505	1950	1335	2450	2805	3388
2	NS	175	380	630	135	665	1040	1300	825	1600	1980	2295
4	NS	115	230	385	90	430	735	830	450	1015	1370	1550
6	NS	85	160	235	50	285	405	545	340	620	890	1095
8	NS	60	105	125	40	185	255	355	270	395	575	710

[a] API Casing tests reflect the fluid time for a cement grout to be placed under normal casing conditions where a large amount of mud and/or water precedes the grout and lowers the formation temperature. Under API squeeze conditions, smaller volumes of cement are used, and the grout is pumped down tubing for squeezing against a formation. Under these conditions high temperatures and shorter thickening times result.
[b] Cured under API curing pressure conditions.

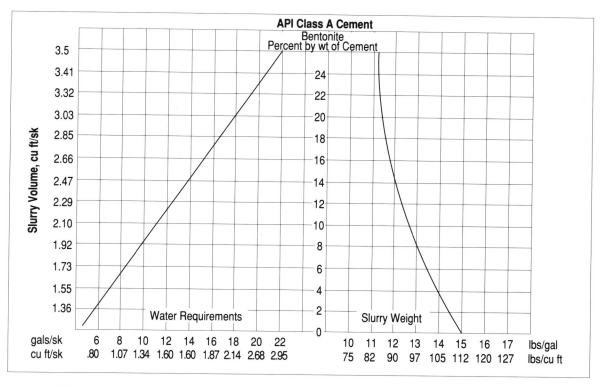

Figure 12.2. Slurry weight, volume, and water requirement with 0–24 percent cement-bentonite slurries (courtesy of Halliburton Services).

TABLE 12.13   Effect of Organic Polymers on the Fluid Loss of API Class A Cement

Polymer Percent by Weight of Cement	API Fluid Loss at 1000 psi (cc/30 min)	Permeability of Filter Cake at 1000 psi (md)	Time to Form 2-in-Cake (min)
0.00	1200	5.00	0.2
0.50	300	0.54	3.4
0.75	100	0.09	30.0
1.00	50	0.009	100.0

After Smith, 1987.

TABLE 12.14   Summary of Commonly Used Cement Dispersants[a]

Type of Material	Usual Cement Amount Used (lb/sack)
Polymer: Blend	0.3 to 0.5
Long chain	0.5 to 1.5
Sodium chloride	1 to 16
Calcium lignosulfonate, organic acid (retarder and dispersant)	0.5 to 1.5

After Smith, 1987.

[a] Turbulence inducers to lower viscosity, yield point, and gel strength.

for expansion. These expanding properties improve the cement bond between the pipe and the formation, thus effecting a better seal against annular migration of formation fluids.

12.5 CONSIDERATIONS DURING THE MIXING AND PLACEMENT OF CEMENT

Downhole Influences on Cement

Pressure, time, and temperature are the basic parameters that influence the performance of cement slurries. Each has its effect on the development of strength necessary for casing support. Temperature has a most pronounced influence upon the performance of a cement slurry. Cement at 40° to 60°F sets very slowly, whereas at 95° to 110°F the slurry sets fast and develops strength very rapidly.

The hydrostatic pressure of well fluids will reduce the pumpability of cement during placement and accelerate the formation of hydration products that aid the development of compressive strength. The cooling effect caused by fluid displacement lowers the circulating temperature considerably during the casing cementing operations.

Strength of Cements to Support Casing

Very little early strength is needed to support a string of casing. Data have shown that an annular sheath of cement 10 ft long possessing only 8 psi tensile strength would support over 200 ft of standard weight casing. This high degree of support was provided even under relatively poor bonding conditions with mud-wet casing.

During the drilling-out process, cemented casing is subjected to additional loads that must be supported by the cement sheath. Since during cement-strength testing the cement is usually in compression rather than in tension, the values must be converted from compressive to tensile strength. As a rule, compressive strength is approximately 8 to 10 times greater than tensile strength. It is generally accepted that a compressive strength of 500 psi is adequate for most operations and waiting on cement (WOC) time can be adjusted to this.

Since the last volume of cement pumped will be placed around the shoe, particular care should be taken to ensure that it has the desired physical properties. Cement slurries may be mixed with a lower water ratio toward the end of pumping for more strength around the shoe.

Viscosity and Water Content of Cement Slurries

To achieve a good bond to the formation and casing, the viscosity of a cement slurry should be designed to give the most efficient drilling-fluid displacement. To serve these functions, most slurries are mixed with that amount of water that will provide a minimum of free-water separation from

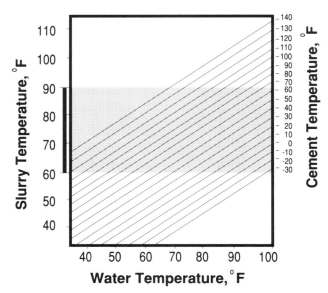

Figure 12.3. Slurry temperature for various temperatures of water and cement (after Smith, 1976).

the slurry, as shown in Table 12.2. Particle size, surface area, and additives influence the amount of mixing water, viscosity, and free water in cement slurry.

Mixing water temperature affects the viscosity of a slurry. Desirable slurry temperatures should range between 60° and 90°F, as shown in Figure 12.3.

Thickening Times of Cement Slurries

The minimum thickening time, or fluid time, required for placement of a cement slurry is that required to mix and displace the slurry down the hole and up the annulus behind the casing. Specific requirements depend upon the type of job (cementing casing, spotting a cement plug or squeeze cementing), the well conditions, and the volume of cement being pumped. Equipment for measuring the thickening time of any cement slurry under laboratory conditions is defined in API Standard RP10B.[3]

Slurry Density and Quality Control

Effective sealing is correlated with the quality of the cement-mixing operation. The measurement of slurry density is used to control the relative amounts of water and cement used, and it should be monitored and recorded to ensure maintenance of the correct water–solids ratio.

In field operations, slurry control is customarily maintained by measuring density with the standard mud balance. For accuracy, samples should be selectively taken and vibrated in the mud cup to remove the finely entrapped air from the

---

3. *Recommended Practice for Testing Oil Well Cements and Cement Additives*, API RP10B, 20th ed., API Division of Production, Dallas (1974).

jet mixer. To avoid the effect of aeration, samples for weighing should be obtained from a location on the discharge side of the displacement pump unless the slurry is being batch mixed. Acceptable density varies but ranges between 100 lb/ft^3 and 118 lb/ft^3.

Effects of Drilling Fluid and Drilling-Fluid Additives on Cements

The effects on cement of several drilling-fluid additives are shown in Table 12.15.

Grouting systems are influenced by such factors as:

- Mud contamination.
- Mud dilution.
- Mud chemicals.
- Filter cake.
- Mixing water which contains contaminants.

The most satisfactory way of combating the effects of drilling-fluid additives on cement grout during the cementing process is to use wiper plugs and spacers, or preflushes. Wiper plugs help to eliminate contamination between the drilling fluid and the cement slurry inside the casing, and a preflush, or water spacer, helps to eliminate contamination in the annular space between the outside of the casing and the formation.

Mixing-water contaminants include fertilizers, decomposed plant life, agricultural products, soil chemicals, and waste effluents.

Preflushes

The preflush functions as a spacer to minimize mixing and interfacial gelation in the annulus. For freshwater-base drilling-fluid systems, water in sufficient volume is an excellent wash because it can be easily pumped under turbulent flow conditions without affecting the setting of cement.

The contact time is the period of time that a fluid flows past a particular point in the annular space. Where turbulent flow prevails, studies indicate that a 10-min or longer preflush provides excellent mud removal. The equation for calculating the volume of fluid needed to provide a specific preflush contact time is as follows:

$$V_t = (T_c)(Q_t)(5.615) ,$$

where

V_t = volume [ft^3],
T_c = desired contact time of spacer [min],
Q_t = displacement rate for spacer [bbl/min]

The calculation of the fluid volume needed for a specific length of contact time is independent of the casing and borehole diameters. The equation holds as long as all of the fluid passes the point of interest.

Displacement

The effective displacement of drilling fluid by cement is a critical factor in successful sealing operations. Cementing failures are predominantly created by channels of drilling fluid bypassed by the cement in the annulus. These channels are highly dependent upon drilling-fluid viscosity and filter-cake deposits on the borehole wall. If channels are eliminated, a variety of cementing compositions will provide an effective seal.

TABLE 12.15   Effects of Drilling Fluid Additives on Cement

Additive	Purposes	Cement Effects
Barium sulfate (BaSO$_4$)	Weighting agent	Increases density, reduces strength
Caustic (NaOH, Na$_2$CO$_3$, etc.)	pH adjustment	Acceleration
Calcium compounds (CaO, Ca(OH)$_2$, CaCl$_2$, CaSO$_4$, 2H$_3$O)	Conditioning and pH control	Acceleration
Hydrocarbons (diesel oil, lease crude oil)	Control fluid loss, lubrication	Decreases density
Sealants (scrap, cellulose, rubber, etc.)	Seal against leakage to formation	Retardation
Thinners (tannins, lignosulfonates, quebracho, lignins, etc.)	Disperse mud solids	Retardation
Emulsifiers (lignosulfonates, alkyl ethylene oxide products, hydrocarbon sulfonates)	Forming oil-in-water or water-in-oil muds	Retardation
Bactericides (substituted phenols, formaldehyde, etc.)	Protect organic additives against bacterial decomposition	Retardation
Fluid-loss-control additives C.M.C., starch, guar, polyacrylamides, lignosulfonate	Reduce fluid loss from mud to formation	Retardation

After Smith, 1987.

Drilling-fluid displacement is primarily influenced by the following factors:

1. Casing centralization significantly aids displacement. Flow velocity in an eccentric annulus is not uniform, and the highest velocity region is on the side of the hole with the largest clearance. If the casing is close to the borehole wall, it may not be possible to pump the cement at a rate high enough to develop uniform flow throughout the annulus.
2. Casing movement, either rotation or reciprocation, helps enormously in drilling-fluid displacement. The limitations of water well drilling equipment generally preclude the use of this technique.
3. Control of drilling-fluid properties (low plastic viscosity and low yield point) greatly increases displacement efficiency.
4. High displacement rates that produce turbulent flow increase displacement efficiency.
5. A minimum contact time of 10 min with turbulent flow aids in displacement.
6. Buoyant force due to the density difference between cement and drilling fluid is a relatively minor factor in displacement.

12.6   PLACEMENT TECHNIQUES

Casing Centralizers

The uniformity of the cement sheath around the casing determines to a great extent the effectiveness of the seal between well bore and casing. Since most boreholes are not straight, the casing will generally be in contact with the wall at several places unless centralized. The requirement for number and spacing of centralizers will vary considerably, depending upon borehole alignment.

Cementing

Most cementing or grouting is performed by pumping the slurry down the casing and up the annulus; however, there are several techniques commonly used:

1. Cementing through casing—the Bradenhead method.
2. Outside or annulus cementing through tubing—the tremie method.
3. Innerstring cementing with float collar or shoe.
4. Cementing through casing—the normal displacement technique using top and bottom plugs.
5. Dump-bailer method.

Pressures during Cementing

Casing is subject to both internal (bursting) and external (collapsing) pressures during certain types of cementing operations, and these pressures should be considered before commencing any cementing operation. Chapter 9 discusses these stresses in detail.

Cementing Conductor Casing

Large-diameter conductor casing is usually cemented using the Bradenhead procedure or through tubing installed in the annulus.

Bradenhead Method.   The Bradenhead method is an effective cementing procedure. As shown in Figure 12.4, a cement head (Bradenhead) is welded to the casing. Cement is pumped under pressure through tubing down the inside of the fluid-filled casing and up the annulus to the surface. There are two major advantages to this method. The possibility

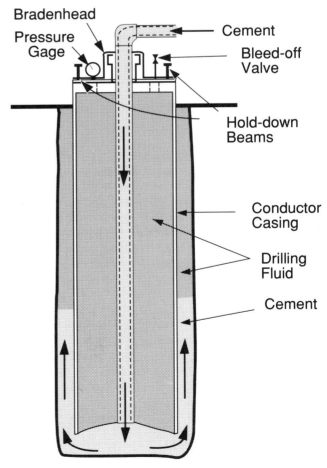

Figure 12.4. Bradenhead method.

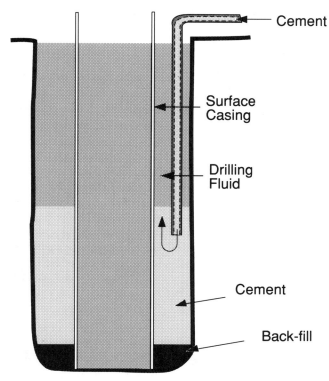

Figure 12.5. Cementing down the annulus.

of mud channels is minimized, and the fluid-filled conductor casing cannot be collapsed since the pressures inside and outside the casing are equalized. However, until the cement sets, there is upward lift on the casing, and it must be anchored at the surface. For example, with 24-in. casing, the lift can exceed 21 tons at a cementing pressure of 100 psi. Cementing pressures can easily and quickly become greater than 100 psi, particularly if positive-displacement cementing pumps are used.

Tremie, or Grout Pipe, Method. Pumping cement through tubing or a small-diameter pipe run in the annulus between casings or between the casing and the borehole is referred to as the "tremie," or "grout pipe," method and is widely used for placing grout near the surface on casing 16 in. in diameter and larger (see Figure 12.5). Generally, if the cement is poured into place, the process is referred to as "tremieing." If the cement is pumped, it is pumped through a grout pipe. This operation can be performed in stages when there are limitations imposed by the casing's collapse strength.

Cementing Pump Housing and Intermediate Casing

Large-diameter pump housing or intermediate casing (10-in. diameter or greater) is cemented through the casing by either

the Bradenhead method, the innerstring method, or with the normal displacement method, using top and bottom plugs.

Innerstring Method. Small-diameter tubing is commonly used for innerstring placement, illustrated in Figure 12.6. This reduces the cementing time, cementing-plug diameter, volume of cement to bump the plug, and the volume of cement that has to be drilled out of the casing. The technique uses a back-pressure valve and packer attached to the tubing. Where the casing is equipped with a back-pressure valve, the tubing can be disengaged and withdrawn from the casing as soon as the plug is seated. Consideration must be given to the casing's collapse strength since internal and external pressures are not balanced with this method.

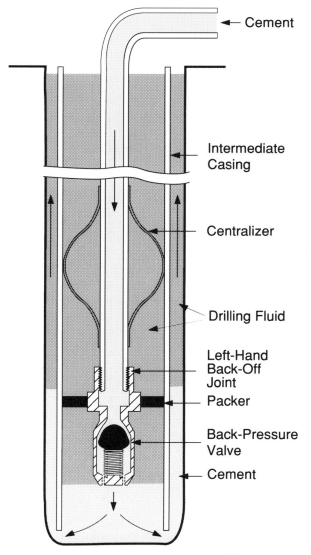

Figure 12.6. Innerstring cementing (courtesy of Halliburton Services).

Normal Displacement Method. In many respects the most effective cementing method for deeper casing strings is normal displacement, using top and bottom plugs (Figure 12.7). This placement technique usually achieves the grouting objectives of displacement and elimination of mud channels better than alternative procedures. If channels remain, even the highest-quality cement slurry will not provide the necessary seal between the casing and the formation.

This operation requires use of a special container for the plugs, which is welded to and becomes an integral part of the casing. Prior to placement of the slurry, the hole should be cleaned by circulating with drilling fluid, and a generous volume of water flush should precede the bottom plug. The bottom plug is released and pumped ahead of the slurry, fulfilling two functions:

1. It provides a barrier between the slurry and the water flush.
2. It cleans the casing wall of drilling fluid.

When this plug reaches the float collar, the differential pressure ruptures the diaphragm at the top of the plug and allows the cement slurry to proceed through the plug and floating equipment and up the annular space behind the casing.

To complete the operation, the top plug is released from the container or inserted at the top of the casing. This plug is designed so that when it is pumped to the bottom it sits on the bottom plug, causing a pressure buildup. The volume of the displacing fluid should be calculated. If the top plug does not "bump" (i.e., sit at the float collar causing a pressure increase) at the calculated displacement volume, pumping should be stopped to prevent overdisplacement of the slurry. The two-plug system requires knowledge of borehole conditions and formations by the cementer since, once the top plug is placed, no additional cement can be added.

Dump-Bailer Method. A dump bailer is sometimes used to place cement slurry, particularly with cable tool well drilling procedures. Since surface or conductor casing is often placed in unstable alluvium, it may be difficult to drill open hole and thus the casing is often driven into place. Cementing is done by underreaming an enlarged borehole under the casing shoe, using a dump bailer to fill this borehole with slurry and then driving the casing into the slurry (Figure 12.8).

Recommended Cements

Casing is usually cemented with accelerated API Class A (ASTM Type I) cement. The WOC (waiting on cement) time is typically 24 to 48 hours. Two-percent calcium chloride may be used as an accelerator.

Lost circulation additives, such as sand, gilsonite, and cellophane, may be added without a significant effect on compressive strength. Where the loss of circulation is severe, a thixotropic cement can be used to prevent fallback.

12.7 SQUEEZE CEMENTING

Squeeze cementing is the process of applying hydraulic pressure to force or squeeze a cement slurry into a formation void or a porous zone to:

1. Prevent fluid migration from other zones.
2. Seal off thief or lost-circulation zones.
3. Correct a defective cementing job.
4. Repair casing leaks.

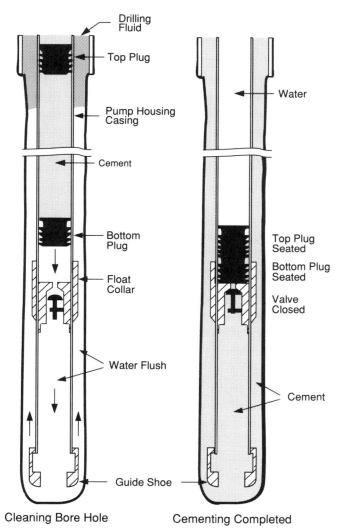

Figure 12.7. Cementing through casing—normal displacement technique using top and bottom plugs (courtesy of Halliburton Services).

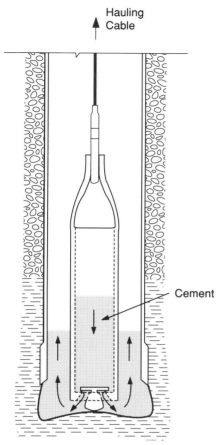

Hauling
Cable

Cement

Figure 12.8. Dump-Bailer method.

Achieving a successful squeeze requires the use of small quantities of grout properly placed at the interval needing repair. An understanding of grout behavior under pressure is very important for good results.

A cement grout is basically composed of cement particles and water. The particle size is too large to allow particles to enter the voids of most formations. However, they can be separated from the water under a low-pressure differential so that the cement filtrate can be pumped into the voids while the particles form a dehydrated filter cake of cement on the formation face. As the filter cake builds, the pump-in pressure increases until a squeeze pressure less than fracturing pressure is attained. The formation permeability must be high enough to accommodate a reasonable pump-in rate before this ideal squeeze condition is attained.

Bradenhead Squeezing

One squeeze technique uses a Bradenhead and is known as "Bradenhead squeezing". A predetermined amount of slurry is pumped through tubing to a specific height in the borehole outside the tubing. The tubing is then pulled back enough to clear the slurry, and the Bradenhead is packed off at the surface. The displacing fluid is pumped down the tubing until the desired squeeze pressure is reached or until a calculated amount of fluid has been pumped. The grout is forced to move into or against zones of weakness, since it can no longer circulate up the annulus. This method is used in plugging and in sealing off zones of partial loss of circulation encountered during drilling.

The Squeeze-Packer Method

Other applications of squeeze cementing include isolating certain aquifers and preventing unsuitable waters from entering the well by cementing the face of the aquifer behind a section of screen. The squeeze-packer method uses either a retrievable or a nonretrievable packer run on tubing to a position near the top of the zone to be squeezed, as shown in Figure 12.9. This technique is generally considered superior to the Bradenhead method since it confines pressures to a specific zone in the borehole.

API Class A cement is suitable for squeeze-cementing water wells. The slurry must be designed to remain fluid long enough to be placed properly and to achieve the desired squeeze pressure. Filtration properties must be controlled to avoid the formation of a thick filter cake that prevents slurry flow into the formation.

12.8 LINERS

Liners are sections of pipe that are run in specific well intervals when it is not desirable to bring the casing back to the surface. They may be used to repair a split or damaged section of casing, or to seal off a contaminated zone by covering the screen. They are installed either by setting them to the bottom of the hole or suspending them with a liner hanger. In the grouting process, cement slurry is mixed at the surface and displaced down tubing and back up the liner to the overlap. The procedure for installing and cementing liners in water wells is discussed in Chapter 17.

The common cements and additives used for grouting liners are:

1. API Class A cement.

2. Bentonite-cement, 2% to 4%, may be used for weight control and to reduce compressive strength and thickening time.

3. Dispersant additives may be used to lower viscosity and to improve flow properties.

4. Quick-set additives may be used to accelerate setting time.

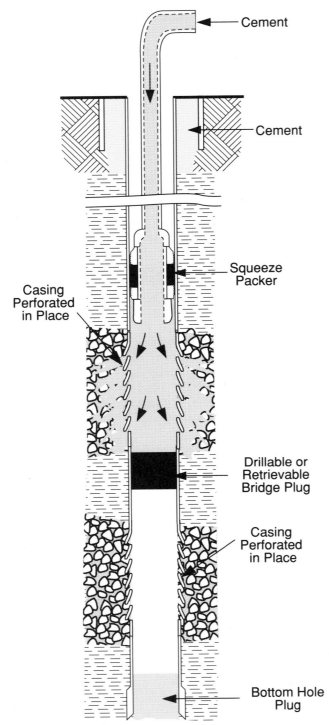

Cement

Cement

Squeeze Packer

Casing Perforated in Place

Drillable or Retrievable Bridge Plug

Casing Perforated in Place

Bottom Hole Plug

Figure 12.9. Squeeze-packer method.

12.9 WELL ABANDONMENT

When abandoning wells, it is essential that all aquifers containing usable-quality water be protected. Most states, or sometimes agencies within the state, have strict regulations regarding well abandonment. Permits and subsequent abandonment reports are often required.

In many instances cement plugs are placed by the squeeze method through tubing between packers or between a packer and a bridge plug. If the formation is to be sealed, cement is squeezed through the screen or perforations. If the formation is outside the casing, it must first be perforated. Some agencies require plugs to isolate aquifers if the water qualities or hydrostatic pressures differ sufficiently to justify separation. A 10-ft cement plug is often placed at the top of the well, and the casing is cut off 3 ft below the ground's surface. Usually a bentonite slurry of at least 9½ lb/gal is placed in all portions of the well not filled with cement.

Some agencies require that the entire well be backfilled with cement. With this requirement the cement is usually tremied into the well.

Downhole Plugging

If a borehole is not completed, regulatory rules generally govern the method and placement of a plug. Each operation presents a problem because a relatively small volume of cement slurry or grout is placed in a large borehole. Mud-contaminated cement often results in a weak, diluted, or unset plug even after a reasonable waiting period. There are three basic factors that influence the condition and effectiveness of a cement plug placed in a borehole:

- The condition of the drilling fluid in the borehole.
- The volume and type of cement used.
- Plug placement techniques.

Well Fluids

The first factor to be considered in setting a plug is the condition of the water in the well at time of placement. If there is any movement of the cement after it is placed, it will not set properly. If water is moving through a section and percolating up through the slurry, the cement could become permeable or honeycombed.

Placement Technique

Regulations for abandonment of a borehole might include a cement plug from the bottom of the hole extending 50 to 100 ft above the static water level, a plug at the bottom of the conductor casing, and a plug at the ground surface. In shallow borehole situations a single plug could extend from the bottom of the borehole to the surface.

Plug-back techniques commonly used in water wells are:

- Balance method (pumping cement down tubing).
- Dump-bailer method.
- Bull plug with ports.

A simple placement method for shallow wells known as the bull-plug technique uses closed-end tubing with 4 to 6 holes of ¾ to 1 in. diameter, as illustrated in Figure 12.10. This technique jets the cement slurry against the wall of the borehole to improve bonding.

With this method it is essential that the drilling fluid be circulated at least once prior to placing the cement plug in the borehole so that fluid density is equalized throughout. This is an important factor in avoiding cement contamination.

The following suggestions should be helpful in placing a successful cement plug in a borehole containing drilling fluid:

1. Select a competent section of borehole to be plugged. Condition the borehole before plugging.
2. Calculate cement, water, and displacement volumes. Always plan to allow for an oversize borehole.
3. Use a densified (low water ratio) slurry, API Type A or C cement.
4. Use a tail pipe with centralizers and scratchers through the interval plugged.
5. Reciprocate the tubing while placing the cement.
6. Use tubing wiper plugs.
7. Place the plug with care and move the tubing slowly out of the cement to minimize mud contamination.
8. Allow ample time for the slurry to set.

A dump bailer can be used to place a measured volume of cement slurry by lowering a bailer on a wire line to the designated depth. A limit plug, cement basket, permanent bridge plug, or gravel fill is usually placed below this depth. The bailer is opened by touching the bridge and raising it to release the cement slurry. This method is primarily limited by the quantity of slurry that can be placed on each run.

Required Volumes of Cement

The volume of cement required to seal the casing should be based on field experience and regulatory requirements. In the absence of specific measurements, a volume equal to 1.5 times the calculated volume should be used. Caliper logs may be necessary to determine the borehole enlargement and the proper location of centralizers or scratchers. Too much cement rather than too little is always advisable, especially where there is a possibility of mud contamination or dilution.

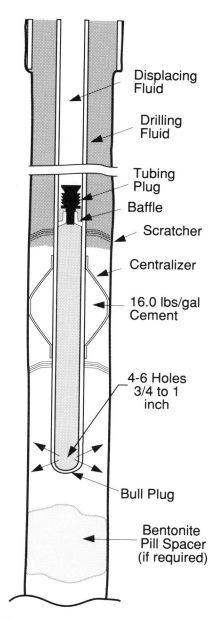

Figure 12.10. Setting down-hole plug (courtesy of Halliburton Services).

The volume of cement slurry may be calculated from data in service company handbooks. These data indicate the slurry yield in terms of cubic feet per sack from cement manufactured to API requirements, based on the amount of mixing water. A standard sack of API Class A cement (94 lb) mixed with the recommended amount of mixing water (5.2 gal) will produce a slurry volume of approximately 1.18 ft³.

12.10 CONSIDERATIONS AFTER CEMENTING

For most water wells, the mixing and placement of the cement slurry will require less than 1 to 2 hr. After placement, internal pressure, time waiting on cement to set, bonding, and quality of the bond are the major considerations.

Waiting on Cement Time

The total waiting time to set is dependent upon well conditions and mixing temperature. A reasonable WOC time (24 to 36 hr) should permit cement to attain sufficient strength to anchor the casing, withstand shock of subsequent operations, and seal permeable zones. With the selection of densified cements and 2% to 3% calcium chloride, WOC time can be reduced to 12 hours during summer operations, or 24 hr during winter operations. WOC time depends on the class of cement, admixtures, placement time, and borehole temperature and pressure. A minimum compressive strength of 500 psi is usually desirable before drilling out a plug. The recommended WOC time prior to running a cement bond log is 24 to 72 hr.

Bonding

In a well bore, the shear and hydraulic bonds are the two forces to be considered for an effective zonal isolation along the cement-casing and cement-formation interfaces. The shear bond mechanically supports the casing in the borehole and is determined by measuring the force required to initiate the casing's movement in a cement sheath. This force, when divided by the cement-casing contact surface area, yields the shear bond.

The hydraulic bond blocks the migration of fluids in a cemented annulus and can be measured by applying pressure at a casing-cement interface until leakage occurs. Hydraulic bonding is of greater significance than shear bonding for zonal isolation since adequate hydraulic bonding provides sufficient mechanical support to hold the casing in place.

While the bond between cement and casing is critical, the bond between the cement and formation normally determines whether or not there will be fluid communication in the annulus. Cement set against a clean formation will provide higher hydraulic bonding than cement set against a filter cake.

READING LIST

Smith, D. K. 1976. "Cementing, Monograph Volume 4." Society of Petroleum Engineers (revised edition, 1987).

Environmental Protection Agency. "Manual of Water Well Construction." EPA-570/9-75-001.

Campbell, M. D., and J. H. Lehr. 1973. *Water Well Technology.* McGraw-Hill, New York.

Anderson, K. E. 1959. "Water Well Handbook." Missouri Water Well Driller's Association.

Halliburton Services. 1983. "Halliburton Cementing Tables." Duncan, OK.

Goins, W. C., Jr. 1952. "How to Combat Circulation Loss." *Oil and Gas J.* pp. 71–92 (June 9).

Herndon, J., and D. K. Smith. 1978. "Setting Downhole Plugs: A State-of-the-Art." *Petroleum Engineer* pp. 56–71 (April).

Moore, P. 1974. "Drilling Practices Manual." Petroleum Publications, Tulsa, OK.

Murphy, W. C. 1976. "Squeeze Cementing Requires Careful Execution for Proper Remedial Work." *Oil and Gas J.* pp. 87–94 (February 16).

Rike, J. L. and E. Rike. 1981. "Squeeze Cementing: State-of-the-Art." 2nd SPE of AIME Prod. Operations Symp. pp. 133–142 (March 1–3).

Smith, R. C. 1982. "Checklist Aids Successful Primary Cementing." *Oil and Gas J.* pp. 72–75 (November 1).

"Specification for Materials and Testing for Well Cements." 1984. Spec. 10, API, Dallas, TX.

CHAPTER 13

Formation Stabilizer and Filter Pack

13.1 INTRODUCTION

Gravel envelope materials may be divided into two categories. The first is commonly known as a "formation stabilizer." Its primary purpose is to fill the annular space between the borehole and the well casing and screen in unstable formations, preventing sloughing. However, if the character of the aquifer indicates sand will be produced with the discharged water, a selected, finer "filter pack" is used. It performs the functions of a formation stabilizer while filtering the formation particles. Installation of a properly designed filter pack extends well life and reduces maintenance costs of wells, pumps, and meters.

Discussion of gravel envelopes requires a standard for grading the granular material. The one used in this chapter is:

d_p The sieve size of particles in a granular material such that $p\%$ of the material is finer than d_p

D_p The sieve size of particles in a granular material such that $p\%$ of the material is coarser than D_p

13.2 GRAVEL ENVELOPE CRITERIA

Filter pack design originated with the soil mechanics pioneer, Karl Terzhagi. Terzhagi addressed the problem of cavities in earthen dams created by seepage washing out the finer earth particles. His solution was to place in the base of the dam a granular filter bed of such gradation that it would retain the fine materials while passing the seepage. The operating principle for a water well gravel envelope is precisely the same.

Terzhagi's patented criteria (1921) for filter design related the size grading of the formation and the filter material such that

$$\frac{d_{15} \text{ (filter)}}{d_{85} \text{ (formation)}} < 4 < \frac{d_{15} \text{ (filter)}}{d_{15} \text{ (formation)}} \ .$$

It is evident Terzhagi regarded the characteristic or representative size of both the filter and formation particles as d_{15} and determined that the filter material should be larger

than four times the characteristic size of the formation. In addition he established that the finest 15% of the filter should be smaller than four times the coarsest 15% of the formation. This is shown in Figure 13.1.

These criteria are based on two fundamental principles. The first is that the characteristic particle size of the filter must exceed that of the formation to provide a significant increase in the filter permeability. This reduces fluid pressure gradients and consequently water velocity within the filter. Since particle mobility is proportional to velocity, movement of fine material is inhibited. However, in the environment of a pumping well, water velocity may never be sufficiently reduced to prevent movement of the finer particles found in many aquifers.

The second principle involves the relationship of the characteristic filter size to the finer segment of the formation. If the ratio of these sizes is too large, some formation material will migrate through the pack. Thus a limitation on the maximum size for the finer segment of the filter pack is imposed.

The Terzhagi principles have been examined many times, and at least 40 studies relating to gravel envelopes have been published. There have been many alternative criteria suggested. Some investigators (Halliburton 1965; Sawaro et al. 1983) have suggested a standard that dictates the maximum size of the smallest material in the filter. The ratio d_{10-15} (filter)/d_{50} (formation) is used, with a generally recommended ratio from 4 to 5. This is not inconsistent with the Terzhagi maximum d_{15} (filter) criterion but is slightly more stringent.

Another common modification of the Terzhagi formula considers the d_{50} (filter)/d_{50} (formation) ratio. Researchers have recommended this ratio be from a low of 4 to a high of 58, a rather wide range. However, the consensus is d_{50} (filter)/d_{50} (formation) = 5 ≈ 6 (Smith 1954; Ahrens 1957).

It should be noted that unlike the Terzhagi criteria, these alternative criteria do not specify any particular grading of the filter material or relate it to the gradation of the formation. Any formula that also specifies a maximum size for the finer portion of the pack may offer more assurance of success than use of the d_{50} ratio alone.

Neither Terzhagi nor these two alternative criteria specify a complete filter gradation. Although this limitation does

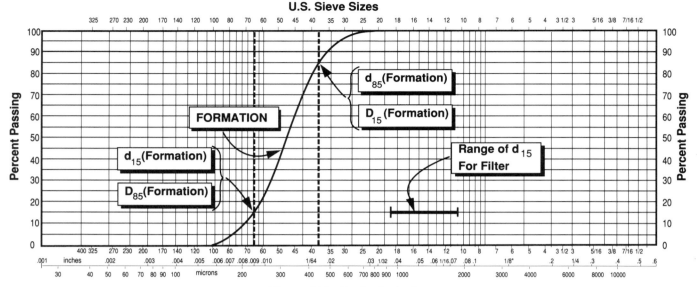

Figure 13.1. Terzhagi criteria.

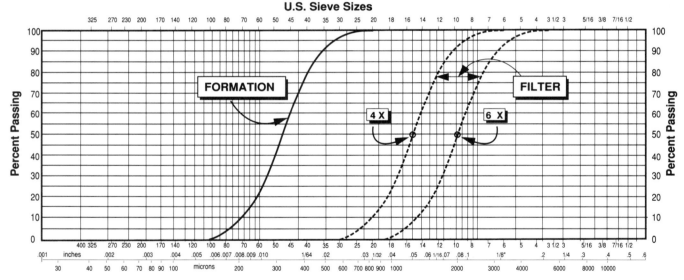

Figure 13.2. Aquifer and filter pack gradation.

not cause problems with earthen dam filters that are compacted in place, it may occasionally create difficulties with gravel envelope water wells. The problem occurs not from the criteria but from possible size segregation resulting from differential settling of the gravel envelope during placement. Gravel envelope material with a broad particle size distribution may settle in the annulus at a nonuniform rate, resulting in layering of finer and coarser particles. This is particularly true in deeper wells when proper drilling-fluid control and installation procedures are not employed.

Some suggest use of a more uniform gravel, the most extreme example being a ratio of largest to smallest particles not greater than 2 (Halliburton 1965; Cecil et al. 1979). A more typical standard is that provided in the *AWWA Standard for Water Wells* (AWWA A100-84), which suggests that the uniformity coefficient[1] of the gravel not exceed 2.5.

Another alternative calls for the plotted gradations of the filter pack and formation to be parallel, with a filter size

1. The definition of uniformity coefficient used here is d_{60}/d_{10}.

four to six times that of the formation (Stow 1962). This criterion is frequently and effectively used by well engineers and contractors despite the fact that a close parallel is seldom possible due to limitations of gravel screening and availability. It is illustrated in Figure 13.2.

Experience of many drilling contractors and engineers has shown that in most circumstances the filter will be satisfactory if the midsize of the pack is four to six times larger than the midsize of the aquifer. To prevent movement of excessive quantities of fine particles through a highly uniform filter, this is modified to include the Terzhagi requirement of d_{15} (filter)/d_{85} (formation) < 4.

13.3   NEED FOR A GRAVEL ENVELOPE

It is generally accepted that a gravel envelope well is not required if 90% of the aquifer is coarser than 0.010 in. and the material has a uniformity coefficient greater than 2 (Bennison 1947; Williams 1981). Similarly, any formation whose uniformity coefficient is greater than 5 does not require a gravel envelope, since a natural filter can be developed. Installation of a gravel envelope under these circumstances makes development more difficult and does not improve well productivity. However, an aquifer whose characteristic size and gradation does not meet these standards generally requires a gravel envelope for filtering purposes.

An exception to these rules occurs where a very poorly sorted formation consists of uniformly large gravel with the pores filled with fine sands or silts. The standards for natural development may be met, but use of a formation stabilizer in a gravel envelope well or construction of a naturally developed well can result in excessive sand production.

Experience has shown that some types of aquifers nearly always require a filter pack. Examples are highly uniform fine-grained sediments, such as beach sand deposits, some river alluvia, and poorly cemented, friable sandstones.

Highly laminated, unpredictable, and nonuniform alluvial formations, often found in the western United States, usually require a filter pack. Gradation of particles in these formations can vary from extremely coarse to extremely fine.

13.4   BASIC PROCEDURE FOR SELECTING GRAVEL GRADATION

Selected filter pack gradations can be no better than the reliability and accuracy of the formation samples collected during drilling. Cuttings samples may not be representative of the formation, regardless of drilling method or the care exercised in obtaining them. Hole erosion caused by fluid flow, mechanical enlargement from rotation of the drill pipe, and sloughing contribute to sample unreliability. In addition finer formation particles usually are lost into the fluid. This

masking of true aquifer gradation often causes potential water-bearing formations to appear coarser and more productive than they eventually prove to be, and may result in design errors. However, careful work will usually provide acceptable samples.

Sampling

Sampling and analysis of cuttings generated by direct or reverse rotary drilling require care in the sampling operation and understanding of the limitations of the results obtained. Direct rotary samples are often obtained from the material retained as the returning drilling fluid crosses the shale shaker screen. The finer segments that pass through the screen customarily are removed from the fluid by a hydrocyclone. However, the formation gradation is difficult to determine from observation of these segments separately. Samples may also be obtained from a sample catcher (a box with several baffle plates placed in the return fluid flow), but again the finer formation particles are usually retained in the drilling fluid.

There is better assurance of representative direct rotary samples when the following procedures are used. From 3 to 5 ft of borehole is drilled at the depth a sample is needed. The bit is then raised from the bottom of the hole, and circulation continued until all cuttings from the interval are cleared from the hole and caught on the shale shaker screen or in the sample catcher. This operation is repeated as required. Samples obtained from each interval are thoroughly mixed and quartered until an approximate 2-qt representative sample remains. The sample is then placed in a 5-gal pail, and water is added to fill it. After thorough stirring, the contents are allowed to settle, and the muddy water is decanted.

With reverse rotary drilling, cuttings are brought to the surface at high velocity, and it is usually not necessary to raise the bit from the bottom of the hole and circulate. However, the general practice of putting a strainer or bucket in the return flow and grabbing a sample is not reliable since the finer portion is lost. Samples should be taken directly from the discharge pipe, which is equipped with a 2½-in. bypass pipe with a quick close manual valve attached. The bypass pipe should penetrate the discharge pipe at an angle to direct it into the flow.

When the samples are obtained, they are dried, and a sieve analysis is run. The finest aquifer to be screened is selected to determine the envelope gradation. The chosen criterion for gravel selection is then applied to this gradation. A recommended method is to multiply the d_{50} of the formation by a factor of 4 to 6 to establish the d_{50} of the filter pack. Through these initial points, two curves parallel to the formation gradation are drawn, as shown in Figure 13.2.

The percentages passing are noted at the points where the filter pack gradation curve intersects appropriate standard sieve sizes to define the material. A commercially available gravel that lies within these curves is then chosen.

Delayed Sand Pumping

Since alluvial aquifer gradation varies in any well and there is no practical method of installing individually graded gravel segments, the filter pack and aperture size are selected for the finest formation encountered. However, wells completed with a filter pack and aperture too large for the finest zones often are initially good sand-free producers. In these cases most production comes from the coarser-grained aquifers with high hydraulic conductivity, and little from the potential sand-producing zones. Thus a satisfactory well results even though the envelope design is theoretically improper. Such wells may eventually produce sand, often in troublesome quantities. This nearly always occurs when a deeper pump setting is required, due to either a lowering of the water table or a decrease in specific capacity.

Sand Production through Filter Packs

Although filter pack gradation is designed to retain the aquifer materials, invariably a few fine-grained particles will migrate with water flow. A properly designed pack permits a few of these fines to move through it and to enter the well during pumping surges. In many wells a small amount of sand produced with each surge aids in maintaining specific capacity. Ideally, it should be limited to a few minutes of pumping, with peak sand production not exceeding 30 to 50 ppm. Sand production should rapidly diminish to zero with time, as shown in Figure 13.3. This will yield a satisfactory sand-free water during the course of a pumping cycle of at least

several hours. Criteria for sand production is discussed in Chapter 14.

13.5 SIMPLIFIED METHOD OF ENVELOPE GRADATION SELECTION

The basic purpose of the gravel envelope is to act as a filter that will stabilize the formation but not clog. Since there are an infinite number of possible aquifer gradations, an infinite number of filter and screen combinations can be designed. However, experience indicates that the selection process may be simplified, and most wells drilled in unconsolidated material can be completed by using one of two gravel gradations and corresponding screen openings.

Formation Stabilizer

In general, when the aquifer(s) is suitable for natural development, a gravel envelope is needed only for formation stabilization. A typical formation stabilizer material, often used with a relatively large screen aperture (3/32 to 1/8 in.), is shown in Table 13.1.

Aquifers Suitable for Formation Stabilizer

Any reasonably well-sorted formation whose gradation approximates or is coarser than line 1 on Figure 13.4 may be stabilized with a gravel envelope material similar to gravel A, and a 3/32-in. screen aperture. It will be noted that this is consistent with the basic rules set forth earlier. Aquifer gradations lying in the area between lines 1 and 2 require more attention, but wells completed in these formations, with gravel A and the 3/32-in. screen opening, have produced water sufficiently sand free for most uses.

Filter Pack

A filter pack must be used in wells where aquifers do not meet the criteria for natural development, or to the left of line 2 on Figure 13.4. In principle, precise filter pack selection can be derived from borehole samples, or in some cases by sieve analysis of sand produced from nearby wells or collected when these wells are cleaned out. However, aquifer char-

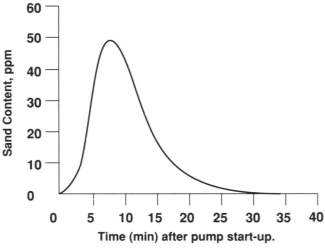

Figure 13.3. Typical sand production curve.

TABLE 13.1 Gravel A

U.S. Standard Sieve Number	3	4	8	16	20
Inches		0.187	0.093	0.047	0.0328
Percentage passing	100	85–95	25–35	5–20	2–10

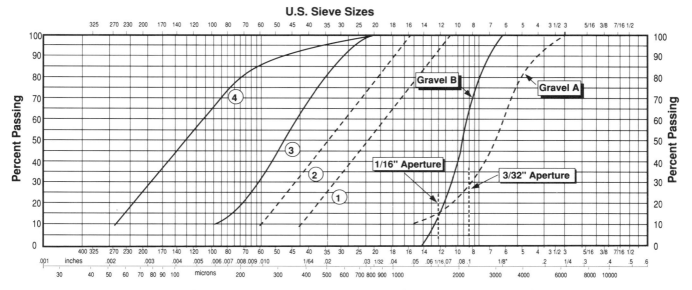

Figure 13.4. Formation stabilizer, filter pack, and formation gradation analysis.

acteristics vary within a well and between wells, and filter pack material is a natural product subject to size variation and segregation during transport and storage. In a practical sense it may be unrealistic to design filter packs and screen openings to excessively close tolerances because of these natural variations.

A gravel gradation typically used with a ¹⁄₁₆-in. screen opening under a broad variety of conditions for wells completed in alluvial formations requiring a filter pack is shown in Table 13.2.

Gradation curves of gravels *A* and *B* (formation stabilizer and filter pack) are also shown in Figure 13.4. These or similar materials are generally commercially available. Gradations 3 and 4 are actual aquifers screened in producing wells. Gradation 3, filtered with gravel *B* and ¹⁄₁₆-in. aperture screen, produces over 2000 gpm with a specific capacity of more than 100 gpm/ft of drawdown, demonstrating the efficacy of this aperture size/filter pack combination. Gradation 4 is a formation of much lower hydraulic conductivity, also completed with the combination of gravel *B* and a ¹⁄₁₆-in. screen opening. Although this well is far less productive (about 800 gpm with a specific capacity less than 10), it has pumped satisfactorily sand-free for many years. This represents the finest formation gradation that is able to use this combination of filter pack and screen.

Very Fine Aquifers

Aquifers are seldom as fine as gradation 4. If one is encountered, the filter pack requires special consideration. Gradations 3 and 4 were successfully filtered with the combination of gravel *B* and a ¹⁄₁₆-in. screen opening because there is a small percentage of relatively coarse material in each of these aquifers. This is generally true of even the finest producing formations.

If the formation were more uniform with the same median grain size, this standard combination would not work. Under such conditions gravel *B* has been successfully modified by adding about 20% of a finer filter material, such as 12-20 (100% passing sieve size No. 12 and retained on sieve size No. 20), again using a ¹⁄₁₆-in. screen opening. Except in the most unusual situations, screen openings smaller than 0.050 in. are not required in filter-packed wells and should be avoided due to potential clogging problems and greater development time and effort.

13.6 PERCENT OF PACK PASSING WELL SCREEN

Considerable analysis has been performed on the relationship of aperture size to envelope gradation. Too small an opening can result in the screen plugging with unnecessarily high head losses and reduction of flow. However, too wide a screen opening allows an excessive amount of finer material from the filter pack and formation to enter the well. Over

TABLE 13.2   Gravel B

U.S. Standard Sieve Number	4	6	8	12	16
Inches	0.187	0.130	0.093	0.068	0.047
Percentage passing	100	95–100	70–80	15–25	0–5

an extended period this will result in discontinuities in the pack, allowing cavities to form. If these cavities do not fill with envelope material, the formation may become structurally unstable, risking loss of the well.

In general, the screen aperture is selected to prevent most of the pack from passing. The combination of gravels and screen openings shown in Figure 13.4 allows the generally recommended range of between 10% and 20% (occasionally as high as 30%) of the envelope material to pass.

The actual percentage of filter pack that will pass a screen is subject to the configuration of the opening as well as the aperture size. For example, it is more difficult to stabilize particles around vertical screen openings. In addition vertical machined slots with parallel sides encourage plugging by finer particles. These characteristics require use of a smaller aperture designed to allow only a small percentage of the envelope to pass, increasing the difficulty of well development.

Horizontal openings are inherently more stable, permitting them to be relatively larger. As shown in Figure 13.5, the down-facing louver aperture of the shutter screen allows the coarser percentage of pack material to form a stable filter at each orifice. Shutter screen is also especially well suited to resist clogging and to stabilize pack material regardless of any minor mismatch between gravel and aperture, and it will retain a filter pack with wider gradation limits. Although generally the best gradation/aperture size relationship allows

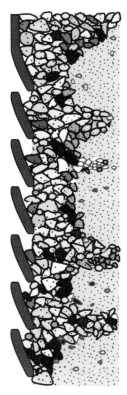

Figure 13.5. Natural filter formed against horizontal louver.

approximately 10% to 20% of the envelope material to pass, shutter screen has been successfully used with 50% or more envelope material passing. The only problem encountered has been a slightly longer period of well development needed.

13.7 PREPACKED SCREENS

Sometimes aquifers are encountered whose character makes the success of conventional gravel-packing procedures very doubtful. Very fine sands or poorly consolidated sandstones are examples. An extreme case is a sandstone aquifer consisting of particles no larger than a U.S. Standard Sieve Number 200 (0.0021 in.), with d_{50} approximately U.S. Standard Sieve Size Number 270 (0.0016 in.). Such zones may be quite productive because of thickness or faulting. Others less productive must be exploited when no alternate source exists.

Filtering of this material by standard analysis would require a filter pack with a d_{50} of U.S. Standard Sieve Number 50 (0.0116 in.) and a screen opening of from 0.005 to 0.010 in. Such extremely fine filter materials with correspondingly small screen openings are difficult to place and develop. Screen clogging or closing by incrustation often occurs due to the very small aperture.

The use of prepacked screens has proved successful under these conditions. They are fabricated with a strainer (often a stainless steel mesh) concentrically placed outside a conventional screen, providing an annular space of 2 to 3 in. This annulus is filled with a relatively coarse sand (Figure 13.6). After the prepacked screens are placed in the borehole, the annulus between the prepacked screen and the borehole wall is filled with a much finer material, designed to filter the formation.

Use of a prepacked screen results in two layers of filter pack. The outer retains the aquifer and the inner (designed to the same criterion) retains the outer layer. In the example just presented, a filter pack designed with a d_{50} of U.S. Standard Sieve Size No. 50 (0.0116 in.) with a pre-packed layer of d_{50} of U.S. Standard Sieve Size No. 10 (0.078 in.) would be appropriate. This allows a 1/16-in. screen opening that is less apt to clog and facilitates development.

13.8 THICKNESS CONSIDERATIONS

With the proper pack/aquifer ratio, filtering is accomplished within a few grains of the interface; thus filtering is not a function of pack thickness. A thin envelope also enhances the effectiveness of development methods in restoring the damage zone.

Although theoretically the greater permeability of the envelope compared to the aquifer increases well production, this relationship is based on the logarithm of the effective

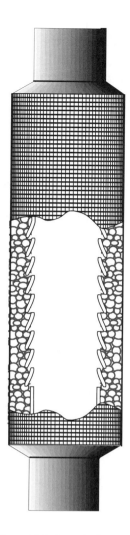

Figure 13.6. Prepacked screen.

shape is less likely to produce fine debris during transport. It is also thought that a rounded material produces an envelope with higher permeability for a given characteristic size, although this may not always be true. Some studies have shown that the angularity of the envelope material helps it to bridge at the aperture, resulting in greater permeability.

Nonetheless, experience indicates that rounded material is less apt to compact during well development operations. In view of this crushed rock is never used for an envelope, and natural material containing more than a minimum percentage (usually from 2% to not more than 10%) of flat surfaces is not recommended.

13.10 ENVELOPE PLACEMENT

Placement is the last step in gravel envelope well construction and the first step in well development.

The primary methods of placing the gravel envelope for rotary drilled wells are:

• Pouring into the annulus from the surface.
• Pumping through a gravel feed (tremie) pipe.
• Pumping through a crossover sub.

These methods can be used with either direct or reverse circulation equipment. It is preferable, however, to provide circulation up the annulus during gravel placement (Figure 13.7), except when a crossover sub is used.

Pouring

Gravel is often placed in shallow and medium-depth wells (less than 1000 ft) by pouring, in combination with the swab and circulation procedures shown in Figure 13.7. Pouring is slow but continuous down the annulus against the return fluid circulation, minimizing gravel segregation. A hopper designed to install up to 10 tons per hour is used to meter the installation rate.

With wells drilled by the reverse rotary method, the gravel is sometimes poured into the annulus with the circulating fluid (Figure 13.8). Circulation is maintained by placing open-end drill pipe near the bottom of the screen section, with pumping performed through this pipe. Gravel segregation is minimized with this procedure, although drilling debris scoured from the borehole walls by the falling gravel may be retained in the pack material if it cannot enter the well through the screen.

Installation through Gravel Feed Pipe

Despite the advantage of scouring the borehole and removing the filter cake, installation by pouring is sometimes not

well radius so that this is not a consideration in determining envelope thickness.

The annular space must be adequate for the pack to be installed without voids. Because of borehole irregularities, it is difficult to ensure placement unless the borehole diameter is at least 8 to 12 in. larger than the casing and screen. To avoid reducing the efficiency of development procedures, it should not exceed the screen's diameter by more than 16 in.

13.9 MATERIAL FOR GRAVEL ENVELOPE

The most suitable material is well-rounded, water-worn siliceous rock washed free of dirt, silt, clay, organic matter, or other objectionable material. Calcareous gravel must be avoided, with most authorities agreeing that it should not exceed 3% to 5% of the total volume.

Smooth and well-rounded material is preferred to angular, sharp-faced materials, primarily because it is believed this

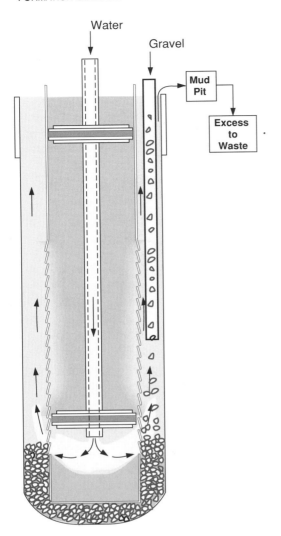

Figure 13.7. Flow diagram for gravel placement.

allowed because of possible size segregation. Segregation seldom occurs when placing the coarser material used for formation stabilization. If good drilling fluid properties are maintained and the gravel is added slowly and consistently, pouring can also be used for the finer-grained filter pack material. Segregation is far more apt to occur during transport and handling than during installation. Nonetheless, pumping through a tremie is often required for gravel installation in wells over 1000 ft deep.

When long sections of pump-housing casing are used, the tremie method may be required on shallower wells because of the tendency of gravel to bridge behind blank casing. The benefits of scouring are lost when this method is used.

Gravel is usually tremied by being metered into the suction of a dredge-type pump capable of handling solid material. The pump should produce a large slurry volume (typically 150 to 200 gpm), consistent with tremie pipe diameter. In a borehole radius at least 6 in. larger than the casing, 2½-

in. upset tubing is practical. This system will install about 5 to 10 tons of gravel per hour.

Although gravel installation rates for pouring and pumping are approximately equal, the total time required for pumping is considerably longer. This is due to the additional setup time and to the inevitable equipment operating delays.

Crossover Method

Another gravel placement procedure occasionally used with smaller-diameter, deeper wells, and those with underreamed borehole in the screened section is performed with a crossover sub or tool. As shown in Figure 13.9 the crossover sub is placed on the bottom of the drill pipe string. The screen and riser casing are attached below the sub by a back-off joint and telescoped into position through the pump housing casing.

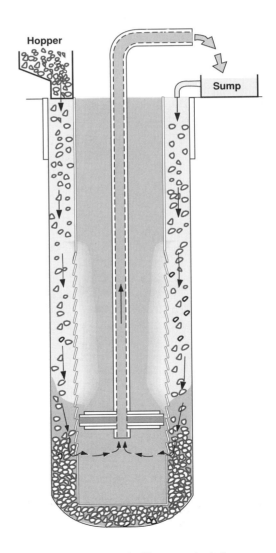

Figure 13.8. Pouring gravel with reverse circulation.

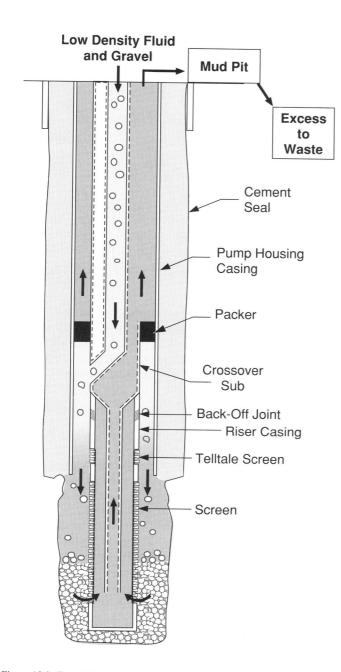

Low Density Fluid and Gravel

Mud Pit

Excess to Waste

Cement Seal

Pump Housing Casing

Packer

Crossover Sub

Back-Off Joint

Riser Casing

Telltale Screen

Screen

Figure 13.9. Gravel placement using a crossover sub.

A gravel–water mixture is pumped down the drill pipe through a port in the sub into the annulus outside the screen. The return fluid flow is carried up through a pipe, whose lower end is near the bottom of the screen section, and then ported through the sub to return to the surface in the annulus between the drill pipe and casing. A packer around the sub prevents the mixing of these two flows. The riser casing laps the pump housing casing a sufficient distance to provide a reserve of gravel above the screen.

A short section of screen is often placed in the riser casing below the crossover sub. When the level of the gravel envelope

rises above this screen (sometimes known as a telltale screen), circulating pressure suddenly rises, alerting the operators to stop gravel installation. This system is complex and expensive, with elaborate equipment and experienced operators required for its use.

13.11 BOREHOLE CONDITIONING

Drilling-fluid conditioning is required prior to installation, varying according to drilling method. Generally, water is added until the fluid properties before and during installation of gravel are:

Weight	Maximum of 68 lb/ft^3
Viscosity	Maximum of 30 sec, API Marsh funnel test
Sand content	Maximum of 1% by volume

Circulation and Mud Conditioning

The filter cake is removed during gravel placement. This is more pronounced with pouring but occurs even when the pack is pumped. Continued conditioning is required to control fluid weight and viscosity. The primary treatment is thinning, and ample quantities of freshwater must be available.

Swabbing and circulating are usually performed during gravel or filter pack placement in wells drilled with direct rotary equipment. After the fluid is properly conditioned and the casing and screen are installed and suspended from the surface, the drill pipe (or 3-in. tubing) with two tight-fitting swabs attached is placed in the casing and screen (Figure 13.7). The lower swab at start of graveling is positioned near the bottom of the screen. The upper swab is located in the blank casing. Fluid is circulated up the annulus to remove particles scoured from the borehole, but at a rate that permits downward flow of the gravel. If this procedure is done with reverse circulation equipment, special pumps not normally available with these rigs are required.

In wells up to 1000 ft deep, the pouring process may be completed without moving the swabs other than to raise them periodically to prevent the sand from locking them. The lower swab should always be below gravel level. In a very deep well with a long screen section, the swabs must be raised periodically so that the desired return flow continues.

As the envelope fills around the blank casing, it becomes increasingly difficult for the flow to circulate up the annulus. In this event the top swab is removed when the pack level is above the top of the screen, and return flow continues inside the well casing.

Cable Tool Gravel Installation

Several well designs are available for cable tool gravel envelope wells (Chapter 6). One calls for placement of a screen section into an open borehole below a conductor casing. The screen can be installed by using a permanent casing string suspended from the surface or by using a temporary pipe that is removed after the screen section is telescoped through the casing and graveled.

If the open hole is supported by drilling fluid during drilling and a pump is available, gravel is usually poured against the circulating fluid, similar to the procedure shown in Figure 13.7. If the well is shallow or if circulating pumps are not available, gravel may be poured without circulation. In both pouring operations the borehole is scored and considerable debris is brought into the well. This debris is removed by bailing concurrently with graveling. Bailing also helps prevent gravel bridging during its installation.

In another design, casing is installed to final depth. The well screen is installed inside this casing, attached either to the pump housing casing or to a temporary pipe. As the pack material is poured in place, the outer casing is withdrawn until it clears the screen.

Gravel is added in approximate 3-ft increments. Following each sequence the outer casing is pulled back until its lower end is just below the gravel level. This avoids locking the outer casing and screen together.

13.12 GRAVEL DISINFECTION

Gravel should be disinfected during its installation. Acceptable methods are described in AWWA Standard C 654-87.

READING LIST

Ahrens, T. 1970. "Basic Considerations of Well Design—Part III, Screens—Slot Size and Materials—Gravel Packs." *Water Well J.* (June), pp. 47–51.

Ahrens, T. 1970. "Basic Considerations of Well Design—Part IV, Various Approaches to Well Development and Testing; the Author's Conclusions on Well Design." *Water Well J.* (August), pp. 35–37.

Bennison, E. W. 1947. *Ground Water, Its Development, Uses, and Conservation.* Edward E. Johnson Inc., St. Paul, MN.

Bertram, G. E. 1940. "An Experimental Investigation of Protective Filters." Soil Mechanics, Series No. 7. Graduate School of Engineering, Harvard University.

Blair, A. H. 1970. "Well Screens and Gravel Packs." *Ground Water* 8, 1, pp. 164–168.

Boulet, D. P. 1979. "Gravel for Sand Control: A Study of Quality Control." *J. Pet. Tech.* (February), pp. 10–21.

Cecil, L. B., J. H. Wilson, and G. L. Birt. 1979. "Gravel Packs and Screens Raise Oil Production 79%." *World Oil* (July), pp. 97–104.

Clark, L., and P. A. Turner. "Experiments to Assess the Hydraulic

Efficiency of Well Screen." *Ground Water* 21, 3, pp. 270–186.

Durrett, J. L., W. T. Golkin, J. W. Murray, and R. E. Tighe. 1977. "Seeking a Solution to Sand Control." *J. Pet. Tech.* (December), pp. 1664–1672.

Gulati, M. S., and G. P. Maly. 1975. "Thin Section and Permeability Studies Call for Smaller Gravels in Gravel Packing." *J. Pet. Tech.* (January), pp. 107–112.

Gumpertz, B. 1940. "Screening Effect of Gravel on Unconsolidated Sands." *J. Pet. Tech.* (May), pp. 1–8.

Gurley, D. G., C. T. Copeland, and J. D. Hendrick. 1977. "Design, Plan, and Execution of Gravel-Pack Operations for Maximum Productivity." *J. Pet. Tech.* (October), pp. 1259–1266.

E. E. Johnson, Inc. 1966. "Developing and Completing Water Wells." *Ground Water and Wells*, ch. 14.

E. E. Johnson, Inc. 1966. "Water Well Design." *Ground Water and Wells*, ch. 10.

Kruse, G. 1960. "Selection of Gravel Packs for Wells in Unconsolidated Aquifers." *Tech. Bull. 66*, March. Colorado State University, Experiment Station, Fort Collins, CO.

Leatherwood, F. N., and D. F. Peterson., Jr. 1954. "Hydraulic Head Loss at the Interface between Uniform Sands of Different Size." *Trans. Am. Geophys. Union*, 35, 4, pp. 588–594.

Monroe, S. A., and W. L. Penberthy, Jr. 1980. "Gravel Packing High-Volume Water Supply Wells." *J. Pet. Tech.* (December), pp. 2097–2102.

Rogers, E. B., Jr. 1971. "Sand Control in Oil and Gas Wells— 1." *Oil & Gas J.* (November), pp. 54–60.

Rogers, E. B., Jr. 1971. "Sand Control 2—Sizing Gravel Pack Line Slots in Oil and Gas Wells." *Oil & Gas J.* (November), pp. 58–60.

Rogers, E. B., Jr. 1971. "Sand Control 4—Combinations, Comparisons, and Costs." *Oil & Gas J.* (November), pp. 64–68.

Shyrock, S. G. 1979. "Tests Show Methods for Improved Gravel Packing." *World Oil* 188, 2, pp. 55–58.

Smith, H. F. 1973. "Gravel-Packing Water Wells." *Water Well J.* (February), pp. 24–26.

Stramel, G. F. 1965. "Maintenance of Well Efficiency." *J. Am. Water Works Assn.* 57, pp. 996–1000.

Terzaghi, K. 1925. *Erdbaumechanik aug Bodenphysikalscher Grundlage.* Deutecke, Leipzig.

Terzaghi, K., and R. B. Peck. 1948. *Soil Mechanics in Engineering Practice.* J. Wiley and Sons, New York.

Todd, D. K. 1959. *Ground Water Hydrology.* J. Wiley and Sons, New York.

U.S. Bureau of Reclamation. 1947. "Laboratory Tests on Protective Filters for Hydraulic and Static Structures." Earth Materials Lab. Report No. EM-132. Denver, CO.

Williams, E. B. 1981. "Fundamental Concepts of Well Design." *Ground Water* 19, 5, pp. 527–542.

Williams, B. B., L. S. Elliott, and R. H. Weaver. 1972. "Productivity of Inside Casing Gravel-Pack Completions." *J. Pet. Tech.* (April), pp. 419–425.

CHAPTER 14

Well Development

14.1 INTRODUCTION

Development is an integral phase of well construction and to some degree will be required for any well. It comprises the systematic procedures followed to ensure the maximum discharge rate at the highest specific capacity with minimum production of particulate matter. The best procedures and equipment to achieve this can vary considerably, reflecting a number of conditions such as aquifer characteristics, sand content standards, well design, and construction method.

14.2 WELL DEVELOPMENT

Water well development may be divided into three phases: predevelopment, preliminary development, and final development.

Predevelopment

Predevelopment consists of the steps followed during well construction to minimize formation damage. This primarily applies to direct or reverse rotary drilling systems using drilling fluid to support an open borehole. Such steps include control of drilling-fluid properties during all phases of construction.

Preliminary Development

Preliminary development comprises those procedures performed with the drill rig and/or a development rig after casing, screen, and gravel installation (if required). These include swabbing, flushing, sand pumping, bailing, jetting, and airlifting.

Final Development

Final development is performed with a pump and requires appropriate pumping, surging, and backwashing techniques.

14.3 TIMELINESS

Wells completed in most types of aquifers benefit from reasonably continuous construction operations, although some drilling methods and aquifers can better tolerate delays or longer construction time. However, after completion of the final borehole with rotary drilled wells, construction should be expeditious through preliminary development. If too much time elapses between the casing and pack installation and the start of preliminary development, drilling fluids may gel and be difficult to remove from outside the screen and within invaded aquifers.

In general, development of cable tool wells need not be initiated as expeditiously as with other construction methods, although occasionally well performance may be impaired when development is delayed.

14.4 PURPOSE OF WELL DEVELOPMENT

There are three basic purposes of well development:

1. To create a stable, effective filter zone between the aquifer and screen from the formation in a naturally developed well, or from the gravel envelope.
2. To increase hydraulic conductivity in the near-well zone.
3. To increase effective well radius in naturally developed wells.

14.5 DEVELOPMENT OBJECTIVES FOR GRAVEL ENVELOPE WELLS

The development of gravel envelope wells requires the following:

1. Removal of drilling fluid and debris from the screen openings, filter pack, borehole wall, and formation.
2. Creation of an effective hydraulic interface between the envelope and the aquifer.
3. Consolidation and stabilization of the envelope.

14.6 PREDEVELOPMENT

Chapters 8 and 13 cover some of the procedures that constitute predevelopment. Careful drilling-fluid control during the final drilling stage, rapid placement of casing and screen, and fluid conditioning prior to gravel installation are fundamental requirements. The degree of emphasis on the various steps of predevelopment depend on the drilling procedure and method of gravel installation. For instance, gravel placed from the surface can remove significant quantities of filter cake by scouring the wall of the borehole.

The screen aperture size is selected to permit a portion of the envelope to pass. Entrance of some pack material into the well facilitates development and indicates favorable progress. Little or no pack material entering the well suggests that it may still be contaminated with drilling fluid, cuttings, or filter cake.

14.7 PRELIMINARY DEVELOPMENT

The relative effectiveness of any of the different preliminary development techniques available for gravel envelope wells is often uncertain. The differing yields of nearby wells of similar design and construction is often attributed to the development procedures used, suggesting that a particular method is always more effective. However, the results may be due to the aquifers' characteristics, the diligence of the crew, or even the sequence of procedures followed. No one method of preliminary development is clearly superior in all situations.

Swabbing and Circulating

Preliminary development of gravel envelope wells should commence immediately after installation of the gravel. The initial activity is usually swabbing and circulating, done with a single swab on the bottom of the drill string, as shown in Figure 14.1. The swab rubbers should fit tightly into the screen section, and be adequately supported by steel plates.

Beginning at the bottom of the screen, the swab is vigorously raised and lowered the length of travel in the mast. This helps break down bridges, consolidate the envelope, fill voids, and flush the filter cake and debris from the envelope for removal by the circulating fluid. After a section has been swabbed, the tool is raised by removing one or two drill pipe lengths, and the process is repeated. Swabbing is performed two or more times through the screen section, or as long as gravel movement is noted.

Records. Records should be maintained during swabbing and circulating and on all subsequent development operations to assess the effectiveness of the work. These should note the time, any movement of the envelope, the quantity of gravel added, as well as the quantity, type, and gradation of material bailed from the well. The effectiveness of swabbing is measured by the quantity of fine particles pulled in for a given length of screen. The presence of viscous drilling

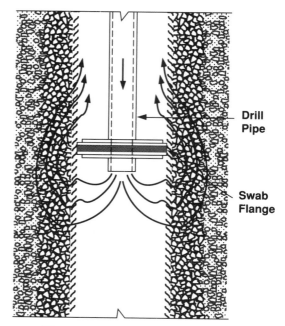

Figure 14.1. Single-swab development.

fluid in the material bailed from the well indicates that filter cake is being removed.

In a gravel envelope well, it is particularly important to compare the quantity of material entering the well with that being added to the envelope. An excess of material being brought in may be an indication of voids in the envelope.

Flushing

Well flushing is a procedure in which the drill rig and mud pump are used to wash the envelope with a double-swab tool. This tool is fabricated in several forms. One version consists of a short section of pipe with two swabs, attached to the bottom of the drill pipe string, with the swabs separated 18 to 24 in. The top swab is attached below the drill pipe tool joint, and the bottom swab is placed at the lower end of the pipe. For circulation, 1-in. holes are drilled into the pipe between the swabs (Figure 14.2).

The efficiency of the flushing procedure is reduced when the formation below the tool is pressurized. The condition in this case is virtually identical to a single-swab operation, and flow losses to aquifer recharge may be significant. If this pressure is relieved, development is faster and more effective, since a much higher fraction of the available energy is expended within the gravel envelope. The flushing tool design incorporating a bypass shown in Figure 14.3 accomplishes this, and its use is preferred.

Flushing effectiveness is dependent upon the amount of water that can be circulated. For higher-capacity wells (12-in. screen or larger), at least 600 to 1000 gpm should be

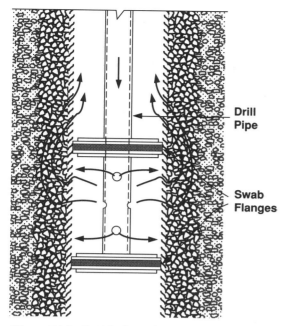

Figure 14.2. Double-flanged swab without bypass.

the well. As the operation proceeds, the amount of material suspended in the return fluid is observed. When the return fluid clears, the drill string is raised and lowered vigorously.

The tools are periodically removed, and the amount of fill measured. If fill has reached the bottom of the screen, the well must be cleaned out before continuing. The process is continued until there is no further envelope settlement, the material flushed into the well is minimal, and return flow is relatively clear.

Wire-Line Swabbing

Wire-line swabbing on higher capacity wells requires use of a medium to large cable tool or wire-line rig with a minimum 80- to 130-hp engine. Prior to swabbing, the well is cleaned out by bailing. The washing action caused by the passage of the bailer through the screen initially flushes considerable drilling debris and fluid into the well.

Cable tool swabs are built in many forms, but most are fabricated on a scow or heavy bailer, as shown in Figure 14.4. The advantage of this type is that the flapper valve

circulated, depending upon screen and borehole diameters. If more than 1000 gpm is circulated, the swab separation is sometimes increased up to 4 ft to provide more exit area. Ideally, freshwater is used for flushing. However, if adequate settling pits are available, the fluid may be recirculated.

Flushing is started at the top of the screen. The return fluid removes the cuttings, mud, and debris brought into

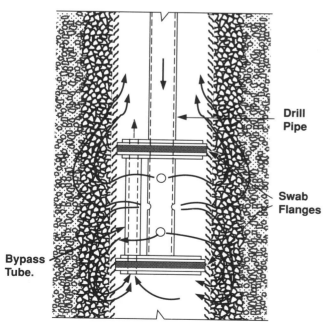

Figure 14.3. Double-flanged swab with bypass.

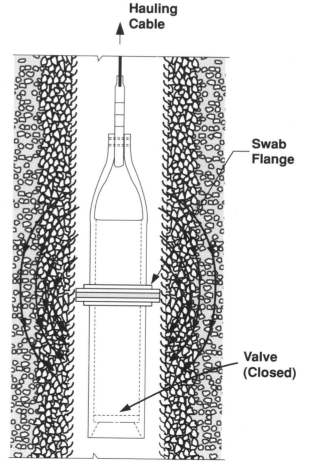

Figure 14.4. Wire-line swab.

allows the swab to fall rapidly. When it is raised, the valve closes, lifting the column of water above the swab. This draws water into the well below the swab. Very effective washing of the envelope takes place because of the reciprocating action of the water through the envelope.

Maximum swabbing efficiency requires a tight-fitting swab. Swab rubbers should have no more than ½-in. clearance when starting this procedure. They are replaced when wear increases the clearance to ½ in. Vigorous swabbing can place considerable stress on the screen, a consideration in selection of screen type.

Wire-line swabbing is far less effective in wire wrap screen, due to the internal rods that support the wrapping wire. The outside diameter of the swab cannot be as large as the diameter inside the rods, and in addition much of the water lifted by the swab bypasses through the inside of the screen. Consequently considerably less swabbing energy is applied to the formation or gravel envelope. This also applies to other systems of swabbing and flushing.

Line swabbing begins at the top of the screen and continues in short intervals (not more than 50 ft) to the bottom. The swab is repeatedly hoisted through each increment with the speed and length of hoisting increased until the swab with full engine horsepower is pulled through the entire screen. The well should be frequently bailed to remove and evaluate the materials drawn in.

By identifying different zones of greater or less resistance to swab movement, an experienced operator can locate voids and intervals requiring additional development. There is less resistance to swab movement opposite voids, while sections requiring additional development because of very fine material in the filter pack are detected by a noticeable increase in resistance.

Envelope Voids

Voids are usually indicated by the presence of fine aquifer material in the well. Collapse of formation against the screen not only permits direct flow of material into the well but also prevents gravel consolidation below this point. When noted, these voids must be filled quickly. This is done by breaking down the bridge in the envelope above the void by swabbing with the spudding beam opposite the bridge. The repeated washing action of the swab usually quickly breaks it down so that envelope material settles into the void. A small amount of water should be continuously added to the annulus while swabbing the well in order to lubricate the gravel.

Caution must be used in consolidating voids. Collapse of the bridged material into a void and subsequent pack settlement may create another void and bridge higher in the annulus. A bridge thus created behind blank casing can be very difficult to break down. If the bridge is below water level, a scow or heavy bailer may be rapidly lowered in the blank casing and quickly braked to a stop. The rapid shutting of the valve creates considerable vibration. Another method for breaking down the bridge is to operate the scow on the beam inside the blank casing, with the scow valve blocked in a half-open position. The valve acts as a vane, forcing the scow barrel to strike the casing on both the up and down strokes. This hammering helps move the gravel.

Alluvial formations often consist of prolific aquifers separated by clay layers. If voids in the gravel envelope of wells completed in these formations are not quickly located and filled, a major loss of well productivity may result. Aquifer material may enter the well through unfilled voids, creating a cavity under a clay layer. If the envelope does not fill the cavity, the clay will eventually slough across the face of the aquifer. When this occurs, redevelopment of that zone is extremely difficult.

14.8   PRELIMINARY DEVELOPMENT PROCEDURES—NATURALLY DEVELOPED WELL

The objectives of development of naturally developed wells are:

1. Remove drilling debris from the aquifer face and formation.
2. Develop a filter on the perforations or screen.
3. Increase productivity by developing a zone of high hydraulic conductivity surrounding the perforations or screen.

Natural well development requires creating a highly permeable zone adjacent to the apertures, allowing maximum flow of water while inhibiting the entrance of material. This is accomplished by moving water in and out of the openings at velocities sufficient to disturb finer formation materials. Successively smaller percentages of these finer granular fractions are drawn out of the aquifer as the distance from the well increases.

Natural well development is usually most effective in cable tool wells drilled in coarse, generally clean (silt-free) aquifer materials, where down-hole casing perforations or screen apertures are appropriately sized relative to the formation (see Chapter 11). Development can be difficult in a highly heterogeneous aquifer where interbedded fine-grained (sand, silt, clay) strata occur.

Although all gravel envelope well development methods may be used with naturally developed wells, generally only sand pumping is required where casing is installed concurrent with drilling and is later perforated-in-place. Since the casing supports and covers the borehole, there is less migration of drill cuttings into the aquifers. The oversize casing shoe shears the formation wall as the casing is installed and may remove much of the damage zone. However, when casing is driven into unconsolidated formations, the vibration can compact the formation adjacent to the well, lowering its hydraulic conductivity.

Sand Pumping (Mechanical)

Typically, the aperture size of the screen or in-place perforations in a naturally developed well will permit 50% (or occasionally even more) of the formation material to pass (Figure 14.5). Development usually necessitates bringing in greater quantities of aquifer materials than with gravel envelope wells, and this is effectively achieved by sand pumping.

Sand pumping uses a scow or large-diameter bailer (approximately 75% of screen's inside diameter) operated with a reciprocating motion opposite the screen or down-hole perforations. Fine particles are efficiently drawn into the well by the washing and surging action. Sand pumping usually begins at the top of the screen, or down-hole perforations, and continues to the bottom. Material brought in is bailed out as required, and the quantity and character of the material removed is noted.

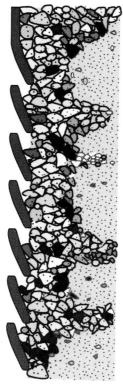

Figure 14.5. Natural filter pack.

Surge Block Development

The effect of a surge block (sometimes called a "surge plunger") is similar to that of sand pumping, although the surge block is usually not operated opposite the screen. A surge block is a tight-fitting solid piston incorporated into a heavy stem. It is operated with a spudding motion so that the water column is raised and then forced back down the well. The surge block is normally operated in the blank casing, starting a few feet below the static water level so that the pulsating action of the water moves throughout the entire screen's length. This minimizes stress on and possible damage to lower-strength screens and minimizes the possibility of sand locking the tool in the well.

As material brought in fills the screen, the surge block action is concentrated in a diminishing section. Therefore the well is bailed frequently, particularly in initial stages of the operation. As the well develops, the surge block is lowered closer to the screen, and the speed of the reciprocating motion is increased.

Use of a surge block is a common and effective development procedure in wells with relatively short screen sections. It is less often used in wells with long or intermittent sections of screen or perforated casing, since these wells usually require that at least some energy be expended directly through the screen. If operated in a screen section, care must be exercised so that the substantial quantities of drawn-in material expected in a naturally developed well do not sand lock the surge block.

Swabbing

Line swabbing naturally developed wells is usually done only if sand pumping has not brought in enough material. Since its effect is more powerful, it is performed less vigorously than in gravel envelope wells. Looser (about ½-in. clearance) swabs are operated with a reciprocating motion on the spudding beam, rather than with full power hoisting.

Indications of Completion. The optimum quantity of formation materials drawn into a naturally developed well during swabbing or sand pumping is related to the lithologic character of the screened or perforated interval and the number and size of openings. Because of such variables, "rules of thumb" have evolved in many areas, depending upon formations completed and well design. A typical rule for wells drilled in unconsolidated alluvial formations calls for a minimum of 1 ft of fill to be brought in for every foot of perforated casing or screen. In practice, all water wells differ, so more or less than that amount of material brought in may be appropriate.

As development proceeds, materials brought in become coarser. Small quantities of fine formation particles present in the material bailed are the best indicator of completion of this stage of development.

Jetting

Jetting can be used in both naturally developed and gravel envelope wells. Equipment required includes a high-pressure pump, hoses and fittings, threaded and coupled pipe, and a jetting tool. Rotary rigs are usually equipped with suitable pressure pumps for jetting.

As shown in Figure 14.6, the jetting tool has two or four nozzles, with orifice size selected to produce velocities between 100 and 200 ft/sec at the pump capacity available. The tool diameter should position the jets close to the inside of the screen.

In theory, the water jet passes through the screen openings into the natural or artificially placed materials, agitating and washing them. If an airlift pump capable of producing at least as much water as is being introduced is simultaneously operated, fine-grained material entering the well can be removed. Airlift pumping while jetting creates a steady flow of water into the well that helps remove dislodged material.

It is important that clean water be used because silt or sand will quickly erode the jet nozzle as well as the screen. For this reason the jet must also not be left stationary. To enhance the operation, chemicals such as polyphosphates may be added to the jetting water, although as discussed later, these chemicals must be used cautiously with certain formations.

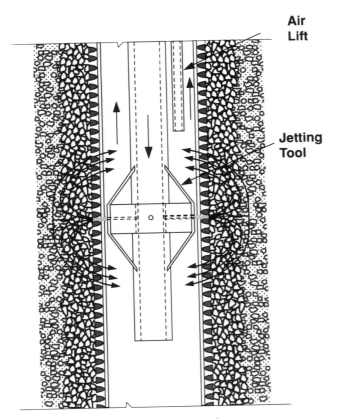

Figure 14.6. Jetting development.

The jetting tool is slowly rotated and lowered through the screen at 1 min/ft of screen until the discharge from the airlift pump is relatively clear of turbidity and fine particles. After the tool reaches the bottom of the screen, it is removed at a rate of up to 5 min/ft. Unless pumping is done concurrently, the well must be periodically bailed. Progress is judged by the amount and character of the material removed, and the process is usually continued until only a negligible amount of material is brought in.

14.9 AIR DEVELOPMENT

Air development can be employed on both gravel envelope and naturally developed wells but is more frequently used where the drilling method requires equipment capable of producing large volumes of high-pressure compressed air. The technique consists of three operations combined in varying fashion:

Surging	Sufficient high-pressure air is injected to create an aerated column of water that is lifted to or near the surface. The air supply is then shut off, allowing the water column to flow back through the screen.
Blowing	A similar operation except that air is injected for a longer period so that the water is pumped or blown from the well.
Airlift pumping	The procedure of pumping water more or less continuously from the well. In development operations, this may be done to clean the well after surging or blowing. Its use avoids the wear that would occur if a turbine pump were used for development.

An alternate development procedure infrequently used is air jetting, which is similar to jetting with water.

Equipment

Most airlift pumps consist of a discharge, or eductor, pipe and a smaller concentric air-injection pipe (air line), both partially submerged in the well. For pumping operations, the air-injection pipe is set slightly shallower than the eductor pipe. The bottom end of the air-injection pipe is usually sealed, with a number of small-diameter holes drilled just above the seal. This introduces the compressed air in a finely divided state, ensuring a mixed column of bubbles of air and water inside the discharge.

A more efficient system, used for airlift pumping only, locates the air-injection pipe alongside the eductor pipe; the air-injection pipe is connected to the eductor at an appropriate depth. This procedure is seldom used in well development because surging cannot be efficiently performed, and there are complications in simultaneously handling and installing two pipe strings.

Separate eductor pipes are required for large-diameter wells, deeper static water levels, or limited quantities of compressed air. In smaller wells the casing may be used as the eductor pipe (casing-air method). However, control of discharge is difficult, and drawdown measurements are not possible. Additionally development work cannot be concentrated in specific intervals. Airlift pump design is described in Appendix I.

Principle of Operation.

Airlift pumping depends on a number of factors, including air volume, submergence (the depth below static level where the air is introduced), total lift, and the cross-sectional area of the discharge or eductor pipe. An airlift pump operates because of the lower specific gravity of a mixture of water and air bubbles compared to water. When air is first injected, the volume of bubbles produced is insufficient to raise the water level inside the discharge above static. Assuming sufficient submergence and air volume, continued injection creates more bubbles, and the water level inside the discharge raises above static level. As still more air is introduced, pumping commences intermittently due to a slug of air lifting a slug of the aerated water. If enough air is available, the discharge rate will be fairly constant.

The usual application of airlift development is known as "open," or "conventional," so called because it affects the entire screen section. Sometimes isolation tools are employed that restrict the development to desired portions of the screen.

Surging and Blowing

The techniques of surging and blowing are usually performed at intervals throughout the screen, depending upon the volume and pressure of air available. When commencing development, the air line is typically placed in or just above the screen. The turbulence of the air bubbles formed agitates the material surrounding the screen. This grades and stabilizes the material while debris and drilling fluid remnants are washed into the well.

When pumping is used in conjunction with surging, the eductor pipe is generally set to the depth where development is to be performed. The air line is lowered a few feet below the eductor pipe, and air is pumped for a brief period of time to surge the well. If the operator is certain the screen is not plugged, the air line is then raised into the eductor pipe, usually while maintaining air flow. When the air line is several feet inside the eductor pipe, pumping of sediment-filled water commences.

As the well develops, the duration of air injection with the air line below the eductor is lengthened until the water is blown from the well if possible. When the discharge clears, the process is repeated until the desired level of development is attained. As air pressure and volume permit, the eductor pipe can be set at various depths in the screen to concentrate the development effects. At the end of development, the entire screen section may be "blown" with the air line set at the bottom of the screen. The well is then cleaned out by pumping.

Isolation Tools.

Isolation tools include air seals or swabs to concentrate pneumatic and hydraulic action on selected screen intervals, as shown in Figure 14.7. They are designed primarily for cleaning and stabilizing rather than pumping. Each screen section starting at the top is selectively developed by surging and pumping.

In a similar procedure, usually done with reverse rotary drilling equipment, the well is mechanically swabbed by raising and dropping drill pipe equipped with a tight-fitting double swab on the bottom. During swabbing, water is airlifted from the well by the same procedure used during reverse circulation drilling. Swabbing is started at the top of the screen section, to minimize the danger of locking the swabs.

Air-Jetting Tools.

Air-jetting tools usually incorporate designs as simple as a 90° bend in the air injection pipe, but more complex multiple nozzle arrangements are possible. Such arrangements are sometimes used when water for jetting is not readily available, and they concentrate the forces of

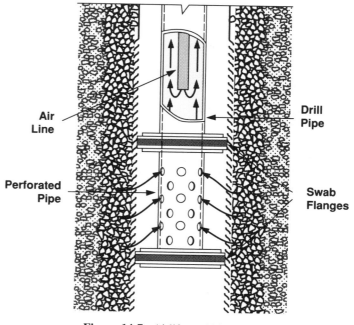

Figure 14.7. Airlift swabbing tool.

injected air at specific points along the screen section being developed.

Formation Limitations

Use of air development techniques is not advisable in certain aquifers, especially fine-grained sands or sandstones, where air locking of the formation may occur. Other formations susceptible to air locking are unconsolidated sands and gravels separated by thin clay aquitards.

14.10 CHEMICAL DEVELOPMENT

Polyphosphates

Since dispersants may cause a breakdown of clays, which in turn may slough and cover the aquifer faces, the use of dispersants is generally considered as a last resort if conventional (and safer) methods have not achieved the desired results. In gravel envelope wells completed in alluvial aquifers, chemical development usually is limited to use of dispersants. Dispersant chemicals include polyphosphates such as hexametaphosphate and sodium acid pyrophosphate. These are referred to as deflocculants, and are effective in removing filter cake from the borehole wall and clay fractions in the formations.

Dispersants are mixed aboveground in concentrations that will provide between 10 to 20 lb of chemical for each 100 gal of water in the well. The solution is tremied into the well, mixed by agitating it with a bailer, and then left overnight. This operation is repeated until the operator is satisfied that maximum benefit has been obtained.

Some polyphosphates may produce a gelled precipitate under certain conditions. Unusually cold water in the well usually encourages the formation of this precipitate, and its removal can be very difficult.

Limestone Development

Acidizing, typically with hydrochloric acid, is sometimes employed as a development procedure to enlarge solution passages adjacent to wells drilled in fissured limestone. A well-head seal or packer is placed on the casing, fabricated so that a string of tubing may be run through it. The bottom of the tubing is placed near the top of the screen. After the appropriate quantity of acid is pumped into the well (usually equal to the borehole volume below the top of the screen), a measured quantity of water, sufficient to displace the acid in the screen section and tubing, is pumped through the tubing.

A valve is placed in the tubing string above the well-head packer so that it can be closed after the acid and displacing water are emplaced. Another valve and a pressure gauge are attached to the packer. This valve is used to release any excess pressure created by gas generated by dissolution of the limestone. The pressure is monitored, with completion of chemical reaction indicated by a noticeable drop.

Limestone development is comparatively a rapid procedure, and some contractors pump the spent acid water from the well in about 1 hour. Experience indicates that fewer precipitates are deposited if the contact time between the acid and limestone is limited. Typically, acid development in limestone is followed by other development procedures such as surging or swabbing.

Turbulence-Producing Chemicals

Turbulence-producing chemicals are sometimes used for development. Dry ice (CO_2) or a combination of a weak acid and base solution separately injected into a well create a short-lived but sometimes violent turbulence. In the case of dry ice, the solid CO_2 changes to gas and combines with water to yield HCO_3 (carbonic acid). Development with dry ice does not usually achieve the results obtained with other, more controlled development methods. Usually little development results from the carbonic acid formed. The injection of weak acid and base solutions generates turbulence similar to dry ice as the chemicals neutralize.

Caution should be exercised in using turbulence-producing chemicals, since occasionally the gas generated will blow nearly all the water from the well. The resulting pressure differential could cause the casing (or occasionally the screen) to collapse. Reactions are generally rapid and of relatively short duration. Cleanout and other development operations complete the work.

Explofracing

In low-production metamorphic and igneous rock wells, explofracing may increase yields. "Explofracing" is a term that covers the use of small explosive charges tied to primacord to produce a rapid series of explosions that enlarge the well borehole and increase the fracturing of the rock. Production is frequently improved by 2 to 10 times with this technique if the rock is already jointed or faulted.

Hydrofracing

Hydrofracing (hydrofracturing) is another method for improving yield from fracture systems in consolidated rock aquifers. The producing zone is pressurized by water to an amount in excess of its overburden pressure or just enough to increase the interconnections between the nearby water-bearing fractures and the borehole.

14.11 FINAL DEVELOPMENT

The effectiveness of preliminary development is a primary factor in the methods used and the time required to complete final well development. If development objectives have been nearly attained, final development pumping can be concluded relatively quickly. However, in many cases, an extensive program of pumping, surging, and backwashing may be necessary. It should be noted that development pumping alone may not be sufficient to fully develop all the aquifers. Zones with higher hydraulic conductivity and piezometric levels will more readily develop with pumping, whereas zones with less favorable hydraulic characteristics will be relatively unaffected. Preliminary development techniques are more selective to each aquifer and are critical to maximizing final well efficiency and specific capacity.

Since it is necessary to pump at varying rates during final development, the use of a line-shaft pump driven by an internal combustion engine is preferred. The pump is installed to a greater depth than the planned production pump setting—often at least 50 ft depending upon the depth of pump housing casing. The development pump is also used for the final pump test.

Development pumping rates usually range from a low percentage of the expected well capacity to at least 150% of the anticipated permanent pump discharge rate. To achieve this performance range, the pump bowl assembly performance curve must be fairly flat.

Development and Test Water Disposal

Development and test water disposal must be considered according to well site and other factors. This subject is covered in Chapter 6.

Development Pumping in Alluvial Formations

Development pumping of wells producing from alluvial aquifers is started at about 20% of expected well capacity (or at the lowest amount the test pump will produce). Usually initial pumping even at this low rate will produce particulate or color, but if none is observed, production is slowly increased. After sand and color in the discharge is observed, the discharge rate is maintained while the water level is monitored. Pumping is continued until the discharge clears, and sand content has dropped considerably. When a fairly stable pumping level and production rate are reached, the well is surged. Surging with a turbine pump is the process of slowing or stopping the pump, permitting the water in the discharge and column pipe to flow back into the well.

The point of discharge is usually near the well, so that a limited amount of water is stored in the discharge line. In some locations a very long pipeline must be used to conduct the water to a remote disposal site. In this situation it may be necessary to install quick-action valves and a vacuum break to save time and avoid potential development problems caused by returning a large quantity of water into the well at an inappropriate time in the development process.

In most wells recovery occurs in two stages after the pump is stopped. The first stage consists of a rapid return to a level often referred to as the 5-min recovery level. The second stage is the time required for full recovery to the initial static water level.

Since the time for full recovery can be as long as that spent pumping, the pump is normally restarted following the first-stage water level recovery. When the discharge clears, the process is repeated several times without increasing the pumping rate. If moderate surging no longer produces much sand, multiple surges at the same rate may be done to increase the washing action.

Moderate surges during the early stages of development pumping are often executed by slowing the pump to gradually lower the water level in the column pipe until it reaches the 5-min recovery level. The original pump speed is then resumed. As development continues, the severity of surging increases. Severity depends not only on number of surges but also on how they are conducted. Severe surges are created by disengaging the engine clutch and allowing the pump to backspin as the column and discharge pipes empty. This results in a faster change in the velocity of the water entering the well and a more vigorous washing action.

The discharge rate is increased only when the well evidences reasonable stability. Pumping should never be stopped, nor the well surged, while large quantities of sand (plainly visible in the well discharge) are being produced. The discharge rate is incrementally increased, with periodic and/or multiple surging at each stage. In gravel envelope wells, a small quantity of water and surfactant should be added to the envelope during the operation to help lubricate and settle it.

There is no merit in severe multiple surging, particularly at higher pumping rates, without improvement in well performance. Improvement consists of a combination of increase in specific capacity, more rapid clearing of the water produced after surging, and settling of the envelope in gravel envelope wells. Multiple surging should be discontinued if it results in high, long-duration sand production. This is particularly important if there is no settlement of the gravel envelope.

Experience has shown that excessive discharge rates during initial stages of final well development can be very damaging, and the production loss suffered may be irreversible. This apparently occurs because high water velocity through the filter zone "locks in" aquifer fines and drilling debris. Maximum discharge rates should evolve from the program of surging and gradual pumping increases described.

These development pumping procedures apply to porous media aquifers. Wells drilled in limestone, basalt, or other rock aquifers usually do not require this program, and high pumping rates may be initiated early.

Completion of Development. It can be difficult to be certain that the development program has achieved the highest possible well production and specific capacity. Comparison with nearby wells may be misleading, because hydrogeologic characteristics can change substantially in a short distance. Also different sections of the aquifer(s) may have been screened. This question can only be answered by careful analysis of the development records.

Greater specific capacity with higher pumping rates suggests insufficient development. A fully developed well has a higher specific capacity at lower pumping rates and time of pumping, due to the combination of additional well losses and the expanding cone of depression (see Chapter 5).

Other indications of incomplete development are fluctuating or excessive sand content, intermittent cloudy discharge, or continued gravel settlement. If an analysis of the pumping records indicates that the preliminary development is incomplete, the pump should be removed and more preliminary development performed.

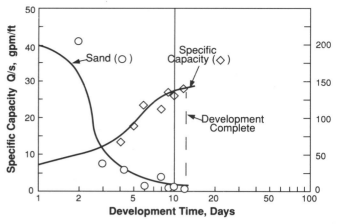

Figure 14.8. Development history.

Specific Capacity as a Measure of Final Well Development

Occasionally specifications require that at the completion of development, the specific capacity remain at a constant value over a designated period of pumping. The specific capacity depends in part on a static water level, and a problem may arise if this level is not clearly defined. When the development and test pump is first installed, the water level is measured and noted. Frequently, the 5-min recovery level may continue to lower throughout development pumping. This is often observed in wells where large quantities of drilling or infiltrated development water have created an artificially high static level, but it may occur for other reasons. Some wells have a higher initial water level than will ever be recorded after pumping. If the specific capacity calculated from the declining 5-min recovery level is used for evaluation, the records will suggest that the development is not complete.

Under these circumstances the original static level should be used for calculating the specific capacity. If the pumping level remains fairly constant, and the completion criteria set forth in this chapter are met, development may be considered complete. Figure 14.8 illustrates a typical development history with clear indications of completion.

Underdeveloped Aquifers. It is sometimes observed that potential producing zones of some wells (usually completed in multiple heterogeneous aquifers) may not initially respond to development. This usually occurs because an aquifer or part of an aquifer (generally an upper zone having higher hydraulic conductivity and/or piezometric level) quickly develops, and the development action concentrates there. However, if these more productive aquifers become depleted, the static water level declines, or a deeper pump setting is required due to a decrease in specific capacity, the underdeveloped aquifer(s) may begin to produce. This is sometimes

noted by unexpected turbidity and increase in sand content. The well's specific capacity may also increase.

If an underdeveloped zone is suspected, isolation development techniques may be employed. This can be particularly effective with air development using isolation tools, or flushing, where development can be concentrated on the designated area. Packers can be installed on the pump suction to isolate deeper zones. However, care must be exercised because packers can become sand locked in the well.

Backwashing

With most wells, pumping and surging at increasing rates is the only final development work performed. However, if the development program was correctly executed but results are not satisfactory because of unexpectedly low well production, a supplemental program may be considered. Pumping and surging is a relatively gentle process. It accomplishes its results through changes in water level which vary the inlet velocity. These variations in velocity break down the bridges of small particles in the aquifer and filter zone, causing fine materials to be pumped into the well.

During a surge the amount of water contained in the pump column and discharge pipes, combined with the well's natural recovery, is generally not sufficient to cause the water level to rise above the static level. The washing action is therefore not as severe as would occur if the flow through the screen were reversed.

A combination of pumping and backwashing is used to achieve this reversal. Supplemental water is added to the well when pumping is stopped for a surge. This water should be free of sand and silt. Backwash is continued up to 10 min after the backspin has stopped, or until the water level stabilizes. Judgment should be exercised concerning backwash duration and the water quantity added, since to some extent success depends upon the frequency of reversing flow rather than on how high a water level is reached. However, it is desirable that the level exceed the 5-min recovery level during each surge.

Backwashing requires a high-capacity water source. Water can be obtained by diverting part of the pumped water into a large tank(s). When the well is surged, the stored water is allowed to flow back down the well through the pump column. A similar technique uses a long elevated discharge line as a reservoir.

As with all development pumping techniques, pumping and backwashing should begin with a low pumping rate and increase gradually as positive results are observed. As long as specific capacity increases, the program should be continued. Although backwashing can be very effective, some formations may require more than 50 to 100 hours of development pumping.

14.12 MEASUREMENTS DURING PUMPING

Complete records must be maintained in order to assess the progress of development pumping. These data are also used to organize the test pumping program. The necessary information includes water levels, production rates, sand content, and turbidity. Other measurements that may be of use later include water conductivity and temperature. Methods for obtaining water level measurements and rate of pump discharge are discussed in Chapter 15.

Sand Content and Turbidity

Standards for sand content and turbidity vary with end use and system design and operation. Excessive sand content in pumped waters can result in plugging of transmission facilities, fouling of irrigation systems, and severe pump wear. Turbidity, usually colloid clay particles, can result in loss of system efficiencies, unpleasant tastes and odors, and occasionally encouragement of growth of slimes and other organisms.

There are no absolute standards for allowable quantities of sand in water. Studies indicate a rather narrow range between acceptable and troublesome quantities, according to end use. Typical sand content limits during continuous pumping are:

1. Wells supplying water for flood-type irrigation, and where the aquifers will not be harmed by producing sand, may produce up to 15 ppm of sand by volume.

2. Wells supplying water to sprinkler irrigation systems, industrial evaporative cooling systems, or other uses where a moderate amount of sand is not harmful may produce up to 10 ppm.

3. Wells supplying water to homes, institutions, municipalities, and most industries may produce up to 5 ppm. Ideally, wells discharging directly into municipal water-supply mains (without passing through reservoirs or settling basins) should produce less than 0.7 ppm of sand to ensure minimum customer complaint and meter fouling (1).

4. Wells supplying water to be in contact with or used for processing food or beverages may produce no more than 1 ppm (2).

Many pump manufacturers state that pumping of sand in any quantity will void bowl efficiency guarantees. However, this is usually impossible to achieve in wells pumping from alluvium. A more realistic limit to avoid excessive wear is a production rate of 5 ppm or less.

Sand production from properly designed and constructed wells will generally average less than 1 ppm during a pumping cycle of at least several hours. Even though inherently stable, many wells will generate markedly more sand during the first few minutes after start-up. Although this will decline to 0 to 1 ppm within 10 to 20 min, depending upon well depth, it may be necessary to pump the first few minutes of production to waste. It is also desirable to pump the well at as constant a rate as possible, consistent with demand and storage capacity and the possible need to cycle the pumping to maintain specific capacity and discharge rate.

During initial development pumping, gross sand content is checked with an Imhoff cone or similar device with appropriate gradations, taking a liter sample of the well discharge from a point of turbulent flow. After settling, the sand content is read in cubic centimeters (cc) and estimated to tenths of a cc, which equates to parts per million. When more accurate measurements of quantities less than 50 ppm are required— for example, during the final testing of a well for acceptance—a Rossum Sand Tester may be used. This device will measure sand content as low as 0.5 ppm when operated over a period of approximately 10 min.

The Rossum Sand Tester (Figure 14.9), is coupled to a small-diameter pipe tapped horizontally in the well discharge close to the pump head or in another turbulent zone. Back pressure should be sufficient to force a 0.5 gal/min sample to flow through the tester. Sand particles are separated from the water and collected in a calibrated tube.

Recording the accumulated quantity of sand in the tube over time enables calculation of the varying sand content of the discharge water. For analysis, measurements are made during steady pumping after a surge.

More detailed analysis requires an automatic particle counter. Currently they are, for the most part, employed to measure very fine fractions of particulate matter contained in liquids where control of granular solids is critical. Such

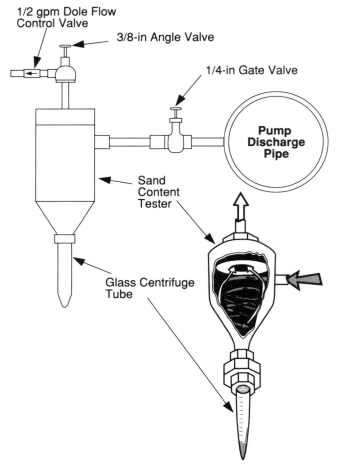

1/2 gpm Dole Flow Control Valve

3/8-in Angle Valve

1/4-in Gate Valve

Pump Discharge Pipe

Sand Content Tester

Glass Centrifuge Tube

Figure 14.9. Rossum sand tester.

liquids include petroleum products, film emulsions, medicines, and water used in demineralization processes.

A particle counter functions by light blockage and light scatter principles. Each particle and its relative size can be recorded in a specific volume of fluid over time. Parts per million of particles by size (mechanical grading analysis) or the total particulate content contained within the fluid is easily calculated.

Excessive Sand Production

Excessive sand production in gravel envelope wells may be due to a void in the filter pack. If so, reconsolidation by swabbing will probably be required. Excessive sand production in naturally developed wells can usually only be controlled by restricting the discharge rate. However, in many of these wells sand production decreases as the well is operated, and the discharge rate can be increased over a period of time.

Turbidity

Turbidity in well water is frequently generated by clay, colloidal, or microorganic particles, which can impart taste, odor, color, or other undesirable characteristics. Turbid water is particularly troublesome in domestic systems and manufacturing processes, but it can also cause problems in irrigation systems where drip equipment is used. In most areas, properly designed, constructed, and developed wells can easily meet turbidity requirements.

Well construction specifications frequently require meeting turbidity standards before acceptance. The usual maximum allowable is five Jackson turbidity units (JTU). Measurement of turbidity in parts per million can also be done. However, this usually requires laboratory analysis.

For accurate turbidity (JTU) measurements one of two instruments, the Jackson candle turbidometer, or the nephelometer, is used. The Jackson turbidometer consists of a long glass tube closed at the bottom end. The tube is placed on a stand under which a lighted candle is placed. Water to be tested is poured slowly into the tube until the candle flame, viewed through the water, can no longer be distinguished. Tables convert the height of the water column into the JTU value.

The Nephelometer is a more sophisticated instrument and has a range of 0.1 to 40 JTU. Measurements are made by placing the water sample between a standard light source and a calibrated photoelectric cell connected to a millivolt meter. The cell and meter are calibrated against a standardized suspension of formazin or other light-diffusing material. Test readings are taken from the millivoltmeter that displays JTU directly or converts to JTU with tables.

Pump operators do not customarily measure turbidity with instruments but record their observations by qualitative visual descriptions, such as cloudy, milky, dirty, and crystal clear.

Safe Pumping Rate

The production characteristics of each well limit the maximum rate at which the well can be safely pumped. The maximum safe pumping rate for both gravel envelope and naturally developed wells may be limited by sand production. In gravel envelope wells, where production is from very fine aquifers, a normally satisfactory filter pack/aquifer ratio may still allow excessive sand production if sufficient drawdown is imposed. Naturally developed wells characteristically pump less sand as they are operated, and the maximum safe pumping rate may then be increased.

The safe pumping rate may be restricted by the depth of the pump housing casing. In some wells a point may be reached where production does not increase even with greater drawdown (critical discharge or breakover pumping rate). Critical discharge occurs when the screened formation is incapable of producing more water because of its internal resistive forces, or when the discharge is equal to or greater than the recharge to the aquifer.

REFERENCES

1. Rossum, J. R. 1954. "Control of Sand in Water Systems." *J. Am. Water Works Assn.* 46, 2.

2. Environmental Protection Agency, Office of Water Supply. *Manual of Water Well Construction Practices.* EPA 570/9-75-001. U.S. Government Printing Office, Washington, DC.

READING LIST

Driscoll, F. G. 1986. *Groundwater and Wells.* Johnson Division, St. Paul, MN.

CHAPTER 15

Well and Aquifer Evaluation from Pumping Tests

15.1 INTRODUCTION

Development of ground water resources frequently includes solving complex problems concerned with both extraction and replenishment. Examples of these problems include optimum spacing of wells, determination of safe yield, artificial recharge, movement of contaminants, and determination of the amount of available storage. These problems can be solved by application of mathematical methods and models. However, the results obtained depend upon use of accurate values of the hydraulic characteristics, and on properly assumed initial and boundary conditions. If they are not correct, computations of ground water flow will be erroneous.

A pumping test is the most useful means of determining hydraulic properties of the well and aquifer system. In 1870 the German scientist Adolph Thiem published the first formula to derive aquifer hydraulic properties through pumping a well and observing the effects in nearby wells (1). Since then, flow of water into pumping wells has been thoroughly researched, and many methods of analysis are now available.

All formulas for analyzing pumping-test data are based on certain assumptions and generalizations. In many cases when erroneous results are obtained, blame is placed on the incorrectness of the formula applied, while the actual cause of error is that field conditions did not satisfy the underlying assumptions which apply to the equation. In this chapter special attention is given to these assumptions and generalizations to allow choosing a method that best fits the specific field condition. Knowledge of the magnitude of deviations from the theoretical condition must be known. In some cases the observed data will have to be corrected before applying the formulas.

This chapter relies on the theories and principles developed in Chapter 5 and is concerned only with pumping tests in porous media. Interpretation examples of the most commonly encountered field conditions are given. A summary of methods discussed is found at the end of this chapter. Further examples are in Kruseman and DeRidder (2).

15.2 TYPES OF PUMPING TESTS

Field pumping tests are used to find aquifer and well parameters and well operational characteristics. In organizing the various combinations of aquifer and test conditions, the following groups are designated:

1. Determination of aquifer parameters
   a. Confined aquifers
      i. Nonsteady-state test methods
      ii. Steady-state test methods
   b. Unconfined aquifers
      i. Nonsteady-state test methods
      ii. Steady-state test methods
   c. Semiconfined (leaky) aquifers
      i. Nonsteady-state test methods
      ii. Steady-state test methods
   d. Corrections and tests for special aquifer conditions
2. Determination of well parameters
   a. Production wells
   b. Injection wells

Within each test group, one or more of the following may be performed:

- Constant rate time-drawdown test
- Recovery test
- Constant rate interference test (use of observation wells)
- Variable rate time-drawdown (step-drawdown) test.

The type of pumping test performed depends on the test purpose, available resources, and site-specific limitations. For example, in an area containing only a single well, it is not possible to calculate aquifer storativity. Similarly, observation wells not screened in the same aquifer as the pumping well cannot provide reliable interference data. Some restrictions may also be imposed by limited water disposal facilities, water quality constraints, or noise restrictions. Obtaining good data from a pumping test involves a carefully planned program, including consideration of the test purpose, assurance of continuous operation, and execution of the necessary steps.

Tests for Well and Aquifer Parameters

Time Drawdown and Recovery Tests. Pumping tests for determination of aquifer parameters fall into two general categories: nonsteady state and steady state. In both categories the pumping rate is held constant throughout the duration of the test. In nonsteady-state (time varying) tests, water-level changes in response to a constant pumping rate are measured over a period of time. Nonsteady-state tests can determine storativity, transmissivity, and the leakage coefficient.

In steady-state tests, pumping is continued until near-equilibrium conditions are reached (i.e., negligible change in water levels with time). Steady-state tests approach only semi- or quasi-steady-state conditions, as true equilibrium may never be obtained because of ever-present aquifer recharge or discharge. Steady-state tests may be used to find aquifer transmissivity and, in some cases, the leakage coefficient.

In constant-rate tests, two time-dependent cycles are recognized: drawdown and recovery. The drawdown cycle is the time period between the beginning and end of pumping, whereas the recovery cycle is measured from end of pumping to completion of all test measurements. Completion of test measurements does not imply full recovery, rather only that measurements have continued until establishment of a predictable trend.

Time-drawdown pumping tests are nonsteady-state tests that measure declining water levels as a function of time after the start of pumping. Drawdown is calculated from the difference between static and dynamic water levels. These tests measure water levels in the pumping well (time drawdown) and/or water levels in observation wells located at varying distances from the pumping well, usually less than 500 ft (distance drawdown).

During the drawdown cycle, water levels decline with time as the cone of depression expands. During recovery, water levels rise (i.e., recover) as formation water replenishes the cone.

Recovery tests are nonsteady-state tests that measure depth to water against time after pumping has stopped. A "residual" drawdown and "calculated" recovery are then determined from the measurements.

Full test evaluation involves the analysis of data obtained from both cycle periods. In general, data obtained during the recovery period are more reliable due to the lack of water-level fluctuations caused by discharge variations.

Interference Tests. The measurements of water-level changes in observation wells are known as "interference measurements." Unlike pumping well measurements, interference measurements contain no turbulent flow components. Since equations for formation parameters (i.e., transmissivity) do not consider turbulent-flow loss, interference data are preferred. Reliable field interference tests dictate that observation wells be completed in the same aquifer(s) as the pumping well. Determination of transmissivity from steady-state testing requires data from at least one observation well in addition to the pumping well. Calculation of storativity and the leakage coefficient also requires data from interference tests. Various pumping test types are illustrated in Figure 15.1.

As a rule, the more observation wells available for measurement during a test, the more reliable the information obtained on aquifer characteristics. If the observation wells are oriented in different directions away from the pumping well, aquifer anisotropy can be determined. Unfortunately, wells that may be used for observation purposes are unavailable in most cases, and piezometers (wells drilled primarily for observation purposes) must be constructed.

Ideally piezometers should be placed in four quadrants surrounding the well at radial distances ranging from 10 to 1000 ft. Spacing between the wells should be closer nearer the pumping well where drawdown changes are greatest.

As the shape of the cone of depression depends upon both time and aquifer parameters, placement of observation wells should consider both. For example, water level changes in unconfined aquifers may be measurable within only relatively short distances from the pumping well (e.g., 100 to 500 ft), whereas confined aquifers may show significant changes in water level hundreds to several thousands of feet away.

Cones of depression in aquifers having low transmissivities are steep and limited, contrasted with the broad, flat cones characteristic of highly transmissive formations. This is illustrated in Figure 15.2. In addition, if hydrologic boundaries are known or suspected, it is desirable to place observation wells between the boundary and the pumping well. Knowledge of aquifer type also is important in observation well placement.

In some cases production wells close enough to the pumping well to show measurable interference effects may be used for observation measurements, provided they are screened in the same aquifer(s). If not, they may reflect different hydraulic effects, leading to erroneous results.

Variable-Rate Test on Pumping Wells. A step-drawdown test measures changes in drawdown corresponding to changes in discharge. It is performed to determine production, drawdown, and well efficiency relationships. Step-drawdown test data are important in the design of the pumping plant, as detailed in Chapter 16.

Injection Well Tests. Injection tests are conducted in areas where use of injection wells is considered (Chapter 19). The test procedures and equations used for analysis are similar to those of drawdown tests, except that injection pressure head (difference between the static water level and the water level during injection) is substituted for drawdown, and injection rate for pumping rate.

15.3 TEST PROCEDURE

Test Pump

Requirements. The pump should be set below the maximum anticipated pumping level, although above the screen.

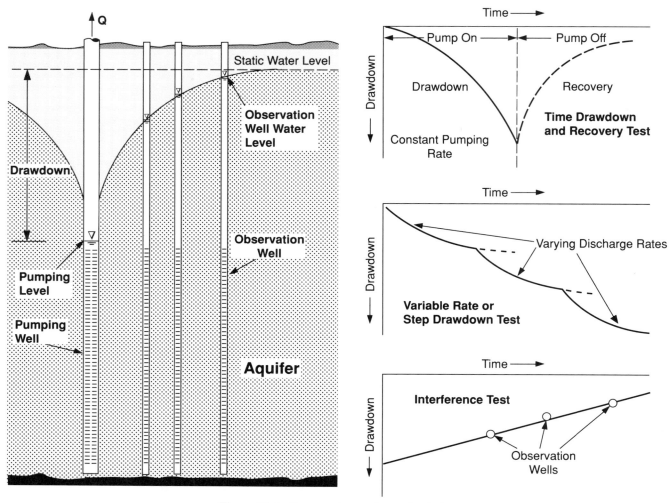

Figure 15.1. Common types of pumping tests.

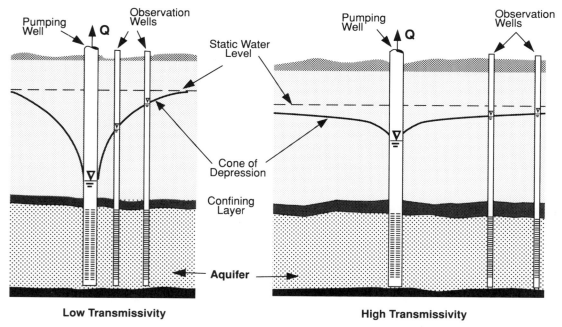

Figure 15.2. Observation well placement.

If it is powered by an internal combustion engine, there should be enough fuel on the site to complete the test. To control discharge, a regulating valve should be provided for minor adjustments. Accurate devices to measure pump discharge and water level in the well must be installed.

Discharge of Pumped Water. When producing from a shallow unconfined aquifer, care should be taken that test water does not re-enter the well and affect test results. It may be necessary to discharge water 300 to 500 ft or more away, depending upon site conditions. In pumping tests where the quality of water discharged exceeds allowable contaminant levels, provision must be made for storage or transport of water pumped during the test.

Personnel Requirements

The number and responsibility of personnel involved in the test depends on its purpose and duration. Tests involving rapidly changing water levels with multiple observation wells require more personnel than periodic measurements at the end of a single-well constant-rate test. Assigning individual responsibilities and performing trial measurements and calculations are done before the test begins. Pre-test trials provide excellent training while working out logistic and other problems.

Pre-Test Measurements

Prior to conducting a pumping test, it is important to understand hydrologic influences in the area that may affect results. In general, all pertinent information related to water-level fluctuations should be noted. This includes the discharge rate, water levels, pumping schedules, interference effects of nearby wells, location of hydrologic boundaries (e.g., surface-water bodies such as canals or streams in hydraulic continuity with the aquifer being tested), and atmospheric pressure changes. Initial measurements record any transient trends that may affect test results. The time period required to establish these trends is related to field conditions and may vary from several hours to several days.

Test Measurements

Test Duration and Time Interval for Measurements. The cone of depression surrounding a pumping well expands at a rate dependent on time since start of pumping, aquifer characteristics, and recharge. The length of a pumping test therefore depends upon both the test purpose and the hydraulic properties of the aquifer.

At the beginning of the test, the cone expands rapidly as aquifer storage in the immediate vicinity of the well is depleted. As pumping continues, horizontal expansion of the cone slows as larger and larger volumes of water become available.

Quantitatively this time and distance expansion may be stated as

$$s \propto \log\left(\frac{t}{r^2}\right) ,$$

where

$s =$ drawdown at distance (r) and time (t) after start of pumping.

To the inexperienced observer, this slowing is often mistaken for the steady-state condition (equilibrium). However, steady-state conditions will not occur until the cone of depression has expanded to the point where recharge to the cone equals the discharge of the well. In some wells equilibrium may be reached within a few hours, whereas others require days, weeks, or longer. It may not be necessary to continue the test until steady-state conditions are reached as nonsteady-state methods are available for analysis. However, in either case, for design of important and costly facilities such as municipal water-supply wells, pumping until equilibrium (or near-equilibrium) conditions are reached is recommended. Long-term pumping may also reveal the presence of nearby hydrologic boundaries that may affect operation of the well.

The essential measurements taken during any pumping test are time, depth to water level, and discharge rate. Measurements of the time of start, stop, and pumping interval must be made with reasonable accuracy (± 0.1 min). Any irregular events, such as pump failure and restart, occurring during the test should be noted. Actual time of day as well as cumulative test times are taken.

Water levels decline or recover most rapidly immediately following a change in pumping rate. For this reason frequent measurements are required during the first few minutes to tens of minutes from the beginning of each step in a step-drawdown test, and after start and stop of pumping in a constant-rate test.

Table 15.1 presents a practical range of time intervals for measurement. The time intervals in the table are for the pumping well and nearby observation wells. Frequency of

TABLE 15.1 Time Intervals for Pumped Well and Nearby Piezometers

Time since Pumping Started or Stopped	Time Interval
0–5 min	1 min
5–60	5 min
60–120	15 min
120–6 hr	60 min
6 hr–end of test	2 hr

measurement is less important in observation wells located considerable distances from the pumped well.

A good field practice during the test period is to plot depths to water level (or drawdown) versus time on semi-logarithmic paper as they are recorded. These plots can be used to determine the frequency of a particular measurement, as well as the length of the test. A similar plot made during recovery will also guide frequency of measurements.

Measuring Pump Discharge Rate. The discharge rate should be carefully regulated during the test to ensure consistency of pumping. When adjustment is necessary, it is more accurately accomplished by regulating a gate valve in the discharge pipe rather than changing the speed of the pump.

Many different types of flow-measuring devices are available; however, only those most commonly used for pumping tests on water wells are discussed here. Discharge-rate accuracy should be within ±2%.

A commonly used discharge-measuring device, the propeller meter, is placed in a straight section of the discharge pipe. Five pipe diameters of straight approach are required to guarantee ±2% accuracy. The meter averages flow by counting propeller revolutions per time period; these revolutions are proportional to velocity. A full pipe is required, as illustrated in Figure 15.3.

Most propeller meters have dials for direct reading of discharge and the cumulative amount pumped. They are calibrated for various pipe sizes. Accuracy is affected by air in the discharge water and sand buildup in the bottom of the discharge pipe.

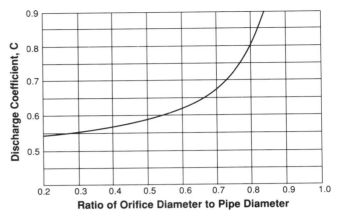

Figure 15.4. Orifice Weir constants.

Another commonly used device to measure discharge during pumping is the circular orifice weir. Flow through the weir is a function of the head as measured in a manometer, and the ratio of pipe to orifice diameter. The equation for flow through an orifice weir with a free discharge is

$$Q = CA\sqrt{2gh} = 8.025\, CA\sqrt{h}$$

where

Q = flow through orifice [gpm],

A = cross-sectional area of the orifice [in.^{2}],

h = water level in piezometer tube [in.],

C = discharge coefficient (a function of orifice and pipe diameter).

g = gravitational constant (32.2 ft/sec^2)

Figure 15.3 shows an orifice installed at the free discharge end of a pipe, and Figure 15.4 a plot of the discharge coefficient versus the ratio of orifice to pipe diameter.

For very small discharge rates (less than 50 gpm), calculation can be made from measurements of the time required to fill a container of known volume. Typically, a 1- or 5-gal bucket is placed under the discharge stream, and the discharge rate is calculated from the time required to fill the bucket.

Measuring Water Level. Accuracy of depth to water-level measurements during a pumping test should be within 0.05 ft in observation wells and 0.1 ft in the pumping well. Accuracy depends primarily upon the measuring device, but sometimes technique and test conditions are a controlling factor, particularly in pumping wells.

A simple and reliable method for measuring depth to water in observation wells is through use of an ordinary steel tape with markings in tenths and hundreths of feet. The approximate depth to water must be known, and ideally

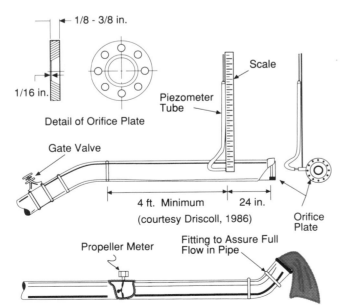

Figure 15.3. Discharge measuring devices.

the lower portion of the tape is covered with an indicator paste which changes color upon contact with water. In most cases, however, dirt is used as an indicator.

Depth to water is measured by lowering the tape to a predetermined level and holding on an even foot mark at the surface reference point. The difference between the reference point reading and the point of water-level contact, as shown by the indicator, is the depth to water. The wetted tape method is seldom used in a pumping well because of fluctuating water levels, cascading water, or pump-column leaks.

A refinement of the steel tape method of measurement uses a two-conductor electrical wire, with an internal supporting cable to prevent stretching. Alternately, some electric sounders utilize one conductor wire, with the second lead grounded to the well casing. When contacted, the water in the well completes an electrical circuit, incorporating a bell, a light, or a meter in series with a battery. The wire is generally calibrated at regular intervals (usually every 5 ft) so that actual depth to water is found by measuring and adding the length from the surface reference point to the uppermost calibration marker in the well. These devices may also be used with a geophysical type reel containing direct reading depth counters.

Electric sounders are also affected by cascading water or pump column leaks. However, these problems can be avoided by use of a small-diameter pipe (stilling well) placed alongside the pump column.

Another technique used for measuring depth to water relies on the water pressure–depth relationship. A pressure head of 1 lb/in.^{2} is exerted by 2.31 ft of freshwater. An airtight tube is set to a known depth that is greater than any anticipated drawdown. Depth to water at any time is calculated from the difference between the length of air line and the pressure required to evacuate the line. This method, commonly used in pumping wells, is calibrated against an electric sounder or steel tape.

Continuous water-level recording devices consist of a float, transducer, or a continuously reading air line (also known as a "bubbler"). These devices are connected to a recorder which may range from light-sensitive film and paper charts to local or remote computer storage. The degree of sophistication of continuous recording devices is limited only by cost, and technology is available for use in a wide range of applications. Advantages are accuracy, continuous records, computer processing of data, and labor savings.

An example of a continuous recording device is the pressure transducer. Changes in transducer submergence cause corresponding changes in electrical output (current or voltage). Typically, current is measured in the 4 to 20 ma (milliamperes) range and is equated to depth of water through an appropriate rating curve. Transducers can be easily interfaced to supervisory control and data acquisition (SCADA) systems.

Figure 15.5 shows commonly used water-level measuring devices.

Recording Test Data

Pumping test forms vary widely in style and format. The example shown in Figure 15.6 is typical. In addition to test data it is important to record any information that might be needed for later analyses. This includes measurement of distances to piezometers, diameters of wells, discharge pipes, etc.

15.4 PUMPING TEST INTERPRETATION EXAMPLES

In this section some important and common methods for evaluation and use of pumping test data are described. It is beyond the scope of this chapter to review all methods available. For descriptions of the mathematical derivations of the formulas, the reader is referred to Chapter 5 and the references cited there.

The examples assume that the aquifer type and initial and boundary conditions are known. Certain sequential steps are common in every analysis procedure:

1. Classification and compilation of data, including converting time, discharge, and water level to conventional units (e.g., min, gpm, and ft).
2. Correcting drawdown data for regional changes in water levels.
3. Graphical plotting of data.
4. Calculating well and aquifer parameters from measurements made from graphs using appropriate equations.

The examples presented in this chapter are based on typical field test data, and in most cases numerical calculations are graphically illustrated. In general, the calculation of parameters requires only solving simple equations. For the more complex examples, step-by-step procedures are presented.

Basic Assumptions

When applying the methods presented in this section, it is important to consider carefully the underlying assumptions from which the equations are derived. Even though most of the assumptions do not occur in nature, in the majority of cases the formulas are adequate and reliable hydraulic characteristics can be obtained.

Some assumptions are more important than others. For example, the assumption of constant aquifer thickness is very important in calculation of transmissivity. Aquifer thickness generally does not change significantly within the cone of depression, but if major variations occur due to lithologic change, they should be recognized.

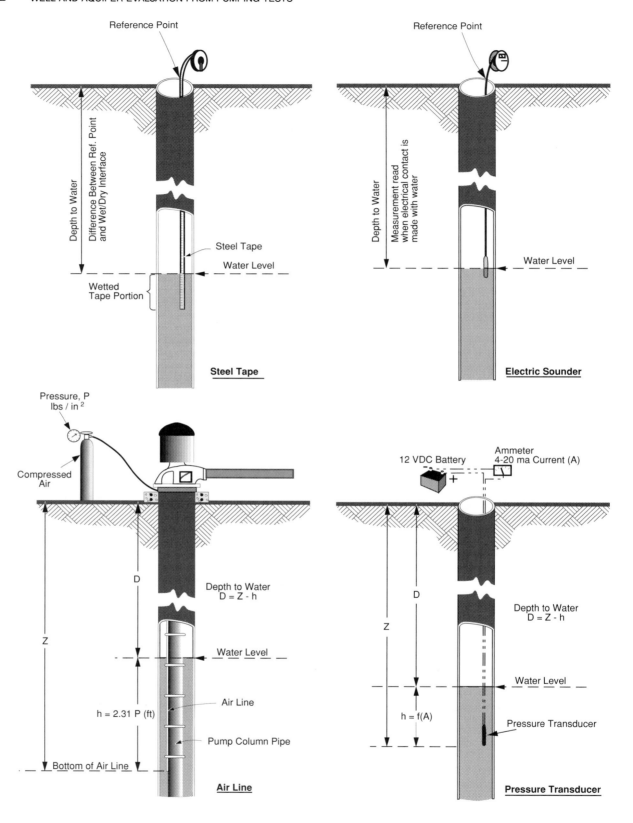

Figure 15.5. Water level measuring devices.

Well / Aquifer Test Data Sheet   Sheet No. _____

Circle one: Test Data for **PUMPING WELL   OBSERVATION WELL**   Well No:_____

Well Location:_____

Well Owner:_____Distance to pumping well: _____

SWL Depth:_____ Reference pt. elev.: _____Test Date:_____

Data File:_____

Circle one: **STEP DRAWDOWN   CONSTANT RATE   RECOVERY**

Time of Day	Time min	Depth to Water, ft	Drawdown ft	Discharge gpm	COMMENTS

Figure 15.6. Well/aquifer test data sheet.

The basic assumptions are:

1. The aquifer material is assumed to consist of porous media with laminar flow obeying Darcy's law.

2. The aquifer is considered to be homogeneous, isotropic, of infinite areal extent, and constant thickness over the area influenced by the pumping test.

3. Unless otherwise accounted for, water is released from (or added to) internal aquifer storage instantaneously upon change in water level.

4. Storage in the well borehole is assumed negligible.

5. The pumping well penetrates the entire aquifer and receives water from the entire thickness by horizontal flow.

6. The slope of the water table or piezometric surface is assumed flat during the test, with no natural (or other) recharge occurring that would affect test results.

7. The pumping rate is assumed constant throughout the entire time of pumping in a constant-rate test—and during each particular discharge step in a variable-rate test.

8. Time measurements are referenced to the start of pumping for drawdown tests, and to the end of pumping for recovery tests.

9. Water levels measured in observation wells are assumed to reflect the same hydraulic conditions encountered in the pumping well.

10. Water-level fluctuations caused by interference from nearby wells, or other causes (e.g., tidal influences) are considered insignificant (or correctable) during the duration of the test.

Calculation of Aquifer Parameters

Theis' Type-Curve Method. When, in addition to the basic assumptions, the aquifer is confined and flow to the well is nonsteady state, the following method may be applied. In 1935 Theis outlined a graphical procedure for evaluating transmissivity and storativity in confined aquifers (3). The method involves matching field data with a "type curve" calculated from the Theis equation. Type curves may be used with field data curves to solve for aquifer parameters. Theis' solution for drawdown in a confined aquifer is written

$$s = \frac{114.6Q}{T}W(u) \ , \qquad (15.1)$$

where

s = drawdown [ft],
Q = discharge rate [gpm],
$W(u)$ = well function of Theis,
T = transmissivity [gpd/ft],

and

$$u = \frac{1.87r^2S}{Tt} \ , \qquad (15.2)$$

where

t = time [days],
r = distance from pumping well [ft].
S = storativity

Rearranging terms of Equation 15.2 and comparing with Equation 15.1 yields

$$\frac{r^2}{t} = \frac{Tu}{1.87S} \ , \qquad (15.3)$$

$$s = \frac{114.6Q}{T}W(u) \ . \qquad (15.4)$$

The similarity between Equations 15.3 and 15.4 is apparent and plots of s versus r^2/t and $W(u)$ versus u differ by only constant factors.

The following example illustrates the graphical procedure suggested by Theis, which requires superimposing field data (s vs r^2/t) upon a plot of $W(u)$ versus u (type curve). The coordinate axes of the two curves are kept parallel. The field data curve is adjusted until a satisfactory match is achieved. An arbitrary match point is then selected and transmissivity and storativity determined. Figure 15.7a illustrates this procedure.

The match point data from Figure 15.7a are:

$W(u)$ = 1.8,
u = 0.1,

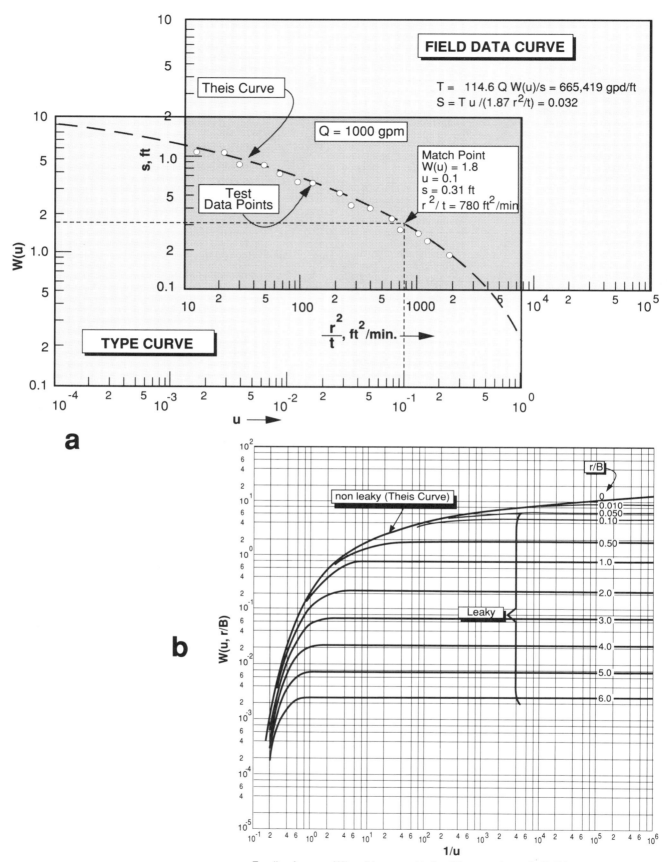

Family of curves W(u, r/b) versus 1/u for different values of r/B (After Walton, 1962).

Figure 15.7. Example of Theis' method.

s = 0.31 ft,
r^2/t = 780 ft²/min (1,123,200 ft²/day),
Q = 1000 gpm.

Substituting the match point data into Equations 15.4 and 15.3 yields

$$T = \frac{114.6QW(u)}{s} = 665,419 \text{ gpd/ft} , \quad (15.5)$$

$$S = \frac{Tu}{1.87r^2/t} = 0.032 . \quad (15.6)$$

Step-by-Step Procedure

1. Construct a type curve of $W(u)$ versus u on log paper, from Table 15.2 where $r/B = 0$.

2. Plot a field-data curve using observed values of s versus r^2/t (field-data curve) on same-scale log paper.

3. Superimpose the field-data curve on the type-curve sheet, keeping coordinate axes parallel to a point of best fit of observed points to the type curve.

4. Select an arbitrary "match point" and record coordinates s, r^2/t, and $W(u)$, and u. Substitution of the coordinates into Equations 15.5 and 15.6 yields values of T and S.

Jacob's Straight-Line Method Applied to Time Drawdown. For the following three examples of Jacob's straight-line method, as in the case of the Theis method the following additional assumptions are required:

- The aquifer is confined.
- Flow to the well is nonsteady state.

TABLE 15.2   Values of the Function $= W(u, r/B) = \int_u^\infty (1/y) \exp(-y - r^2/4B^2 y) \, dy$ (11)

u	\multicolumn{12}{c}{r/B}
	0	0.01	0.05	0.10	0.5	1.0	2.0	3.0	4.0	5.0	6.0
0	∞	9.4425	6.2285	4.8541	1.8488	0.8420	0.2278	0.0695	0.0223	0.0074	0.0025
.000001	13.2383										
.000002	12.5451										
.000003	12.1397	9.4425									
.000004	11.8520	9.4422									
.000005	11.6289	9.4413									
.000006	11.4465	9.4394									
.000007	11.2924	9.4361									
.000008	11.1589	9.4313									
.000009	11.0411	9.4251									
.00001	10.9357	9.4176									
.00002	10.2426	9.2961									
.00003	9.8371	9.1499									
.00004	9.5495	9.0102									
.00005	9.3263	8.8827									
.00006	9.1440	8.7673									
.00007	8.9899	8.6625	6.2285								
.00008	8.8563	8.5669	6.2284								
.00009	8.7386	8.4792	6.2283								
.0001	8.6332	8.3983	6.2282								
.0002	7.9402	7.8192	6.2173								
.0003	7.5348	7.4534	6.1848	4.8541							
.0004	7.2472	7.1859	6.1373	4.8539							
.0005	7.0242	6.9750	6.0821	4.8530							
.0006	6.8420	6.8009	6.0239	4.8510							
.0007	6.6879	6.6527	5.9652	4.8478							
.0008	6.5545	6.5237	5.9073	4.8430							
.0009	6.4368	6.4094	5.8509	4.8368	1.8488	0.8420	0.2278	0.0695	0.0223	0.0074	0.0025

(*Table continues on p. 286.*)

TABLE 15.2 (*Continued*)

u	r/B										
	0	0.01	0.05	0.1	0.5	1.0	2.0	3.0	4.0	5.0	6.0
.001	6.3315	6.3069	5.9711	4.8292	1.8488	0.8420	0.2278	0.0695	0.0223	0.0074	0.0025
.002	5.6394	5.6271	5.4516	4.7079							
.003	5.2349	5.2267	5.1084	4.5622							
.004	4.9482	4.9421	4.8530	4.4230							
.005	4.7261	4.7212	4.6499	4.2960							
.006	4.5448	4.5407	4.4814	4.1812							
.007	4.3916	4.3882	4.3374	4.0771							
.008	4.2591	4.2561	4.2118	3.9822	1.8488						
.009	4.1423	4.1396	4.1004	3.8952	1.8487						
.01	4.0379	4.0356	4.0003	3.8150	1.8486						
.02	3.3547	3.3536	3.3365	3.2442	1.8379						
.03	2.9591	2.9584	2.9474	2.8873	1.8062	0.8420					
.04	2.6813	2.6807	2.6727	2.6288	1.7603	8418					
.05	2.4679	2.4675	2.4613	2.4271	1.7075	8409					
.06	2.2953	2.2950	2.2900	2.2622	1.6524	8391					
.07	2.1508	2.1506	2.1464	2.1232	1.5973	8360					
.08	2.0269	2.0267	2.0231	2.0034	1.5436	8316					
.09	1.9187	1.9185	1.9154	1.8983	1.4918	8259					
.1	1.8229	1.8227	1.8200	1.8050	1.4422	0.8190	0.2278				
.2	1.2227	1.2226	1.2215	1.2155	1.0592	7148	2268	0.0695			
.3	0.9057	0.9056	0.9050	0.9018	0.8142	6010	2211	694			
.4	7024	7024	7020	7000	6446	5024	2096	691			
.5	5598	5598	5595	5581	5206	4210	1944	681	0.0223		
.6	4544	4544	4542	4532	4266	3543	1774	664	222		
.7	3738	3738	3736	3729	3534	2996	1602	639	221		
.8	3106	3106	3105	3100	2953	2543	1436	607	218	0.0074	
.9	2602	2602	2601	2597	2485	2168	1281	572	213	73	
1.0	0.2194	0.2194	0.2193	0.2190	0.2103	0.1855	0.1139	0.0534	0.0207	0.0073	0.0025
2.0	489	489	489	488	477	444	335	210	112	51	21
3.0	130	130	130	130	128	122	100	71	45	25	12
4.0	38	38	38	38	37	36	31	24	16	10	6
5.0	11	11	11	11	11	11	10	8	6	4	2
6.0	4	4	4	4	4	4	3	3	2	2	1
7.0	1	1	1	1	1	1	1	1	1	1	0
8.0	0	0	0	0	0	0	0	0	0	0	

Jacob noted that for small values of u ($u \leq 0.05$), the higher-order terms in the Theis equation become negligible, resulting in the following equation (4):

$$s = \frac{264Q}{T} \log\left(\frac{0.3Tt}{r^2 S}\right) \quad \text{(Jacob's equation)} \ . \quad (15.7)$$

Jacob's approximation formula is valid if $t \geq 5r^2 S/T$. This is the case for most field problems of practical interest.

The Jacob method is one of the most commonly used analytical procedures to find transmissivity in confined aquifers and, with a correction, in unconfined aquifers. The solution is based on measuring drawdown in an observation well at two different times along the best-fit straight line within the valid time range. Jacob's equation (Equation 15.7) may be written in general form as

$$s = \beta + \alpha \log t \ ,$$

where

$$\beta = \alpha \log(0.3T/r^2 S),$$
$$\alpha = 264Q/T.$$

The difference (Δs) between two drawdowns (s_1, s_2) taken at times (t_1, t_2) can be expressed as

$$\Delta s = s_2 - s_1 = (\beta + \alpha \log t_2) - (\beta + \alpha \log t_1)$$
$$= \alpha \log(t_2/t_1) \ .$$

Substituting α,

$$\Delta s = \frac{264Q}{T} \log\left(\frac{t_2}{t_1}\right) \ . \qquad (15.8)$$

A semilogarithmic plot of drawdown versus time will yield the transmissivity (T) from Equation 15.8. If the times of measurement are taken a log cycle apart (e.g., 10, 100 or 30, 300, etc), Equation 15.8 reduces to

$$T = \frac{264Q}{\Delta s} \ ,$$

where

Δs = difference in drawdown taken one log cycle apart [ft].

As can be seen from Equation 15.8, calculation of transmissivity is independent of the radial distance (r) and thus data from the pumping well itself (e.g., the first step of a step-drawdown test) may be used. However, caution is recommended as comparisons of field test analyses generally show lower transmissivities obtained from pumping-well data than from observation-well data. The difference is related to the influence of turbulent components in pumping-well drawdowns.

Calculation of aquifer storativity can be made by rearranging Jacob's equation (Equation 15.7) with the following assumption:

$$t = t_0 \quad \text{when} \ s = 0 \ ,$$

and

$$\frac{264Q}{T} \log\left(\frac{0.3Tt_0}{r^2 S}\right) = 0 \ .$$

Solving for storativity,

$$S = \frac{0.3Tt_0}{r^2} \ ,$$

where

t_0 = time where straight-line extrapolation intercepts at $s = 0$ [days].

Figure 15.8 illustrates the use of Jacob's straight-line method to calculate transmissivity and storativity from a pumping test in a confined aquifer.

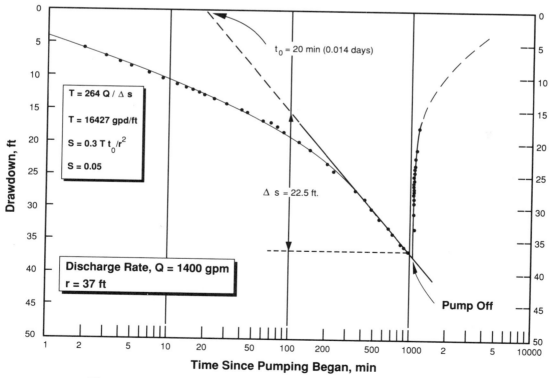

Figure 15.8. Example of Jacob's straight-line method for time-drawdown.

Jacob's Straight-Line Method Applied to Distance Drawdown.

Analysis similar to that used with time-drawdown data is used with distance-drawdown (observation-well) data to derive transmissivity and storativity. Solving Equation 15.7 in the same manner as was done for time duration, the field equations for distance-drawdown analysis are

$$T = \frac{528Q}{\Delta s} ,$$

$$S = \frac{0.3Tt}{r_0^2} ,$$

where

t = time at which drawdown measurements were taken [days],

r_0 = intercept where drawdown is zero [ft].

Figure 15.9 illustrates the use of multiple observation well data on a distance-drawdown plot to calculate formation parameters.

The distance-drawdown plot also may be used to estimate the drawdown just outside the pumping well, thus showing the formation loss and enabling calculation of well efficiency, as discussed later.

Jacob's Straight-Line Method Applied to Recovery.

The procedure used for recovery data analysis is identical to that used during the pumping cycle except that time is measured after the pump has stopped and "calculated recovery" is used instead of drawdown. Calculated recovery is the difference between the extrapolated time-drawdown curve and the residual drawdown. Residual drawdown is the difference between the static water level and depth to water after the pump is stopped. This procedure applied to an observation well is shown in Figures 15.10 and 15.11.

Chow's Method.

Chow proposed a method of deriving aquifer parameters for confined aquifers, similar to Jacob's method, but not limited to the restriction of ($u \leqslant 0.05$) (5). In addition to the basic assumptions,

- The aquifer is confined.
- Flow to the well is nonsteady state.

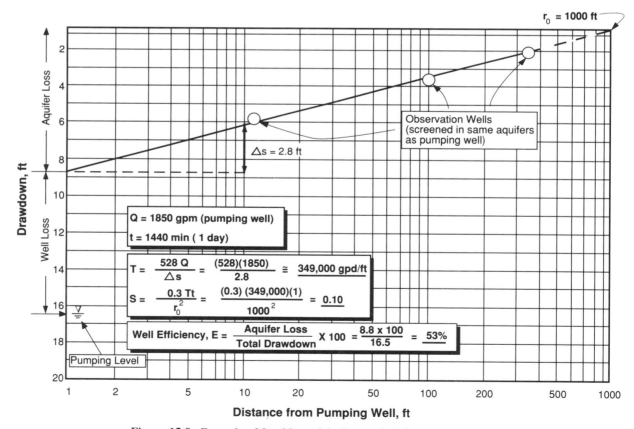

Figure 15.9. Example of Jacob's straight-line method for distance-drawdown.

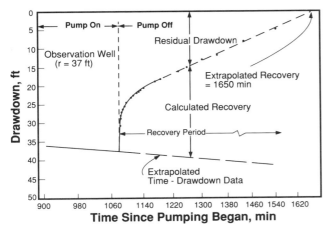

Figure 15.10. Calculated recovery.

Chow's method requires simple computations and involves plotting drawdown against time on semilogarithmic paper as in the Jacob approximation method. An arbitrary point on the best-fit curve is then chosen, and time (t) and drawdown (s) noted (see Figure 15.12). The slope of the tangent of the curve at this point gives drawdown per log cycle of time. A function $F(u)$ is calculated from

$$F(u) = \frac{s}{\Delta s},$$

where

s = drawdown at arbitrary point [ft],

Δs = difference in drawdown per log cycle. [ft]

$W(u)$ and u are found from the Chow type curve (Figure 15.12), and T and S are determined from Equations 15.5 and 15.6, which are the Theis method of solution. In the example of Figure 15.12, the arbitrary point of the time-drawdown field plot requires:

t	= 200 min = 0.14 day,
s	= 28 ft,
Δs/log cycle	= 17 ft,
$F(u)$	= 1.65.

From Figure 15.12,

Q	= 1000 gpm,
u	= 0.02,
r	= 50 ft,
$W(u)$	= 3.3,
T	= $(114.6Q/s)W(u)$ = 13,506 gpd/ft,
S	= $Tut/1.87r^2$ = 0.008.

Thiem's Steady-State Method. In 1906 Thiem proposed a method utilizing two or more piezometers to determine

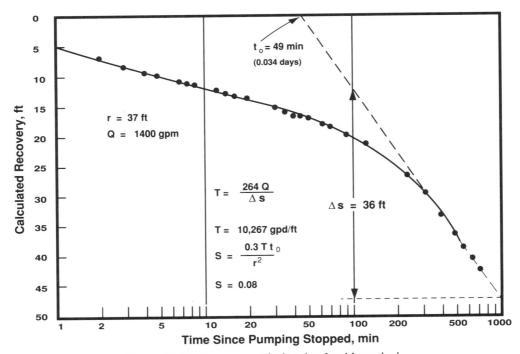

Figure 15.11. Recovery analysis using Jacob's method.

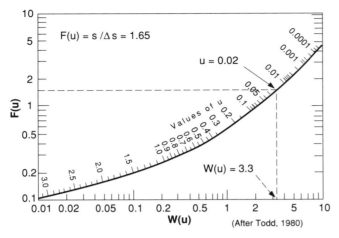

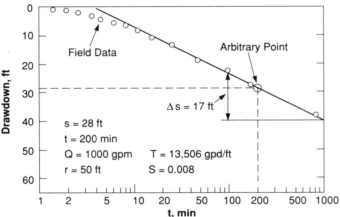

Figure 15.12. Chow's method.

states that, unless accounted for, the release of water from storage occurs instantaneously with the decline in head. In unconfined aquifers this is not true because the drop in the water table is usually faster than the rate at which pore water is released. Pore water drains very slowly by gravity until eventually it reaches the water table, a phenomenon known as "delayed yield," or sometimes "slow drainage."

Pumping tests in unconfined aquifers typically exhibit a three-phase water table decline. Initially, the water level drops rapidly, followed by a period of nearly constant level, and then it resumes a normal decline.

Care should be taken in use of and interpretation of early data before the effects of delayed yield are complete. Transmissivity calculations should not be made until the slow drainage phase is in effect, usually after several hours of pumping. Boulton outlined a type-curve procedure for determining transmissivity and storativity from time-drawdown data in unconfined aquifers, correcting for the delayed yield phenomenon (7, 8). The procedure is similar to that of the Theis curve method. Figure 15.13 shows Boulton's family of type curves for delayed yield.

The solution for early time (before equilibrium from gravity drainage) may be written

$$s = \frac{114.6Q}{KD} W(u_A, r/B) , \qquad (15.10)$$

where

K	=	hydraulic conductivity [gpd/ft^2]
D	=	average saturated thickness [ft],
$W(u_A, r/B)$	=	well function of Boulton,
B	=	drainage factor [ft],

and

$$u_A = \frac{1.87r^2S_A}{KDt} , \qquad (15.11)$$

where

S_A = storativity for early time (first segment),
t = time [days].

For later time, the equations reflecting gravity drainage in equilibrium with pumping are

$$s = \frac{114.6Q}{KD} W(u_y, r/B) , \qquad (15.12)$$

where

$$u_y = \frac{1.87r^2S_y}{KDt} \qquad (15.13)$$

aquifer transmissivity (6). For confined aquifers in steady-state conditions, the discharge can be expressed as

$$Q = \frac{T(h_2 - h_1)}{528 \log(r_2/r_1)} , \qquad (15.9)$$

where

Q	=	discharge rate [gpm],
T	=	transmissivity [gpd/ft],
h_1, h_2	=	elevations of water levels in piezometers located at distances r_1 and r_2 from the pumping well. [ft]

Equation 15.9 is known as the "equilibrium" or "Thiem equation." Substitution of the measured values of Q, r, and h into Equation 15.9 gives the value for aquifer transmissivity.

Boulton's Method with Delayed Yield. When in addition to the basic assumptions the aquifer is unconfined but shows delayed yield and flow to the well that is nonsteady state, Boulton's method may be used. One of the basic assumptions

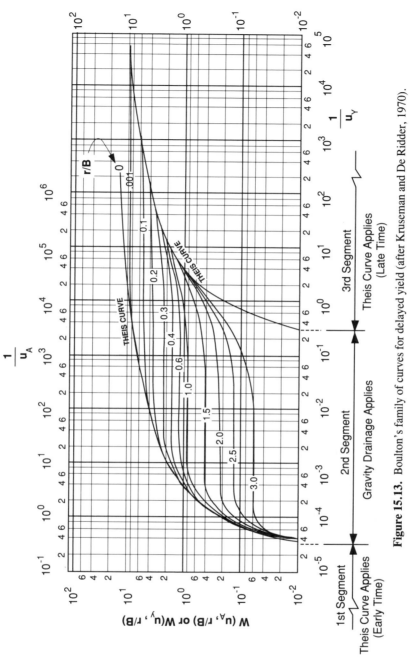

Figure 15.13. Boulton's family of curves for delayed yield (after Kruseman and De Ridder, 1970).

and

S_y = storativity during late time (third segment)

= specific yield.

For the second segment (i.e., start of gravity drainage),

$$1 + \frac{S_y}{S_A} \rightarrow \infty \ ,$$

and the drawdown is expressed as

$$s = \frac{Q}{2\pi KD} K_0(r/B) \ , \qquad (15.14)$$

where

B = drainage factor ($\sqrt{KD/\alpha S_y}$) [ft],

K_0 = modified Bessel function,

$1/\alpha$ = Boulton delay index [days].

For early time data at match point A (Figure 15.14),

r/B	= 0.6,
$1/u_A$	= 10,
$W(u_A, r/B)$	= 1,
s_A	= 0.068 ft,
t_A	= 16 min = 0.0111 days.

Substituting these values in Equation 15.10 yields

$$KD = \frac{(114.6)(160)(1)}{0.068} = 269,647 \ \frac{\text{gpd}}{\text{ft}} \ .$$

From Equation 15.11,

$$S_A = \frac{(0.10)(269647)(0.0111)}{(1.87)(295)^2} = 0.0018 \ .$$

For late time data, match point Z is chosen with

r/B	= 0.6,
$1/u_y$	= 1,
$W(u_y, r/B)$	= 1,
s_Z	= 0.12 ft,
t_Z	= 250 min = 0.174 days,

$$KD = \frac{(114.6)(160)(1)}{0.12} = 152,800 \ \frac{\text{gpd}}{\text{ft}} \ ,$$

$$S_y = \frac{(1)(152800)(0.174)}{(1.87)(295)^2}$$

$$= 0.163 \ .$$

The drainage factor (B) may be calculated as

$$B = \frac{r}{r/B} = \frac{295}{0.6} = 492 \text{ ft} \qquad (15.15)$$

and

$$\frac{1}{\alpha} = \frac{S_y B^2}{KD} = \frac{(0.163)(492)^2(7.48)}{152800} = 1.9 \text{ days} \ . \qquad (15.16)$$

The time (t_{wt}) in which delayed yield effects are negligible may be calculated from

$$t_{wt} \simeq \frac{2.5r/B + 2}{\alpha} \quad \left(\text{if } \frac{r}{B} > 0.5 \right)$$

$$\simeq \frac{(2.5)(0.6) + 2}{0.53} \simeq 7 \text{ days} \qquad (15.17)$$

Step-by-Step Procedure

1. Construct the family of "Boulton type curves" by plotting $W(u_{Ay}, r/B)$ versus $1/u_A$ and $1/u_y$ for a practical range of values of r/B on double logarithmic paper (see Figure 15.13). The left-hand portion of this figure shows the "type A" curves, early time [$W(u_A, r/B)$ vs. $1/u_A$], and the right-hand portion shows the type Y curves, late time [$W(u_y, r/B)$ vs. $1/u_y$].

2. Prepare the observed-data curve on another sheet of double logarithmic paper of the same scale as that used for the type curves, by plotting the values of the drawdown (s) against the corresponding time (t) for a single piezometer at a distance (r) from the pumped well, Figure 15.14.

3. Superimpose the observed-data curve on the early time curves (type A) and, while keeping the coordinate axes of the two curves parallel, adjust until as much as possible of the early time field data fall on one of the type A curves. Note the r/B value of the selected type A curve.

4. Select an arbitrary point A on the overlapping portion of the two sheets of graph paper, and note the values of s, t, $1/u_A$, and $W(u_A, r/B)$ for this point A.

5. Substitute these values into Equations 15.10 and 15.11, and since Q is also known, calculate KD and S_A.

6. Move the observed-data curve until as much as possible of the later time field data falls on the type Y curve with the same r/B value as the selected type A curve.

7. Select an arbitrary point Z on the superimposed curves, and note the values of s, t, $1/u_y$, and $W(u_y, r/B)$ for this match point Z.

8. Substitute these values into Equations 15.12 and 15.13 and, because Q is also known, calculate KD and S_y.

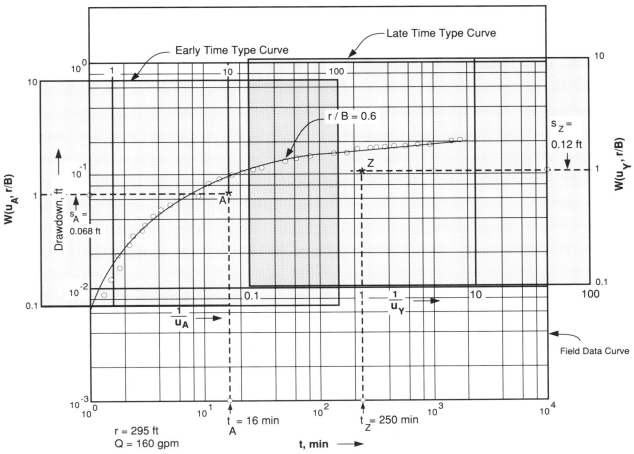

Figure 15.14. Bolton's method for unconfined aquifers (after Kruseman and De Ridder, 1970).

9. Calculate B, α, and t_{wt} from Equations 15.15, 15.16, and 15.17 by first finding the value of B from the value of r/B and the corresponding value of r.

Dupuit's Method. Determination of transmissivity during steady-state pumping in unconfined aquifers is based on the Dupuit equation for unconfined aquifers. The method is similar to the Thiem method and consists of measuring drawdown from two or more observation wells or from an observation well and the pumping well. However, in applying the latter procedure, errors may be introduced if turbulent-loss components are present in the pumping well drawdown.

In an unconfined aquifer the transmissivity (T) may be calculated using the following equation (2, 9):

$$ T = \frac{528Q \, \log(r_2/r_1)}{[s_1 - (s_1^2/2h_0)] - [s_2 - (s_2^2/2h_0)]} , \left[\frac{\text{gpd}}{\text{ft}} \right] $$

where

Q = discharge rate [gpm]

r_1, r_2 = radial distances [ft],

s_1, s_2 = drawdowns at radii r_1 and r_2 [ft],

h_0 = full saturated thickness of aquifer [ft].

Hantush's Inflection Point Method for Leaky Aquifers. When in addition to the basic assumptions, the aquifer is semiconfined and flow to the well is nonsteady state, Hantush's method may be used. In some aquifer tests time-drawdown data may exhibit recharge trends that are indicative of a leaky aquifer system, although some effects of partial penetration (the screened portion of the well less than the full thickness of the aquifer) show similar trends. If partial penetration may be discounted, then analysis for leakage may be done by a simplified graphical method and transmissivity, storativity, and leakance calculated using a straightforward graphical method (10, 11).

Figure 15.15 shows a time-drawdown plot of a well exhibiting typical leakage effects. The shape of the plot shows the characteristic early time data curve ($t < 5r^2S/T$), and an additional decline in the rate of drawdown at later times suggests leakage. The resulting curve assumes a reverse S-shape, with an inflection point between the concave and convex portions.

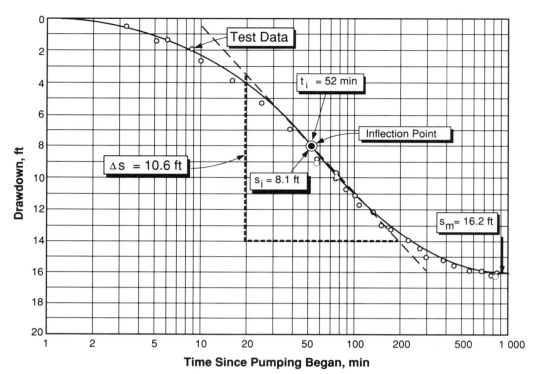

Figure 15.15. Hantush inflection point method for leaky aquifers.

The drawdown at the inflection point (s_i) is calculated as

$$s_i = 0.5s_m = \frac{Q}{4\pi T} K_0\left(\frac{r}{B}\right)$$

$$= \frac{114.6Q}{T} K_0\left(\frac{r}{B}\right) ;$$

further this is written

$$\frac{2.3s_i}{\Delta s} = f\left(\frac{r}{B}\right) = \exp\left(\frac{r}{B}\right) K_0\left(\frac{r}{B}\right) ,$$

where

Q = discharge rate as measured (1995 gpm)

r = distance from pumping well as measured (75 ft)

s_i = drawdown at inflection point (8.1 ft),

s_m = maximum drawdown (16.2 ft),

Δs = slope of straight line at inflection point s_i per log cycle of time (10.6),

K_0 = modified Bessel function of the second kind and order zero.

B = leakage factor = $\sqrt{Kb/K'/b'}$ [ft]

Calculation of $2.3(s_i/\Delta s])$ yields $f(r/B)$ as 1.76. From Table 15.3 r/B and $K_0(r/B)$ are obtained as 0.34 and 1.26, respectively.

The aquifer parameters are then

Transmissivity

$$T = \frac{229Q}{s_m} K_0\left(\frac{r}{B}\right)$$

$$= \frac{(229)(1995)(1.26)}{16.2}$$

$$\approx 36,000 \ \frac{\text{gpd}}{\text{ft}} .$$

Storativity

$$S = \left(\frac{r}{B}\right) \frac{0.27Tt_i}{r^2}$$

$$= \frac{(0.34)(0.27)(36000)(0.036)}{75^2}$$

$$= 0.02 .$$

Leakage Factor

$$B = \frac{r}{0.34} = \frac{75}{0.34} = 221 \text{ ft} .$$

Leakance

$$\frac{K'}{b'} = \frac{T}{7.48B^2} \left(\frac{1}{\text{day}}\right)$$

$$= \frac{36000}{(7.48)(221)^2}$$

$$= \frac{0.10}{\text{days}} .$$

Walton's Type-Curve Method for Leaky Aquifers.
When the aquifer is semiconfined and flow to the well is nonsteady state, Walton presents a solution for the leaky aquifer problem.

Walton's solution involves a type-curve matching technique similar to the Theis and Boulton methods (12). Using a semilogarithmic plot of drawdown against time and imposing this plot on the appropriate r/B type curve of Figure 15.7b yields values of $W(u, r/B)$, $1/u$, s, and t. Transmissivity and storativity are found from

$$T = \frac{114.6Q}{S} W\left(u, \frac{r}{B}\right) \left(\frac{\text{gpd}}{\text{ft}}\right) ,$$

$$S = \frac{Ttu}{1.87r^2} .$$

The leakance (K'/b') may be calculated from

$$\frac{K'}{b'} = \frac{T}{7.48B^2} \left(\frac{1}{\text{days}}\right) .$$

Hantush-Jacob Method for Leaky Aquifers.
When the aquifer is semiconfined and flow to the well is steady state, the Hantush-Jacob method is used.

Hantush and Jacob noted that if $r/B < 0.05$, then the maximum drawdown in a leaky aquifer may be approximated by (13)

$$s_m = \frac{528Q}{T} \log\left(1.12 \frac{B}{r}\right) ,$$

where

s_m = maximum drawdown [ft],
r = distance from pumping well [ft],
B = leakage factor [ft].

The procedure involves constructing a semilogarithmic plot of s versus r and drawing the best-fit straight line through the data points (Figure 15.16). The slope (per log cycle) is equal to $528Q/T$ and the r-intercept at zero drawdown (r_0) equals $1.12B$. Transmissivity (T) and leakance (K'/b') may be calculated from

$$T = \frac{528Q}{\Delta s} \left[\frac{\text{gpd}}{\text{ft}}\right]$$

$$= \frac{(528)(1250)}{2.25} \simeq 293,000 \, \frac{\text{gpd}}{\text{ft}} ,$$

$$\frac{K'}{b'} = \frac{(1.94)10^{-6}T}{r_0^2} \text{ sec}^{-1}$$

$$= \frac{(1.94)10^{-6}(293,000)}{2500^2}$$

$$= \frac{9.1 \times 10^{-8}}{\text{sec}} ,$$

where

r_0 = distance to zero drawdown [ft].

Corrections for Special Test or Aquifer Conditions

Unconfined aquifers and some other conditions that are exceptions to the basic assumptions outlined earlier may be analyzed (with limitations) using the standard equations (14, 15). These exceptions include partially penetrating wells, perched aquifers, hydrologic boundaries, or wells in nonporous media (e.g., fractured rock or cavernous formations).

Corrections for Regional Changes in Water Levels during the Test.
For proper analysis of pumping test data, it may be necessary to correct observed data for effects not accounted for in the analytical methods. An example of such an effect would be tidal influence on a piezometer in a coastal aquifer.

The correction methods involves adding or subtracting water-level deviations measured during the test period to a reference water level (usually taken as the water level at start of the test). Figure 15.17 shows an example of the correction procedure as applied to changes in water level caused by tidal influence.

Unconfined Aquifer Correction.
When drawdown is small relative to saturated thickness, unconfined aquifers may be analyzed using the methods for confined aquifers (15). Specifically, if

$$\frac{s_m}{h_0} \leq 0.02 ,$$

where

s_m = maximum drawdown measured during the test [ft],
h_0 = initial depth of saturation [ft].

TABLE 15.3 Values of the Functions e^x, $K_0(x)$, $e^x K_0(x)$, $-Ei(-x)$, and $-Ei(-x)e^x$ (11)

x	e^x	$K_0(x)$	$e^x K_0(x)$	$-Ei(-x)$	$-Ei(-x)e^x$	x	e^x	$K_0(x)$
0.010	1.0101	4.7212	4.7687	4.0379	4.0787	0.10	1.1052	2.4271
11	1.0111	4.6260	4.6771	3.9436	3.9874	11	1.1163	2.3333
12	1.0121	4.5390	4.5938	3.8576	3.9044	12	1.1275	2.2479
13	1.0131	4.4590	4.5173	3.7785	3.8282	13	1.1388	2.1695
14	1.0141	4.3849	4.4467	3.7054	3.7578	14	1.1503	2.0972
15	1.0151	4.3159	4.3812	3.6374	3.6925	15	1.1618	2.0300
16	1.0161	4.2514	4.3200	3.5739	3.6317	16	1.1735	1.9674
17	1.0171	4.1908	4.2627	3.5143	3.5746	17	1.1853	1.9088
18	1.0182	4.1337	4.2088	3.4581	3.5209	18	1.1972	1.8537
19	1.0192	4.0797	4.1580	3.4050	3.4705	19	1.2093	1.8018
0.020	1.0202	4.0285	4.1098	3.3547	3.4225	0.20	1.2214	1.7527
21	1.0212	3.9797	4.0642	3.3069	3.3771	21	1.2337	1.7062
22	1.0222	3.9332	4.0207	3.2614	3.3340	22	1.2461	1.6620
23	1.0233	3.8888	3.9793	3.2179	3.2927	23	1.2586	1.6199
24	1.0243	3.8463	3.9398	3.1763	3.2535	24	1.2713	1.5798
25	1.0253	3.8056	3.9019	3.1365	3.2159	25	1.2840	1.5415
26	1.0263	3.7664	3.8656	3.0983	3.1799	26	1.2969	1.5048
27	1.0274	3.7287	3.8307	3.0615	3.1452	27	1.3100	1.4697
28	1.0284	3.6924	3.7972	3.0261	3.1119	28	1.3231	1.4360
29	1.0294	3.6574	3.7650	2.9920	3.0800	29	1.3364	1.4036
0.030	1.0305	3.6235	3.7339	2.9591	3.0494	0.30	1.3499	1.3725
31	1.0315	3.5908	3.7039	2.9273	3.0196	31	1.3634	1.3425
32	1.0325	3.5591	3.6749	2.8965	2.9908	32	1.3771	1.3136
33	1.0336	3.5284	3.6468	2.8668	2.9631	33	1.3910	1.2857
34	1.0346	3.4986	3.6196	2.8379	2.9362	34	1.4050	1.2587
35	1.0356	3.4697	3.5933	2.8099	2.9101	35	1.4191	1.2327
36	1.0367	3.4416	3.5678	2.7827	2.8848	36	1.4333	1.2075
37	1.0377	3.4143	3.5430	2.7563	2.8603	37	1.4477	1.1832
38	1.0387	3.3877	3.5189	2.7306	2.8364	38	1.4623	1.1596
39	1.0398	3.3618	3.4955	2.7056	2.8133	39	1.4770	1.1367
0.040	1.0408	3.3365	3.4727	2.6813	2.7907	0.40	1.4918	1.1145
41	1.0419	3.3119	3.4505	2.6576	2.7688	41	1.5068	1.0930
42	1.0429	3.2879	3.4289	2.6344	2.7474	42	1.5220	1.0721
43	1.0439	3.2645	3.4079	2.6119	2.7267	43	1.5373	1.0518
44	1.0450	3.2415	3.3874	2.5899	2.7064	44	1.5527	1.0321
45	1.0460	3.2192	3.3673	2.5684	2.6866	45	1.5683	1.0129
46	1.0471	3.1973	3.3478	2.5474	2.6672	46	1.5841	0.9943
47	1.0481	3.1758	3.3287	2.5268	2.6483	47	1.6000	.9761
48	1.0492	3.1549	3.3100	2.5068	2.6300	48	1.6161	.9584
49	1.0502	3.1343	3.2918	2.4871	2.6120	49	1.6323	.9412
50	1.0513	3.1142	3.2739	2.4679	2.5945	50	1.6487	.9244

$e^x K_0(x)$	$-Ei(-x)$	$-Ei(-x)e^x$	x	e^x	$K_0(x)$	$e^x K_0(x)$	$-Ei(-x)$	$-Ei(-x)e^x$
2.6823	1.8229	2.0147	1.0	2.7183	0.4210	1.1445	.2194	.5964
2.6046	1.7371	1.9391	1.1	3.0042	.3656	1.0983	.1860	.5588
2.5345	1.6595	1.8711	1.2	3.3201	.3185	1.0575	.1584	.5259
2.4707	1.5889	1.8094	1.3	3.6693	.2782	1.0210	.1355	.4972
2.4123	1.5241	1.7532	1.4	4.0552	.2437	0.9881	.1162	.4712
2.3585	1.4645	1.7015	1.5	4.4817	.2138	.9582	.1000	.4482
2.3088	1.4092	1.6537	1.6	4.9530	.1880	.9309	.0863	.4275
2.2625	1.3578	1.6094	1.7	5.4739	.1655	.9059	.0747	.4086
2.2193	1.3098	1.5681	1.8	6.0496	.1459	.8828	.0647	.3915
2.1788	1.2649	1.5295	1.9	6.6859	.1288	.8614	.0562	.3758
2.1408	1.2227	1.4934	2.0	7.3891	.1139	.8416	.0489	.3613
2.1049	1.1829	1.4593	2.1	8.1662	.1008	.8230	.0426	.3480
2.0710	1.1454	1.4273	2.2	9.0250	.0893	.8057	.0372	.3356
2.0389	1.1099	1.3969	2.3	9.9742	.0791	.7894	.0325	.3242
2.0084	1.0762	1.3681	2.4	11.0232	.0702	.7740	.0284	.3135
1.9793	1.0443	1.3409	2.5	12.1825	.0623	.7596	.0249	.3035
1.9517	1.0139	1.3149	2.6	13.4637	.0554	.7459	.0219	.2942
1.9253	.9849	1.2902	2.7	14.8797	.0493	.7329	.0192	.2854
1.9000	.9573	1.2666	2.8	16.4446	.0438	.7206	.0169	.2773
1.8758	.9309	1.2441	2.9	18.1742	.0390	.7089	.0148	.2693
1.8526	.9057	1.2226	3.0	20.0855	.0347	.6978	.0131	.2621
1.8304	.8815	1.2018	3.1	22.1980	.0310	.6871	.0115	.2551
1.8089	.8583	1.1820	3.2	24.5325	.0276	.6770	.0101	.2485
1.7883	.8361	1.1630	3.3	27.1126	.0246	.6673	.0089	.2424
1.7685	.8147	1.1446	3.4	29.9641	.0220	.6580	.0079	.2365
1.7493	.7942	1.1270	3.5	33.1155	.0196	.6490	.0070	.2308
1.7308	.7745	1.1101	3.6	36.5982	.0175	.6405	.0062	.2254
1.7129	.7554	1.0936	3.7	40.4473	.0156	.6322	.0055	.2204
1.6956	.7371	1.0779	3.8	44.7012	.0140	.6243	.0048	.2155
1.6789	.7194	1.0626	3.9	49.4025	.0125	.6166	.0043	.2108
1.6627	.7024	1.0478	4.0	54.5982	.0112	.6093	.0038	.2063
1.6470	.6859	1.0335	4.1	60.3403	.0100	.6022	.0033	.2021
1.6317	.6700	1.0197	4.2	66.6863	.0089	.5953	.0030	.1980
1.6169	.6546	1.0063	4.3	73.6998	.0080	.5887	.0026	.1941
1.6025	.6397	.9933	4.4	81.4509	.0071	.5823	.0023	.1903
1.5886	.6253	.9807	4.5	90.0171	.0064	.5761	.0021	.1866
1.5750	.6114	.9685	4.6	99.4843	.0057	.5701	.0018	.1832
1.5617	.5979	.9566	4.7	109.9472	.0051	.5643	.0016	.1798
1.5489	.5848	.9451	4.8	121.5104	.0046	.5586	.0014	.1766
1.5363	.5721	.9338	4.9	134.2898	.0041	.5531	.0013	.1734
1.5241	.5598	.9229	5.0	148.4132	.0037	.5478	.0011	.1704

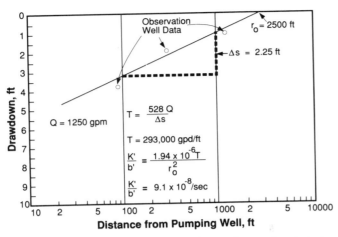

Figure 15.16. Hantush-Jacob method for steady-state flow in leaky aquifers.

When $s_m/h_0 > 0.02$, the confined aquifer equations may be used, provided that the observed drawdown (s) is replaced by s' and that T and S in the confined aquifer equations are replaced with Kh_0 and S_y (specific yield):

$$s' = \text{corrected drawdown [ft]} = s - \frac{s^2}{2h_0}.$$

Hantush's Correction for Partial Penetration in Confined Aquifers. When the pumped well does not penetrate the full aquifer thickness, the aquifer is confined, and flow to the well is nonsteady state, the Hantush correction applies. Wells that do not penetrate the full saturated aquifer thickness require more drawdown for a given discharge than do fully penetrating wells. Data from partially penetrating wells may

be analyzed in the same manner as fully penetrating wells if the observation well is located at a distance away from the pumping well equal to at least 1.5 times the saturated aquifer thickness (16).

Hantush states that for long pumping times,

$$t > \frac{3.7bS}{K},$$

where

t = time after which data apply [days],
b = saturated aquifer thickness [ft],
K = hydraulic conductivity [gpd/ft^2],

the drawdown equation is

$$s = \frac{114.6Q}{T}\left[W(u) + f\left(\frac{r}{b}, \frac{D}{b}, \frac{D'}{b}, \frac{z}{b}\right)\right],$$
(15.18)

where

$W(u)$ = well function of Theis, and

$$f = \frac{4b}{\pi(D - D')}\sum_{n=1}^{\infty}\left(\frac{1}{n}\right)K_0\left(\frac{\pi nr}{b}\right)\cos\left(\frac{\pi nz}{b}\right)$$

$$\times \left[\sin\left(\frac{\pi nD}{b}\right) - \sin\left(\frac{\pi nD'}{b}\right)\right],$$
(15.19)

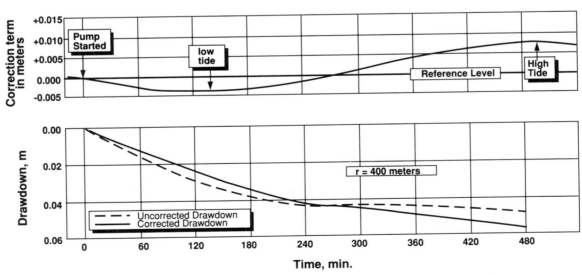

Figure 15.17. Corrections for tidal influence (after Kruseman and De Ridder, 1970).

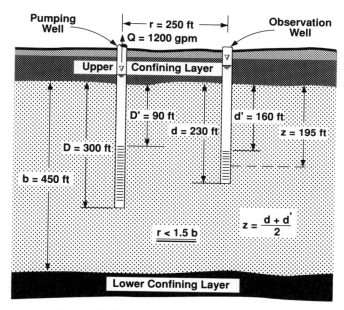

Figure 15.18. Partial penetration well example.

K_0 = modified Bessel function of second kind order zero.

All angles are in radians and other terms as annotated on Figure 15.18.

Step-by-Step Procedure

Figure 15.19 shows test data from the partially penetrating configuration of Figure 15.18.

1. An estimate of K from samples and logs yields approximately 500 gpd/ft^2. Storativity estimates (from nearby wells) are approximately 0.001. The time after which Equation 15.18 applies is calculated as

$$t > \frac{(3.7)(450)(0.001)}{500} = 0.0033 \text{ days} \simeq 5 \text{ min} .$$

2. The slope (Δs) of the best fit line (after 5 min) is 2.0 ft per log cycle.
3. Calculate transmissivity from:

$$T = \frac{264Q}{\Delta s} = \frac{(264)(1200)}{2} = 158,400 \text{ gpd/ft} .$$

4. Calculate f from Equation 15.19. After three iterations, f converges to $\simeq 0.0707$.
5. Calculate storativity from

$$S = \frac{0.3t_0 T e^f}{r^2}$$
$$= \frac{(0.3)(6.25 \times 10^{-4})(158,400)e^{0.0707}}{250^2}$$
$$= 5.1 \times 10^{-4} .$$

Hantush's Partial Penetration Correction for Unconfined Aquifers. Hantush has shown that wells partially penetrating unconfined aquifers may be analyzed using methods for fully penetrating wells if the aquifer is unconfined, the pumping well penetrates to a depth (D) that is less than

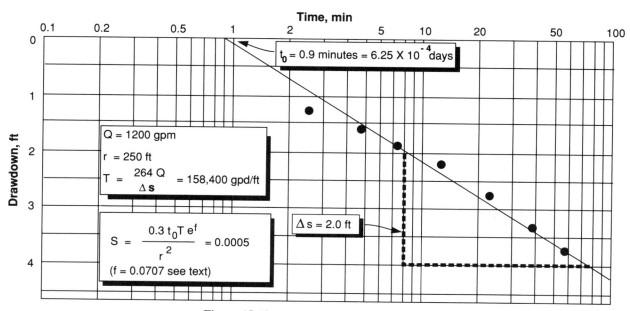

Figure 15.19. Partial penetration test data.

the full aquifer thickness (b), and the observed drawdown (s) is replaced by $s - s^2/2h_0$ (Jacob's correction) (10).

Aquifers Influenced by Boundaries.
In the analysis of pumping test data, a definite change in slope of the time-drawdown graphs may sometimes be observed. When the slopes steepen, this often reflects interception of a geohydrologic boundary by the cone of depression (e.g., limited aquifer conditions, impermeable bedrock, or fault barriers). When slopes flatten, the cone may have reached a condition of equilibrium, have encountered a recharge boundary (e.g., stream), or be receiving recharge from leakage.

Recharge Boundary—Hantush's Method.
Wells pumping in the vicinity of, and in hydraulic continuity with, a stream or river may be analyzed utilizing image wells and the method proposed by Hantush (17). The following is assumed:

- The aquifer is transected by a straight recharge boundary in hydraulic continuity with the aquifer.
- The recharge boundary (i.e., stream) has a constant water level.
- Flow to the well is nonsteady state.
- The observed drawdown is replaced by an adjusted drawdown, $s' = s - s^2/2h_0$.
- Storativity remains constant.

Figure 15.20 shows the relationship of the stream and pumping well.

The drawdown in the vicinity of the well/stream system is given by

$$s(r, t) = \frac{114.6Q}{T} M(u, \beta) ,$$

where

$$M(u, \beta) = W(u) - W(\beta^2 u) \text{ (see Table 15.4),}$$
$$\beta = r_i/r_r \text{ (see Figure 15.20),}$$
$$r_r = \text{distance between the pumping well and observation well [150 ft],}$$
$$r_i = \text{distance between the observation well and recharging image well [960 ft].}$$

Figure 15.21 shows a time-drawdown plot of corrected drawdowns for the test example.

$f(\beta)$ is defined as

$$f(\beta) = \frac{s_m}{\Delta s} ,$$

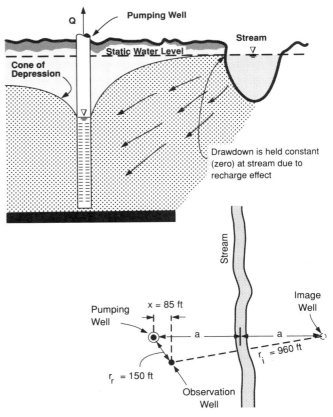

Figure 15.20. Well pumping near a stream.

and from Figure 15.21,

$$f(\beta) = \frac{5.0}{2.75} = 1.818 ,$$

where

$$s_m = \text{maximum drawdown (5.0 ft),}$$
$$\Delta s = \text{drawdown change at inflection point (2.75 ft),}$$
$$s_i = s_m/2, \text{drawdown at inflection point (2.5 ft),}$$
$$t_i = \text{inflection point time (0.125 days).}$$

β is then calculated as 6.4, using Table 15.4. Also

$$u_i = 0.0930 ,$$
$$M(u_i, \beta) = 1.988 .$$

The distance between the observation and image well (r_i) may be calculated from

$$r_i = r_r\beta = (150)(6.4) = 960 \text{ ft} .$$

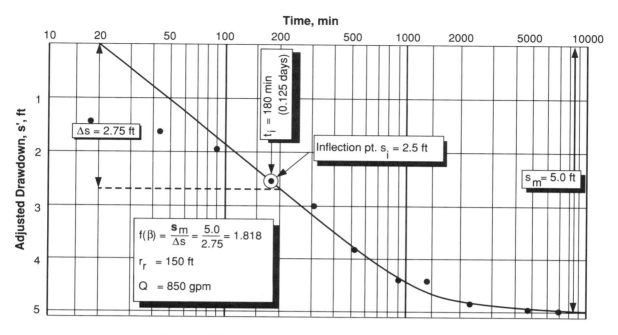

Figure 15.21. Test data from a well pumping near a stream.

TABLE 15.4   Values of the Functions u_i, $M(u_i, \beta)$, and $f(\beta)$ (17)

β	u_i	$M(u_i, \beta)$	$f(\beta)$	β	u_i	$M(u_i, \beta)$	$f(\beta)$	β	u_i	$M(u_i, \beta)$	$f(\beta)$	β	u_i	$M(u_i, \beta)$	$f(\beta)$
1.0	1.000	0.000	1.179	5.0	0.134	1.553	1.667	10	0.0466	2.534	2.115	35	0.00582	4.576	3.109
1.1	0.909	0.070	1.183	5.2	0.127	1.604	1.688	11	0.0400	2.680	2.188	36	0.00554	4.624	3.134
1.2	0.830	0.135	1.188	5.4	0.120	1.653	1.710	12	0.0348	2.815	2.251	37	0.00528	4.671	3.155
1.3	0.761	0.195	1.194	5.6	0.114	1.703	1.731	13	0.0306	2.940	2.312	38	0.00505	4.717	3.178
1.4	0.702	0.252	1.203	5.8	0.108	1.750	1.752	14	0.0271	3.057	2.367	39	0.00483	4.761	3.199
1.5	0.649	0.306	1.214	6.0	0.102	1.796	1.770	15	0.0241	3.172	2.423	40	0.00462	4.805	3.221
1.6	0.603	0.357	1.223	6.2	0.0976	1.840	1.794	16	0.0218	3.271	2.472	41	0.00443	4.847	3.242
1.7	0.562	0.407	1.235	6.4	0.0930	1.988	1.814	17	0.0203	3.342	2.520	42	0.00424	4.889	3.262
1.8	0.525	0.456	1.247	6.6	0.0888	1.927	1.833	18	0.0179	3.462	2.564	43	0.00407	4.930	3.282
1.9	0.492	0.502	1.262	6.8	0.0848	1.969	1.852	19	0.0164	3.551	2.609	44	0.00391	4.969	3.301
2.0	0.462	0.548	1.273	7.0	0.0812	2.010	1.871	20	0.0150	3.637	2.647	45	0.00376	5.008	3.321
2.2	0.411	0.635	1.301	7.2	0.0777	2.050	1.889	21	0.0138	3.716	2.687	46	0.00362	5.046	3.339
2.4	0.368	0.717	1.329	7.4	0.0745	2.089	1.908	22	0.0128	3.793	2.725	47	0.00349	5.084	3.357
2.6	0.332	0.796	1.357	7.6	0.0715	2.127	1.925	23	0.0119	3.867	2.761	48	0.00336	5.120	3.375
2.8	0.301	0.872	1.385	7.8	0.0687	2.165	1.943	24	0.0111	3.938	2.796	49	0.00325	5.156	3.393
3.0	0.275	0.945	1.413	8.0	0.0661	2.202	1.960	25	0.0103	4.007	2.837	50	0.00313	5.191	3.410
3.2	0.252	1.016	1.435	8.2	0.0636	2.238	1.977	26	0.00966	4.072	2.862	55	0.00265	5.358	3.491
3.4	0.232	1.083	1.467	8.4	0.0613	2.273	1.994	27	0.00906	4.135	2.893	60	0.00228	5.510	3.565
3.6	0.214	1.149	1.493	8.6	0.0590	2.308	2.010	28	0.00852	4.196	2.923	65	0.00198	5.650	3.634
3.8	0.199	1.212	1.500	8.8	0.0570	2.342	2.026	29	0.00803	4.256	2.952	70	0.00174	5.781	3.697
4.0	0.185	1.273	1.545	9.0	0.0550	2.376	2.041	30	0.00757	4.313	2.980	75	0.00154	5.903	3.757
4.2	0.173	1.333	1.571	9.2	0.0531	2.408	2.057	31	0.00716	4.369	3.008	80	0.00137	6.017	3.812
4.4	0.162	1.390	1.597	9.4	0.0513	2.441	2.072	32	0.00678	4.423	3.034	85	0.00123	6.124	3.864
4.6	0.152	1.447	1.619	9.6	0.0497	2.472	2.087	33	0.00643	4.475	3.059	90	0.00111	6.226	3.913
4.8	0.142	1.500	1.642	9.8	0.0481	2.503	2.102	34	0.00611	4.526	3.085	95	0.00102	6.311	3.960
5.0	0.134	1.553	1.667	10.0	0.0466	2.534	2.115	35	0.00582	4.576	3.109	100	0.00092	6.412	4.004

For values of $t > 4t_i$ (i.e., $t > 720$ min), the drawdown approaches the maximum drawdown (s_m) and transmissivity (T) may be calculated from

$$T = \frac{528Q}{s_m} \log \beta = \frac{(528)(850)}{5.0} \log (6.4)$$

$$= 72,363 \text{ gpd/ft} ,$$

where

Q = pumping rate = 850 gpm,
s_m = maximum drawdown = 5.0 ft.

Storativity is calculated from

$$S = \frac{Tt_i u_i}{1.87 r_r^2} = \frac{(72,363)(0.125)(0.0930)}{(1.87)(150)^2} = 0.02$$

The distance (a) to the boundary can be found from

$$a = \frac{4x + \sqrt{16x^2 + 16r_r^2(\beta^2 - 1)}}{8} ,$$

where

x = reflection of r_r as measured perpendicular to stream (see Figure 15.20) = 85 ft.

$$a = \frac{(4)(85) + \sqrt{(16)(85)^2 + (16)(150)^2(6.4^2 - 1)}}{8}$$

$$= 519 \text{ ft}$$

Impermeable Boundary. Recharge boundaries in the vicinity of a pumping well are normally visible. However, impermeable boundaries may have no surface expression (e.g., buried bedrock or fault barriers)(Figure 15.22). The following example illustrates location of an impermeable boundary from pumping test data.

The times of occurrence of equal drawdown in observation wells vary directly as the squares of distances between the observation and production well. This principle, analogous to the law of times, is stated as (12)

$$\frac{t_i}{r_i^2} = \frac{t_r}{r_r^2} \qquad (15.20)$$

where

t_r = arbitrary time after start of pumping but before the effect of the image well is felt [min],

r_r = distance between pumping well and observation well,

t_i = time after start of pumping where divergence of drawdown (due to image well) equals arbitrary drawdown (s_A) measured at time t_r [min].

From Equation 15.20,

$$r_i = r_r \sqrt{\frac{t_i}{t_r}} .$$

For observation well 1, $r_r = 100$ ft from Figure 15.23, $t_r = 15$ min and $t_i = 160$ min; thus

$$r_1 = 100 \sqrt{\frac{160}{15}} = 327 \text{ ft} .$$

Similarly, for observation well 2, $r_r = 140$ ft, $t_r = 100$ min, $t_i = 700$ min:

$$r_2 = 140 \sqrt{\frac{700}{100}} = 370 \text{ ft} .$$

The discharge image well is located at the intersection of the arcs of radii r_1 and r_2 (see Figure 15.22). The barrier is located perpendicular to and halfway between the line connecting the real and image well. Obviously, the more observation wells available, the more precise the barrier location would be.

Perched Aquifer Condition. Analysis of pumping test data in perched aquifers shows effects similar to wells pumping near impermeable hydrologic boundaries. As a result recharge to the cone of depression in a perched aquifer is limited with the effect being similar to pumping from a closed tank. In most cases long-term sustained production is unobtainable unless well production is decreased to balance the limited recharge. Figure 15.24 shows steepening slopes of time-drawdown data from a well pumping from a perched aquifer. Pumping test data indicating a perched aquifer may be distinguished from impervious or semipervious barrier data by knowledge of local hydrogeologic conditions.

Aquifer Zone Testing for Water Quality. Although traditionally pumping test analyses have been used to determine hydraulic characteristics of wells and aquifers, measurement of water quality is increasingly important. An example of this is selective aquifer zone testing to determine the vertical extent of contamination. The procedure involves isolating aquifers by placement of impermeable material. The isolated

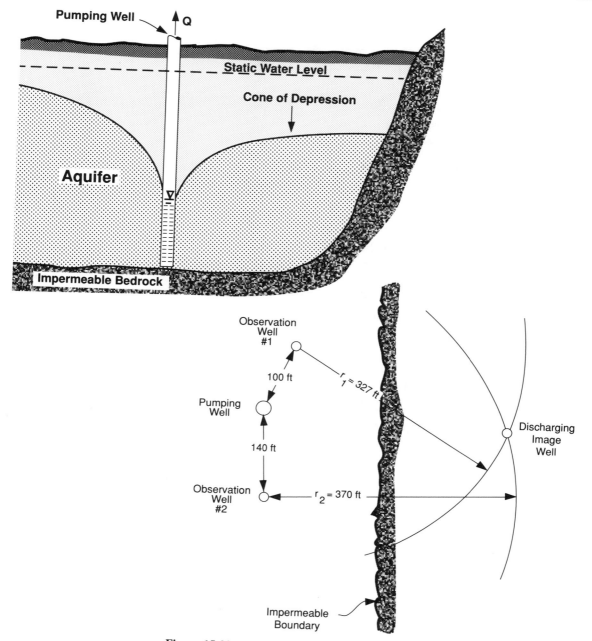

Figure 15.22. Impermeable boundary example.

zone is then pumped for several hours or longer, and the water samples are analyzed for contaminants. In this way estimates of both yield and water quality for various aquifer zones are procured, with only acceptable zones completed for production. Figure 15.25 shows results of a zone test in Southern California.

Calculation of Well Parameters

Step-Drawdown Tests. Calculation of well characteristics are necessary to determine specific capacity and efficiency

relationships. These relationships are used to help design production pumps, gauge the degree of development, and determine well maintenance programs.

Well parameters are found using variable rate pumping tests, commonly called "step-drawdown" tests. The following example assumes that the well is fully developed and that no near-well turbulent flow losses occur.

The drawdown in a pumping well satisfying these assumptions may be written (18)

$$s_w = BQ + CQ^2 \ldots , \qquad (15.21)$$

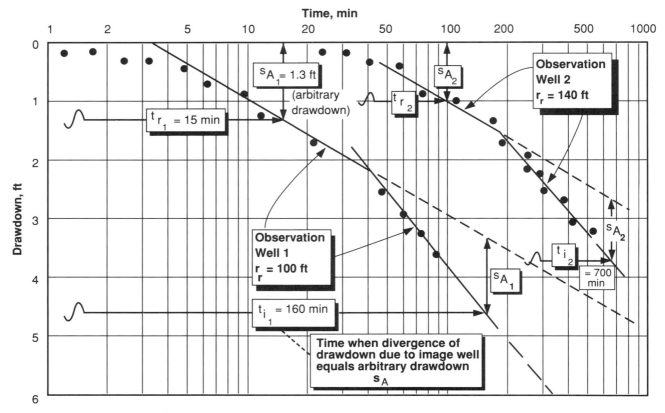

Figure 15.23. Observation well data for a well pumping near an impermeable boundary.

where

s_w = drawdown in the pumping well [ft],
Q = pumping rate [gpm],
B = formation-loss coefficient [ft/gpm],
C = well-loss coefficient [ft/gpm²].

Calculation of the parameters B and C makes use of the principles developed in Chapter 5.

Step-by-Step Procedure

1. The time-drawdown data for a three-step variable rate test is shown in Figure 15.26.

2. Incremental drawdowns (δs_i), occurring 200 min after the start of each step, are measured as the distance between the drawdown at the particular step and the extrapolated drawdown from the previous step. The t^* time of 200 min was selected as the minimum time required for the linear portion of the data to occur from the start of the last step (see Figure 15.26).

The total drawdown for each step s_m is obtained from

$$s_m = \sum_{i=1}^{m} \delta s_i \; .$$

Specific drawdown is obtained by dividing total drawdown by the pumping rate for each step. A plot of specific drawdown versus pumping rate is shown in Figure 15.27. The data from Figure 15.26 is summarized in Table 15.5.

3. A best-fit straight line is drawn through the data points on Figure 15.27, and the formation loss coefficient (B) is measured from the zero discharge intercept. The well-loss coefficient (C) is calculated from the slope of the line. The specific capacity diagram (Figure 15.28) was constructed by substituting the values of B and C into Equation 15.21.

Calculation of Well Efficiency from Step-Drawdown Test Data.
Due to higher energy costs and the expanding

dependence on ground water in many areas, well efficiency has become of increasing interest to water system managers, engineers, and owners. Although all agree on the essential concept, analysts have defined a number of ways of measuring the parameters necessary to determine its value. The step-drawdown procedure described in this chapter offers the advantage of determining well efficiency by delineating the turbulent and laminar flow components of drawdown from a test on the pumping well itself.

Well efficiency may be written as (18)

$$E = \frac{100}{1 + CQ/B}$$

and is plotted as the small dashed line in Figure 15.28 (uppermost curve).

A design pumping rate and drawdown were selected based on operational requirements and well efficiency. The resulting drawdown and efficiency were

$$s_w = 0.09Q + 1.9 \times 10^{-5}Q^2 \;,$$

$$E = \frac{100}{1 + 2.1 \times 10^{-4}Q} \;.$$

At the design pumping rate of 1250 gpm, the well drawdown is 142 ft with a well efficiency of 79%.

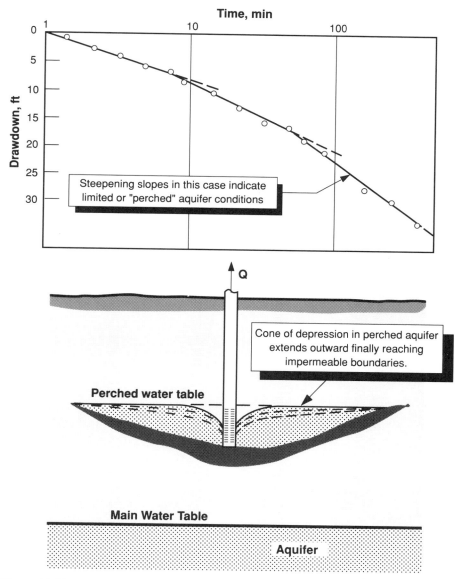

Figure 15.24. Example of time-drawdown data from a well completed in a perched aquifer.

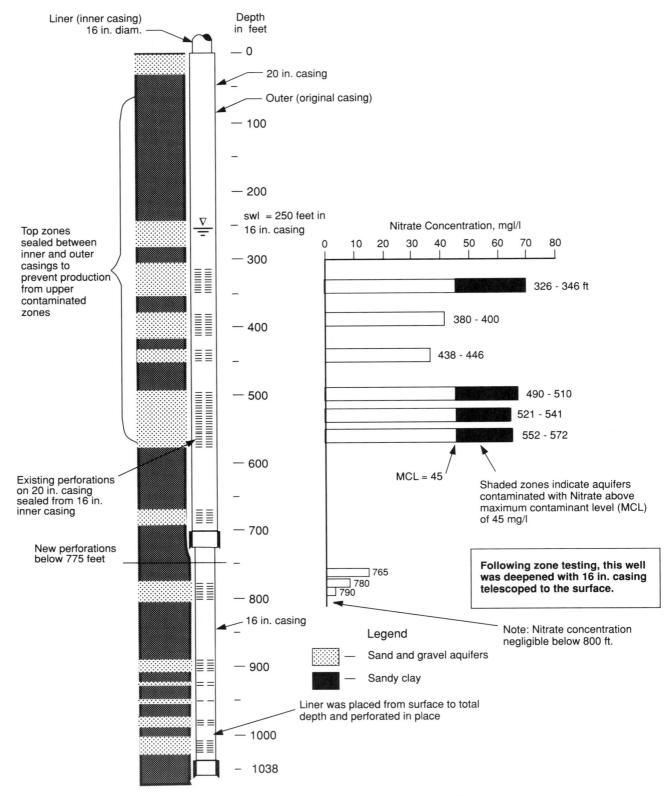

Figure 15.25. Aquifer zone testing for water quality.

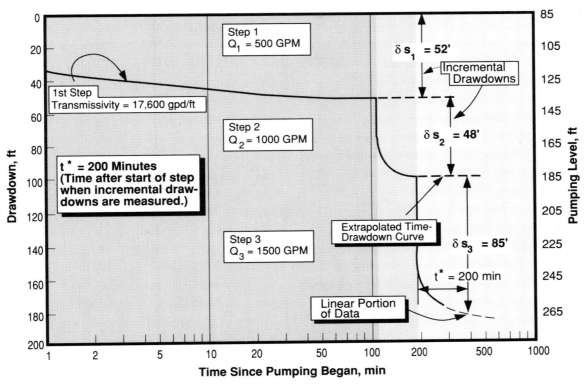

Figure 15.26. Step-drawdown test example.

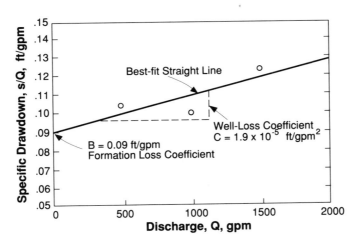

Figure 15.27. Specific drawdown plot.

TABLE 15.5 Summary of Data from Figure 15.26

Step, m	Pumping Rate, Q_m [gpm]	Incremental Drawdown, δs_i [ft]	Total Drawdown, s_m [ft]	Specific Drawdown, $(s/Q)_m$ [ft]
1	500	52	52	0.104
2	1000	48	100	0.100
3	1500	85	185	0.1233

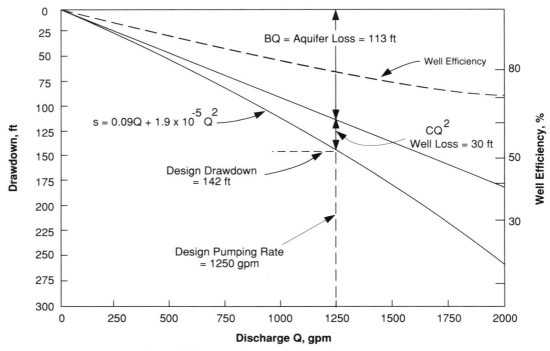

Figure 15.28. Specific capacity and efficiency diagram.

Other Methods of Calculating Well Efficiency. The most reliable method of measuring efficiency, other than having piezometers in the near-well zone, is through the distance-drawdown method. In this method water levels are measured in multiple observation wells and extrapolated on semi-logarithmic paper to the face of the pumping well.

Figure 15.29 shows an example of well efficiency calculated using this method.

The following example compares the step-drawdown method with a method based on estimating theoretical specific capacity. Figure 15.30 shows the specific drawdown plot for the example well. From the plot, B and C are obtained,

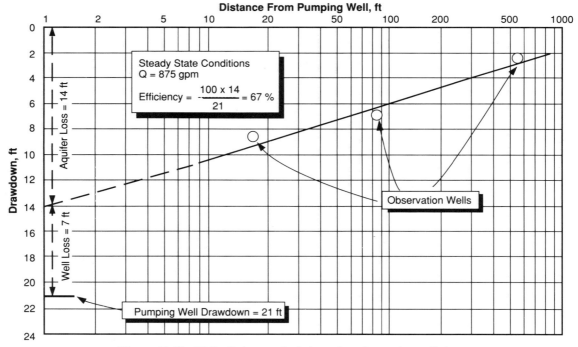

Figure 15.29. Well efficiency calculation using observation well data.

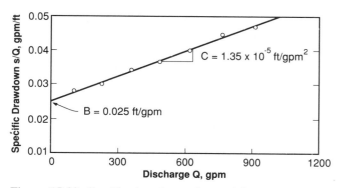

Figure 15.30. Specific drawdown plot used in well efficiency example.

and well efficiency is calculated for a discharge rate of 900 gpm:

$$E = \frac{100}{1 + CQ/B}$$

$$= \frac{100}{1 + (1.35 \times 10^{-5})(900)/0.025}$$

$$= 67\% .$$

Figure 15.31 shows a constant rate pumping test on the same well. Applying the Jacob straight-line method, transmissivity is calculated as 47,500 gpd/ft. Assuming an effective well radius r_e of 1 ft and a storativity of 0.001, the aquifer drawdown after 1 day of pumping can be calculated from Jacob's equation:

$$s_{1\ day} = \frac{264Q}{T} \log\left(\frac{0.3Tt}{r_e^2 S}\right)$$

$$= \frac{(264)(900)}{47,500} \log\left(\frac{(0.3)(47,500)(1)}{(1^2)(0.001)}\right)$$

$$= 35.8\ ft .$$

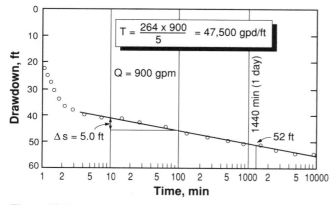

Figure 15.31. Time-drawdown plot of efficiency comparison well data.

The theoretical specific capacity (i.e., only considering aquifer loss) is calculated as

$$\frac{Q}{s} = \frac{900}{35.8} = 25 \frac{gpm}{ft}$$

The actual 1-day specific capacity as measured in the well (see Figure 15.31) was 900 gpm/52 ft, or 17.3 gpm/ft.

Well efficiency is also defined as the ratio of the actual to the theoretical specific capacity, or

$$\frac{(Q/s)_{actual}}{(Q/s)_{theoretical}} \times 100 = \frac{17.3}{25.1} \times 100 = 69\% .$$

As can be seen, this value is very close to the value of 67% obtained from the step-drawdown analysis.

Each method of determining well efficiency has its proponents. However, within the reliability of the data and the simplifying assumptions made to permit development of these equations, the efficiencies calculated from all methods will be close in most cases.

Critical Discharge and Recommended Production Rate. Based on results from the pumping test program, a maximum safe pumping rate can be recommended. This long-term rate is generally the highest pumping rate at which equilibrium conditions can be maintained.

When the equilibrium rate is exceeded, the pumping level continues to decline at a rate proportional to both well and aquifer parameters. When this decline is fairly rapid (several feet per minute), the well may dewater in a short period of time with the pump "breaking suction." This latter condition is potentially serious and could result in casing collapse.

The point of discharge at which this rapid decline occurs is called the "critical discharge." This condition varies widely from area to area and from well to well. In areas having high transmissivities and storativities, the full capacity of the pump may never be able to create a critical discharge situation. However, in low-yielding aquifers, rapid dewatering of the well may be achieved quite easily.

Water quality may also limit the safe rate. For example, in wells pumping directly into the distribution system, sand or air in the discharge may be critical factors requiring reduced pumping below the well's potential.

Calculation of Effective Well Radius. In a fully developed well, the hydraulic conductivity in the near-well zone is higher than that of adjacent aquifer material. This results in a slightly lower drawdown and a larger effective radius. The following example shows how to calculate the effective well radius from step-drawdown test data.

Having found the formation-loss coefficient (B) from Figure 15.27 as 0.09 ft/gpm and transmissivity (T) from the slope of the first step in Figure 15.26 (Jacob's straight-line method) as 17,600 gpd/ft, the effective well radius is calculated as follows (19):

$$r_e = \sqrt{\frac{0.3Tt^*}{10^{0.0038TB}S}} \quad (15.22)$$

where

r_e = effective well radius [ft],
T = tranmissivity [17,600 gpd/ft],
t^* = time from beginning at each step [0.139 days],
B = formation-loss coefficient [0.09 ft/gpm],
S = storativity (estimated as 0.0005).

Substituting the values of B, T, t^*, and estimating (S) to be 0.0005, r_e is calculated as

$$r_e = \sqrt{\frac{(0.3)(17,600)(0.139)}{10^{(0.0038)(17,600)(0.09)}(0.0005)}}$$

$$= 1.19 \text{ ft} = 14 \text{ in.}$$

The effective radius of 14 in. is greater than the nominal radius of 10 in, confirming increased conductivity in the near-well zone, and complete development.

Injection Well Tests. Methods for analysis of injection wells are similar to pumping well methods. A description of injection well hydraulics is found in Chapter 19.

The equation for pressure head in an injection well may be written

$$P = BQ + CQ^2 ,$$

where

P = injection well pressure [psi],
Q = injection well rate [cfs],
B, C = formation- and well-loss coefficients, respectively [psi/cfs], [psi/cfs^2].

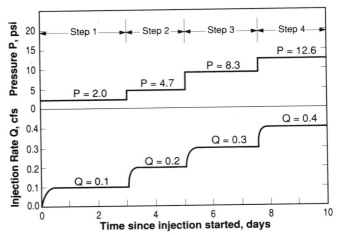

Figure 15.32. Example of a variable rate injection test.

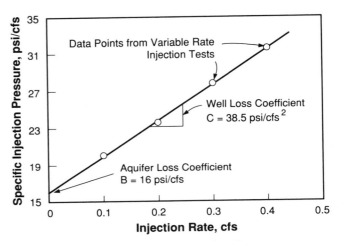

Figure 15.33. Specific injection diagram.

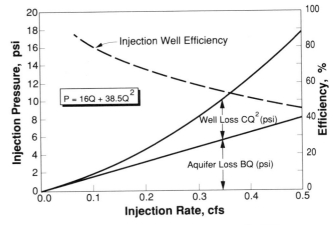

Figure 15.34. Injection pressure, rate and efficiency.

TABLE 15.6   Injection Well Test Data

| | Injection (Well Head) | | |
Step	Injection Rate, Q [cfs]	Pressure, P [psi]	Specific Injection P/Q [psi/cfs]
1	0.1	2.0	20.0
2	0.2	4.7	23.5
3	0.3	8.3	27.7
4	0.4	12.6	31.5

TABLE 15.7 Summary of the Methods Used in This Chapter

Aquifer Type	Flow Type	Parameters Calculated	Method Name	Reference
		Calculation of Aquifer Parameters		
Confined	Nonsteady	T, S	Theis' type-curve	3
Confined	Nonsteady	T, S	Jacob's straight line	4
Confined	Nonsteady	T, S	Chow	5
Confined	Steady	T	Thiem	6
Unconfined	Nonsteady	T, S	Boulton's delayed yield	7, 8
Unconfined	Steady	T	Dupuit	9
Semiconfined	Nonsteady	$T, S, K'/b'$	Hantush's inflection point	10
Semiconfined	Nonsteady	$T, S, K'/b'$	Walton's type-curve	12
Semiconfined	Steady	$T, S, K'/b'$	Hantush-Jacob	13
		Corrections and Special Conditions		
Confined	Nonsteady	T, S	Hantush's partial penetration	10
Unconfined	Nonsteady	T, S	Hantush's partial penetration correction	10
Unconfined	Nonsteady	T, S	Jacob's unconfined aquifer correction	15
Confined	Nonsteady	T, S, a	Hantush's recharge boundary	17
Confined	Nonsteady	a	Impermeable boundary	12
		Calculation of Well Parameters		
Confined	Nonsteady	T, B, C, E	Step-drawdown	18
Confined	Nonsteady	T, B, C, E	Step-injection	This chapter

T = transmissivity, S = storativity, K'/b' = leakance, a = distance between well and boundary, B = formation-loss coefficient, C = well-loss coefficient, E = well efficiency.

Injection well efficiency is identical to pumped well efficiency with the clarification that well losses are those turbulent flow losses associated with injecting water into the aquifer system.

Determination of the aquifer- and well-loss coefficients B and C involves conducting a variable rate injection test. This test uses a procedure similar to the step-drawdown test. At least three separate rates are required; however, more "steps" (injection rates) result in better accuracy. The specific injection pressure (P/Q) is plotted against the injection rate (Q) on arithmetic paper. The best-fit straight line through these points determines the parameters B and C (the intercept of the line at $Q = 0$ is B, and the slope of the line is C).

Table 15.6 illustrates use of this technique to find injection well efficiency of a seawater barrier well. The injection well test data are contained therein.

Figure 15.32 is a plot of injection pressure and rate against time, and Figure 15.33 is specific injection (P/Q) against the injection flow rate (Q). The best-fit straight line through the four data points defines the well coefficients (B) and (C). The aquifer-loss coefficient (B) is obtained from the intercept of the line at $Q = 0$ and equals 16 psi/cfs. The well-loss coefficient (C) is the slope of the line, and for the example is 38.5 psi/cfs^2.

The resulting injection pressure equation is

$$P = 16 \times Q + 38.5 \times Q^2 .$$

A plot of this equation for various injection rates is seen in Figure 15.34.

15.5 SUMMARY OF METHODS

Table 15.7 is a summary of the methods used in this chapter.

REFERENCES

1. Thiem, A. 1870. "Die Ergiebigkeit Artesischer Bohrlocher, Schachtbrunnen und Filtergallerien." *J. F. Gasbel. und Wasservers.* 14, pp. 450–567.

2. Kruseman, G. P., and N. A. De Ridder. 1970. "Analysis and Evaluation of Pumping Test Data." *Bulletin 11.* International Institute for Land Reclamation and Improvement, Wageningen, The Netherlands.

3. Theis, C. V. 1935. "The Relation Between the Lowering of the Peizometric Surface and the Rate and Duration of Discharge of a Well Using Ground-Water Storage." *Am. Geophys. Union Trans.*, 16th Ann. Mtg., pt. 2, pp. 519–524.

4. Cooper, H. H., and C. E. Jacob. 1946. "A Generalized Graphical Method for Evaluating Formation Constants and Summarizing Well Field History." *Am. Geophys. Union Trans.* 27, pp. 526–534.

5. Chow, V. T. 1952. "On the Determination of Transmissivity and Storage Coefficients from Pumping Test Data." *Am. Geophys. Union Trans.* 33, pp. 397–404.

6. Thiem, A. 1906. *Hydrologische Methoden*. Gebhard, Leipzig.

7. Boulton, N. S. 1963. "Analysis of Data from Non-Equilibrium Pumping Tests Allowing for Delayed Yield from Storage." *Proc. Inst. Civ. Eng.* 26, pp. 469–482.

8. Boulton, N. S. 1963. "Analysis of Data from Non-Equilibrium Pumping Tests Allowing for Delayed Yield from Storage: A Discussion" *Proc. Inst. Civ. Eng.* 28, pp. 603–610.

9. Dupuit, J. 1863. *Etudes theoriques et pratiques sur le mouvement des eaus dans les canaux decouvert et a travers les terraines permeables*. 2d ed. Dunot, Paris.

10. Hantush, M. S. 1964. "Hydraulics of Wells." In *Advances in Hydroscience*. Vol. 1. Academic Press, New York.

11. Hantush, M. S. 1956. "Analysis of Data from Pumping Tests in Leaky Aquifers." *AGU* 37, 6, p. 431.

12. Walton, W. C. 1962. *Groundwater Resource Evaluation*. McGraw-Hill, New York.

13. Hantush, M. S., and C. E. Jacob. 1955. "Non-Steady Radial Flow in an Infinite Leaky Aquifer." *Am. Geophys. Union Trans.* 36, pp. 95–100.

14. Jacob, C. E. 1963. "Correction of Drawdowns Caused by a Pumped Well Tapping Less than the Full Thickness of an Aquifer." In *Methods of Determining Permeability, Transmissibility and Drawdown*. U.S. Geological Survey Water-Supply Paper 1536-I, pp. 272–282.

15. Jacob, C. E. 1950. *Engineering Hydraulics*. J. Wiley and Sons, New York.

16. Hantush, M. S. 1962. "Aquifer Tests on Partially Penetrating Wells." *Am. Soc. Civ. Eng. Trans.* 127, pt. I, pp. 268–283.

17. Hantush, M. S. 1959. "Analysis of Data from Pumping Wells near a River." *J. Geophys. Res.* 65, pp. 3713–3725.

18. Williams, D. E. 1985. "Modern Techniques in Well Design." *J. Am. Water Works Assoc.* (September).

19. Jacob, C. E. 1947. "Drawdown Test to Determine Effective Radius of Artestian Well." *Trans. Amer. Soc. Civil Engrs.* 112, pp. 1047–1070.

PART III

MANAGEMENT AND USE

■■■■■■ CHAPTER 16

Vertical Turbine Pumps

16.1 INTRODUCTION

The history of the deep-well turbine pump spans more than 100 years. The original basic concept of vertically stacked centrifugal impellers and vane diffusers allowed the development of small-diameter, multistage pumping units capable of delivering large quantities of water from wells under high heads. Vertical turbine pumps have many diverse applications. This chapter, however, will discuss only those turbine pumps designed for use in high-capacity agricultural, industrial, and municipal wells.

16.2 PRINCIPLES OF TURBINE PUMP OPERATION

The operating characteristics of a vertical turbine pump are a function of the design of the impeller and bowl, and the speed of rotation. The impeller-bowl assembly operates on a modified radial-flow centrifugal principal. Kinetic energy is imparted to the fluid as a result of the speed differential between the eye of the impeller and its outer edge. The directional vanes of the bowl surrounding the impeller convert this from velocity to pressure as the vane passage area increases and the fluid is redirected axially.

16.3 SYSTEM ASSEMBLIES

Line-Shaft Pump

A line-shaft powered vertical turbine pump consists of three separate but co-dependent elements. They are the bowl, column and shaft, and discharge head assemblies, as shown in Figure 16.1.

Bowl Assembly. The bowl assembly consists of single or multiple stage units, each with impeller, bowl case, pump shaft, and bearings.

Column and Shaft Assembly. The column pipe suspends the bowl assembly from the discharge head assembly and conducts water between them. In addition the column houses the drive shaft (line shaft), or shaft-enclosing tube and drive shaft, and shaft bearings as appropriate for the pump type.

Discharge Head Assembly. The head assembly consists of the driver and discharge head. The discharge head supports the driver and column pipe, and directs the pumped water from the column to the external piping.

Oil Lubricated versus Water Lubricated

The majority of high-capacity vertical turbine pumps are oil lubricated. This type has a shaft-enclosing tube that contains the drive shaft bearings and conducts oil from the surface to the bearings. Oil lubrication allows a much simpler seal in the discharge head assembly. The drive shaft has better support because it has bearings at 5-ft intervals, which is particularly important with deep pump settings. No prelubrication is necessary because residual oil is retained on the shaft bearings. Its main disadvantage is accumulation in the well of surplus oil discarded through bypass ports. The flow is small, but in time it accumulates into a floating mass several feet thick. This will not create problems unless oil is drawn into the pump suction, but it should be removed by bailing when the pump is removed for maintenance. Food-grade oil may be used as a precaution against contamination.

Use of water-lubricated pumps is advantageous in some applications where static water levels are high and settings are shallow, or where oil cannot be tolerated. In this type, lubrication of the shaft and bearings is provided by the pumped water. Its major disadvantage is higher maintenance cost, especially if sand is being pumped. If the column is not filled with water during start-up, the line shaft must be prelubricated from an external-pressurized water source to protect the bearings. Typically, this is controlled by an automatic timer and solenoid valve.

Submersible Pump

This vertical turbine pump type consists of column pipe, a discharge head, a bowl assembly close-coupled to a squirrel-cage induction electric motor, and a submersible power cable.

315

1 Impeller
2 Shaft, Pump
3 Ring, Impeller
4 Shaft, Head
5 Shaft, Drive
6 Packing
7 Gland
8 Ring, Lantern
9 Bushing, Bearing
10 Bell, Suction
11 Bushing, Stuffing Box
12 Collar, Protecting
13 Nut, Shaft Adjusting
14 Coupling, Shaft
15 Lubricator
16 Bracket, Lubricator
17 Stuffing Box
18 Collet, Impeller Lock
19 Tube, Shaft Enclosing
20 Pipe, Column
21 Bearing, Line Shaft,
    Enclosed
22 Nut, Tubing
23 Plate, Tension, Tubing
24 Head, Surface Discharge
25 Flange, Top Column
26 Coupling, Column Pipe
27 Retainer Bearing, Open
    Lineshaft
28 Adapter, Tubing
29 Case, Discharge
30 Bowl, Intermediate
31 Case, Suction
32 Strainer
33 Pipe, Suction

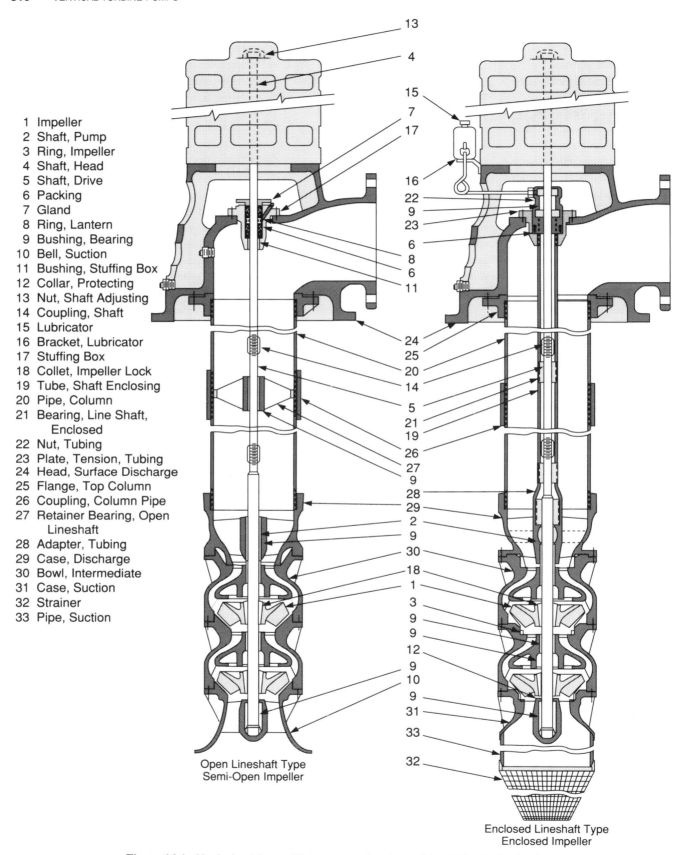

Open Lineshaft Type
Semi-Open Impeller

Enclosed Lineshaft Type
Enclosed Impeller

Figure 16.1. Vertical turbine multi-stage pump (courtesy of the Hydraulic Institute).

The motor and bowl assembly are suspended from the discharge head on the column pipe. Electrical energy is supplied to the motor by a submersible power cable attached to the outside of the column pipe.

No line shaft is required since the directly connected motor is located just beneath the bowl assembly. Water enters the intake screen between the motor and bowl assembly, passes through the pump stages, and is discharged through the column pipe. The discharge head, which does not have to support the prime mover, consists of a surface plate from which the column pipe is suspended, contains a passage for the submersible cable, and may include a heavy-duty pipe elbow which directs the water into the external piping. A typical submersible pump is shown in Figure 16.2.

Submersible pumps have become the major type used in domestic wells and increasingly are installed in high-capacity wells, particularly when they are crooked or pumping levels are very deep. Design, installation, and maintenance of line-shaft pumps becomes difficult when settings exceed 700 ft below ground surface.

Submersible pumps have several advantages:

1. The motor is directly coupled to the impellers, eliminating line-shaft losses.
2. The motor is easily cooled because of complete submersion.
3. Ground-surface noise is eliminated.
4. The pump can be installed and operated in crooked wells.
5. A pump house is not necessary if a pitless adapter (underground discharge) is used.

Their disadvantages are:

1. Electrical problems that may be caused by the submerged cable and its splice at the motor.
2. Motor efficiency is generally lower, due to diameter constraints on motor design.
3. Motor seals cannot tolerate sand pumping.
4. The motor is not accessible for repairs.
5. The motor requires a higher degree of protection against voltage fluctuations.
6. Larger-diameter pump housing casing is required in most instances to accommodate the submersible motor and cable.

16.4 COMPONENTS

The following identify and describe the primary components of line shaft and submersible pumps, as applicable.

The Driver

The driver is mounted on the discharge head and transmits power to the top shaft. It contains the means for impeller adjustment and provides a bearing to carry the thrust load. When the prime mover (vertical hollow shaft motor) is directly connected to the head shaft, it is also the driver. Engine prime movers coupled with right-angle gear drives are also used, particularly when electricity is not available or electric outages cannot be tolerated.

Motor Standards

The National Electrical Manufacturers Association's (NEMA) *Specifications MG-1* contains standards for electric motor selection and application to line-shaft turbine pumps.

Motor Types

The standard vertical turbine motor is a squirrel-cage ac induction motor that rotates under full load at about 1% to 3% lower rpm than the synchronous speed of the electric input, due to the slip of the rotor relative to the rotating field in the stator. The synchronous speeds (also known as nominal speed) generally used in well pumps are 1200, 1800, or 3600 rpm with 60 cycle electric power.[1] Under full load the typical nominal 1800 rpm motor will rotate at 1750 rpm to 1790 rpm, depending on its size and design. Manufacturers' bowl-performance curves are usually represented at about the full-load rpm of the most common motors of that size.

Vertical hollow shaft (VHS) motors are designed for line-shaft pumps. The motor's hollow shaft slips over the head shaft and is attached by a coupling and top nut that controls vertical adjustment of the impellers. Vertical solid shaft (VSS) motors are not used for deep settings since shaft and impeller adjustment are virtually precluded with this type.

Submersible motors are also squirrel-cage induction motors. A shaft end seal prevents water from entering between the stator and rotor.

Service Factor. The service factor (SF) refers to the amount a motor can be overloaded and still operate within the design limit of temperature rise for the insulation system used. SF for a typical well motor is 1.15 up to 200 hp size and 1.0 for larger motors. For example, a 100 hp motor with a 1.15 SF may be loaded to 115 hp, whereas a large motor with a 1.0 SF should not be loaded beyond its rated hp. Motors are seldom loaded beyond rated (name plate) horsepower,

---

1. Synchronous speed is a function of the number of motor poles. For example with 60 cycles/second AC power, two-pole motor synchronous speed is [(60 cycles/sec × 60 sec/min)/2 poles] × 2 = 3600 rpm. Similarly 4-pole and 6-pole motors would have synchronous speeds of 1800 rpm and 1200 rpm, respectively.

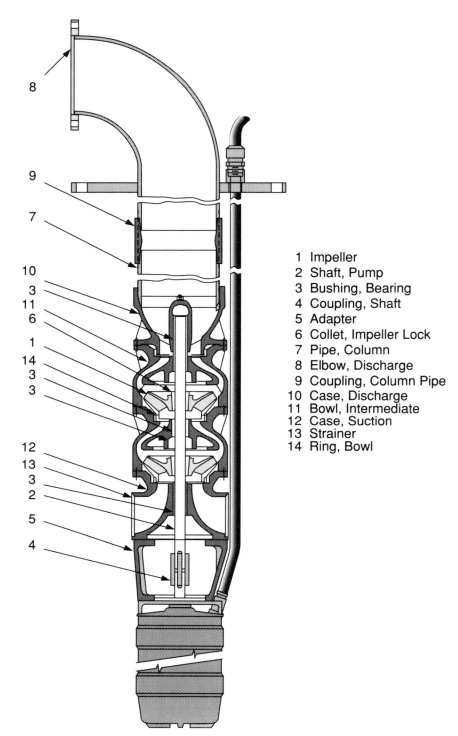

1 Impeller
2 Shaft, Pump
3 Bushing, Bearing
4 Coupling, Shaft
5 Adapter
6 Collet, Impeller Lock
7 Pipe, Column
8 Elbow, Discharge
9 Coupling, Column Pipe
10 Case, Discharge
11 Bowl, Intermediate
12 Case, Suction
13 Strainer
14 Ring, Bowl

Figure 16.2. Vertical turbine multi-stage submersible pump (courtesy of the Hydraulic Institute).

regardless of the SF, since the excess heat generated by extended periods of overload operation greatly reduces motor life.

Efficiency. The motor's efficiency is an important concern because the cost of energy to operate a motor over its life usually materially exceeds the initial cost. Manufacturers' brochures express efficiency values as "NEMA nominal efficiency," which means that the motor has met criteria established by NEMA *Specifications MG-1*, Sections 12.53a and 12.53b. Motor efficiency is the ratio of useful power output (brake horsepower) to the total input electrical power, expressed as a percentage. Manufacturers offer "normal efficiency" (NE) and "high efficiency" (HE) motors. The typical NE motor is 85% to 92% efficient, depending on factors such as horsepower. The HE motor is more expensive but is 2% to 5% higher in efficiency. HE motors tend to be physically larger than NE motors because of the additional material used in construction. They also run much cooler and are less noisy, due to the reduction of losses that would otherwise be converted to heat and noise. In most cases economic analysis confirms that a HE motor is the better choice to power a vertical turbine pump.

Submersible Motors

NEMA standards have not been developed for submersible motors. These motors are extremely compact and generally do not withstand overheating and fluctuations in voltage. They are cooled by water passing the motor casing into the pump intake, so a free flow of water must be maintained. Overheating may occur if the well has cascading water or if water enters the well above the pump intake. This problem can be overcome by placing a shroud over the pump intake that directs the water past the motor before it enters the pump.

In most cases submersible motors are equipped with 2- to 3-ft cable leads. These must be spliced to a submersible electrical cable from the ground surface. This splice must be watertight to keep the unit operating successfully.

Operating voltage is especially important for submersible motors that are designed to operate within 10% of motor rated (name-plate) voltage. Voltage variation in excess of this range, or power fluctuations due to electrical storms, may cause motor failure unless necessary overload protection is provided.

Right-Angle Gear Drives

A right-angle gear drive is used when a vertical turbine pump is driven by a horizontal prime mover such as an internal combustion engine. Consideration must be given to engine rpm and required pump speed when selecting the appropriate gear ratio. Right-angle gear drives are built in VSS and VHS configurations, with VHS usually used.

Space Heaters

Space heaters keep the electric motor warm and dry when it is not operating, preventing failure due to low temperatures or high humidity during start-up.

Head-Shaft Subassembly

This subassembly, which connects the driver and line shaft, consists of the head shaft, shaft adjusting nut, lock screw, gib key, and top drive coupling. The top drive coupling connects the head shaft with the VHS motor rotor, transmits driver torque to the line shaft, and vertically aligns the head shaft. It also protects the pump from turning in the direction opposite normal rotation, which may be caused by motor phase reversal. Reverse rotation (backspin) may cause pump shaft couplings to unscrew.

Two types of couplings are used in VHS motors: the nonreverse ratchet (NRR) and the self-release coupling (SRC). Both couplings consist of two parts, one connected to the motor and the other connected to the head shaft. The NRR prevents pump backspin by use of a ratchet device between the two parts that engages automatically when the motor is turned off and the torque is reversed. NRR is useful when there is no time-delay control to limit pump re-start, and with water-lubricated pumps when the line-shaft bearings might otherwise run dry. The disadvantages of NRR are noisy operation, a particular problem with wells located in residential neighborhoods, and extra maintenance cost.

The SRC allows backspin by separating the two coupling components, thereby releasing the motor from the head shaft. It should only be installed with time-delay restart control to prevent re-starting during backspin. The SRC has lower initial and maintenance costs than the NRR coupling.

Thrust Bearing

The motor thrust bearing carries the axial downthrust on the shaft. It is supported by the frame of the motor, or right-angle gear drive, and the discharge head. There are three types of thrust bearings, which vary primarily in their load-carrying capability and life expectancy.

Kingsbury bearings, known as "pivoted segmental-thrust bearings," ride on a thin film of oil and have very long life if the oil is clean and the bearing rating is not exceeded. Their drawbacks are high cost, low-energy efficiency due to friction loss in the plates, and lack of ready availability.

There are two types of antifriction bearings. Spherical roller bearings are used when more thrust capacity and reasonably long life are needed. They are cheaper and more energy efficient than Kingsbury bearings and will carry as large a load, but have a shorter life. Water cooling of the oil bath is almost always required with Kingsbury and spherical roller bearings. Angular contact antifriction ball bearings have the lowest thrust capacity and shortest life but are the cheapest and most energy efficient due to lower friction.

Antifriction bearings are rated by the Anti-Friction Bearing Manufacturers Association (AFBMA) for their life expectancy. A typical rating is "AFBMA continuous B-10 1-year minimum and 5-year average life," which means that under continuous operation with the rated thrust load applied, average bearing life will be 5 years, but 10% will fail before one year. The minimum life is equal to 20% of the average life, which is adequate for most applications. Higher ratings can be specified, such as B-10 2-year minimum and 10-year average life. Thrust bearings are specified for thrust at the bowl-assembly design point. However, the bearings should be capable of carrying momentary thrust at shutoff head, the maximum load to which they will be subjected.

Upthrust is not normally a problem in well pumps because the weight of the rotating elements usually counteracts the weak upthrust force during start-up. When upthrust is of concern, the two halves of the motor coupling are bolted together, and a split inner face is installed on the thrust bearing to hold it in place. For severe upthrust problems the shaft is restrained by mounting two angular-contact bearings back-to-back in the motor.

Discharge Head

The discharge head serves a number of essential purposes:

1. The top of the discharge head, dimensioned to fit the driver, supports and aligns the rotating elements of the pump.
2. The bottom of the discharge head supports the complete pump and driver assemblies on the foundation.
3. A column flange or coupling attaches the column pipe to the discharge head. This fitting must carry the entire weight of the column, shaft, tubing, and bowl assembly.
4. The discharge head must incorporate a water seal for the extension of the head shaft into the discharge head from the column pipe.
5. The discharge head must provide support and tensioning of the shaft-enclosing tube.
6. The discharge head must receive the flow from the column pipe and discharge it to external piping.

Discharge heads are constructed of cast iron or fabricated steel, normally with an ANSI 125 or 250 flange. The base should enclose the well casing and have sufficient bearing surface on the pad to support the weight of the pump. Some discharge heads are manufactured with a recess on the bottom to allow use over casing extended above the pad, or with a plate that raises the pump a few inches above the pad.

Tube Tension Nut Subassembly. This subassembly in oil-lubricated pumps is located in the discharge head and exerts the required tension on the shaft-enclosing tube to align the shaft. It consists of a tubing nut and tube-tension plate, and miscellaneous accessories.

Stuffing Box and Mechanical Seals. In water-lubricated drive-shaft-driven pumps, the discharge head must have a seal where the line shaft extends through the discharge head. This minimizes leakage while not restricting the free rotation of the shaft. The stuffing-box packing seal consists of rings of sealing material retained in a stuffing box around the shaft held in place by a packing gland. The packing gland is adjusted to allow sufficient leakage to cool the shaft and minimize shaft scoring. The water leakage should be collected and piped to waste.

The pressure capacity of a packing seal may be increased, up to a point, by stacking more packing material and lantern rings into the stuffing box. Advantages of a stuffing box are low initial cost and ease of maintenance. Disadvantages are the required water leakage, and shaft scoring that necessitates frequent replacement. Mechanical seals are seldom, if ever, used in water wells since sand, even in minute amounts, will destroy the seal.

Water-Lubricated Column and Shaft Assembly

Column pipe is manufactured in 10-ft lengths according to AWWA E101 Section A4.2.7. Pipe ends are machined square and straight threaded to allow the ends to butt, facilitating alignment and ensuring a tight seal at each coupling. A bearing retainer complete with a long, fluted, synthetic-rubber bearing-bushing is threaded into the center of the column coupling, with the adjacent column lengths securely butted against it. These retainers and bushings align the shaft throughout the column length.

The drive shaft, extending inside the column pipe from the bowl shaft to the head shaft, transmits drive torque to the impellers and transfers the axial hydraulic thrust to the thrust bearings. Shafting is also supplied in 10-ft lengths. The ends are machined square, with straight threads tightened until they butt against the next shaft. Line shaft and shaft bearings are manufactured in accordance with AWWA E101, Sections A4.3.4 and A4.3.5. When carbon steel shafting is used, stainless-steel shaft sleeves at the bearing surfaces are usually specified to reduce line-shaft wear.

Oil-Lubricated Column and Shaft Assembly

The column pipe used for oil-lubricated pumps is manufactured to the same standards as for water-lubricated pumps, except in 20-ft lengths. Oil-lubricated line shaft and shaft bearings are manufactured in accordance with AWWA E101, Sections A4.2.4. and A4.2.5. The shaft-enclosing tubing contains the drive-shaft bearings and conducts oil to the bearings. Oil tubing is made from Schedule 80 steel pipe in 5-ft sections according to AWWA E101, Section A4.2.6. The sections are connected by bronze couplings that serve as the drive-shaft bearings. The bearings are machined with axial or spiral grooves for passage of lubricating oil. Stabilizers, placed at intervals within the 20-ft column pipe, support the shaft-enclosing tube.

The shaft-enclosing tube terminates in the discharge head on top of the water-passage elbow. A tubing-nut assembly keeps the tube in tension, ensuring proper shaft alignment. This assembly incorporates a seal that prevents water from leaking out of the discharge head. An oil-feed system, located at the discharge head, drips oil into the tube while the pump is running. It consists of a lubricator that regulates the flow of oil, a window for the operator to visually inspect the flow and a solenoid valve to stop the flow when the pump is not running. A 1-gal oil reservoir is normally sufficient.

Bowl Assembly

Cavitation is usually not a problem in a pump that has sufficient submergence and does not produce water in excess of designed bowl rating. Therefore the three factors generally affecting the life and efficiency of the bowl assembly are abrasion, wear, and corrosion. Abrasion occurs when there is sand in the water and high water velocity in the impellers and water passages of the bowl cases. Wear, which is due to mechanical friction, occurs at the bearing surfaces and mating surface of the impeller seals. Wear increases over time, occurs more rapidly at higher rpm, and results in decreased pump efficiency. The typical 1800-rpm pump will not wear as quickly as a 3600-rpm unit but faster than a 1200-rpm unit. Most bowl-assembly designs in a typical well environment will require bowl replacement because of wear before corrosion materially affects performance, although galvanic action may require replacement of parts.

Impellers. The pump impeller imparts velocity head to the water, which the bowl case chamber converts to pressure head. Impellers are particularly susceptible to erosion from sand, which sharpens and thins the ends of the vanes, reducing pump efficiency. They are available in two configurations: semiopen and enclosed.

The semiopen configuration has a shroud (skirt) only on top of the impeller. Positioning of a shaft-adjusting nut maintains a seal by the close tolerance between the bottom of the impeller and the bowl surface. This design offers the advantage of partially restoring efficiency and capacity of a worn impeller by adjustment, allowing longer-term production of sand-carrying water without bowl replacement. Its primary disadvantage is the difficulty, due to drive-shaft stretch, of maintaining a close running fit between impeller and seal area in deep pump settings. For this reason semiopen impellers are seldom used in pump settings greater than 100 ft, or when the pump discharges into a variable-pressure

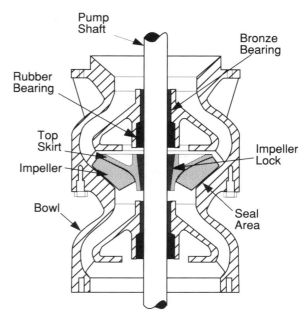

Figure 16.3. A semiopen impeller arrangement.

system (which causes either inefficiency or excessive wear). Semiopen impellers are illustrated in Figure 16.3.

Enclosed impellers are selected for most deep well pumps. This configuration also has a lower shroud (skirt) that serves as a seal, limiting bypass and recirculation (Figure 16.4). The seal contact is a peripheral mating of the bottom impeller skirt and the bowl case. Since some vertical lateral is provided in the bowl case, the impeller position relative to the bowl is less critical. With this seal, impeller-position readjustment

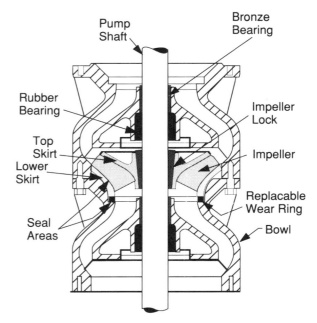

Figure 16.4. An enclosed impeller arrangement.

is normally not required. Advantages of the enclosed impeller include less field adjustment, less downthrust, and use of replaceable wear rings that may be incorporated in the bowl. Disadvantages include a slightly lower bowl efficiency (due to increased friction in the seal) and the fact that, unlike the semiopen system, efficiency can only be restored by pump removal and repair.

Impellers can be trimmed to match the bowl-assembly performance characteristics to system demand. The most common trimming method reduces the impeller's diameter and vane length by machining. In Figure 16.5, curve 1 represents the performance of a full diameter impeller. Reducing the impeller diameter lowers performance accordingly (curves 2–4). Reduced efficiency usually limits the amount of trim.

Bowl-Assembly Shaft, Bearings, and Lubrication.
Bowl-assembly shafts are usually manufactured from stainless steel, to limit corrosion and abrasion. Lubrication of the

pump shaft below the tube-adapter bearing (in the case of oil lubrication) or its entire length (in the case of water lubrication) is by the pumped water. The exception is the suction-case bearing which is prepacked with waterproof grease at assembly. The bowl bearings, functioning primarily as a shaft guide, are bronze, synthetic rubber, or a combination of both, depending on anticipated service. Rubber bearings have proved best in sand-producing wells due to their resistance to abrasion.

Materials and Coatings.
The typical bowl assembly is bronze-fitted cast iron, which consists of cast iron bowls, bronze impellers and bearings, and stainless-steel pump shaft. This material combination is relatively inexpensive and offers long life even when some sand is pumped.

Other material combinations are available at varying cost. In all-iron assemblies the bowls and impellers are cast iron with stainless-steel shaft and bronze bearings. In all-bronze assemblies, bowls, impellers, and bearings are bronze and

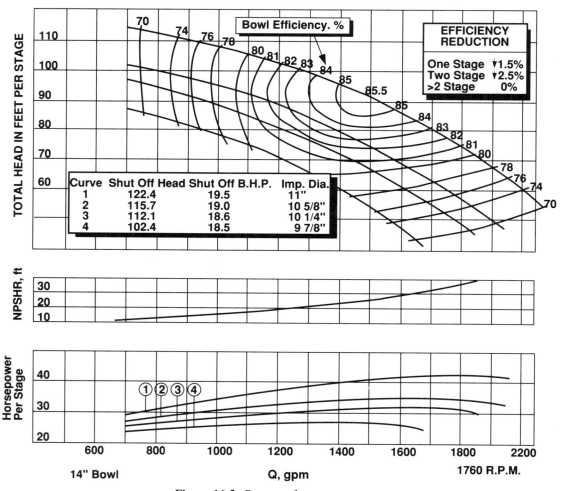

Figure 16.5. Pump performance curves.

the shaft is stainless steel. A stainless-fitted assembly consists of cast-iron bowls with stainless steel impellers and shafting, and bronze bearings. Where sand is a severe problem, an all-stainless bowl, impeller, and shafting combination may be used.

Porcelain-lined bowls and cast-iron impellers are available, and they offer corrosion and abrasion protection as well as some increase in efficiency. Other coatings with these advantages are available, but a cost-benefit analysis is needed for selection.

Steel and cast iron act as anodes to the more cathodic bronze and stainless steel. Galvanic corrosion is reduced in bowls by using combinations of metals that are close in the galvanic series. Anodic metals for small parts should be avoided, particularly in areas with aggressive waters.

In highly conductive well water, carbon-steel bolts connecting bowl cases will corrode. This is aggravated by the small size of the bolts compared to the larger size of the cathodic impellers and bowls. Corrosion may be so severe that bowl disassembly requires removal with a cutting torch. Failure during pump operation causes the bowl cases to separate, destroying the bowl assembly.

Plated bolts provide some protection; however, a better choice is bolts made of annealed stainless steel such as AISI Type 316 or 304. These bolts will not fail and will likely be easy to disassemble.

As a general precaution, bowl bolts should be coated with an antiseize compound that insulates both the bolt and bowl case threads, and seals out water. Otherwise, galvanic cells will develop at the threads due to moisture and the stress applied by tightening. Stainless cap screws will not corrode, but the threads in the anodic cast-iron cases will, eventually causing failure.

Suction Pipe and Strainer

A suction pipe connected to a strainer is a common accessory for deep-well turbine pumps, although its usefulness is questionable. It directs incoming flow to the bowl assembly at a velocity less than the flow in the column pipe, which ensures that any sand particles will be carried to the surface and not accumulate in the column pipe. Normally, the suction pipe is the same diameter as the column pipe. A suction case, located below the bottom bowl, connects to the suction pipe and directs flow to the eye of the first impeller. If a suction pipe is not used, this case takes the form of a suction bell.

Strainers are supplied in various configurations: the most common for deep wells is the cone type. Its open area of three to four times that of the pump suction is helpful in preventing relatively large particles from entering the pump, but does not protect it from fine sand.

Occasionally, conditions will dictate that the pump suction be installed in the well screen. The problem of turbulence can be minimized by equipping the pump with a long perforated suction closed at the bottom. Typically, this suction is 20-ft long, fabricated from well screen or standard column pipe. If column pipe is used, a large number of ½-in. holes are drilled throughout its length. The number should provide total inlet area several times the cross-sectional area of the column pipe. Damage caused by the water originating from above is minimized by allowing it to enter the pump over a long vertical distance. The backwash effect when the pump is stopped is also minimized.

Discharge Case

The discharge case serves as the adapter to accommodate the column and shaft connections, and to redirect the water as it passes from the top bowl. Velocity head is reduced by the discharge case. With oil-lubricated pumps it has bypass ports and accommodates the tubing adapter bearing.

16.5 SPECIFIC SPEED

There are three types of pumps within the general classification of centrifugal pumps: vertical turbine, mixed flow, and axial flow (propeller). Differentiation between these types depends on the proportional distribution in radial and axial components of the kinetic energy imparted to the pumped liquid by the differing impeller designs. This differentiation can be established by computing the specific speed of the pump by the equation

$$N_s = \frac{nQ^{0.5}}{T_H^{0.75}} \tag{16.1}$$

where

N_s = specific speed of the pump,
n = rotating speed [rpm],
Q = discharge rate at optimum efficiency [gpm],
T_H = total bowl assembly head per stage [ft] (for vertical turbine pumps).

The specific speed of vertical turbine pumps ranges from 1000 to 4000, mixed flow from 4000 to 10,000, and axial flow higher than 10,000 with some overlap between types. Mixed and axial flow pumps, which generally deliver large capacities at relatively low heads, are usually inappropriate for water wells.

16.6 OPERATING PARAMETERS

To select pump components, a number of parameters are considered.

Bowl-Performance Curves

Pump manufacturers supply bowl-performance curves for each of their standard units, covering a full range of bowl and impeller combinations for a single-stage unit operating at a given rotating speed. These curves give data, including discharge rate against total dynamic head, efficiency variations, horsepower, and required submergence. Multiple curves depicting performance with different impeller trims are also available.

Figure 16.5 shows performance characteristics for a particular bowl-impeller combination.

Rated Capacity. This bowl and impeller combination offers a pumping range from 1100 gpm to 1800 gpm at 80% and 81% efficiency, respectively, with peak efficiency of 85.5% at 1485 gpm. For purposes of analysis, a rated capacity of 1485 gpm with full (untrimmed) impellers is assumed (curve 1).

Rated Head. A bowl-assembly head per stage of 75 ft to 105 ft, with 92 ft at rated capacity is shown.

Total Bowl-Assembly Head. Most deep-well turbine pumps consist of multistage units. The bowl-assembly head is the total of the heads developed per stage. As an example, the bowl-impeller combination in Figure 16.5, if assembled as an eight-stage unit, develops a total bowl-assembly head of 736 ft at rated capacity. This is the working head available from the pump-bowl assembly.

Shutoff Head. Shutoff head is the total head at zero discharge with the pump running. For this bowl-impeller combination, shutoff head would be 122.4 ft per stage or 979.2 ft for eight stages. This head condition is an important factor in calculating required driver horsepower and thrust-bearing capacity. Only in very special cases should a well pump be allowed to operate against its shutoff head, for this can cause severe damage if sand in the column settles into the bowls. Continuous operation at the shutoff head causes overheating, and the impeller operation in steam will lead to bearing or wear ring failure.

Efficiency. A family of iso-efficiency curves is included with the bowl-performance data of Figure 16.5, varying in this case from 70% to a peak efficiency of 85.5%. A table of efficiency reductions for single or multistage configurations is also included.

Bowl-Assembly Horsepower. The bowl-assembly performance curves define required horsepower delivered at the bowl assembly. Bowl-assembly horsepower requirements may also be calculated from the formula

For this example with eight stages, the calculation would be

$$hp = \frac{(1485)(736)}{(3960)(0.855)} = 322.8 \text{ hp}$$

The solution is 40.35 hp per stage, which corresponds with the curve value.

For this particular bowl-impeller configuration, horsepower requirements continue to increase as capacity increases and head decreases (as shown on Figure 16.5). This is not always the case. The relationships of these operating parameters are crucial in final pump and prime mover selection, because of head variations that may be imposed on the pumping unit by the system.

Net Positive Suction Head (NPSH). A turbine pump will operate satisfactorily only if sufficient water can freely enter the first-stage impeller. This condition will be satisfied when the pressure available to deliver it is equal to or exceeds the net positive suction head required (NPSHR). NPSHR is usually shown with the bowl-performance curve data. For this particular bowl-impeller combination it varies from 10 to 40 ft, with 26 ft required at rated capacity and peak efficiency.

Net positive suction head available (NPSHA) is that available from the pump's environment. It consists of the head of water above the impeller plus atmospheric pressure less vapor pressure and friction loss in the suction pipe. NPSHA should always exceed NPSHR to avoid cavitation. Cavitation occurs when the absolute pressure of moving water drops below the vapor pressure of the water. Small bubbles of water vapor form at the impeller suction and rapidly collapse as pressure increases when the water flows to the impeller discharge. Impeller metal fatigue occurs, resulting in pitting. When heard, it sounds like rocks are passing through the pump. Cavitation is usually worse at higher discharge rates.

The equation for calculation of available NPSH is

$$NPSHA = H_a + H_s - H_f - H_{vp} \qquad (16.3)$$

where

H_a = atmospheric pressure [ft of water],

H_s = elevation of the water above the impeller eye while pumping [ft of water],

H_f = friction loss in suction piping [ft of water],

H_{vp} = absolute vapor pressure of the water at pumping temperature [ft of water].

$$hp = \frac{\text{capacity (gpm)} \times \text{total bowl-assembly head (TBH)(ft of water)}}{3960 \times \text{bowl efficiency}} . \qquad (16.2)$$

Rated Speed. Each performance curve is based on a specific rotational speed and is accurate in expressing performance only at that speed. The most common drivers for deep-well turbine pumps are 60-Hz four-pole electric induction motors, with a synchronous speed of 1800 rpm and full-load speed varying between 1760 and 1790 rpm. The performance curves shown in Figure 16.5 are for 1760 rpm, the rated speed of this particular combination. It is important to correct for true full-load speed because power consumption varies with the cube of the speed. The motor manufacturer will furnish data on full-load speed of the selected motor under actual load and field power conditions.

Pump Speed and Diameter. The relationship between pump capacity, head, horsepower, and rotating speed is described by the pump affinity laws. These laws state that capacity (q) varies directly with rotating speed (n), head (h) varies with the square of the rotating speed, and horsepower (hp) varies with the cube of the rotating speed:

$$\frac{q_1}{q_2} = \frac{n_1}{n_2} , \quad \frac{h_1}{h_2} = \frac{n_1^2}{n_2^2} , \quad \frac{hp_1}{hp_2} = \frac{n_1^3}{n_2^3} .$$

Since efficiency does not change with speed for any point on a given bowl-performance curve, these relationships may be used to redraw a bowl-performance curve for other speeds.

However, speeds should not exceed the limits indicated by the manufacturer.

High speed can generate high output from a relatively small-diameter pump. This lowers initial cost at the expense of maintenance cost and reliability. Noise can be a problem at higher speed. Pumps operating at a nominal 3600 rpm tend to be noisy and have high maintenance costs, pumps operating at 1200 rpm tend to be too large and expensive for most well uses, and pumps operating at 1800 rpm usually are a good compromise. The discharge rate of a pump operating at a given rate is a function of the diameter of the impeller and the bowl.

16.7 PUMP LOSSES

The following hydraulic and mechanical losses must be considered in design computations and pump selection.

Column Losses

Friction loss in the column pipe is computed from data such as those presented in Figure 16.6. This figure is applicable to column pipe and shaft-enclosing tube up to and including 16-in. column with 4-in. tube. These data may also be used for water-lubricated pumps since, for equivalent shaft and

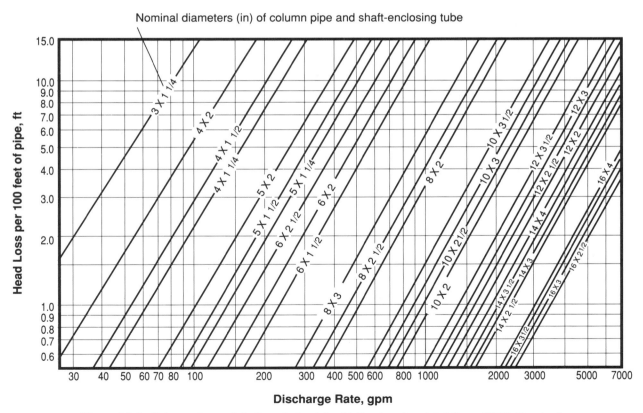

Figure 16.6. Column pipe and shaft-enclosing tube friction loss (reprinted from *AWWA Standard for Vertical Turbine Pumps-Line Shaft and Submersible Types*). By permission. Copyright © 1988, American Water Works Association.

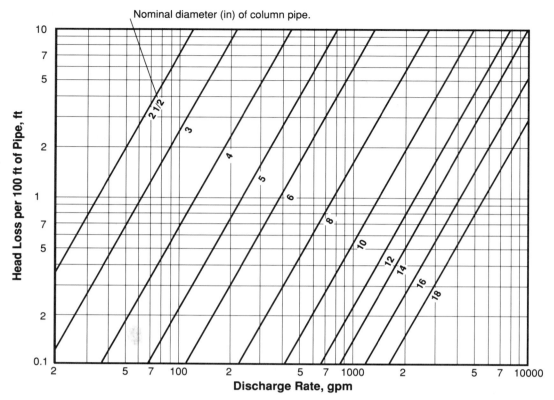

Figure 16.7. Column friction loss (reprinted from *AWWA Standard for Vertical Turbine Pumps-Line Shaft and Submersible Types*). By Permission. Copyright © 1988, American Water Works Association.

column sizes, the turbulence loss caused by the rotating shaft is roughly equal to the friction loss from the tube.

In submersible pumps, column friction loss may be computed from Figure 16.7. Suction pipe and strainer losses are not usually significant except where a long suction pipe is installed.

Discharge Head Losses

In most applications, discharge head losses are insignificant. If necessary, losses may be taken from Figure 16.8.

Shaft Losses

Shaft losses, due to rotation of the shaft within the bearings, are mechanical losses measured in horsepower per 100 ft of shaft. They may be computed from the data contained in Figure 16.9.

Load Capacity. Shaft diameter and material specifications must be selected prior to shaft-loss calculations. Required torque, hydraulic thrust, and the combined shear stress of these forces must be taken into account. In addition the shaft diameter selected must limit the elongation resulting from the total maximum thrust, to conform to bowl-lateral constraints. Table 16.1 defines the maximum recommended horsepower for a given shaft diameter, taking into account

the effect of hydraulic thrust and the weight of the shaft and suspended rotating parts. This table applies to shaft steel

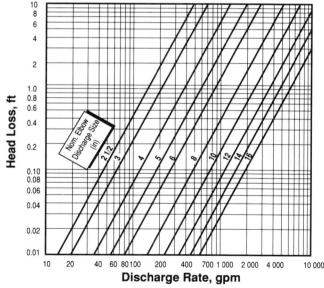

Figure 16.8. Discharge head loss (reprinted from *AWWA Standard for Vertical Turbine Pumps-Line Shaft and Submersible Types*). By Permission. Copyright © 1988, American Water Works Association.

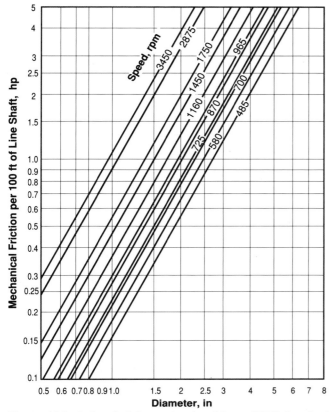

Figure 16.9. Drive shaft losses (reprinted from *AWWA Standard for Vertical Turbine Pumps-Line Shaft and Submersible Types*). By Permission. Copyright © 1988, American Water Works Association.

having a minimum yield strength of 40,000 psi and tensile strength of 67,000 psi.

The shaft diameter required to transmit the torque and to withstand the combined thrust can also be found from the formula

$$D^3 = \frac{16}{\pi S} \sqrt{\left(\frac{TD}{8}\right)^2 + \left(\frac{396,000P}{2\pi n}\right)^2} \,, \quad (16.4)$$

where

D = shaft diameter at root of thread [in.],

T = total axial thrust (hydraulic thrust plus weight of shaft and impellers) [lb],

P = power transmitted through the shaft [hp],

n = rotational speed of the shaft [rpm],

S = allowable combined sheer stress (limited to 30% of the elastic limit or 18% of the tensile strength of the shafting steel) [psi].

Elongation. Shaft elongation under operating conditions consists of two components: the weight of the shaft and impellers, and the axial hydraulic thrust developed by the impellers. The first of these elongations will occur at installation and is of interest to ensure adequate top-shaft-thread length. It may be computed by the formula

$$e_1 = \frac{(L_s)(12)(W)}{EA} \,, \quad (16.5)$$

TABLE 16.1 Shaft Selection Chart for Type B Material[a]

Shaft Diameter in.	Speed rpm	Pump Thrust–*1000 lb*								
		1	2	3	5	7.5	10	15	20	30
		Power Rating–*hp*			(*kW* = *hp* × *0.746*)					
¾	3500	39.7	38.8	37.4	32.4					
	2900	32.9	32.2	31.0	26.9					
	1760	20.0	19.5	18.8	16.3					
	1460	16.6	16.2	15.6	13.5					
1	3500	94.5	93.8	93.0	89.5	82.5				
	2900	78.3	77.7	77.0	74.2	68.4				
	1760	47.5	47.2	46.7	45.0	41.5				
	1460	39.4	39.1	38.7	37.3	34.4				
1³⁄₁₆	3500	167.0	167.0	166.0	163.0	157.0	149.0			
	2900	138.4	138.4	137.5	135.1	130.1	123.5			
	1760	84.0	84.0	83.5	82.0	79.0	75.0			
	1460	69.6	69.6	69.2	67.9	65.5	62.1			
1⁷⁄₁₆	3500			296.0	294.0	289.0	283.0	264.0		
	2900			245.3	243.6	239.5	234.5	218.7		
	1760			149.0	146.0	145.0	142.0	133.0		
	1460			123.5	121.0	120.1	117.7	110.2		
	1160			98.3	97.6	96.0	94.0	87.6		
	960			81.4	80.8	79.5	77.8	72.5		

(*Table continues on p. 328.*)

TABLE 16.1 (*Continued*)

Shaft Diameter in.	Speed rpm	Pump Thrust—*1000 lb*								
		1	2	3	5	7.5	10	15	20	30
		Power Rating—hp				(kW = hp × 0.746)				
1½	3500			336.0	334.0	330.0	324.0	306.0		
	2900			278.4	276.7	273.4	268.5	253.5		
	1760			169.0	168.0	166.0	163.0	154.0		
	1460			140.0	139.2	137.5	135.1	127.6		
	1160			111.2	110.7	109.2	107.2	101.4		
	960			92.0	91.6	90.4	88.7	83.9		
1¹¹⁄₁₆	1760			252.0	251.0	248.0	246.0	239.0	227.0	
	1460			209.1	208.2	205.7	204.1	198.3	188.3	
	1160			166.0	165.0	164.0	162.0	157.0	150.0	
	960			137.4	136.6	135.7	134.1	129.9	124.1	
	860			123.0	122.0	121.0	120.0	117.0	111.0	
	710			101.6	100.7	99.9	99.1	96.6	91.6	
1¹⁵⁄₁₆	1760				393.0	392.0	390.0	382.0	373.0	345.0
	1460				326.0	325.2	323.5	316.9	309.4	286.2
	1160				259.0	258.0	257.0	252.0	246.0	228.0
	960				214.3	213.5	212.7	208.6	203.6	188.7
	860				192.0	192.0	191.0	187.0	182.0	169.0
	710				158.5	158.5	157.7	154.4	150.3	139.5
2³⁄₁₆	1760				578.0	577.0	576.0	570.0	562.0	538.0
	1460				479.5	478.7	477.8	472.8	466.2	446.3
	1160				382.0	381.0	380.0	376.0	371.0	355.0
	960				316.1	315.3	314.5	311.2	307.0	293.8
	860				283.0	282.0	281.0	279.0	275.0	263.0
	710				233.6	232.8	232.0	230.3	227.0	217.1
2⁷⁄₁₆	1760					816.0	815.0	810.0	802.0	781.0
	1460					676.9	676.1	671.9	665.3	647.9
	1160					537.0	537.0	533.0	529.0	515.0
	960					444.4	444.4	441.1	437.8	426.2
	860					398.0	398.0	395.0	392.0	381.0
	710					328.6	328.6	326.1	323.6	314.6
2¹¹⁄₁₆	1760						1070.0	1062.0	1055.0	1035.0
	1460						887.6	881.0	875.2	858.6
	1160						703.0	700.0	696.0	682.0
	960						581.8	579.3	576.0	564.4
	860						520.0	518.0	515.0	505.0
	710						429.3	427.7	425.2	416.9

[a]Steel with a minimum elastic limit of 40 000 psi and a minimum tensile strength of 67 000 psi (after AWWA Standard E101-7).

where

e_1 = initial shaft elongation [in.],

L_s = total shaft length [ft],

12 = conversion from feet to inches,

W = weight of shaft and impellers (from manufacturer's data) [lb],

E = modulus of elasticity [psi],

A = cross-sectional shaft area at full diameter [in.^{2}].

The second of these elongations, which occurs during a full-load operation, is of interest to head-shaft thread length but, more important, to impeller adjustment within the available bowl lateral. This is particularly true with semiopen impellers that depend on maintenance of precise bottom clearance for a seal. This elongation is less critical with closed impellers that seal equally anywhere within the bowl lateral. However, particularly with deep settings, there must be sufficient bowl lateral to allow for downthrust. Column stretch, which also affects these adjustments, may also be computed but is normally insignificant.

Dynamic shaft elongation may be computed by the formula

$$e_2 = \frac{12(\text{TBH})(L_s)(K_t)}{EA} , \qquad (16.6)$$

where

e_2 = dynamic shaft elongation [in.],
12 = conversion from feet to inches,
TBH = total pump bowl assembly head [ft],
L_s = total shaft length [ft],
K_t = thrust constant (furnished by manufacturer) [lb/ft],
A = cross-sectional area of shaft at full diameter [in.^{2}],
E = modulus of elasticity [psi].

The total shaft elongation during the operation is the sum of both elongations.

Thrust-Bearing Loss

Prior to computing thrust-bearing loss, total thrust (T) must be determined. As in the case of elongation, it consists of two components: the weight of the shaft and impellers or static thrust, and the hydraulic thrust developed by the impellers.

The weight of shaft and impellers can be found from data supplied by the pump manufacturers. Hydraulic thrust is calculated by multiplying the thrust constant (K_t) value for the selected impeller-bowl combination by the total bowl-assembly head. The K_t value for the impeller-bowl combination is derived by the manufacturer through laboratory tests.

The thrust-bearing loss may then be approximated by the following empirical equation

$$\text{hp}_t = \frac{0.0075(n)}{100} \left(\frac{T}{1000} \right) , \qquad (16.7)$$

where

hp_t = thrust-bearing loss [hp],
n = operating rotating speed [rpm],
T = total hydraulic and static thrust [lb].

Submersible Cable Losses

Although there will be no line-shaft losses with a submersible pump, there will be electrical-resistance cable losses from the ground surface to the motor leads. This loss can be found from Figure 16.10 after selection of pump setting and wire size.

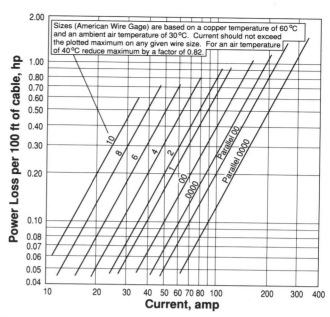

Figure 16.10. Power loss for three-conductor copper cable (reprinted from AWWA *Standard for Vertical Turbine Pumps-Line Shaft and Submersible Types*). By permission. Copyright © 1988, American Water Works Association.

16.8 SYSTEM EFFICIENCIES

A number of pumping-system component efficiencies determine overall pump efficiency and system operational cost.

Bowl-Assembly Efficiency

Bowl-assembly efficiency is the ratio of bowl-assembly output to bowl-assembly input.

$$E_b = \frac{Q(\text{TBH})}{3960(\text{hp}_b)} , \qquad (16.8)$$

where

E_b = bowl-assembly efficiency [%],
Q = pump bowl discharge rate [gpm],
TBH = total bowl head [ft],
hp_b = power input to bowl assembly [hp].

Pump Efficiency

The complete assembled and installed pump is the bowl assembly plus components needed to deliver water from the suction to the outlet of the discharge head, not including the driver. Pump efficiency is computed as follows:

$$E_p = \frac{H_1(Q)}{3960(\text{hp}_1)} , \qquad (16.9)$$

where

E_p = pump efficiency [%],

H_1 = total bowl-assembly head minus column and discharge head losses [ft],

Q = pump discharge rate [gpm],

hp_1 = bowl-assembly horsepower input plus shaft and thrust-bearing losses [hp].

Motor Efficiency

Motor efficiency is given by the manufacturer from test data.

Overall Efficiency

Overall efficiency is defined as the combined efficiency of the entire assembly, including all pump components and the driver. It is calculated as

$$E_o = E_p \times E_m , \qquad (16.10)$$

where

E_o = overall efficiency,

E_p = pump efficiency,

E_m = motor efficiency.

Wire-to-Water Efficiency

Wire-to-water efficiency is a measure of the efficiency of the complete facility considering external piping losses, external elevation head, and all internal pump losses. It is an index of the effectiveness of the entire system in delivering water from its pumping level to its final delivery point and is the efficiency most easily determined in the field.

Wire-to-water efficiency is obtained from

$$E_w = \frac{0.7457(Q)(Z_{st})}{3960(kW_i)} , \qquad (16.11)$$

where

E_w = wire-to-water efficiency,

Z_{st} = total static lift at the specified discharge rate and equals the sum of the drawdown, depth to static water level, and lift to the top of the reservoir (Figure 16.11) [ft],

Q = discharge rate [gpm],

kW_i = electrical power input measured at the power meter or motor terminals [kW].

Analysis of wire-to-water efficiency is used to budget and evaluate alternative system designs.

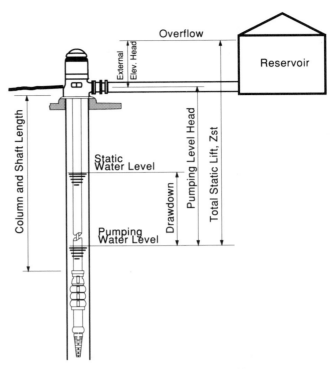

Figure 16.11. External head conditions.

16.9 SYSTEM CONSIDERATIONS

The proper design and selection of pumping equipment requires consideration of many well-, pump-, and system-operating parameters, including

1. Inside diameter and alignment of casing.
2. Static water level.
3. Well test data.
4. Seasonal and long-term hydrologic variations.
5. Availability and cost of energy sources.
6. External pressure variations.
7. System demand characteristics (especially peak demand).

Computation of the external head must take into account variations in static lift (Z_{st}) imposed by elevation change and loss from pipe friction external to the pump. It is desirable to impose a low external head when possible. A relatively constant external head is desirable, since variations will cause corresponding variations in the pumping capacity. Wide pressure variations require caution in planning direct discharge from a well pump into a distribution system. Direct discharge may also cause system operating difficulties if sand is produced. Accumulation and velocity-induced movement of the sand within the system will wear meters, valves, and intermediate booster pumps. Initial cost is less with direct-system connection, although there is risk of excess pump cycling. Overcycling can be a major cause of well deterioration and decreased operating efficiency. Direct discharge of the well pump into adequately sized storage is recommended. This allows relatively constant external head, long-term pumping cycles and settling time for any sand produced.

The velocity head developed in the system is dissipated when the discharge is vented to the atmosphere, and this is therefore a loss to the overall system. Under most conditions this loss is small and can safely be neglected; however, its relative magnitude should be understood. It is expressed as

$$h_v = \frac{v^2}{2g}, \qquad (16.12)$$

where

h_v = velocity head [ft],

v = average velocity in the pipe at the point of pressure head measurement [ft/sec],

g = acceleration due to gravity [32.2 ft/sec^2].

Miscellaneous

Well test-data collection is discussed in Chapters 14 and 15. Casing diameter and alignment are discussed in Chapter 6.

16.10 DESIGN POINT

The design point is the selected operating discharge rate and head on the pump-performance curve. It is a combination of the discharge rate and bowl-assembly head, selected from total head-loss calculations.

Bowl Selection

There are many sizes and types of pump bowls available. They vary in operating speed, diameter, efficiency, steepness of performance curve, and output. Among different manufacturers, however, the performance curves are generally similar for a given speed and diameter. The designer initially selects the desired bowl characteristics and usually several options are found.

The objective of bowl selection is to obtain the desired operating characteristics at highest efficiency. The total head required is achieved by stacking the appropriate number of bowl stages. Bowls that are 2 to 4 in. smaller than the casing's inside diameter should be selected, and they should have the highest efficiency throughout the expected head-operating range as well as at design point.

Steep versus Flat Pump Curve

Bowls with a steep pump curve have a shutoff head that is much higher than peak efficiency head. With a flat curve the shutoff head may be close to the peak efficiency head.

The choice of a steep or flat curve for bowl design is based on the external head range. Flat curves are desirable in systems requiring a large change in discharge rate for a small change in head, such as in a system where constant pressure is desirable.

Steep performance curves are typical and desirable for well pumps that usually operate over a wide head range. They display a small change in discharge rate for a large change in head. A constant discharge rate is beneficial in minimizing sand production. As the impellers wear, the discharge will remain more constant than with a flat performance curve.

Surface Booster Pumps

When the external head is high and the pump-housing casing inside diameter too small to accommodate an appropriate bowl assembly, surface booster pumps can be added to the system. Usually, the pumped water is discharged into a tank, which is used as a source for the booster. The disadvantages of boosters are the cost of equipment and the higher maintenance required by additional facilities. Advantages are a smaller well pump and a tank that can be used as a reservoir.

16.11 PUMP-SELECTION CALCULATIONS

One method of pump selection begins with the construction of a system head curve from a plot of system head loss against flow rate. A selected pump-performance curve can then be superimposed on the system head curve, and the point of intersection will be the design point for head and capacity. This method is effective when system head is primarily a function of losses from pumping directly into an elevated tank, but it becomes cumbersome if other head variables control.

A second method involves first establishing the head range in which the pump is most likely to operate during its useful life. The design point is found from the midpoint of this head range together with the desired discharge rate. A pump curve may be chosen for a bowl assembly that will be most efficient at this design point but will also operate safely and efficiently at the high and low limits established.

With design head and capacity established, an algorithm for line-shaft pumps is used to optimize pump-component design. Figure 16.11 sets forth various external and static head conditions, and Figure 16.12 lists the steps of the algorithm. This concept is developed in the sections that follow. The numbering system used corresponds to that of the algorithm.

Items	High Conditions	Low Conditions
1. System Condition		
a. Desired Capacity (gpm)	_____	_____
b. Current Pumping Level at Desired Capacities (ft)	_____	_____
c. Seasonal Pumping Level Changes (ft)	_____	_____
d. Long Term Pumping Level Changes (ft)	_____	_____
e. External Elevation Head (ft)	_____	_____
f. External Friction Loss (ft)	_____	_____
g. Velocity Head (ft)	_____	_____
h. Estimated Column Loss (ft)	_____	_____
i. Estimated Total Bowl Assembly Head (ft)	_____	_____
j. Total External Head (ft)	_____	_____

Items	Base Conditions
2. Selection Parameters	
a. Pump Capacity (gpm)	_____
b. External Head (ft)	_____
c. Speed (rpm)	_____
d. Pump Setting (ft)	_____

Items	Base Data	Interim Data	Final Data
3. Pump Selection Calculations			
a. Column Size (in)	_____	_____	_____
b. Column Friction Loss (ft)	_____	_____	_____
c. Discharge Head Selection	_____	_____	_____
d. Discharge Head Loss (ft)	_____	_____	_____
e. Total Bowl Assembly Head (ft)	_____	_____	_____
f. Bowl-Impeller Selection	_____	_____	_____
g. Bowl Head per Stage (ft)	_____	_____	_____
h. Bowl Efficiency (%)	_____	_____	_____
i. Bowl Assembly Horsepower (hp)	_____	_____	_____
j. Shaft Selection	_____	_____	_____
k. Shaft-Impeller Total Weight (lb)	_____	_____	_____
l. Hydraulic Thrust (lb)	_____	_____	_____
m. Total Thrust (lb)	_____	_____	_____
n. Shaft Stretch (in)	_____	_____	_____
o. Thrust Bearing Loss (hp)	_____	_____	_____
p. Shaft Loss (hp)	_____	_____	_____
q. Pump Horsepower Required (hp)	_____	_____	_____
r. Motor Efficiency (%)	_____	_____	_____
s. Prime Mover Required (hp)	_____	_____	_____
t. Pump Efficiency (%)	_____	_____	_____
u. Overall Efficiency (%)	_____	_____	_____
v. Wire-to-Water Efficiency (%)	_____	_____	_____

Figure 16.12. Pump-selection algorithm.

System Conditions

1a. With an analysis of maximum safe pumping rate (Chapters 14, 15) and system demand, choose the pump discharge rate as the lesser of the two. This is the maximum capacity desired from the pumping facility throughout its useful life. From selected available bowl-performance curves, estimate the most likely range of capacity at which the pump will operate with design capacity at the midpoint.

1b. Based on these upper and lower limits, select the corresponding pumping levels from the step draw-down test.

1c. Seasonal pumping level changes are estimated from data based on knowledge of seasonal variations of the static level of other wells in the area.

1d. Long-term variations in pumping levels are evaluated from available data and ground water models.

1e. External elevation head, as shown in Figure 16.11, is the elevation difference from the centerline of the pump discharge to the overflow of the receiving reservoir. High and low limits would be the overflow and minimum anticipated level respectively. If the well discharges directly into a pressurized system, pressure recorders will show maximum and minimum expected external heads.

1f. External friction loss is that head loss resulting from the external piping and accessories.

1g. Velocity head will be negligible under most conditions as discussed, but it may be computed from Equation 16.12.

1h. With estimated depth of pump setting, column friction loss can be estimated from Figure 16.6. At the high range of anticipated discharge rates friction loss should not exceed 5 ft per 100 ft of column.

1i. Determine the preliminary values of probable high and low conditions of total-bowl-assembly head as the sum of 1b through 1h. The design point is the midpoint between these values and represents the most likely head that the pump will operate at over its life. This value should be used in the initial selection of the bowl and impeller combination, with head and discharge rate at a point of maximum efficiency.

1j. The sum of all heads, not including velocity head (1g) and column loss (1h), constitute the high and low conditions of the total external head. The design point relative to external conditions is the midpoint of these perceived conditions. The sum of these heads is sometimes referred to as the "total dynamic head" (TDH).

Selection Parameters

The following basic design parameters are required for use in pump selection.

2a. Design pump capacity as derived from the step-drawdown test and the system conditions analysis.

2b. External head as computed from item 1j.

2c. Pump speed predicated on the characteristics of the prime mover and the selected pump performance curve.

2d. Pump setting based on items 1b, 1c, and 1d as well as the NPSHR of the bowl-impeller combination initially selected.

Calculations and Selection

Selection of pump components is an iterative process interdependent on preliminary and subsequent selection and computations. Listed in Figure 16.12 are the base data (first data iteration), interim data (the second iteration if required), and final data when all calculations and selection have been finalized.

The pump selection algorithm shown in Figure 16.12 continues:

3a. *Column Size*. With the data from Figure 16.6 and selected pumping capacity (2a), the column size may be selected predicated on a maximum allowable 5 ft of column head loss per 100 ft. Depending on well conditions, a sufficiently high annular velocity must be maintained in the column to ensure that all sand particles passing through the pump will be transported to the surface. Enter this selection under base data until column friction loss and fluid velocity have been calculated and the adequacy of the final shaft diameter has been checked. After final revision, enter under final data.

3b. *Column Friction Loss*. With the column size selected (3a) and the pump setting (2d) known, column friction loss may be computed from Figure 16.6. Enter under base data.

3c. *Discharge Head*. Select the discharge head based on column diameter, anticipated motor or gear head size, and discharge pipe size, and enter under base data. After all computations enter under final data.

3d. *Discharge Head Loss*. From Figure 16.8 enter the head loss through the discharge head under base data.

3e. *Total Bowl-Assembly Head*. This item is the sum of items (2b) external head, (3b) column friction loss, (3d) discharge head loss, and (1g) velocity head. Enter under base data until all computations and selections have been checked; then enter under final data.

3f. *Bowl-Impeller Selection.* Using pump capacity (2a), selected speed (2c), and total bowl-assembly head (3e), select a bowl-impeller combination to provide insofar as possible the following:

- High efficiency.
- Minimum number of stages.
- Preferred curve shape (steep or flat), depending on probable system conditions.

Enter manufacturer's bowl and impeller designation under base data. When all calculations and selections are confirmed, enter under final data.

3g. *Bowl Head per Stage.* Divide total-bowl-assembly head (3e) by available head per stage from the selected bowl curve. Round off to the next higher number of stages, and enter under base data. Compare with the selected bowl-performance curve. This will indicate impeller trim required, the required head per stage, and whether selection has been appropriate relative to efficiency. If trim has excessively reduced efficiency, other bowl-impeller selections may be superior, and steps 3f and 3g may need repeating.

3h. *Bowl Efficiency.* From final review of steps 3f and 3g enter the bowl efficiency under base data until all calculations and selections are completed; then enter under final data.

3i. *Bowl-Assembly Horsepower.* Compute bowl-assembly horsepower from Equation 16.2, using pump capacity (2a), total bowl-assembly head (3e), and bowl efficiency (3h). Enter under basic data.

3j. *Shaft Selection.* With Figure 16.9, Equation 16.4, and Table 16.1, make initial shaft material and diameter selection. For this use the bowl-assembly horsepower (3i) and estimated total thrust. This selection must be reviewed against Figure 16.6, as well as final total thrust (3m) and final pump horsepower requirements (3q), with appropriate corrections if required. Enter under base data until final computations and selection are made and confirmed; then enter under final data.

3k. *Shaft and Impeller Total Weight.* From manufacturers tables find the total weight of shafting and impellers. Enter this value under base data.

3l. *Hydraulic Thrust.* This value is the K_t value of the selected impeller-bowl combination times the total bowl-assembly head (3e). Enter under base data.

3m. *Total Thrust.* Enter under base data the sum of items 3k and 3l. Review against the estimated value used in 3j, and correct if required.

3n. *Shaft Stretch.* From Equation 16.6 with the hydraulic thrust value (3l) and total bowl-assembly head (3e), compute anticipated shaft stretch (shaft stretch resulting from hydraulic thrust must be accommodated by available lateral within the selected bowl). This value must be reviewed relative to alternate shaft selection or bowl modification. Enter under base data.

3o. *Thrust-Bearing Loss.* Compute the horsepower loss in the thrust bearing from Equation 16.7. Enter under base data.

3p. *Shaft Loss.* With confirmed shaft size (3j) and Figure 16.9, find the shaft horsepower loss. Enter under base data.

3q. *Pump Horsepower Required.* This value is the sum of bowl-assembly horsepower (3i), thrust-bearing loss (3o), and shaft loss (3p). Enter under base data.

3r. *Motor Efficiency.* This depends on size, anticipated loading, and type and must be obtained from the motor manufacturer. Enter under base data.

3s. *Prime Mover Required.* This results from all previous reviewed and confirmed calculations and selections and is based on the pump horsepower required (3q). Depending on the operating system characteristics, it may be worthwhile to compute the maximum anticipated horsepower the pumping facility may require and select the next larger standard motor size. Predicated on the horsepower determined, it may be advisable to recalculate all relevant parameters to allow selection of a standard motor size without overload or excessive unused capacity.

3t. *Pump Efficiency.* This item is computed with Equation 16.9 with data from items 2a (pump capacity), 3q (pump horsepower required) and 3e (total-bowl-assembly head) minus items 3b (column friction loss), and 3d (discharge head loss). Enter under final data.

3u. *Overall Efficiency.* Compute this item from Equation 16.10 with data from items 3r (motor efficiency) and 3t (pump efficiency). Enter under final data.

3v. *Wire-to-Water Efficiency.* This item is computed from Equation 16.11, using item 2a (pump capacity), total lift Z_{st} (from Figure 16.11), and the power demand (kW_i) measured at, or as close as possible to, the motor terminals. Readings may be taken from the power meter at the service location but they will include power line losses and transformer losses when present.

When power input readings are unavailable, wire-to-water efficiency may be computed from the formula

$$E_w = \frac{Q(Z_{st})}{(3960)\dfrac{\text{Pump Horsepower Req. (3q)}}{\text{Motor Efficiency (3r)}}}.$$

16.12 INSTRUMENTATION AND PROTECTION

Control and instrumentation are essential for the safe and efficient operation of a pumping system and accumulation of well and pump operational data. Measuring and recording the well discharge are necessary for analysis of any system failure and supply long-term data for well and aquifer analysis.

The well should be equipped with water-level-sensing and recording instrumentation. This may be through use of an air line or a pressure-sensing device. Either is facilitated by installing a separate tube outside but attached to the casing, hydraulically connected to the well below any anticipated pumping level. This allows the instrumentation to be placed and retrieved without interference with the pumping unit.

Current and voltage instrumentation is normally housed in the motor-starter cabinet. These data are essential since any wear or defect in the prime mover will be reflected in the current load, assuming voltage has remained constant.

Usually monthly power-consumption readings will be furnished by the power company. When compared with the recorded quantity of water produced for that period, a rough check of operating efficiency is given.

Each pumping unit should be equipped with minimal fail-safe provisions for protection. These may include the following as appropriate for the particular installation:

1. Phase failure and phase reversal protection of the motor, to shut off the motor and prevent it from starting until the condition has been corrected. Phase failure, also known as a single-phase condition, is the absence of one phase in a three-phase system. Phase reversal is a condition that would cause the motor to rotate in the reverse direction from normal.

2. High-pressure sensing and lock-out, taking effect considerably above any normal operating pressure, to protect against system misoperation and pump damage that may occur from very high thrust.

3. Low pressure sensing and lock-out, to prevent overproduction from the well. Overproduction may damage the well and overload the motor, depending on pump performance characteristics.

4. A time delay sequence, to prevent restarting until counterrotation has ceased. As discussed earlier, serious damage will occur to a pump if the motor or alternative drive unit is started during backspin.

16.13 CONTROLS

It is necessary to provide some type of pumping plant control. This can be as simple as hand control with a manual on–off switch or as sophisticated as an automatic computer-based control system. If automatic control is selected, the on–off control should be activated through the most dependable system available. Reservoir level would be the best control variable for a system with a reservoir. Pressure control can be used if the system demand is fairly constant, or if the system is large enough to allow the pump to run for fairly long periods of time. Care should be taken when using pressure control to avoid excessive cycling of the pump due to pressure variations. If this becomes a problem, a pressure tank can be installed.

Remote manual control is used with systems that have central, manned monitoring. This applies especially to large geographical areas where it becomes impractical for hand control to be efficiently used.

Automatic control responds to measured variables and initiates control action without operator intervention. It provides a higher level of service than manual control by constraining system variables (flow, pressure, etc.) within narrower limits. Automatic control minimizes operator intervention under emergency conditions by instantly reacting to unusual system conditions. Labor costs are reduced since fewer operators are required, and since the need for a continuous monitoring system is eliminated.

Automatic control can be either open or closed loop. With open-loop control, the control action is independent of the controlled variable. A "pump-on-time" control would be an example of open-loop control. Closed-loop automatic control systems measure a variable that is dependent on the controlled variable. Pressure, tank level, and flow control are examples.

Storage-tank level control has several benefits. It maintains system pressure within narrower limits. Pumping can be confined to periods of reduced power rates, and storage used to meet demands during peak periods. Tank-level sensors can be float switches, electrodes in the tank, or an electronic pressure transducer. Usually the pump that supplies the tank is some distance away, necessitating transmission of the start–stop signal or the tank level across a communication medium. This medium can be a dedicated local cable, telephone lines (either leased or dial-up), or radio link.

Tank-level control can be implemented through several modes of operation. The simplest instructs the pump(s) to turn on when the level reaches a single pre-established point. If multiple pumps are involved, multiple-start points will improve operational efficiency, starting pumps to meet system demand. Level control can be integrated with time control to minimize energy costs and take advantage of favorable time-of-use power rates. The rate of tank level change can be used in more sophisticated control systems to further improve operating efficiency when time-of-use power rates are available.

Pressure control will be the most efficient option if a system has no elevated storage or if storage is remote to the supply pump. Wells are only started as required by system demand. Pressure-controlled pumps must have a pressure sensor. This can be a Bourdon tube pressure switch or an electronic pressure transducer. Pressure can be sensed either at the pump or at a location that is more representative of the general condition of the distribution system. The control system can vary from a pressure sensor turning on a single pump to sensors integrated with a computer controlling multiple pumps in an area.

Maintaining constant pressure can also be accomplished by varying the pump speed. This is done by use of an internal combustion engine, a variable-speed coupling, or an electronic control that varies the rotational speed of an electric motor.

Flow control can either maintain constant pressure or constant flow. In either case it is necessary to have some type of variable-speed pump controller.

A computer-based Supervisory Control and Data Acquisition System (SCADA) can be used to integrate these control systems into a water system. It is most economical on a per-point read basis when multiple points must be monitored and controlled. SCADA systems consist of a central computer, remote terminal units (RTUs), and a communication link.

Several variables can be read and controlled with a single remote unit. System expansion is easier with computer-based control than with a conventional wired system. It is easy to add functions at a particular site or add sites. Modern RTUs are programmable and can offer both computer control and an emergency local backup control in the event of central computer failure.

A computer-based control system generates complete data acquisition (historical and instantaneous). The summaries reported are superior to traditional circular or strip charts. Critical variables and current data for operators are summarized for quick reference. These data can also be used for special system studies and system expansion. Computer-control systems are flexible. The operators can react to changing conditions such as power rates, pumping water levels, pump efficiencies, and water quality.

Complex operational programs may be implemented with computer-controlled systems. Several well pumps can be controlled from a single pressure sensor installed at a location within the influence of the wells. If the pressure drops below the sensor setting, for example, 40 psi, the computer will start one of the pumps assigned to the sensor. If, after a sufficient waiting period, system pressure remains below 40 psi, another pump will be started. This grouping program can be used to conserve power by using different start and stop points for day and night operations. Pressures can be lowered in the evening to conserve energy. The pumps assigned to a sensor can be changed by the operator for efficient operation. If the system is large, it can contain several pressure-control groups, each with start and stop points independent of each other. The program can be expanded to select the most efficient pumps to run the longest.

Alternately, operators can assign maximum running hours per month to each well in the group. When another pump is to be started, the computer checks the list. If the next pump has exceeded its maximum allotted hours, the computer continues down the list until it reaches a well that has not.

Modern remote-terminal equipment can monitor flow, pressure, power input, and well pumping level so that a continuous record of pump and well performance can be maintained. These data can also be used to plan maintenance programs.

READING LIST

Dicmas, J. L. 1987. *Vertical Turbine, Mixed Flow, and Propeller Pumps.* McGraw-Hill, New York, N.Y.

"American National Standard for Vertical Turbine Pumps—Line Shaft and Submersible Types," AWWA E101. American Water Works Association, Denver, CO.

CHAPTER 17

Well and Pump Operation and Maintenance

17.1 INTRODUCTION

Well efficiency, pumping plant efficiency, and operating relationships between well and distribution systems are a function of well design and construction, pumping plant design, installation, and operation, and maintenance procedures. Operation and maintenance are particularly crucial because annual energy demands for pumping may be as much as 70% of capital cost. This chapter discusses record keeping and its importance for wells and pumps, the major causes of deteriorating well performance, and general approaches to well maintenance. Although emphasis has been placed on high-capacity wells, the fundamental and general information presented has widespread application.

17.2 PRELIMINARY DATA

Well construction entails collection of samples, logs, and records. This information forms the basis for evaluation of well performance.

For wells drilled by the rotary method, the following records should be kept:

1. The final well report showing a complete description of the character and depths of all formations; casing diameter, wall thickness, and material; depths of setting and lengths of casings and screens installed; details of reducing sections; screen type, aperture size, and pattern; borehole diameters; cemented conductor casing; gradation of gravel envelope; quantity of gravel initially installed, quantity of gravel added during development operations, and quantity of material removed during development operations; and all other pertinent details.
2. Geophysical borehole logs.
3. Development and test data showing production rates, static water level, pumping levels, drawdown, sand production, and other relevant information concerning development.
4. The final pump test report.
5. Two plots of plumbness and alignment in planes oriented at 90° with respect to each other.

The following records should be prepared for wells drilled by the cable tool method:

1. Full descriptive notes of all conditions found while drilling. These notes should include explanations for delays, reasons for any casing reductions, description of any swaging done, and all other pertinent information.
2. The height to which water naturally rises or falls as each aquifer is encountered.
3. The final well report showing complete description of the character and depths of all formations; a description of all casings, including starters and shoes, complete with diameter, wall thickness, and depth placed in the well; any reductions in well diameter; the method of well completion; and the location, type, dimensions, aperture size, and pattern of screens or downhole perforations.
4. Development and test data showing production rates, static water level, pumping levels, drawdown, sand production, and other pertinent information concerning development.
5. The final pump test report.
6. Two plots of plumbness and alignment in planes oriented at 90° with respect to each other.

Examples of final well reports for gravel envelope and naturally developed wells are shown in Tables 17.1 and 17.2.

For the permanent pumping plant, the following information should be recorded:

1. Prime mover type (electric motor or engine).
2. Prime mover data (electrical—voltage, frequency, phase, and rpm; mechanical—engine type and operating rpm range).
3. Driver (vertical hollow-shaft motor, vertical solid-shaft motor, vertical hollow-shaft right-angle drive, combination drive, submersible motor).
4. Drive-shaft lubrication required.
5. Type of discharge (surface or below base).

TABLE 17.1 Final Report, Gravel Envelope Well

Well No. _____ Drilled for _____

Name _____

Address _____

_____

Location _____

_____

_____

Started Work _____

Completed Work _____

Total Depth Drilled _____

Total Depth Completed _____

Drilled by Direct Hydraulic, Reverse Rotary _____

	DIAMETER	FROM	TO
PILOT BORE	in.	ft	ft
	in.	ft	ft
CONDUCTOR BORE	in.	ft	ft
	in.	ft	ft
COMPLETED WELL BORE	in.	ft	ft
	in.	ft	ft
	in.	ft	ft

CASING AND SCREEN SCHEDULE

Conductor Casing

Material _____

Diameter _____ in. Wall Thickness _____ in.

Installed From _____ ft. To _____ ft.

Cemented From _____ ft. To _____ ft.

Well Casing

DIAMETER	WALL	MATERIAL	FROM	TO

Screen

Type _____

Material _____

DIAM.	WALL	APERTURE SIZE	FROM	TO

FORMATION
Include Size of Aquifer Material

ft	to	ft
" "	"	"
" "	"	"
" "	"	"
" "	"	"
" "	"	"
" "	"	"
" "	"	"
" "	"	"
" "	"	"
" "	"	"
" "	"	"
" "	"	"
" "	"	"
" "	"	"
" "	"	"
" "	"	"
" "	"	"
" "	"	"

Development Record

Method of Swabbing _____

No. of Hours _____

Total Material Removed _____

Gravel Added _____

Rig No. _____ Developer _____

Test Record

Water Level When Test Started _____ ft.

Drawdown from standing level _____ ft.

No. of Gallons per Minute Pumped When Test Started _____

No. of Gallons per Minute Pumped When Test Completed _____

Drawdown at Completion of Test _____ ft.

Hours Testing Well _____

Tons of Gravel Installed _____

Gravel Gradation _____

Give any additional data that may be of future value _____

_____

_____

Date of Report _____

_____
                                              Driller

Type and Rig No. Used _____

TABLE 17.2 Final Report, Naturally Developed Well

Well No. _____ Job No. _____

Owner _____

Address _____

_____

Location _____

_____

_____

_____

Started Work _____

Completed Work _____

Total Depth Drilled _____

Depth Water First Encountered _____

MATERIALS
Conductor Casing

Material _____

Diameter _____ in. Wall Thickness _____ in.

Installed From _____ ft To _____ ft

Cemented From _____ ft To _____ ft

Well Casing

DIAMETER	WALL OR GAUGE	MATERIAL	FROM	TO

Starter Used _____ wall or gauge

Size Shoe _____

DOWN-HOLE PERFORATIONS OR SCREEN

Type of Perforator or Screen Used _____

FROM	TO	WIDTH	LENGTH	HOLES PER FOOT	SQ. INCH PER FOOT

FORMATION
Include Size of Aquifer Material

	ft	to		ft	
_____	"	"	_____	"	_____
_____	"	"	_____	"	_____
_____	"	"	_____	"	_____
_____	"	"	_____	"	_____
_____	"	"	_____	"	_____
_____	"	"	_____	"	_____
_____	"	"	_____	"	_____
_____	"	"	_____	"	_____
_____	"	"	_____	"	_____
_____	"	"	_____	"	_____
_____	"	"	_____	"	_____
_____	"	"	_____	"	_____
_____	"	"	_____	"	_____

If Well Is Reduced, Indicate:

Amount of Lap at Reduction _____ ft

Amount of Lap at Reduction _____ ft

Amount of Lap at Reduction _____ ft

Method of Sealing at Reduction _____

_____

Development & Test Record

Was Well Swabbed or Sand Pumped? _____

No. of Hours _____

Total Material Removed _____

Water Level When Test Started _____ ft

Drawdown from Standing Level _____ ft

No. of Gallons per Minute Pumped When Test Started _____

No. of Gallons per Minute Pumped When Test Completed _____

Drawdown at Completion of Test _____ ft

Hours Testing Well _____

Give any additional data that may be of future value _____

_____

_____

_____

_____

_____

_____

Driller _____

Date of Report _____

Type and Rig No. Used _____

6. Pump design capacity (gpm).

7. Datum (elevation from which the weight of the pump is supported and from which vertical distances are measured).

8. Pumping level below datum at design capacity (ft).

9. Total head above datum at design capacity (ft).

10. Total pump head at design capacity (ft).

11. Operating head range (minimum total pump head and maximum total pump head) (ft).

12. Overall pump length (datum to inlet of pump suction case) (ft).

13. Length and design of suction pipe and strainer installed.

14. Pump bowl diameter and number of stages.

15. Major pump components and materials.

16. Field pump performance data.

17.3 WELL AND PUMP OPERATION PROCEDURES AND RECORDS

Proper well operation, maintenance, and rehabilitation require periodic inspection and record keeping. Records will show variations of pump and well performance as well as changes in hydrogeologic conditions and their influence on well performance. A yearly summary of the well and pump operation should be prepared in addition to other records.

A complete set of records includes a detailed catalog of every component in the well. This facilitates ordering of parts for pump repairs as well as identification of existing parts that may be wearing at an unusually fast rate. Every aspect of engineering design and materials incorporated in the well and pumping system should be covered.

Operators' Daily Checklist

The following maintenance checklist was developed by the San Jose Water Company from years of operating experience. Its use serves much of the record-keeping requirements discussed earlier.

The items listed are checked daily (unless specified otherwise) for each well station, booster pump station, and reservoir site, as applicable.

General

1. Record the production of each well and booster pump.

2. Record the standing water level and/or pumping water level of each well monthly.

3. Take a water sample (frequency as directed). Deliver the sample to the laboratory promptly.

Line-Shaft Pumps

1. For oil-lubricated pumps, replenish the oil in the reservoir.

2. For oil-lubricated pumps, verify that the oiler rate is correct. Adjust as required.

3. For a pump with mechanical packing, verify that a minimal flow of water is escaping to cool the packing. Tighten the stuffing box studs to decrease leakage if necessary. Remedy any excessive packing leakage or clogged pump head drain.

4. Inspect the vertical hollow shaft motor for signs of distress, including:
   a. Leaking oil.
   b. Unusually high motor surface temperature.
   c. Unusual noise.
   d. Smoke or soot.
   e. Excessive vibration.
   f. Motor windings space heater not operating.

Submersible Pumps

1. Inspect the pump for signs of distress, such as:
   a. Excessive vibration.
   b. Cracked power cable.
   c. Constant release of air through the release valve.
   d. Excessive movement of the check valve arm due to fluctuating discharge.

Horizontal Booster Pumps

1. For the mechanical packing, verify that a minimal flow of water is escaping to cool the packing. Tighten the packing bolts to decrease leakage if necessary. Correct excessive packing leakage or a clogged pump case drain.

2. Inspect the pump and motor for signs of distress, including:
   a. Leaking grease or oil.
   b. Unusually high surface temperature.
   c. Unusual noise.
   d. Smoke or soot.
   e. Excessive vibration.
   f. Motor windings space heater not operating.

3. Replenish grease in the bearing cups, as necessary.

Motor Control Center

1. Change the circular chart for each booster pump recorder (weekly).

2. Change the circular chart for each well pump recorder (weekly).

3. Change the circular chart for the pressure recorder (weekly).

4. Wind up any mechanical time clocks in the recorders.

5. Inspect for signs of distress, including:

   a. Any breaker in the "tripped" position.

   b. Any selector switch in the "hand" or "off" position.

   c. The cooling fan or space heater not functioning (seasonal).

   d. Recorder pens not inking.

   e. Soot or burned spots on the mechanical switch gear.

   f. Overhead lighting not functioning.

   g. Burnt-out indicator light bulbs.

   h. Flow meter and recording equipment not functioning.

   i. Chart drive not functioning.

Tanks and Reservoirs

1. For steel tanks, verify that cathodic protection system potential is approximately 0.9 V.

2. For wood tanks, check for any leakage or evidence of overflow.

3. Check the water-level transmitter mechanism.

4. Check the water-level control floats and weights.

5. Check the underdrain for leakage.

6. Check the reservoir berm for displacement.

7. Check the pavement for cracks.

Miscellaneous

1. Walk the site's perimeter and observe anything unusual, including:

   a. Hole in the cyclone fence.

   b. Trash.

   c. Weeds.

   d. Tree limbs down.

   e. Graffiti on tank.

   f. Standing water or water bubbling out of the ground.

2. Inspect all piping above ground for:

   a. Leaks.

   b. Leaking valve packings.

   c. Leaking dresser pipe couplings.

3. Inspect the station lights.

4. Clean up as required inside the pump houses.

Well Inspection. The San Jose Water Company has developed the following checks and rationale for periodic well inspection:

1. Check each Rossum Sand Tester daily, and record the amount of sand in the glass. Report any significant increase in sand production (an accumulation of 4.0 cc in 24 hours is considered unacceptable). Clean or replace the glass if it is contaminated with algae or scale. (Perhaps sooner than any other indication, an increase in the amount of sand produced warns of the need for remedial work.)

2. With gravel envelope wells, check pack level whenever a pump test is made. (Some settlement may be expected, but any rapid decline in level or increase in the amount of sand pumped by the well without any settlement is an immediate cause for concern. This is particularly important in situations where only a small reserve is available above the screen. Settlement can expose the well screen to direct contact with the aquifer, and excessive sand production could occur.)

3. If possible, check the inside of the well for fill material every six months. (Normal operation of many wells completed in alluvium generates some fill inside the well, and a foot to a few feet a year is inconsequential. However, an increase in the rate of fill suggests a void in the envelope, a casing or screen break, or enlargement of screen openings.)

4. Inspect the well with down-hole television or camera equipment whenever the pump is pulled. Visual inspections may point out the need for remedial work before it is otherwise apparent.

Pump Test. The test report form in Table 17.3, which is based on the AWWA standard field test report form may be used to calculate wire-to-water efficiency (see Chapter 16). Performance tests should be run on an annual basis to document well and pump performance to help decide if replacement or repairs are necessary.

17.4   EVALUATION OF PUMP AND WELL PERFORMANCE

Pump Maintenance Criteria

Typically pumps should be pulled for maintenance when one or more of the following occur:

1. A line-shaft-driven pump has been operating without maintenance for 12 to 15 years.

2. A submersible pump has been operating without maintenance for 10 years.

TABLE 17.3   Annual Pump Test

Date _____

Owner: Name _____

         Address _____

Pump: Location _____ Type _____ Size _____ Stages _____

         Make _____ Serial No. _____

Motor: Make _____ Serial No. _____

         Rated hp: _____ rpm _____ vss _____ vhs _____ subm _____

Power supply: Nominal Voltage _____ Phase _____

Column: Pipe Size _____ Shaft Size _____ Discharge Pipe Size _____

         Length _____ Cable Size _____

Test Conducted by _____ Witnessed by _____

Flow meter used _____

Test Readings and Calculations

All readings except No. 1 are taken when pumping

No.		Units	1	2	3
1	Static water level	ft			
2	Drawdown	ft			
3	Head below datum	ft			
	Datum to centerline discharge gage	ft			
4	Pressure head reading	ft or psi			
5	Pressure head above datum	ft			
6	Velocity head in discharge pipe[a]	ft			
7	Head above datum[a] = (5) + (6)	ft			
8	Total head[a] (3) + (7)	ft			
9	Discharge rate	gpm			
10	Current Line A	amp			
	Current Line B	amp			
	Current Line C	amp			
11	Voltage Phase AB	V			
	Voltage Phase BC	V			
	Voltage Phase AC	V			
12	Revolutions of watt-hour meter disc (constant)				
13	Time	sec			
14	Wattmeter reading				
15	Electrical input[a] from (12 & 13) or (14) or calculated from (10 & 11 with power factor)	kW			
16	Electric horsepower input[a] = (15)/0.746	hp			
17	Water horsepower = (8) × (9)/3960	hp			
18	Wire-to-water efficiency[a] = (17)/(16)	%			
19	Motor efficiency	%			
20	Pump field efficiency[a] = (18)/(19)	%			
21	Specific capacity = (9)/(2)	gpm/ft			

[a]Calculated.

Note: Results will be in hp only if head measurements are in feet of liquid (hp × 0.746 = kw).

3. The wire-to-water efficiency of the unit decreases:[1]
   a. To 65% for key units (24 hr/day operation).
   b. To 60% for pumps that run at least 1000 hr per year.
4. The motor insulation check, or megger reading, changes by more than 50% from the previous year.
5. The megger reading is less than 1 megohm.
6. Field observation of the unit reveals signs of degradation (bearing noise, excessive vibration, hot motor, etc.).

Well Maintenance Criteria

In order to ensure that well performance does not excessively deteriorate, periodic tests for well efficiency should be run and the results compared with the value attained when the well was new. The complete procedure for conducting well efficiency testing is found in Chapter 15. The well can be considered to be maintaining reasonable performance if well efficiency has not declined more than 15% from the original value. When performance falls below an acceptable standard, redevelopment should be considered.

In most wells, well efficiency is never determined and is difficult and expensive to monitor. In these cases specific capacity values are used to evaluate well performance over time. Seasonal and yearly influences on the pumping level must also be considered.

An example of a practical methodology for well and pump evaluation is shown at the end of the chapter.

Economic Analysis for Scheduling Well and Pump Repair and Replacement

In addition to practical guidelines there are also economic guidelines that may be useful for the repair of wells and pumps. An economic analysis for scheduling pump repair, pump replacement, well rehabilitation, and well replacement is presented in "Improving Well and Pump Efficiency" (Helwig, Scott, and Scalmanini) (1). According to Helwig et al., a pump should be repaired or replaced when the present value of the improvements in annual net benefits is greater than the cost of repair or replacement. The well should be rehabilitated when the present value of the improvements in annual net benefits is greater than the cost of rehabilitation.

17.5 MAJOR CAUSES OF DETERIORATING PUMP PERFORMANCE

The major causes of deteriorating pump performance can be summarized as follows:

1. Improper pump installation. Leakage from the column pipe and horsepower losses due to crooked shafts and improper tightening are examples.
2. Changes in system conditions that force the pump to operate in an inefficient range.
3. Insufficient line-shaft lubrication that causes horsepower loss and premature wear of line-shaft bearings.
4. Motor overloading and/or overheating that decreases efficiency and breaks down insulation.
5. Improper pump adjustment causing increased wear and horsepower losses.
6. Cavitation either from entrained air or from insufficient NPSH.
7. Abrasion from sand or silt produced from the well.
8. Wear from rubbing mechanical parts. This can be normal wear expected over time or abnormal wear caused by deformed or bent parts.
9. Corrosion and incrustation of pump parts.
10. Mechanical plugging of the impellers or the pump suction.

A complete guide to pump troubleshooting can be found in *The Manual of Water Well Maintenance and Rehabilitation Technology*, and in *Groundwater and Wells* (2, 3).

17.6 MAJOR CAUSES OF DETERIORATING WELL PERFORMANCE

Well performance can deteriorate through one or a combination of factors principally involving the following:

1. Change in hydrogeologic conditions, leading to declining water tables.
2. Excessive pumping of sand causing deterioration of the filter zone.
3. Plugging of the filter zone by fine particles.
4. Reduction in well efficiency and production due to incrustation or bacterial growth in the well screen and/or filter zone.
5. Degradation of water quality due to contamination.
6. Structural damage of well casing and/or screen.

Deterioration of Well Performance due to Hydrogeologic Factors

Excessive Interference. When pumping wells are too close, interference creates additional drawdowns with corresponding higher pumping lifts. These higher lifts result in a decline in production, the amount dependent on the performance characteristics of the pump.

1. The wire-to-water efficiency as used here is defined in Table 17.10.

The production rate in a pumping well may also decline where basin production exceeds natural recharge. This results in a "mining" condition with lowering water levels.

Proximity to Impermeable Barriers.

In a production well the cone of depression expands to the point where the flow to the cone balances the well discharge. When the expanding cone is limited due to impermeable or semipervious boundaries, drawdown increases to provide the necessary flow to the well, or production declines.

Reduction in Saturated Aquifer Thickness.

When a portion of the aquifer becomes dewatered, the total transmissivity is reduced, and the well requires more drawdown to maintain the discharge rate. Continued operation may cause further dewatering.

Effect on Pump Operation of Excessively Low Pumping Level.

A decline in the pumping level leads to an increase in the pumping head and reduced discharge. If the pump breaks suction, the flow will fluctuate with air in the discharge. Unless action is taken, pumping of air will lead to damaged bowl bearings and pump vibration. This damage can be avoided if a new satisfactory head and discharge rate can be established. This can be done by increasing the back pressure on the pump, modifying the pump by trimming impellers, or replacing the bowl assembly with a higher-head lower-capacity unit.

Excessive Pumping of Sand

Well conditions revealed by analysis of sand production curves are discussed in Chapter 14.

Gravel Envelope Wells.

Major sand problems in gravel envelope wells are caused by improper filter pack design or voids in the pack. In the case of a generally improper pack/aquifer ratio, rehabilitation by redevelopment would be ineffective. Reducing pumping rates with extended pumping cycles is sometimes helpful.

If a well starts to produce excessive sand after an extended period of satisfactory operation, an analysis of sand production versus time must be made. The dashed line on Figure 17.1 represents a well with voids in the filter pack. This erratic curve results from movement of bridged particles of sand that break down periodically and enter the screen through voids.

Techniques to reconsolidate a gravel envelope are similar to cable tool swabbing described in Chapter 14. If the pump suction is in the screen, the void is usually found at the same depth. For pump suctions set in the pump housing casing, the void is frequently located at the top of the screen. Cable tool swabbing is usually effective in locating the void,

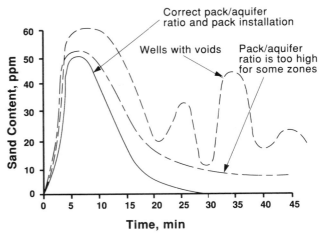

Figure 17.1. Analysis of sand content vs. time from pump start-up.

breaking down the bridge and reconsolidating the gravel envelope.

When sand is being pumped through voids, it is important to rehabilitate promptly. Probabilities for successful rehabilitation decline with continued operation of the well.

Gravel envelope wells can also pump sand when the pack/aquifer ratio is generally correct but too high for one or more sand layers in the well (see Figure 17.1). The well designer may have failed to recognize an isolated layer of uniform sand or silt that cannot be stabilized by the filter pack. If this occurs, the driller's logs, geophysical logs, and any information available about the material being pumped (color, size, shape) should be analyzed to locate its source. These data, coupled with a video inspection, provide guidance in sealing off problem zones. The same procedure may be used when formation is pumped through holes or breaks in the casing and/or screen. The problems described next for naturally developed wells may also occur in gravel envelope wells.

Naturally Developed Wells.

Naturally developed wells can produce excessive quantities of sand for a variety of reasons. Redevelopment will usually stabilize the natural filter.

1. Enlargement of screen aperture through corrosion, allowing sand to enter the well.
2. Incrustation of screen or downhole perforations, causing excessive entrance velocities and movement of sand into the well.
3. Declining water table, resulting in the overpumping of previously underdeveloped aquifers.
4. Inadequate initial development.

5. Pumping rates greater than the original developed well capacity.
6. Excessive pump cycling.

Loss of Efficiency Due to Plugging by Fine Particles

Certain well-operating practices may have an effect on long-term performance. Sometimes a new well is pumped at a rate considerably less than the developed capacity. If this continues for several years, the well may not be able to produce at the higher production rate achieved during development. In some cases redevelopment is not effective, and the loss is permanent.

Another similar phenomenon occurs when a pump operates continuously for long periods of time. It has been observed that many wells operated in this fashion will eventually (sometimes in less than one year) experience a decline in specific capacity. If the pumping pattern is not changed, decline may be very rapid. Continuous flow without backwash fills the interstices of the filter zone with fine particles, reducing permeability. Rehabilitation efforts rarely result in complete restoration of specific capacity.

The precise mechanism by which the natural or artificial filter zone becomes plugged is not understood, but its existence is acknowledged. It is generally associated with production from alluvial aquifers composed of material some of which is as fine as 200 mesh. Magnitude of developed specific capacity does not appear to be a factor since wells with this problem can have specific capacities ranging from 10 to over 100 gpm/ft. Other well design features that appear to be irrelevant include method of drilling, type of completion (naturally developed or gravel envelope), and type of screen. However, the operating cycle appears to have a pronounced effect. Wells that are pumped continuously for months evidence the greatest tendency to plug. Wells that discharge into reservoirs controlled by level sensors usually operate for many years or indefinitely without loss of specific capacity. These wells generally cycle daily, or several times a day.

This evidence suggests that some wells require periodic surging. Pumps in these wells should not be equipped with foot valves or column pipe check valves. Experience is useful in setting optimum operating cycles. In some cases cycling the pump weekly effectively maintains capacity. In others one or more cycles daily are required. Excessive cycling must be avoided to prevent destabilization of the filter zone.

Manual backwashing with supplemental water may be required if decline in specific capacity is observed. It is important to apply remedial measures before decline becomes too severe. When losses reach 50% of original specific capacity, redevelopment rarely effects full recovery.

It is also reported that extending the pump suction to the bottom of the screen may increase the time interval between maintenance by improving the flow distribution through the length of the screen. However, this effect may be negligble with short screens. Plugging may also be ameliorated by reducing the pumping rate and/or increasing the effective well radius by redevelopment.

Loss of Efficiency due to Incrustation or Bacterial Growth

The chemical reactions involved in incrustation were discussed in Chapter 10. Incrustation can occur on the inside or outside of the well screen, in the filter zone, or in crevices in consolidated rock formations. For wells operated in areas known to have these problems, remedial work should take place periodically to prevent significant deterioration in well performance. If necessary, samples of the incrusting materials can be obtained from the screen interior and analyzed along with the well water to choose the most suitable chemical and/or mechanical treatment.

Contamination

Ground water quality and contamination is discussed in Chapter 18. Methods for dealing with unusable-quality water vary and depend upon the source, nature and extent of contamination. They include:

1. Well abandonment.
2. Protecting the well from the source of surface contamination.
3. Sealing or lining off the aquifer or part of an aquifer that is contaminated.
4. Water treatment or blending to reduce contaminants to acceptable levels.

Since locating the precise source of contamination can be difficult, the cost of repair or protection should be weighed against the long-term benefits of well redesign and replacement.

Structural Damage

Causes of Damage. Structural damage of well casing and screen generally occur as a result of:

1. Corrosion.
2. A hole generated in the pump column or bowl assembly by corrosion or sand, allowing water to jet a hole in the casing.
3. Hydrostatic pressure differential on the casing or screen (discussed in Chapter 9).
4. Subsidence (discussed in Chapter 9).

5. A dropped pump or tools, or the improper operation of down-hole well tools.

Evaluation of Structural Failure. Usually a structural failure will be revealed during well operation or pump inspection. Careful record keeping frequently can identify the problem promptly. Some common indicators are:

1. Gravel or sand in the discharge.
2. A sharp drop in production with substantial fill inside the screen.
3. Material in the well larger than the screen or perforation aperture size. Slightly larger particle sizes suggest enlargement of the aperture. Large gravel or formation material indicates ruptured casing or screen.
4. Difficulty in removing or lowering the pump, which suggests a collapse or obstruction.
5. Difficulty in installing or operating clean-out tools.

It is common practice in some areas to make a video or photographic inspection of a well whenever the pump is pulled. This will usually reveal existing or developing structural problems. Videos are available in color or black and white and in vertical or side-looking pictures. If a video cannot be taken, adjustable cages or impression blocks are useful in defining obstructions.

The extent of damage must be carefully appraised. Many factors must be considered, including risk of inability to complete the repair, estimated cost, age and general condition of the well, and whether the repair can be expected to materially augment well life.

17.7 WELL MAINTENANCE AND REHABILITATION

Once the cause of deteriorating well performance is known, one or more maintenance and rehabilitation techniques may be effective in restoring specific capacity and discharge rates or protecting the well from further deterioration:

1. Redevelopment with well-development techniques.
2. Chemical redevelopment with acid or dispersing agents.
3. Mechanical cleaning by wire brushing or high-pressure jetting.
4. Cleaning the screen with vibratory explosives.
5. Structural repairs by setting liners, complete relining, or screen replacement.
6. Deepening the well.

Redevelopment

Redevelopment can often be accomplished by reapplying the development techniques described in Chapter 14. They include combinations of sand pumping, swabbing, airlift pumping and swabbing, and surging and backwashing with an engine-driven turbine pump. Improper application of these techniques may result in damage to the formation or the gravel envelope.

Chemical Redevelopment

Acid Treatment. Incrustations often consist of calcium or iron compounds. Chemical redevelopment involves the use of chemicals, usually acids, to dissolve them. The most effective acids are hydrochloric (HCl), sulfamic (NH_2SO_3H), and hydroxyacetic ($HOCH_2COOH$). If the incrustation contains appreciable amounts of manganese dioxide, it may be necessary to use very strong hydrochloric acid or to add hydrogen peroxide to the acid as a reducing agent. Inorganic acids such as hydrochloric and sulfamic are commercially available with chemical inhibitors to minimize corrosion of carbon steel.

Acid can be introduced in the well via a small-diameter plastic pipe. Enough 15% acid solution should be added to displace the volume of water inside the screen section. The acid is dispersed by surging or swabbing. Two or three hours of contact time should be sufficient to dissolve most incrustations. Longer contact times are desirable for ferric compounds; however, this may not be practical if there is any movement of ground water within the well. This typically occurs in a well that is screened in two or more aquifers with water moving between them due to differences in piezometric levels. The acid may disperse into the aquifer and reduce the effectiveness of the treatment.

Before the treated well is returned to service, the residual acid solution must be pumped from the well and properly disposed. Applicable health standards and environmental regulations must be observed. These regulations typically require neutralization of acid discharges. Environmental and safety considerations dictate that rehabilitation with acids be done by experienced, qualified personnel. See Driscoll (3), for a more detailed discussion of acid treatment techniques and safety precautions.

Incrustations resulting from growth of iron bacteria are also treated with hydroxyacetic acid because it is considered to be a good disinfectant as well as an incrustation-removing acid. Historically, iron bacteria problems have been treated with the disinfectant, chlorine. A plausible explanation for chlorine's effectiveness in reducing incrustation is the dissolution of iron precipitate by the acids (hydrochloric and hypochlorous) formed upon the reaction of chlorine and water. It is generally recognized that iron bacteria are ubiquitous and that disinfection by shock chlorination is only temporarily successful. However, it has been observed that the productivity of wells treated with chlorine or hydroxyacetic acid may be sustained for two or three years before additional treatment is required.

Whenever frequent treatments with acids are anticipated, the use of stainless steel casings and screens should be considered.

Dispersing Agents.

A sodium polyphosphate such as sodium hexametaphosphate is sometimes used in well rehabilitation. In redevelopment, this chemical is primarily used as an agent to disperse residual drilling fluid and aid in the removal of fine particles of silt and clay that have lodged in the filter zone. This technique must be used cautiously since the polyphosphates can also break down formation clays and cause them to swell and slough.

Polyphosphates are pre-mixed on the surface, often in a ratio of about 1 to 1½ lb of polyphosphate to 10 gal of water, and tremied into the well. After placement, the well is agitated, with either a swab or a bailer or by surging, to ensure complete mixing. The solution is left in the well for 24 hours and then pumped to waste. Backwashing and surging may help remove silt and clay dislodged by the chemicals. The well should be chlorinated after treatment.

Mechanical Redevelopment

The mechanical techniques most suited for removal of incrustation will depend on the type of screen installed. Wire brushing and high-pressure water jetting are frequently used. These methods enhance the effectiveness of the acid when used in combination with acid treatment.

Wire Brushing.

Wire brushing cleans the inside surface of well casing and screen with a tool fabricated of pipe, wire rope, and cement. The tool is made by drawing wire rope (⅝- to 1-in. diameter) through holes in a pipe 15 to 20 ft long and approximately 8 in. smaller in diameter than the well to be cleaned. The wires should be separated by approximately 8 in. and extend ½ to 1 in. larger then the inside diameter of the well. With the wires in place, the pipe is filled with cement to hold the wires and to add weight to the tool.

The brush is lowered into the well and alternately dropped and raised through a 10-ft section of casing or screen for several minutes. It then is lowered to the next section and the procedure is repeated. After brushing, debris at the bottom of the well is removed with a bailer or a sand pump and inspected to see if chemical cleaning is desirable. Wire brushing is an economical process due to simplicity and speed.

The brush can be operated with a pump rig or crane; however, it is more effective to use the spudding action of a cable tool rig. Use of a swivel or KAK ratchet rope socket will improve the cleaning action.

Wire brushing should be done by experienced, qualified personnel. Improper use may damage the well; it should not have to be forced up or down. It is not advisable to use this method in a well with a wire wrap screen or where the well design is unknown.

High-Pressure Water Jetting.

High-pressure water jetting was originally developed to remove hard scales from piping, penstocks, and other industrial equipment (4). The water jet must have sufficient energy to penetrate the deposit. Once penetrated, a fluid wedge is driven between the deposit and the surface to which it is affixed and strips it off.

Frequently deposits are porous and layered. Even though very hard, they usually dislodge easily. The jet will enter a pore, build up pressure below the surface, and rupture the pressured area. In some cases dislodged particles entrained in the jetting stream help dislodge other particles.

The rate at which an incrusted surface can be cleaned depends on the speed at which the jet stream can penetrate, dislodge, and remove the deposit, which is proportional to the hydraulic horsepower applied. However, without properly designed nozzles, most of the hydraulic power can be dissipated by line losses, with little cleaning action achieved.

An orifice nozzle assembly consists of a stainless-steel body with male pipe threads and a tungsten carbide nozzle insert. The body is designed to accept inserts with a wide range of orifice sizes. A tapered inlet section further extends the jet stream.

Multiorifice nozzle assemblies are used for cleaning the inside of well screens. The inserts are set at the angle required to direct the jet into the screen aperture (see Figure 17.2). Orifice number, size, and arrangement vary with requirements.

The relationship between velocity and nozzle pressure drop is given in Table 17.4. The assembly should have sufficient nozzles for efficient cleaning. Power output is

$$P = 0.0223 NAp^{1.5} , \qquad (17.1)$$

where

P = hydraulic horsepower,
N = number of jet nozzles,
A = nozzle area [in.^{2}],
p = pressure drop across the nozzle [psi].

The high-pressure jetting described here is not the same as that described for developing in Chapter 14, and the pressures and power utilized could damage or perforate casing or screen if the work is not carefully performed by a professional oil field pumping service.

High-pressure jetting is expensive, but it may be the best way to revitalize a well where redevelopment, chemical treatment, and wire brushing are ineffective. The structural integrity of the well should be a major consideration in evaluating the use of this method.

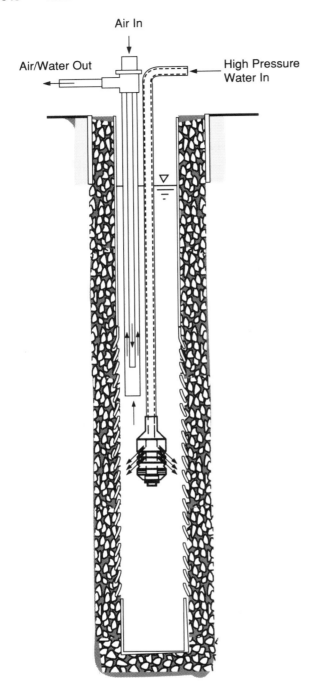

Figure 17.2. Multi-orifice nozzle jetting.

crustation. The charges should be centralized to reduce risk of damage to the screen. After firing, the bottom of the well is cleaned of debris. Explosives are sometimes used in conjunction with acid treatment in particularly stubborn cases of incrustation.

Structural Repair

A variety of techniques have been developed to facilitate repair of damaged wells. Some of the more common are described here.

Liners. For wells with peeled or collapsed casing or screen, repairs must first be made by swaging the casing or screen as close to the original inside diameter as possible. This is achieved with the use of drive swages operated by cable tool equipment or by hydraulically or electrically operated swages (10-in. or larger). The successful use of these tools depends upon the severity of collapse, condition of casing or screen, age of the well, and any existent passageway through the collapsed section. In general, the collapsed area can be restored as long as an opening exists to permit the guiding of the swage. A hydraulically collapsed section that extends for several hundred feet usually cannot be economically restored.

Collapsed sections that have been swaged can be reinforced by placing a liner (a section of casing) across the damaged area. Liners are also used to repair a split or damaged section of casing, or to seal off a contaminated zone by covering the screen. They may be cemented in place or expanded against the repaired section by a hydraulic or electric swage. If expanded inside the original casing, liners are initially 1 to 2 in. smaller in diameter than the inside diameter of the casing or screen. If cemented, they are 2 to 4 in. smaller. The most common cement used is API Class A. Two to 5% bentonite can be used for shrinkage control and lubricity, and up to 2% calcium chloride can be used for setting-time acceleration.

In a typical liner-cementing procedure applicable to larger-diameter casing, the liner is equipped with a cement basket near the bottom. The length of the liner is sufficient to cover the area to be repaired or covered, plus 4 to 6 ft of overlap at each end. The liner is suspended in the well either by hanging at the selected elevation, or by equipping the top of the liner with a left-hand back-off joint so that it can be suspended on a string of threaded-and-coupled pipe at the desired position until cementing is complete.

As shown in Figure 17.3, the liner is also provided with an apparatus to permit the cement to be pumped in place through holes cut in the sides of the liner. After the liner is in place, a string of tubing is lowered into the well (or telescoped through the threaded-and-coupled pipe), properly centered to enter the top bell of the cement-feed-through

Well-Cleaning Explosives. A cleaning procedure using explosive charges sometimes referred to as "vibratory explosives" may be used in wells where incrustation has plugged the screen openings and surrounding filter zone. It involves use of sequentially fired explosive charges placed in the screen section. The quantity of explosive in each charge can be varied according to well diameter and amount of in-

TABLE 17.4 Discharge of Water to Atmosphere through Insert-Type Nozzles (Flow in gpm and Hydraulic Horsepower)

Pressure (psi)	Velocity (ft/sec)	Diameter of Orifice (in.)						
		1/32	3/64	1/16	5/64	3/32	7/64	1/8
2000	535	1.27	2.89	5.11	8.00	11.5	15.6	21.0
		1.48	3.37	5.96	9.34	13.4	18.4	24.5
4000	756	1.80	4.08	7.24	10.3	16.3	22.2	28.8
		4.20	9.51	16.9	24.0	38.0	51.7	67.0
6000	925	2.21	4.98	8.85	13.8	19.8	27.0	35.3
		7.72	17.4	30.9	48.3	68.2	94.5	123
8000	1070	2.55	5.75	10.25	16.0	23.0	31.3	40.9
		11.9	26.9	47.8	74.6	107	146	190
10,000	1180	2.81	6.36	11.3	17.6	25.4	34.6	44.8
		16.4	37.0	66.0	103	148	202	261

Source: Courtesy of Hydro Manufacturing.
Note: Top number = gpm; bottom number = hhp.

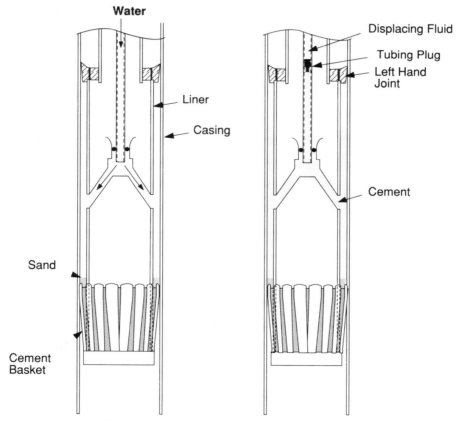

Figure 17.3. Typical liner cementing procedure.

apparatus. The bottom of the tubing is equipped with an entering socket that seals into the bell, usually by means an appropriate O-ring.

After the tubing is in place, a small quantity of sand is flushed down the tubing to seal the cement basket. The calculated volume of cement required to fill the annulus around the liner is then pumped down the tubing, followed by a tubing wiper to flush the cement out of the tubing. After the cement has set, the tubing is withdrawn. The feed-through apparatus can either be hooked and returned to the surface or broken off with a scow or heavy bailer, and cleaned out from the bottom of the well.

Complete Relining. In some instances, particularly in urban areas where sites are difficult and costly to acquire, wells that are old but in reasonably good condition can be rehabilitated by complete relining. Such wells were frequently cased with relatively large-diameter pump housing casing to accommodate low rpm pumps and can tolerate a reduction in diameter of at least 4 in. The annular space between the new and old well screen may be filled with filter pack material, depending upon the nature of the aquifer. The filter pack must be designed either to permit the movement of fine particles into the well or to prevent them from invading and sealing the pack between the two screens.

This technique works best with naturally developed wells that have not experienced significant loss of performance over many years of operation. Relining can double well life at a fraction of the cost of a replacement well.

Replacement of Screens. For relatively shallow wells with short lengths of telescoped screen, the extraction and replacement of the screen may be justified. Conditions encouraging this procedure include:

1. Insufficient and/or declining well yields requiring deepening.
2. Incrustation problems that cannot be resolved by chemical or mechanical treatment.
3. Corrosion damage to the screen.

Procedures to remove screens are set forth in Driscoll (3).

Well Deepening

Well deepening has been used to offset decline in production from lowering water tables. It may achieve this at a substantially lower cost than replacement with a deeper well. Results depend upon the age, condition, and design of the existing well, and the character of the deeper aquifer(s).

Deepening is relatively easy in naturally developed wells where the lower aquifers can also be naturally developed. It is more difficult to deepen gravel envelope wells since

the bottom plug must be drilled out and the gravel envelope stabilized by pressure grouting. In either case the deeper string of casing and screen is usually 4 in. smaller in diameter than the original.

Use of hydraulic rotary equipment for deepening has generally been unsuccessful because of major production losses from the upper aquifers due to large quantities of drilling fluid invading them. For this reason, cable tool equipment is used with three basic variations:

1. With a stable formation, an open hole is drilled followed by installation of screen.
2. With loose unconsolidated formations, blank casing equipped with top and bottom drive shoes is driven to the total depth. Casing is then perforated in place.
3. A machined vertical slotted screen equipped with top and bottom drive shoes is installed by driving.

Generally the original well is lapped with 20 ft of blank casing. Where conditions in the upper casing and screen require it, the well can be relined to surface.

17.8 TYPICAL WELL PROBLEMS OCCURRING IN VARIOUS TYPES OF AQUIFERS

Some difficulties in maintaining the specific capacity of wells result from the nature of the producing aquifers. Solutions include correct well design and construction and appropriate operating and rehabilitation techniques. Each situation must be evaluated individually. (The analyses of causes, prevention, and solution to these problems are discussed in various chapters of this book.) The following lists those problems frequently associated with distinct aquifers.

Alluvium

1. Sand production
2. Plugging of filter zone by fine particles
3. Corrosion
4. Incrustation
5. Bacterial infestation
6. Subsidence
7. Contamination

Sandstone

1. Sand production
2. Corrosion
3. Turbidity with unconsolidated zones containing clay platelets
4. Fissure plugging by mineral deposition

Limestone

1. Plugging by mineral deposition in low-yielding zones.
2. Borehole instability in open-hole completions.
3. Bacterial infestation in shallow permeable zones.
4. Contamination from seawater intrusion.
5. Plugging by clay or silt in fine-fractured shallow zones.

Basalt

Wells completed in basalt are relatively trouble free; however, some extraneous problems have been reported:

1. Contamination from seawater intrusion.
2. Plugging of shallow zones with intruding clays or silts from the overlying formations.
3. Naturally occurring gas, causing problems in the operation of the pumping plant.

Granite and Metamorphic Rock

1. Plugging of shallow zones with intruding clays or silts from overlying formations.
2. Mineralization of fissures.

17.9 METHOD FOR WELL AND PUMP EVALUATION

This section summarizes the methodology set forth in this chapter for well and pump evaluation by use of an example. The guidelines are those developed and used by the San Jose Water Company (SJW). The example selected is their well and pump at Twelfth Street Station Well 8. The pertinent information available for review is presented in the accompanying tables and figures.

Table 17.5 lists the items used by SJW for the analysis of the pumping unit in 1988. The terms in columns 15 through 19, Repair Selection Criteria, are defined as follows:

1. Motor and Pump ages are found by subtracting the year of the last motor or pump repair, respectively, from the year 1988.
2. Meg Low means that the motor megged 0.99 megohms or less. Meg OK means that the motor megged 1.0 megohms or greater.
3. dMEG Low means that the motor megged at least 50% below last year's reading.

   dMEG OK means that the motor megging changed less than 50% since last year. In some cases, megging actually increases.

4. Efficiency Low means that the wire-to-water efficiency of a key pump, at the operating point, is 65% or less (or 60% or less for all other pumps). The pump will be pulled for maintenance.

   Efficiency OK means that the wire-to-water efficiency of a key pump at the operating point is greater than 65% (or 60% for all other pumps). Maintenance is not required.

In reviewing the information in Table 17.5 and the pump maintenance criteria presented in this chapter, the following evaluations can be made:

1. The pump has been operating without maintenance for 22 years, which is beyond the recommended interval of 12 to 15 years. The pump should be pulled.
2. The wire-to-water efficiency of the pumping unit is 51.1%, which is below the recommended minimum of 60%. The pump should be pulled.
3. The motor insulation has a megger reading of 0.05 megohms, which is less than the minimum of 1 megohm. The motor should be repaired while the pump is out of service.
4. The pump discharge rate is only 786 gpm compared to the original design capacity of 1750 gpm. The pump should be pulled and information collected and analyzed with respect to declining water levels and decreased specific capacity.

After reviewing Table 17.5, a recommendation can be made to pull the pump. However, additional information is available to make more specific decisions for the repair of the pump and well.

Table 17.6 gives the original casing schedule and driller's log for this well, which was drilled in 1956. The original test pumping data (Table 17.7) shows a specific capacity of approximately 65 gpm per foot of drawdown and a static water level of 136 ft. In the 1960s the Twelfth Street Station well field was pumped very heavily, and subsidence damaged many of the wells. An artificial recharge program was initiated to stop subsidence and replenish ground water storage. In 1962 liners were expanded opposite telescoped sections at 284'–289', 337'–342', 384'–389', 421'–424', 482'–485', 559'–563', 619'–622', and 678'–688', and to exploit the newly introduced water recharged into the shallow zones several intervals from 280 to 531 ft were perforated. The well was tested and the specific capacity had increased to 80.

Water levels recovered, but wide seasonal variations continue to be experienced. Because of this the well pumps at this station are designed to operate efficiently over a wide range of discharge rates.

TABLE 17.5 SJW Corp.'s Well and Pump Evaluation for 1988 Maintenance

Station/Unit (1)	Date Pump Tested (2)	Design Discharge Head		Operating Discharge Head		Wire-to-Water Test Efficiency (%) (7)	Year Bowls Repaired (8)	Year Motor Repaired (9)
		gpm (3)	ft (4)	gpm (5)	ft (6)			
Twelfth St. W–8	8/11/87	1750	240	786	252	51.1	1966	1962

[a] 1 meg = 1 million ohms.

TABLE 17.6 SJW Corp.'s 12th Street Station Well 8, Well Log and Casing Schedule

Location: 12th Street and Martha, San Jose, CA
Driller: Roscoe Moss Company, Los Angeles, CA
Drilling Method: Cable Tool
Started Work: January 31
Completed Work: April 4, 1956
Total Depth: 714 ft

Casing

48 ft of 20 in., double 10-gauge casing
716 ft of 16 in., double 10-gauge casing

Perforations by Down-Hole Hydraulic Louver Perforator

508 ft to 531 ft, 8 holes per 4 in.
554 ft to 569 ft, 8 holes per 4 in.
574 ft to 579 ft, 8 holes per 4 in.
586 ft to 635 ft, 8 holes per 4 in.
649 ft to 667 ft, 8 holes per 4 in.
688 ft to 697 ft, 8 holes per 4 in.

Perforations $1/4 \times 2^{1/2}$ in.

Water first encountered at 21 ft
Water level before perforating, 128 ft
Water level after perforating, 135 ft

Driller's Log:

0 ft to 58 ft	Fill
58 ft to 92 ft	Brown clay
92 ft to 108 ft	Clay, gravel embedded
108 ft to 124 ft	Brown sandy clay
124 ft to 141 ft	Clay, gravel embedded
141 ft to 188 ft	Blue clay, gravel embedded
188 ft to 196 ft	Brown clay, gravel embedded
196 ft to 238 ft	Blue clay
238 ft to 242 ft	Brown clay, gravel embedded
242 ft to 254 ft	Clay, mixed blue and brown
254 ft to 258 ft	Sand and gravel to 3 in.
258 ft to 268 ft	Brown clay, gravel embedded
268 ft to 288 ft	Sand and gravel to 4 in.
288 ft to 310 ft	Brown clay, gravel embedded
310 ft to 316 ft	Sand and gravel to 3 in.
316 ft to 324 ft	Brown clay
324 ft to 350 ft	Blue clay
350 ft to 364 ft	Brown clay, gravel embedded
364 ft to 380 ft	Blue clay
380 ft to 386 ft	Sand, gravel to $1^{1/2}$ in.
386 ft to 404 ft	Blue clay
404 ft to 414 ft	Brown clay
414 ft to 421 ft	Sand and gravel to 3 in.
421 ft to 446 ft	Brown clay, gravel embedded
446 ft to 461 ft	Sand and gravel to 3 in.
461 ft to 479 ft	Brown clay, gravel embedded
479 ft to 486 ft	Sand and gravel to 3 in.
486 ft to 506 ft	Brown clay, gravel embedded
506 ft to 513 ft	Sand and gravel to 3 in., very tight
513 ft to 514 ft	Brown clay, gravel embedded
514 ft to 528 ft	Sand and gravel to 3 in., tight
528 ft to 538 ft	Brown clay, gravel embedded
538 ft to 554 ft	Blue clay, gravel embedded
554 ft to 567 ft	Sand and gravel to 3 in.
567 ft to 574 ft	Brown clay, gravel embedded
574 ft to 577 ft	Sand and gravel to 3 in.
577 ft to 586 ft	Brown clay, gravel embedded
586 ft to 603 ft	Sand and gravel to 5 in.
603 ft to 606 ft	Brown clay
606 ft to 611 ft	Sand and gravel to 4 in., very tight
611 ft to 614 ft	Brown clay, gravel embedded
614 ft to 633 ft	Sand and gravel to 3 in.
633 ft to 650 ft	Brown clay, gravel embedded
650 ft to 658 ft	Cemented clay and gravel
658 ft to 665 ft	Sand and gravel to 2 in. with clay
665 ft to 674 ft	Brown clay, gravel embedded
674 ft to 678 ft	Blue clay, gravel embedded
678 ft to 682 ft	Blue clay
682 ft to 688 ft	Gray clay, gravel embedded
688 ft to 695 ft	Sand and gravel to 2 in. occasional trace of clay
695 ft to 711 ft	Blue clay
711 ft to 713 ft	Brown clay, gravel embedded
713 ft to 716 ft	Blue clay

Type Unit (10)	Motor hp (11)	Motor bhp 1987 (12)	Motor Megger Test 1987 (13)	Motor Megger 1986 (14)	Repair Selection Criteria				
					Motor (15)	Pump (16)	Meg[a] (17)	dMEG (18)	Efficiency (19)
vhs	125	90.9	Low	0.05	26	22	Low	Low	Low

TABLE 17.7   12th Street Station Well 8, Original Test Pump Record, 1956

Static Water Level, 136 ft

gpm	Pumping Level	Drawdown	Specific Capacity
650	146	10	65.0
1000	150	14	71.4
1600	159	23	69.6
1825	163	27	67.6
2190	170	34	64.4

Note: Length of test 26 hours.

Table 17.8 shows this seasonal variation in water levels. Also indicated is a specific capacity of 78 in 1988, approximately the specific capacity developed in 1962. The water levels and specific capacity, along with the desired discharge rates and the horsepower of the existing motor, are used in Table 17.9 to calculate new design conditions. Although the principles used for pump selection in Chapter 16 are valid in cases such as this when a decision has been made to continue use of most existing components, the algorithm must be modified. Reused here will be the motor and this establishes the limits of the combination of discharge rate and total head.

The primary design point was selected to give the highest wire-to-water efficiency during the peak summer production periods when water levels are at their lowest. The secondary design point is selected so the pump will operate efficiently when water levels are highest. These two points define the operating range for a new bowl assembly. If possible, the design points should fall in the middle or to the right of the highest efficiency point of the performance curve for the new bowl assembly. Since ground water levels generally decline over time and thus add to the pump head, the pump-

operating conditions will move into the high-efficiency range (as water levels decline).

The total head for a selected discharge rate is calculated by adding the static water level, drawdown, friction loss in the pump and aboveground piping, and the external elevation head. The actual discharge rate for the calculated total head is then found by using the limiting values of horsepower and pump efficiency. In this example the pump horsepower available is limited to 125 to utilize the existing motor. In new installations the discharge rate would not be limited by motor-rated horsepower; the appropriate motor size would be selected to match the design discharge rate.

Bowl-assembly efficiencies for the design points are set at 83% and 80%. These are the highest efficiencies that can be reasonably expected from vertical turbine pumps operating at these primary and secondary points. This selection process is an iterative process that is continued until the selected discharge rate matches the calculated discharge rate. This discharge rate, and corresponding total head and efficiency for the primary and secondary design points, were used to select a new bowl assembly.

A downhole video survey was particularly important for this well because of its age, liners added, additional perforations and changes in hydrologic conditions. Specific items of concern were corrosion in the "splash" zone, liners that may be peeling or dislodged, enlargement of downhole perforations, casing breaks, or other damage that may have been caused by subsidence.

The video survey showed the well to be in good general condition with open perforations and 16 ft of fill. There was no significant evidence of sand production, as confirmed by a Rossum sand test. The survey also showed a horizontal break in the casing at 422.5 ft. The break was around the casing circumference evidencing a compressive loading due to subsidence. There was also a protruding lip on one end of a liner installed in 1962.

TABLE 17.8 SJW Corp.'s Well Pump Design Conditions

Station/Unit	Normal Year Static Water Level Low	Normal Year Static Water Level High	Original Design Point Head Discharge (ft)	Original Design Point Head Discharge (gpm)	Summer Head Discharge (ft)	Summer Head Discharge (gpm)	Winter Head Discharge (ft)	Winter Head Discharge (gpm)	Tank Height (ft)	Specific Capacity Original (gpm/ft)	Specific Capacity Current (gpm/ft)	Design Conditions (from Table 17.9) Primary TDH[a] (ft)	Design Conditions (from Table 17.9) Primary Q (gpm)	Design Conditions (from Table 17.9) Secondary TDH (ft)	Design Conditions (from Table 17.9) Secondary Q (gpm)
Twelfth St. W–8	204	100	240	1750	218	1017	114	1525	19	80	78	256	1600	172	2300

[a]TDH = Total dynamic head.

TABLE 17.9   SJW Corp.'s 1988 Design Point Calculations for 12th Street Station Well 8

Calculate the Primary Design Point	*Calculate the Secondary Design Point*

Calculate the Primary Design Point

Discharge rate desired at low static water level is 1600 gpm
Static water level, normal year low, is 204 ft
Column pipe and above-ground pipe friction is 12 ft
Tank height is 19 ft
Limiting horsepower is 125 hp to reuse the existing motor

$$\text{Drawdown} = \frac{\text{discharge rate}}{\text{specific capacity}}$$

$$= \frac{1600 \text{ gpm}}{78 \text{ gpm/ft drawdown}}$$

$$= 21 \text{ ft}$$

Total head = 204 + 19 + 12 + 21 = 256 ft

$$\text{hp} = \frac{\text{gpm} \times \text{TDH}}{3960 \times \text{pump efficiency}}$$

Assume pump efficiency at 83% and calculate the design flow rate

$$\text{gpm} = \frac{\text{hp} \times 3960 \times \text{pump efficiency}}{\text{TDH}}$$

$$= \frac{125 \times 3960 \times 0.83}{256}$$

$$= 1605 \text{ gpm}$$

(If the desired discharge rate and the answer are more than 5% off, select another discharge rate and repeat the process until a discharge rate is found to match the limiting factors, static water level, specific capacity, and the motor horsepower.)

Calculate the Secondary Design Point

Discharge rate desired at high static water level is 2300 gpm
Static water level, normal year high, is 100 ft
Pipe friction is 24 ft
Tank height is 19 ft
Limiting horsepower is 125 hp

$$\text{Drawdown} = \frac{\text{discharge rate}}{\text{specific capacity}}$$

$$= \frac{2300 \text{ gpm}}{78 \text{ gpm/ft drawdown}}$$

$$= 29 \text{ ft}$$

Total head = 100 + 19 + 24 + 29 = 172 ft

$$\text{hp} = \frac{\text{gpm} \times \text{TDH}}{3960 \times \text{pump efficiency}}$$

Assume pump efficiency at 80% and calculate the design flow rate

$$\text{gpm} = \frac{\text{hp} \times 3960 \times \text{pump efficiency}}{\text{TDH}}$$

$$= \frac{125 \times 3960 \times 0.80}{172}$$

$$= 2302 \text{ gpm}$$

Select New Pump to Operate Efficiently at Primary Design Point (1600 gpm at 256 ft) and at Secondary Design Point (2300 gpm at 172 ft)

Following the video inspection, the lip was repaired, a liner was expanded at 421 to 424 ft and the well was cleaned out by bailing. The well is now in good condition and should operate without problems for at least 20 years.

The pump was reinstalled with a new bowl assembly that conforms to the new design points. The column pipe and tubing were sandblasted and painted, and new line-shaft bearings and centering spiders were installed. The performance curves and the field test curves of the new and old pumps are shown in Figure 17.4, and the field test data are given in Tables 17.10 and 17.11. The new pump operates at an optimum wire-to-water efficiency of 70%. The differences in the design performance curve and the field test curve for the new pump occur because the design performance curve is for the bowl assembly only and the field test covered the entire pump, including the column and the discharge

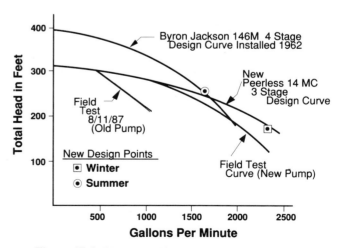

Figure 17.4. New and old pump performances curves.

TABLE 17.10 SJW Corp.'s Old Pump Performance Test

Location of Pump: 12th Street, Well 8
Date: 8/11/87   Pump: 4-stage, 14GM Byron Jackson Bowls   Motor: General Electric VHS

Design Specifications:	gal/min	Total Dynamic Head (ft)	Brake Horsepower	Operating amps	Wire-to-Water Efficiency (%)
	1750	240	125	160	71.8

Test Points	gal/min	TDH[a]	Water hp[b]	Brk hp[c]	Elec hp[d]	Pump Efficiency (%)[e]	Wire-to-Water Efficiency (%)[f]
Pt. No. 1	786.00	251.7	49.96	90.88	97.72	55.0	51.1
Pt. No. 2	687.00	270.3	46.89	87.46	94.05	53.6	49.9

[a] Total dynamic head = Z_{ST} + friction losses

[b] Water horsepower = $\dfrac{Q \times \text{TDH}}{3960}$

[c] Brake horsepower = $\dfrac{1.73\ (I)(E)(PF)(ME)}{746}$

where I = current in amps
   E = voltage
   PF = power factor
   ME = motor efficiency

[d] Electric horsepower = $\dfrac{1.73\ (I)(E)(PF)}{746}$

[e] Pump efficiency = $\dfrac{\text{water hp}}{\text{brk hp}}$

[f] Wire-to-water efficiency = $\dfrac{\text{water hp}}{\text{electric hp}}$

TABLE 17.11 SJW Corp.'s New Pump Performance Test

Location of pump: 12th Street, Well 8
Date 1/14/88   Pump: 3-stage, 14 MC Peerless Bowls   Motor: General Electric VHS

Design Specifications:	gal/min	Total Dynamic Head (ft)	Horsepower	Operating Amps	Wire-to-Water Efficiency (%)
	1600	256	125	148	70.0

Test Points	gal/min	TDH	Water hp	Brk hp	Elec hp	Pump Efficiency (%)	Wire-to-Water Efficiency (%)
Pt. No. 1	2043.00	176.1	90.85	131.36	141.25	69.2	64.3
Pt. No. 2	1697.00	231.3	99.12	131.62	141.53	75.3	70.0
Pt. No. 3	1233.00	272.4	84.82	119.23	128.20	71.1	66.2

head assemblies. The new pump appears to be functioning properly since the design performance and field test curves converge as they approach the shutoff head, and losses in the rest of the pump system are minimized.

REFERENCES

1. Helwig, O. J., V. H. Scott, and J. C. Scalmanini. 1983. *Improving Well and Pump Efficiency*. AWWA, Denver, CO.

2. Gass, T. E., T. W. Bennett, J. Miller, and R. Miller. 1980. *The Manual of Water Well Maintenance and Rehabilitation Technology*. U.S.E.P.A. Ada, OK.

3. Driscoll, F. G. 1986. *Groundwater and Wells*. Johnson Division, Saint Paul, MN.

4. Halliburton Industrial Services, Bulletin 1976. Halliburton Company, Duncan, OK.

CHAPTER 18

Ground Water Quality and Contamination

18.1 INTRODUCTION

"Pure" liquid water does not exist in nature. Even rainwater is at best a very dilute solution containing other dissolved substances. Water exhibits unique properties that enable it to dissolve a variety of other materials. These properties result in the wide ranges of water quality found in nature.

Properties of Water

Ground water quality is greatly influenced by the dissolution of materials existing in the aquifer. The process of dissolving usually involves the separation of ions from a parent compound such as a salt or other mineral. In the parent compound, ions are held in place by the attractive forces between opposite electrical charges. An example is common table salt (sodium chloride), composed of sodium ions (Na^+) and chloride ions (Cl^-). The positively charged sodium ions are bound to the negatively charged chloride ions by electrostatic attractive forces. The result is a solid crystal of sodium chloride composed of alternating positively charged sodium and negatively charged chloride ions (Figure 18.1).

Water is well suited to interfere with these attractive forces. To see how this might occur, the chemical structure of the water molecule needs to be examined. Water is itself a chemical compound of hydrogen (H) and oxygen (O). The molecular formula is H_2O. The bonds between the hydrogens and oxygen form an angle of approximately 105° (Figure 18.2). Since both hydrogens are on the same side of the molecule, that side of the molecule has a net positive charge relative to the other side. One side of the molecule is the positive pole and the other is the negative pole. The molecule is said to be dipolar. This is also discussed in Chapter 1.

The dipolar water molecules are strongly attracted to ions bound on mineral surfaces. As the water molecules approach the mineral ions, the attractive forces that bind the ions are partially neutralized. Ions are released from the mineral surface (dissolved) when neutralization is sufficient to overcome the electrical attraction between them.

When ions are dissolved in water, they are immediately surrounded by the dipolar water molecules. This process, termed "hydration," stabilizes the ions in solution (Figure 18.3). Because the dipolar water molecules are also strongly

attracted to each other, ions in solution are further insulated from the attracting forces of other oppositely charged ions. Thus water makes an excellent solvent for many of the ionic compounds found in the earth's crust.

The chemistry of water is much more complex than the process of dissolving minerals. There are many chemical reactions that occur within solutions. In a number of these, water is a reactive component rather than just a solvent. The different types of reactions are important in determining the chemical composition of natural waters. Many of these will be discussed in the sections that follow.

Impurities in Water

Pure water is defined chemically as H_2O. The term "impurity" is used to describe any constituent dissolved or suspended in water. Impurities can be dissolved chemicals, suspended particles, or even living organisms. As used in this text, the term "impurity" applies to any natural or artificial constituent found in water. These constituents largely determine the chemical, physical, and biological properties of a water—or the "quality" of that water.

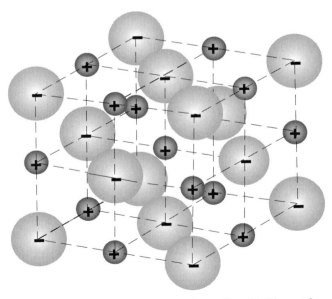

Figure 18.1. Arrangement of ions in a sodium chloride crystal (courtesy of Johnsen, R.H., and E. Grunwald, 1965).

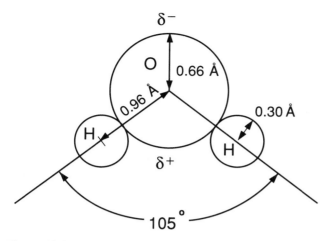

Figure 18.2. Dimensions and spatial arrangement of atoms in a water molecule (Montgomery, 1985).

The quality of most natural waters is constantly changing. In Chapter 1, the hydrologic cycle is depicted as a mechanism for renewing and redistributing pure water. However, even as water droplets form in the atmosphere, their chemical composition is changing. A water droplet passing through the atmosphere comes in contact with a variety of soluble gases. Some of these, such as oxygen (O_2) and nitrogen (N_2), exist in water as simple dissolved gases. Other gases, such as carbon dioxide (CO_2) and sulfur dioxide (SO_2), participate in a series of chemical reactions with water and air, resulting in the formation of new chemical substances such as carbonic and sulfuric acids (H_2CO_3 and H_2SO_4, respectively).

These acids can affect the solution pH and its reactivity toward other substances. For example, carbonic acid (H_2CO_3) is very aggressive toward silicate and carbonate minerals in aquifers. In ground water, carbon dioxide levels may rise as a product of microbial oxidation of organic matter. The resulting attack of carbonic acid on aluminosilicate minerals may be written:

$$\text{Cation-Al-silicate} + H_2CO_3(aq) + H_2O = \text{Cation}$$
$$+ \; HCO_3^- + H_4SiO_4(aq) + \text{Al-silicate} , \quad (18.1)$$

illustrating the release of cations (Na^+, K^+, Mg^{2+}, Ca^{2+}) and dissolved silica in the form of silicic acid (H_4SiO_4) (1). Carbonate minerals, such as calcium carbonate ($CaCO_3$), are similarly attacked by carbonic acid:

$$CaCO_3(s) + H_2CO_3(aq) = Ca^{2+} + 2HCO_3^- , \quad (18.2)$$

resulting in the release of cations, in this case calcium ion (Ca^{2+}), and bicarbonate ion (HCO_3^-).

By the time a water droplet strikes the earth's surface, it will already contain a myriad of constituents that reflect the composition of the atmosphere through which it falls. As the water runs over the earth's surface or percolates through the soil, it acquires additional constituents, and the quality of the water changes further. It follows that ground water generally has higher dissolved mineral concentrations than surface water. This is due in part to the extended contact time between ground water and rocks and soils.

Contamination of Ground Water

In reference to water, the term "contamination" is defined as the degradation of water quality. Both surface and ground waters are subject to contamination, but there is an important distinction. Contamination in surface water may be transient, lasting only a few days or weeks. Conversely, contamination

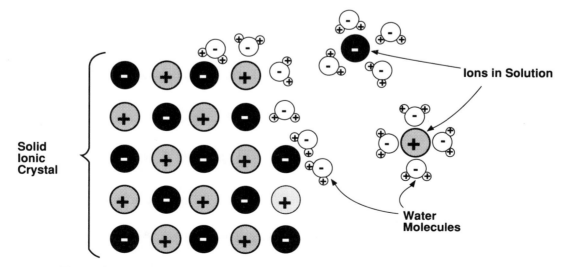

Figure 18.3. Hydration of ions in solution (from *Chemistry: A Conceptual Approach* by Charles E. Mortimer). © 1967 Reinhold Publishing Corporation. Reprinted by permission of Wadsworth, Inc.

in ground water may persist for decades or centuries. If a contaminant is not readily decayed or immobilized underground, it may remain a degrading influence indefinitely.

There are two principal sources of contamination; natural and artificial. Natural contamination is a result of the chemical and physical processes that transfer impurities to water from the atmosphere, biosphere, or lithosphere (earth's crust). Artificial contamination (pollution) results from human activities. Much of this chapter is devoted to the natural processes in aquifers that affect ground water quality. For example, the mineral content of ground water is altered by the leaching of soluble salts from the soils and rocks that surround it. Natural sources of impurities that are external to an aquifer can also significantly alter water quality. These include interchanges between aquifers. For example, highly mineralized connate water or water associated with volcanic or other special geological formations may contaminate aquifers by moving through fault zones or other natural conduits. Seawater intrusion is also a natural source of contamination, although overdrafting of coastal freshwater aquifers accentuates the problem.

Waste disposal practices are the most notable human activities affecting ground water quality. Impurities associated with domestic and industrial wastes range from simple inorganic ions, such as nitrate (feedlots) and chloride (municipal wastewater), to heavy metals, such as chromium (plating operations), and to synthetic organic chemicals, such as tetrachloroethylene (dry cleaning and metal-degreasing wastes). Other impurities that may be associated with waste disposal activities are listed in Table 18.1.

Domestic and industrial wastes are often deposited or stored on or below land surfaces. This includes but is not limited to individual sewage disposal systems (septic tanks, cesspools), land disposal of sludge or solid waste (landfills), industrial wastewater impoundments, disposal of mine wastes and brines, disposal of animal feedlot wastes, deep-well injection of liquid wastes, and the collection, treatment, and disposal of municipal wastewater. Because of the sheer number of waste processing and disposal sites, the potential for contamination of ground water is great. As illustrated in Figure 18.4, uncontained wastes from a variety of sources may enter underlying ground water by percolation.

Sources of contamination not related to waste disposal practices include accidental spills and leaks, infiltration from polluted surface waters, mining, oil and gas exploration, and agriculture. The relative importance of these activities in affecting ground water resources is geographically quite variable. For example, regions located in the snowbelt may be vulnerable in infiltration of surface water contaminated with road-deicing salts, whereas coastal regions may have serious problems with seawater intrusion.

In locating new water wells, it is important to consider existing and potential sources of contamination in addition to the local geology and hydrology. A general rule is to

TABLE 18.1 Major Contaminants Associated with Waste Disposal Practices

Source	Possible Major Contaminants
Landfills	
Municipal	Heavy metals, chlorides, sodium, calcium
Industrial	Wide variety of inorganic and organic constituents
Hazardous waste disposal sites	Wide variety of inorganic (particularly heavy metals) and organic compounds (pesticides, priority pollutants, etc.)
Liquid waste storage ponds (lagoons, leaching ponds, recharge basins)	Heavy metals, solvents, inorganic compounds
Subsurface sewage disposal systems	Organic compounds (degreasers, solvents), nitrogen compounds, sulfates, sodium, microbiological contaminants
Deep-well waste injection	Variety of inorganic and/or organic compounds
Agricultural activities	Fertilizers, herbicides, pesticides
Land application (sludge, wastewater)	Heavy metals, inorganic compounds, organic compounds
Urban runoff infiltration	Inorganic compounds, heavy metals, petroleum products
Deicing activities	Chlorides, sodium, calcium
Radioactive wastes	Radioactivity and radionuclides

Source: Hess, A. F. and J. E. Dyksen. Utility Experiences Related to Existing and Proposed Drinking Water Regulations. Table 3, p. 23 found in Annual Conference Proceedings, AWWA, 1984. Reprinted from 1984 Annual Conference Proceedings, American Water Works Association, by permission. Copyright © 1984, American Water Works Association.

locate water wells upgradient and as far away from potential sources of contamination as is practical. There is no universally defined "safe distance" between a well and a source because there are so many factors that affect the transport of water and contaminants. Mathematical models are used to predict the movement of ground water (see Chapter 20).

The movement of ground water and contaminants is greatly affected by the nature of soils and aquifers. Water and dissolved substances move more slowly in dense strata, such as clay and heavy loam, than in sands and gravels. Fractured rock aquifers can be particularly troublesome because they are channeled or creviced and readily transmit water over great distances. In addition contaminants may be dispersed by diffusion or mechanical mixing, retarded by ion exchange or sorption to soils, and degraded by natural chemical and biological processes. The behavior of a particular contaminant in ground water (and the impact on water quality) is dependent

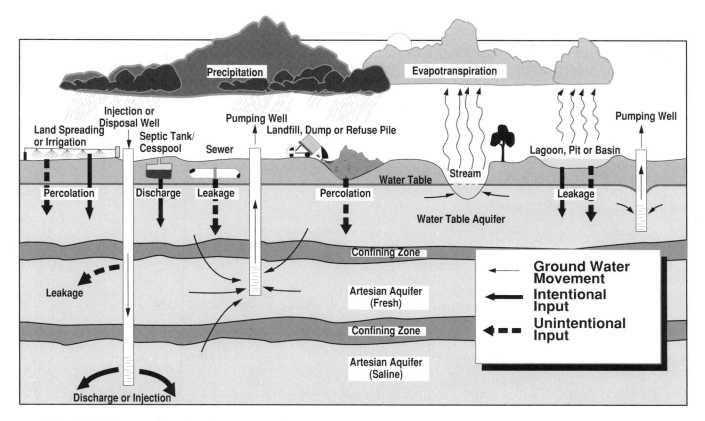

Figure 18.4. How waste disposal practices may contaminate ground water (from *E.P.A. Journal*, Volume 10, No. 6, "Sources of Ground Water Contamination", July-August, 1984).

on its physical and chemical nature and the processes that influence its transport and environmental persistence.

A discussion of the movement of individual contaminants in ground water is beyond the scope of this book. Nevertheless, there are a few generalizations that can be made. Once released, contaminants tend to move in the direction of ground water flow. They generally move in a "plume," the size and shape of which depends on local geology, flow, type, concentration, release rate, and attenuation of contaminant. Where ground water moves rapidly, a plume from a point source may be long and thin. Lesser flows may allow contaminants to spread laterally, resulting in a wider plume. Irregular plumes may be created by local influences such as well pumping or variations in aquifer permeability. It should be noted that plumes of contaminants are not necessarily stable. Figure 18.5 illustrates some of the possible changes and the factors causing them (2).

18.2 WATER-QUALITY CRITERIA AND STANDARDS

Water-Quality Criteria

All ground waters contain impurities from natural sources, and many usable aquifers contain pollutants resulting from

human activities. If all waters contain impurities, then what constitutes good-quality water? The answer to this question is related not only to the quantities and types of impurities but also to the intended use of the water. Table 18.2 lists some of the most common natural impurities found dissolved in ground water and the undesirable effects of high concentrations. In high concentrations, most of the impurities listed in Table 18.2 would render a water unsuitable for domestic use. For example, iron and manganese can cause water to appear colored. Although colored water may be unacceptable to domestic consumers, it may pose no problem for agricultural uses. Therefore the usefulness of a water supply depends on intended use as well as the composition and characteristics of the water. Water quality criteria define desirable characteristics and acceptable levels of constituents for water of various intended uses.

Water intended for domestic use should be aesthetically pleasing as well as safe for human consumption. In the United States legally enforceable drinking water standards exist to protect the public health and welfare. These standards guide water purveyors in producing water that is aesthetically acceptable and free from disease-causing organisms or harmful levels of other constituents.

The quality requirements for industrial use are highly variable. For some uses, such as single-pass condensing or

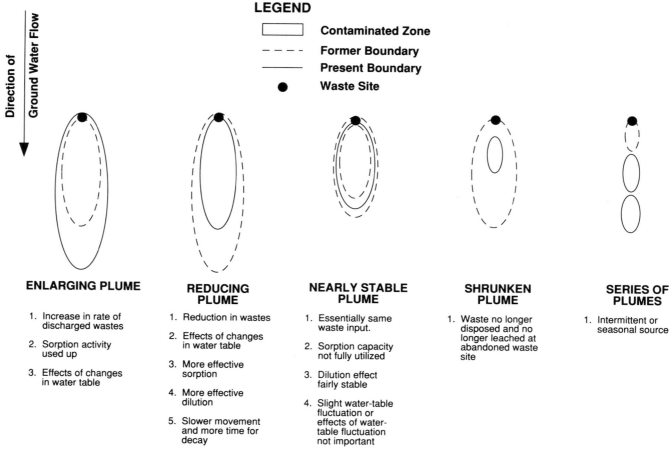

Figure 18.5. Changes in plumes and factors causing the changes (from Miller, D. W., *Waste Disposal Effects on Ground Water*). Copyright © 1980, by the American Geophysical Union.

cooling, even salt and brackish waters may be acceptable. This is not the case for most industrial uses, however, and almost every industrial process has its own quality requirements. Table 18.3 identifies some of the water quality requirements for selected industries and processes. Quality for most industrial process waters would be satisfied if the water supply meets drinking water standards. Exceptions include waters used for processing certain foods and beverages and boiler feedwater. In addition highly purified water is necessary for certain technical operations such as the manufacture of silicon chips for the semiconductor industry. It is technically possible to treat any water to the desired quality requirements. However, the cost of treatment may preclude the use of some water sources for certain industrial applications.

Agricultural use includes irrigation of crops and consumption by livestock. Livestock are generally able to consume waters higher in mineral content than recommended for human consumption. Table 18.4 lists recommended upper limits for total dissolved solids (TDS) for livestock. These limits are much higher than the 500 mg TDS per liter recommended for human consumption. Domestic animals, however, are sensitive to individual toxic substances just as

are humans. Table 18.4 also lists limits for some of these. Incidents where animals have been poisoned by drinking polluted water are not uncommon.

Crops exhibit sensitivity to both total dissolved solids and selected elements. Citrus crops, for example, are very sensitive to the element boron in water. Carrots, lettuce, and cabbage, however, are somewhat tolerant to boron. Soils are also impacted by the chemical quality of water. A soil irrigated with water that contains a high proportion of sodium relative to calcium and magnesium can develop serious drainage problems and may not support the growth of crops. Table 18.5 suggests restrictions on the use of irrigation water containing various dissolved constituents (3).

There have been several efforts to establish water quality criteria for the various beneficial uses of water. In 1963 the State of California published a book entitled *Water-Quality Criteria* (4). In this document the most important beneficial uses of water are identified:

1. Domestic water supply
2. Industrial use
3. Agricultural use

TABLE 18.2 Common Impurities Found Dissolved in Ground Water

Impurity	Undesirable Effects
Gases	
Oxygen	Corrosive to metals
Carbon dioxide	Corrosive to metals
Hydrogen sulfide	Corrosive to metals, odor
Solids	
Sodium carbonate	Alkalinity
Sodium bicarbonate	Alkalinity
Sodium chloride	Taste
Sodium fluoride	Mottles teeth
Calcium carbonate	Hardness, alkalinity
Calcium bicarbonate	Hardness, alkalinity
Calcium sulfate	Hardness
Calcium chloride	Hardness, corrosive
Magnesium carbonate	Hardness, alkalinity
Magnesium bicarbonate	Hardness, alkalinity
Magnesium sulfate	Hardness
Magnesium chloride	Hardness, corrosive
Iron	Color, hardness, taste
Manganese	Color

Source: Reprinted from Manual M21, *Groundwater*, American Water Works Association, by permission. Copyright © 1973, American Water Works Association.

4. Fish and aquatic life

5. Recreational use

The various impurities found in water are also identified and referenced according to their effects on these uses.

In 1972 amendments to the Federal Water Pollution Control Act, P.L. 92-500, mandated the U.S. Environmental Protection Agency (EPA) to publish water quality criteria identifying all known impacts on health and welfare that may be expected from the presence of pollutants in water. In compliance with this requirement, EPA published *Quality Criteria for Water* (5). This document recommends maximum levels for water quality constituents that provide for the protection and propagation of fish and other aquatic life. Criteria are also provided for domestic water. Since these uses require the highest quality, this water will also be suitable for most agricultural and industrial purposes.

Public Health Standards

Many countries establish formal drinking water standards to ensure that the public health is protected from waterborne disease. Two sets of standards have achieved international status: the World Health Organization (WHO) European

TABLE 18.3 Suggested Water-Quality Tolerances for Industrial Use

Industry or Use	Turbidity (units)	Color (units)	Hardness (as CaCO₃) (ppm)	Fe (ppm)	Mn (ppm)	Total Solids (ppm)	Alkalinity (as CaCO₃) (ppm)	Odor, Taste	H₂S (ppm)	Other Requirements
Air conditioning	—	—	—	0.5[a]	0.5	—	—	Low	1	No corrosiveness or slime formation
Baking	10	10	—	0.2[a]	0.2	—	—	Low	0.2	Potable[b]
Brewing										
Light beer	10	—	—	0.1[a]	0.1	500	75	Low	0.2	Potable. NaCl less than 275 ppm. pH 6.5–7.0.
Dark beer	10	—	—	0.1[a]	0.1	1000	150	Low	0.2	Potable. NaCl less than 275 ppm. pH 7.0 or more.
Canning										
Legumes	10	—	25–75	0.2[a]	0.2	—	—	Low	1	Potable
General	10	—	—	0.2[a]	0.2	—	—	Low	1	Potable
Carbonated beverages	2	10	250	0.2 (0.3)[a]	0.2	850	50–100	Low	0.2	Potable. Organic color plus O₂ consumed less than 10 ppm.
Confectionery	—	—	—	0.2[a]	0.2	100	—	Low	0.2	Potable. pH above 7 for hard candy, low for fondant.
Cooling	50	—	50	0.5[a]	0.5	—	—	—	5	No corrosiveness or slime formation
Distilling										
Gin and spirits	10	—	—	0.1[a]	0.1	500	75	Low	0.2	Potable. NaCl less than 275 ppm. pH 6.5–7.0.

(Table continues on p. 364.)

TABLE 18.3 (*Continued*)

Industry or Use	Turbidity (units)	Color (units)	Hardness (as CaCO$_3$) (ppm)	Fe (ppm)	Mn (ppm)	Total Solids (ppm)	Alkalinity (as CaCO$_3$) (ppm)	Odor, Taste	H$_2$S (ppm)	Other Requirements
Whiskey	10	—	—	0.1[a]	0.1	1000	150	Low	0.2	Potable. NaCl less than 275 ppm. pH 7.0 or more.
Blending	—	—	—	—	—	—	—	—	—	Distilled or special treatment
Food, general	10	—	—	0.2[a]	0.2	—	—	Low	—	Potable
Ice	5	5	—	0.2[a]	0.2	170	—	Low	—	Potable. SiO$_2$ less than 10 ppm.
Laundering	—	—	50	0.2[a]	0.2	—	—	—	—	
Plastics, clear, uncolored	2	2	—	0.02[a]	0.02	200	—	—	—	
Paper and pulp										
Groundwood	50	200	180	1.0[a]	0.5	—	—	—	—	No grit or corrosiveness
Kraft pulp	25	15	100	0.2[a]	0.1	300	—	—	—	
Soda and sulfite	15	10	100	0.1[a]	0.05	200	—	—	—	
High-grade light papers	5	5	50	0.1[a]	0.05	200	—	—	—	No slime formation
Rayon (viscose)										
Pulp production	5	5	8	0.05[a]	0.03	100	Total 50, OH 8	—	—	Al$_2$O$_3$, SiO$_2$, Cu less than 8, 25 and 5 ppm, respectively
Manufacture	0.3	—	55	0.0	0.0	—	—	—	—	pH 7.8 to 8.3
Tanning	20	10–100	50–135	0.2[a]	0.2	—	Total 135, OH 8	—	—	
Textiles										
General	5	20	—	0.25	0.25	—	—	—	—	
Dyeing	5	5–20	—	0.25[a]	0.25	200	—	—	—	Constant composition, Al$_2$O$_3$ less than 0.5 ppm
Wool scouring	—	70	—	1.0[a]	1.0	—	—	—	—	
Cotton bandage	5	5	—	0.2[a]	0.2	—	—	Low	—	

Source: Camp, T. R. and R. L. Meserve, *Water and Its Impurities*, 2nd Edition, 1974, Table 7.6 on p. 160. Reprinted by permission Van Nostrand Reinhold.
[a]Limit applies both to iron alone and to the sum of iron and manganese.
[b]Potable means complying with drinking water standards of the U. S. Environmental Protection Agency.

Standards and the WHO International Standards (6, 7). The WHO International Standards were intended as minimal standards and were considered at the time to be achievable by all countries throughout the world. The European Standards were more rigorous and reflected the water treatment technologies available to industrial nations.

Recognizing the many social, economic, and cultural constraints of implementing uniform international standards, in 1984 WHO developed a comprehensive set of guidelines for drinking water quality (8). These supersede the formal European and International standards and instead provide a basis on which individual countries can develop standards and regulations of their own. Selected international drinking water standards are listed in Table 18.6.

The first national drinking water standards in the United States were established by the U.S. Public Health Service (USPHS) in 1914. These were enforceable only for water systems serving interstate carriers such as trains, airplanes, and buses. The USPHS standards were updated in 1925, 1946, and 1962, but for most public water systems were considered voluntary guidelines rather than enforceable standards. The Safe Drinking Water Act of 1974, P.L. 93-523, mandated that the Environmental Protection Agency (EPA) establish national drinking water standards that are legally enforceable. These standards, the National Primary Drinking Water Standards, became effective in June 1977. At present (1989), the primary standards specify maximum contaminant levels (MCLs)—the highest allowable concentration of a specific substance—for ten inorganic chemicals, coliform bacteria, turbidity, five radioactive materials, six pesticides, eight volatile organic chemicals, and total trihalomethanes. All of these substances may have adverse

TABLE 18.4 Recommended Concentration Limits for Water Used for Livestock

	Livestock: Recommended limits (mg/L)
Total dissolved solids	
Small animals	3000
Poultry	5000
Other animals	7000
Nitrate	45
Arsenic	0.2
Boron	5
Cadmium	0.05
Chromium	1
Fluoride	2
Lead	0.1
Mercury	0.01
Selenium	0.05

Source: Freeze, R. A. and J. A. Cherry, *Groundwater*, © 1979, Table 9.2, p. 388. Reprinted by permission of Prentice-Hall.

effects on human health if consumed in sufficient amounts. The MCLs are mandatory standards and are enforceable by EPA or states with designated authority. The current (1989) National Primary Drinking Water Standards are shown in Table 18.7.

Secondary drinking water standards are established for drinking water constituents that affect the aesthetic qualities of water (Table 18.8). They are intended to set limits for substances that affect the taste, odor, or physical appearance of water. Secondary standards are generally not legally enforceable by EPA. However, several states have passed legislation allowing enforcement by the state.

The Safe Drinking Water Act was significantly amended in June 1986. The amendments require EPA to regulate a list of 83 constituents by June 1989 and to subsequently regulate 25 new constituents at three-year intervals. The constituents specified for regulation are listed in Table 18.9.

TABLE 18.5 Guidelines for Interpretation of Water Quality for Irrigation

		Degree of Restriction on Use		
Potential Irrigation Problem	Units	None	Slight to Moderate	Severe
Salinity (Affects availability of water to crop)				
EC	μmho/cm	<700	<700–3000	>3000
TDS	mg/L	<450	450–2000	>2000
Permeability (Affects infiltration rate of water into the soil. Evaluate using EC and SAR together)				
SAR = 0–3		and EC = >700	700–200	<200
= 3–6		= >1200	1200–300	<300
= 6–12		= >1900	1900–500	<500
= 12–20		= >2900	2900–1300	<1300
= 20–40		= >5000	5000–2900	<2900
Specific ion toxicity (Affects sensitive crops)				
Sodium (Na)				
Surface irrigation	SAR	<3	3–9	>9
Sprinkler irrigation	meq/L	<3	>3	
	mg/L	<70	>70	
Chloride (Cl)				
Surface irrigation	meq/L	<4	4–10	>10
	mg/L	<140	140–350	>350
Sprinkler irrigation	meq/L	<3	>3	
	mg/L	<100	>100	
Boron (B)	mg/L	<0.7	0.7–3.0	>3.0
Miscellaneous effects (Affects susceptible crops)				
Nitrogen (total-N)	mg/L	<5	5–30	>30
Bicarbonate (HCO₃)				
(overhead sprinkling only)	meq/L	<1.5	1.5–8.5	>8.5
	mg/L	<90	90–500	500
pH	unit	Normal range 6.5–8.4		
Residual chlorine				
(overhead sprinkling only)	mg/L	<1.0	1.0–5.0	>5.0

Source: Asano, T. and G. S. Pettygrove, Using Reclaimed Municipal Wastewater for Irrigation, *Journal of California Agriculture*, March–April 1987. By permission, California Agriculture, Univeristy of California.
Note: EC = electrical conductivity; SAR = sodium adsorption ratio.

TABLE 18.6 Selected International Drinking Water Standards

Parameter	Units	U.S. EPA (1980)		WHO, International (1971)	
		Recommended	Maximum Contaminant Level	Recommended	Maximum Permissible Level
Aluminum	mg/L	—	—	—	—
Arsenic	mg/L	—	0.05	—	0.05
Chloride	mg/L	250	—	200	600
Copper	mg/L	1	—	0.05	1.5
Fluoride	mg/L	—	1.4–2.4[a]	—	0.6–1.7[a]
Iron	mg/L	0.3	—	0.1	1.0
Lead	mg/L	—	0.05	—	0.1
Magnesium	mg/L	—	—	—	150[c]
Manganese	mg/L	0.05	—	0.05	0.5
Mercury	mg/L	—	0.002	—	0.001
Nitrate	mg/L-N	—	10	—	—
Phosphorus	mg/L	—	—	—	—
Sodium	mg/L	20	—	—	—
Sulfate	mg/L	250	—	200	400
Zinc	mg/L	5	—	5	15
TDS	mg/L	500	—	500	1500
Total hardness	mg/L CaCO$_3$	—	—	100	500
Selenium	mg/L	—	0.01	—	0.01
Color	mgPt/L	15	—	5	50
Turbidity	TU	—	1	5	25
pH	units	6.5–8.5	—	7.0–8.5	6.5–9.2
Coliform	org/100 mL	—	1	—	1
Foaming agents (MBAS)	mg/L	0.5	—	0.2	1.0
Other organochlorine compounds	μg/L	—	—	—	—
Total organic carbon (TOC)	mg/L	—	—	—	—
Trihalomethane (THM)	μg/L	—	100	—	—

Source: Montgomery (1985).
[a] Concentration is described as a function of ambient temperature.
[b] After 16-hr lead or copper pipe.
[c] If there is less than 250 mg/L of sulfate.
[d] The reason for any increase in the usual concentration must be investigated.
[e] Concentration of THM as low as possible.

18.3 WATER-QUALITY CONSTITUENTS AND CHARACTERISTICS

Physical Characteristics

Suspended Solids. Natural waters may contain dissolved, suspended, or settleable solids. The dissolved state is characterized by individual atoms or molecules of solute completely surrounded by solvent molecules. If the particles are large enough to be removed by filtration or to settle out of the solvent by gravity, then the particles are said to be suspended or settleable rather than dissolved. The term "colloidal material" is used to describe extremely small particles that are difficult to filter and remain suspended due to repulsive forces of like electrical charges.

Natural suspended solids include silts, clays, and organic materials. Most ground waters contain relatively small amounts of suspended solids because the formations through which the water flows effectively remove particles by filtration. Once the ground water is brought to the surface, however, chemical precipitation may result in the formation of suspended solids from dissolved minerals. For example, some ground waters are devoid of dissolved oxygen and contain iron largely in the ferrous (reduced) state. Ferrous iron is relatively soluble and stays in solution. When this water is pumped from the ground and contacts air, the ferrous iron

WHO, European (1970)	European Community		The Netherlands (1981)	USSR (1975)	Norway
	Guide Levels	Maximum Admissible			
—	0.05	0.2	—	0.5	<0.1
0.05	—	0.05	0.2	0.05	<0.01
200	25	>200	—	—	—
0.05	0.1	—	3.0[b]	—	<0.05
0.7–1.7[a]	—	0.7–1.5[a]	1.2	0.7–1.5[a]	<1.5
0.1	0.05	0.2	1	0.3	<0.1
0.1	—	0.05	0.3	0.1	<0.05
125[c]	30	50	—	—	<10
0.05	0.02	0.05	0.05	0.1	<0.1
—	—	0.001	—	—	<0.0005
23	6	11	23	10	<2.5
—	0.15	2.0	—	0.0035	—
—	20	175	—	—	—
250	25	250	—	—	—
5.0	0.1	5	1.5[b]	—	<0.3
—	—	1500	—	—	—
100–500	—	—	—	—	—
0.01	—	0.01	0.05	0.001	<0.01
—	1	20	20	—	<5
—	0.4	4	0.5	—	<0.5
—	6.5–8.5	9.5	—	—	8.0–8.5
1	—	<1	0–2	—	<1
0.2	—	0.2	—	—	—
—	1	—	—	—	—
—	—	d	—	—	—
—	e	—	—	—	—

reacts with available oxygen and comes out of solution as particles of ferric oxide.

Suspended solids can have a significant impact on the suitability of a ground water for both domestic and industrial use. Even in small amounts, solids can interfere with the disinfection of drinking water by providing a surface for microbial growth and insulation from chemical disinfectants. In greater amounts solids can physically interfere with machinery and sensitive industrial processes. Additionally, suspended particles may cause water to appear cloudy (turbid), and many types of particles, such as iron oxide, can result in water with a colored appearance. Either of these conditions can make water aesthetically unacceptable. Effects on agriculture include formation of crusts or films on soil and plant surfaces that impede soil aeration and infiltration of water or reduce the marketability of some leafy crops such as lettuce. Suspended solids may be removed by filtration.

Settleable Solids/Sand. "Settleable solids" is the term applied to material that will settle out of suspension within a defined period. The time period, usually 1 hour, is specified by the method used to measure the solids. Settleable solids can be measured either gravimetrically or volumetrically (9).

One of the most troublesome settleable materials found in well water is sand. Its deleterious effects on pumps and systems are discussed in Chapter 17. For continuous measurement, a centrifugal sand separator produces reliable results (see Chapter 14) (10).

Turbidity. Turbidity is an expression of the optical property that causes light to be scattered as it passes through water. It is closely related to the amount and type of suspended solids. Water that is crystal clear has low turbidity and usually contains very few suspended solids. Water that is cloudy in appearance has high turbidity. The measurement of turbidity depends not only on the amount of suspended matter but also on the size, shape, and optical properties of the individual particles. Therefore turbidity cannot be measured in milligrams per liter (mg/L) or other concentration

TABLE 18.7 EPA National Primary Drinking Water Standards (1988)

Constituent	Unit	Maximum Contaminant Level
Inorganics		
Arsenic	mg/L	0.05
Barium	mg/L	1.0
Cadmium	mg/L	0.01
Chromium	mg/L	0.05
Fluoride	mg/L	4.0
Lead	mg/L	0.05
Mercury	mg/L	0.002
Nitrate (as N)	mg/L	10.0
Selenium	mg/L	0.01
Silver	mg/L	0.05
Microbials		
Coliforms		1/100 mL
Turbidity	ntu	1–5
Organics		
2,4-D	mg/L	0.1
Endrin	mg/L	0.0002
Lindane	mg/L	0.0004
Methoxychlor	mg/L	0.1
Toxaphene	mg/L	0.005
2,4,5-TP silvex	mg/L	0.01
Trihalomethanes (chloroform, bromoform, bromodichloromethane, dibromochloromethane)		0.10
Radionuclides		
Beta particle and photon radioactivity	mrem	4 (annual dose equivalent)
Gross beta particle activity	pCi/L	50
Strontium-90	pCi/L	8
Tritium	pCi/L	20,000
Gross alpha particle activity	pCi/L	15
Radium-226 + radium-228	pCi/L	5
Volatile organic chemicals		
Benzene	mg/L	0.005
Carbon tetrachloride	mg/L	0.005
1,2-Dichloroethane	mg/L	0.005
1,1-Dichloroethylene	mg/L	0.007
1,1,1-Trichloroethane	mg/L	0.20
para-Dichlorobenzene	mg/L	0.075
Trichloroethylene	mg/L	0.005
Vinyl chloride	mg/L	0.002

Source: Sayre, I. M., International Standards for Drinking Water, Table 1 on p. 54. Reprinted from *Journal* American Water Works Association, Vol. 80, No. 1, by permission. Copyright © 1988, American Water Works Association.

TABLE 18.8 EPA National Secondary Drinking Water Standards (1979)

Constituent	Maximum Contaminant Level	Effect on Water Quality
Chloride	250 mg/L	Salty taste
Color	15 color units	Objectionable appearance
Copper	1 mg/L	Undesirable taste
Corrosivity	Noncorrosive	Stains, dissolution of metals, economic losses
Foaming agents	0.5 mg/L	Undesirable appearance
Iron	0.3 mg/L	Bitter, astringent taste, stained laundry
Manganese	0.05 mg/L	Impaired taste, discolored laundry
Odor	3 Threshold odor number	Undesirable smell
pH	6.5–8.5	Corrosivity, impaired taste
Sulfate	250 mg/L	Detectable taste, laxative at high concentrations
TDS	500 mg/L	Objectional taste
Zinc	5 mg/L	Undesirable taste, milky appearance at high concentrations

Source: Montgomery (1985).

be measured with an instrument (nephelometer). The standard is based on health considerations related to the effect of turbidity (suspended particles) on the disinfection process. A secondary standard of 5.0 NTU is based on the level at which the human eye can detect turbidity. Drinking water with a turbidity greater than 5.0 NTU is generally not acceptable for consumption.

Color. Two types of color may be present in natural water. "True color" results from the presence of dissolved substances. In natural waters these substances usually originate from the decomposition of soils or vegetation. True color by definition can only be determined in samples that have been filtered or centrifuged to remove suspended solids. "Apparent color" results from the presence of certain types of suspended or colloidal materials. For example, suspended particles of iron oxide can give water a rust-colored appearance. Iron and manganese in the oxidized state can impart a yellow-to-brown color to water. Apparent color is always measured in unfiltered water samples. Color may be measured by

units used to describe suspended solids. Rather, turbidity is reported in turbidity units.

The present (1989) primary drinking water standard for turbidity is 1.0 nephelometric turbidity units (NTU). This level of turbidity is not visible to the naked eye and must

TABLE 18.9 Constituents Specified for Regulation under the Safe Drinking Water Act of 1986

Volatile Organic Chemicals

Trichloroethylene	Benzene
Carbon tetrachloride	Dichlorobenzene(s)
1,1,1-Trichloroethane	1,1-Dichloroethylene
1,2-Dichloroethane	Vinyl chloride

Inorganics

Arsenic	Asbestos
Barium	Sulfate
Cadmium	Copper
Chromium	Nickel
Lead	Thallium
Mercury	Beryllium
Nitrate	Cyanide
Nitrite	Selenium
Fluoride	Antimony

Organics

Tetrachloroethylene	2,3,7,8-TCDD (Dioxin)
Methylene chloride	1,1,2-Trichloroethane
Chlorobenzene	Vydate
Trichlorobenzene(s)	Simazine
trans-1,2-Dichloroethylene	PCB's
cis-1,2-Dichloroethylene	Phthalates
Endrin	Acrylamide
Lindane	Dibromochloropropane (DBCP)
Methoxychlor	1,2-Dichloropropane
Toxaphene	Pentachlorophenol
2,4-D	Picloram
2,4,5-TP	Dinaseb
Aldicarb	Ethylene dibromide
Chlordane	Ethylbenzene
Atrazine	Xylene
Dalapon	Hexachlorocyclopentadiene
Diquat	Styrene
Endothall	Aldicarb sulfone
Glyphosate	Aldicarb sulfoxide
Carbofuran	Alachlor
Epichlorohydrin	Toluene
Adipates	Heptachlor
Heptachlor epoxide	PAH's

Microbiology and Turbidity

Total Coliforms	Viruses
Turbidity	Standard Plate Count
Giardia lamblia	Legionella

Radionuclides

Radium 226 and 228	Uranium
Beta Particle and Photon	Gross Alpha
Radioactivity	Radon

Source: Tables 2-3 & 2-4, p. 12, found in *New Dimensions in Drinking Water*, 1987. Reprinted from *New Dimensions in Drinking Water*, American Water Works Association, by permission. Copyright © 1987, American Water Works Association.

visual or photometric comparison of the water sample to standard color solutions or permanently colored glass discs. The unit of measurement is the color unit, a rating of color intensity. Color in natural water has no direct correlation with any specific dissolved chemical concentration.

The significance of color in ground water is dependent on the chemical nature of the color-causing agent. Color may result from industrial contamination (chromate wastes) as well as from natural decomposition of plants and soils (humic substances). The materials responsible for natural color in water are not known to be harmful to health. However, some of them may react with chlorine to form by-products such as trihalomethanes that have health implications. Aside from specific chemical concerns, color is most significant in terms of aesthetics. Color below ten units is barely noticeable. The secondary drinking water standard for color is 15 color units (apparent). The constituents that cause water to appear colored can generally be removed by chemical precipitation and/or filtration. Many of them are removed incidentally during lime-soda softening.

Taste and Odor. Distilled or deionized water with few dissolved impurities usually tastes "flat." Some mineral substances improve the taste of water for the average person. Other dissolved substances can impart objectionable tastes and odors. A common problem in ground water is dissolved hydrogen sulfide gas, which causes water to smell like rotten eggs. In addition, some metal ions can cause metallic or astringent tastes. For example, secondary drinking water standards for copper and zinc are based on taste thresholds. A chemical's taste or odor threshold is defined as the minimum concentration of the compound that is definitely perceived by the tester.

The majority of taste or odor-causing substances are organic compounds. Some organic chemicals have taste or odor thresholds at part per billion (microgram per liter) levels. Many of these are reported in the literature (11). Some are listed in Table 18.10. There are a few naturally occurring organic substances that can be detected by taste or smell at the part per trillion (nanogram per liter) level. Some of these are produced by algae, bacteria, or other microorganisms.

Treatment to reduce taste and odor in drinking water generally involves the destruction or removal of the taste or odor-causing compounds. Many of them can be destroyed (oxidized) by the same chemical oxidants used to disinfect drinking water. The levels of many taste or odor compounds can also be reduced by aeration or adsorption. Aeration is best suited to remove volatile compounds, whereas adsorption is more effective for less volatile compounds and those with higher molecular weights. Activated carbon is widely accepted as the most effective adsorbent.

Temperature. Water temperature has a significant impact on its beneficial uses as well as on the other chemical and

TABLE 18.10 Concentrations of Some Chemicals Causing Odor or Taste

Compound	Threshold (mg/L)	Characteristic Odor or Taste
Odor thresholds of various chemicals in water at 60°C		
Benzene	0.072	Sweet
Ethylbenzene	0.0024	Sweet
Ethylene glycol	197	Mild, sweet
Hydrochloric acid (32%)	420	Weak chlorinous
Naptha	0.016	Gasoline
Tetrachloroethylene	0.24	Chlorinated solvent
Trichloroethylene	0.56	Chlorinated solvent
Vinylidene chloride	1.6	Perfume
Taste thresholds of various chemicals in water at 40°C		
Chloroform	12	Sweet
Ethylene glycol	1067	Puckery
Hydrochloric acid (32%)	200	Lemony
Methylene chloride	24	Dry
Tetrachloroethylene	2.8	Sweet, dry
Trichloroethane	3.2	Bitter

Source: Alexander, H. C., W. M. McCarthy, E. A. Bartlett, and A. N. Syverud, "Aqueous Odor and Taste Threshold U.S. Environmental Protection Agency. Values of industrial Chemicals, Tables 1, 2, pp. 596, 598, *Journal* American Water Works Association, November, 1982. Reprinted from *Journal* American Water Works Association, Vol. 74, No. 11, by permission. Copyright © 1982, American Water Works Association.

physical properties exhibited by the water. Although there are no drinking water standards for temperature in the United States, many countries have established nonenforceable guidelines for the temperature of the domestic water supply. In Norway the guideline is a temperature less than 10° Celsius, or centigrade (C). In Germany the guide level is greater than 5°C but less than 10°C. These levels are related to the aesthetics of drinking water and local consumer acceptance. In temperate climates, use of ground water with temperatures approaching 25°C has been found to be generally acceptable. Ground water with temperatures approaching 30°C results in consumer complaints. The suitability of a water for direct industrial use may also be influenced by temperature, although the temperature of most natural waters can be controlled for industrial applications.

Thermal ground waters (hot springs) may exhibit unusual chemical properties in addition to elevated temperatures. Because the solubility of most mineral substances increases with temperature, the dissolved solids content of such water can be much higher than that of cooler waters located near the earth's surface. Conversely, dissolved gases are less soluble in warmer waters resulting in further chemical differentiation.

Biological Characteristics

Biological organisms play an important role in a variety of chemical phenomena that affect water quality. For example, certain types of bacteria are critical links in the biogeochemical cycles of carbon, sulfur, and nitrogen. Although the organisms and chemical processes associated with them ultimately affect the quality of natural waters, only a few organisms have been shown to directly impact the development of ground water resources. These include organisms having health-related impacts (disease-causing organisms) and organisms resulting in nuisance conditions.

Disease-Causing Organisms. The occurrence of disease-causing organisms in natural waters is well documented. Bacteria, viruses, protozoans, and even certain types of worms and fungi have been implicated in animal and human disease outbreaks. Historically, the most important source of these organisms is contamination by domestic waste. In general, surface water is more vulnerable to this type of contamination than ground water. Surface water used for domestic supply is usually disinfected and filtered to ensure that disease-causing organisms are removed or destroyed.

Under certain circumstances ground water may also become contaminated by disease-causing organisms. Domestic wastes are often deposited or stored on or below land surfaces. Examples include septic tanks and landfills. If not properly contained, these wastes may percolate to underlying ground water. Highly porous soils and fluctuating water tables contribute to poor filtration of wastewater. A number of episodes of ground water transmission of typhoid fever have been documented (12). Additionally, improper construction or inadequate sealing of a well may allow surface water or shallow ground water to flow downward and contaminate deeper aquifers. Nevertheless, ground waters derived from deep aquifers are generally of good microbiological quality because the percolation of water through soil results in the removal of many organisms. Possible mechanisms for their elimination include adsorption, filtration, changes in the chemical environment, and extended times required for ground water to move great distances. All waters (surface and ground water) intended to be used for domestic supply should be tested to ensure that disease-causing organisms are not present.

At the present time (1989), primary drinking water standards exist for only one biological parameter: coliform bacteria. The coliform group of bacteria comprises all aerobic and facultative anaerobic, gram-negative, nonspore-forming, rod-shaped bacteria that ferment lactose with gas formation within 48 hours at 35°C (9). Coliforms are found in the intestines of humans and other warm-blooded animals. These bacteria, though not known to cause disease themselves, are often associated with disease-causing organisms (pathogens). Thus the presence of coliforms in a water is an indication that pathogens may be present and a good index of the sanitary quality of the water.

For public water systems, the National Primary Drinking Water Standards establish a minimum number of samples that must be tested each week and a maximum number of coliform bacteria that may be present in 100 mL of drinking water. Compliance with the regulation is based on the average number of coliforms detected on a monthly basis. However, the presence of any coliform bacteria is an indication that some type of treatment is necessary. The treatment method most commonly used for disinfection of ground water is chlorination. However, more extensive treatment, such as filtration, may be appropriate if coliforms are present in combination with high turbidity.

Nuisance Organisms. The presence of biological organisms in water does not always represent a health-related problem. For example, algae growing in public water supply reservoirs frequently produce substances that result in nuisance tastes and odors but present no health risk to consumers. In ground water there are a number of nuisance organisms, usually bacteria, that result in similar aesthetic or operational problems. Three such groups of bacteria, the iron bacteria, the sulfate-reducing bacteria, and the sulfur bacteria, are described in Chapter 10 in regard to their impacts on corrosion of metal surfaces.

Iron bacteria, such as *Crenothrix*, *Leptothrix*, and *Gallionella*, occasionally become established in wells. These bacteria obtain energy for growth from the oxidation of ferrous iron and may have some impact on the concentrations of dissolved iron in ground water. If growth is prolific, they can impart taste and discoloration to water.

The iron bacteria require dissolved oxygen and a source of reduced (soluble) iron (Equation 10.22). They grow well in waters containing as little as 0.02 mg/L soluble iron but are most prolific in waters that provide a continuous supply at levels greater than 1 mg/L. Since reduced iron is not stable in the presence of oxygen, iron bacteria are expected to flourish where oxygenated waters mix with waters containing reduced iron. A well that is completed or screened in two different aquifers will provide such an environment, provided that one aquifer is oxygenated and the other exhibits reducing characteristics. The growth of iron bacteria may occur on the well casing or other surface exposed to the mixed water.

Many iron bacteria utilize manganese ion more readily than they do iron.

$$2Mn^{2+} + O_2 + 2H_2O \rightarrow 2MnO_2(s)$$
$$+ 4H^+ + energy . \quad (18.3)$$

When only iron is present, the bacteria are the yellow-brown color of iron rust. If traces of manganese are present, the growth is darkened by the black color of manganese dioxide (MnO_2) (13).

Excessive growth of filamentous iron bacteria, such as *Crenothrix*, results in gelatinous slimes that may seriously reduce water yield from wells. This problem is more likely to occur in a well that is inactive or intermittently operated than in one that operates continuously. When a well is inactive,

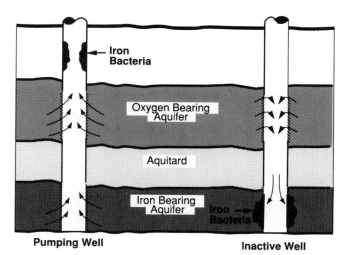

Figure 18.6. Distribution of iron bacteria as a result of pumping and static conditions in a well.

there is an opportunity for water from different aquifers to mix slowly. The movement of water from one aquifer to another is driven by pressure differences. Under these conditions growth may occur in the well screen slots or adjacent aquifer interstices (see Figure 18.6).

The growth of iron bacteria in wells can be controlled by disinfection with chlorine. Chlorine may affect the bacteria in two ways: by inactivation of the organism or by oxidation of ferrous iron. Experience suggests that the control of iron bacteria by chlorine is accomplished by the oxidation of ferrous iron which deprives the bacteria of their energy source.

The sulfate-reducing bacteria are ubiquitous in many ground waters and are capable of prolific growth under favorable environmental conditions. This growth can impact the chemical quality of the water and may result in physical interference with ground water extraction. In the absence of dissolved oxygen, the growth of sulfate-reducing bacteria results in the reduction of sulfate (SO_4^{2-}) to hydrogen sulfide gas (H_2S) as evidenced by a "rotten egg" odor.

Chemical Characteristics

Total Dissolved Solids. Total dissolved solids (TDS) is defined as the residue of a filtered water sample after evaporation. In ground water most of the residue consists of inorganic salts with lesser amounts of organic matter. The TDS is found by weighing the dried residue after a known volume of the filtered water has been evaporated. Alternatively, an approximate value for TDS can be calculated by summing the concentrations of the individual constituents in solution. TDS is expressed in milligrams per liter (mg/L) or other concentration units.

All natural waters contain some dissolved solids. Many rain and snow waters contain dissolved solids in concentrations above 5 mg/L. The major dissolved constituent in rainwater is usually carbonic acid (H_2CO_3) resulting from the dissolution of atmospheric carbon dioxide. Carbonic acid makes water more aggressive toward calcareous minerals. As might be expected, both surface and ground waters are quite variable in dissolved solids content.

Factors determining the TDS of a ground water include the solubility of the minerals it contacts, the length of time the water remains in contact with mineral formations, and the concentrating and diluting effects of other hydrologic cycle processes. The principle cations comprising TDS in ground water are usually sodium, calcium, and magnesium. The major anions are carbonates, bicarbonates, chlorides, sulfates, and sometimes nitrates.

A water high in dissolved solids can be unsuitable for many beneficial uses. Excessive solids are objectionable in drinking water because of unpalatable mineral tastes, adverse physiological effects, and the potential for increased corrosion of plumbing materials used in the water supply system. Physiological effects include laxative effects induced by magnesium salts and the adverse effects of sodium related to certain cardiac disorders. Because of these problems, a secondary drinking water standard for TDS has been set at 500 mg/L.

Many industrial uses require even lower TDS. For example, less than 300 mg/L is desirable for the dyeing of textiles and the manufacturing of plastics, paper, and synthetic fibers. Although the specific mineral composition is important, generally a water that is high in TDS is more corrosive to metals than a water with low TDS. Well screens and other metal components of well structures may be adversely affected by high TDS in ground water.

TDS is an important characteristic of irrigation water because of the potential osmotic effects of dissolved salts and the action of specific ions on plant physiology and soil structure. Table 18.11 provides a general indication of the limitations of using irrigation water with various levels of TDS in arid and semiarid regions. Detrimental effects of TDS on crops are normally not significant until the TDS exceeds 500 to 1000 mg/L. However, much higher levels of TDS may be successfully applied to tolerant crops in very permeable soils.

TABLE 18.11 Dissolved Solids Problems for Irrigation Water (mg/L)

No detrimental effects usually noticed	500
Detrimental effects on sensitive crops	500–1000
Adverse effects on many crops, requiring careful management practices	1000–2000
Only for tolerant plants on permeable soils with careful management practices	2000–5000

Source: USEPA, "Quality Criteria for Water," July, 1976, Table 16 on p. 208. U.S. Environmental Protection Agency.

Treatment to remove almost all dissolved solids is possible (distillation, deionization) but not practical or desirable for most uses. Removal of dissolved solids in municipal water treatment is usually limited to partial reduction of hardness or removal of iron and manganese.

pH. The pH of a solution is an expression of the hydrogen ion (H^+) activity. Activity is the effective concentration of ions in solution. In very dilute solutions the activity of an ion is approximately equivalent to the molar concentration. In more concentrated solutions ionic interactions may affect the availability of ions such that the activity is some fraction less than the molar concentration. As a rule, the approximate equivalence of hydrogen ion activity and molar concentration can only be presumed for solutions with ionic strength[1] less than 0.1 (9). When pH is measured potentiometrically, using a standard pH meter, the hydrogen ion activity is measured.

Even theoretically "pure" water contains some H^+ due to the dissociation, or ionization, of the water molecule itself. This is frequently written:

$$H_2O = H^+ + OH^- . \qquad (18.4)$$

In a "pure" water solution, the dissociation occurs only slightly so that the activity of H^+ is about 0.0000001 equivalents per liter, or 10^{-7} eq/L. For convenience, the term pH is defined to be the negative base-10 log of the hydrogen ion activity. Thus the pH of "pure" water is the negative log of 10^{-7}, or 7. It follows that the hydrogen ion activity in a solution of pH 6 is 10^{-6} eq/L, or 10 times as great as in a solution of pH 7.

Most ground waters in the United States have pH values ranging from 6.0 to 8.5 (14). Hydrogen ions are present in water as a result of the dissociation or hydrolysis of solutes, as well as from the dissociation of water itself. Some of the most important chemical equilibria affecting the pH of natural waters involve the reaction of dissolved carbon dioxide and water to form carbonic acid (H_2CO_3). When carbonic acid dissociates to bicarbonate ion (HCO_3^-), hydrogen ion is released. Further dissociation of bicarbonate to carbonate (CO_3^{2-}) releases additional hydrogen ion. Other chemical reactions in natural waters result in a decrease of hydrogen ion in solution. For example, calcium carbonate ($CaCO_3$) reacts with hydrogen ion (H^+) to form bicarbonate ion (HCO_3^-) and calcium ion (Ca^{2+}). These equilibria are described in more detail later. The pH of ground water is affected by

1. Ionic strength (I) is a measure of the strength of the electrostatic field caused by the ions in a solution. It is computed by the formula

$$I = \sum_{1-i} \frac{m_i z_i^2}{2} ,$$

where m is the molar concentration of a particular ion and z is its charge. To calculate I, the concentrations of the major ionic species and their charges must be known for the solution.

many such chemical reactions and is ultimately determined by a combination of many interrelated chemical equilibria.

pH is an important characteristic of water for all beneficial uses. The National Secondary Drinking Water Regulations recommend that pH be no less than 6.5 and no greater than 8.5 for domestic water supplies. The pH of a solution can affect the toxicity of other elements and has a pronounced effect on many chemical reactions important to industry, agriculture, and domestic water treatment. The pH of most waters can be adjusted by addition of acid or alkaline (basic) substances.

The terms "acidity" and "alkalinity" are frequently misused in relation to the term pH. Waters are incorrectly said to be "acid" only if they have a pH value below 7 and "alkaline" only if the pH is above 7. A more accurate definition of acidity is the capacity of a solution to react with hydroxyl ions (OH^-) or strong base. Since free hydrogen ion (H^+) is not the only constituent in ground water that will react with strong base, it is possible for the water to exhibit measurable acidity even at low H^+ concentrations (pH > 7). Acidity is therefore an aggregate property of water and cannot be interpreted in terms of a single solute such as H^+.

Alkalinity (basicity) is defined as the acid-neutralizing capacity or ability to react with hydrogen ions (H^+). Again, alkalinity cannot be attributed only to the presence of hydroxyl ion (OH^-), so there is no quantitative relationship between pH (or pOH) and alkalinity. In most waters alkalinity is primarily attributed to carbonate (CO_3^{2-}) and bicarbonate (HCO_3^-) ions. These ions are present over a wide range of pH, so it is possible for a water to have considerable alkalinity even at pH well below 7. The term "buffer capacity" is a characteristic of water related to its ability to neutralize additions of either acids or bases. Acidity, alkalinity, and buffer capacity are all dependent on the chemical equilibria of substances dissolved or suspended in water.

Redox Potential.

The oxidation-reduction potential, or redox potential (Eh), is a measure of the relative intensity of the oxidizing or reducing conditions in solutions such as natural waters. In Chapter 10 the significance of Eh was discussed with regard to corrosion. In addition the value of Eh is useful for predicting the relative quantities of oxidized and reduced elements in water.

In waters with a positive value for Eh, oxidizing conditions dominate and most multivalent elements are predicted to be in the oxidized state. A negative value of Eh indicates reducing conditions. In the case of iron and manganese this is significant because the oxidized forms are insoluble while the reduced forms are soluble. Many ground waters contain iron and manganese in the reduced state. When this water contacts air, it becomes oxygenated, and the Eh becomes more positive. The change in Eh correlates with a shift in the oxidation state of iron and manganese and the formation of chemical precipitates.

Plots of Eh versus pH, called "Pourbaix diagrams," are useful for predicting the stability of various ions and compounds in natural waters. They were developed by M. J. N. Pourbaix in the 1950s as a tool for understanding the chemistry of corrosion (15). Based on electrochemical equilibria, boundaries may be plotted that represent limiting Eh and pH conditions for the stability (and hence occurrence) of dissolved and solid chemical species.

Figure 18.7 is a Pourbaix diagram for a water that contains iron, dissolved carbon dioxide species, and dissolved sulfur species. The principal equilibria involved are the oxidation and reduction of dissolved iron species, and the solution and precipitation of iron hydroxides, carbonates, and sulfides. It can be seen that the range of Eh in water is about 1.25, but the absolute values of the upper and lower boundaries vary with pH. These boundaries represent the stability field of water itself. The area above the upper boundary represents

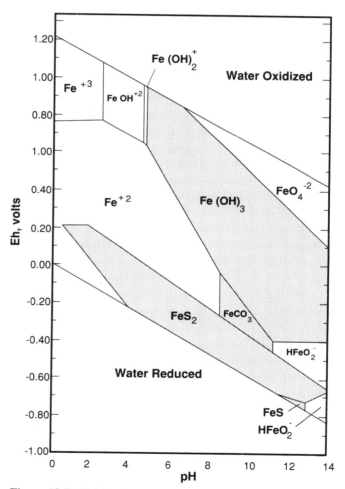

Figure 18.7. Fields of stability for solid and dissolved forms of iron as a function of Eh and pH at 25°C and 1 atmosphere pressure. Activity of sulfur species 96mg/L as sulfate, carbon dioxide species 1000 mg/L as bicarbonate, and dissolved iron 0.0056 mg/L (from Hem, J. D., "Study of Interpretation of the Chemical Characteristics of Natural Water," *Geological Survey Water Supply Paper*, 1970).

conditions in which water is oxidized to oxygen gas (O_2). Below the lower boundary, water is reduced to hydrogen gas (H_2). For any Eh and pH within the boundaries, the diagram indicates the predominant dissolved or solid species of iron.

It should be noted that Pourbaix diagrams have limitations when applied to natural water systems. Their construction depends on the availability of accurate thermodynamic data for the important chemical equilibria in the system. The data are frequently unavailable, especially for the various solids that might form in a solution containing many different inorganic ions. Hem (14) describes three conditions that must be met when the diagrams are applied to natural waters:

1. The system should be in equilibrium.
2. The solids in the system should be reasonably pure and the same as those specified in the diagram.
3. Complexes not allowed for in the computation should be absent or negligible.

Despite these limitations, Pourbaix diagrams are useful for understanding the general behavior of certain dissolved and solid species in water.

Hardness. The term "hardness" is used to describe the properties of water that result in precipitation of ordinary soap. The same properties result in the formation of scales or mineral deposits on plumbing fixtures. Although hardness cannot be attributed to any single constituent, calcium and magnesium are the major contributors in natural waters.

There are two types of hardness that can be described according to the anions found in solution with calcium and magnesium. Carbonate (temporary) hardness results from the presence of calcium and magnesium in combination with carbonate (CO_3^{2-}) or bicarbonate (HCO_3^-). The term "temporary" comes from the fact that this type of hardness can be greatly reduced by boiling the water. Boiling drives off carbon dioxide (CO_2) and causes calcium carbonate to precipitate. This in turn reduces the amount of CO_3^{2-} and HCO_3^- in solution. Noncarbonate (permanent) hardness results from the combination of calcium and magnesium with sulfate (SO_4^{2-}), chloride (Cl^-), nitrate (NO_3^-), or other minor anions. These anions cannot be removed by boiling. Total hardness is the sum of carbonate and noncarbonate hardness.

Hardness is usually expressed as mg/L $CaCO_3$ equivalent. If the concentrations of calcium and magnesium ions are known (in mg/L), total hardness can be determined using the expression:

$$\text{Total hardness as } CaCO_3 = 2.5\,(Ca^{2+}) + 4.1\,(Mg^{2+})\,,$$

where each concentration is multiplied by the ratio of the formula weight of $CaCO_3$ to the atomic weight of the ion (15).

The terms "hard water" and "soft water" are relative expressions of hardness. Because consumer acceptance of hardness varies widely, there are no formal drinking water standards for hardness. In general, water with less than 75 mg $CaCO_3$/L is considered "soft," and more than 75 mg $CaCO_3$/L is considered "hard." Ground waters from limestone or dolomite formations commonly have total hardness exceeding 200 to 300 mg/L. In these cases many households choose to install water softeners. Lime-softening and ion exchange processes may be used to reduce hardness in municipal supplies.

Conductivity. Conductivity is a measure of the ability of an aqueous solution to carry an electrical current. The conductivity of water is due to the migration of ions (electrolytes) in solution. This is vastly different from the electrical conductance of a solid metal, which is the result of electron flow from atom to atom.

The ability of a water to conduct electricity depends on several factors, including the concentrations and types of dissolved materials present. Waters containing dissolved inorganic salts and the resulting ions are relatively good conductors. Dissolved organic compounds that do not dissociate contribute very little to conductivity.

The conductivity of a solution is determined from a measurement of the specific resistance or resistivity. Resistivity is usually expressed as ohms-centimeter. Conductivity is the reciprocal of resistivity and is expressed as mhos per centimeter. For convenience, the conductivity of water is expressed in micromhos per centimeter (μmhos/cm) or, using the International System of Units, in millisiemens per meter (mS/m); 1 mS/m = 10 μmhos/cm. When reporting conductivity, the temperature must also be reported, since the conductivity of solutions increases with temperature at a rate of about 2% per degree Celsius. It is customary to report the conductivity of water measured at 25°C.

For most ground waters, conductivity is a good measure of the relative mineralization. It may also be used to estimate another important property of water related to mineral content. Total dissolved solids (TDS) in milligrams per liter may be estimated by multiplying conductivity (μmhos/cm at 25°C) by an empirical factor. This factor may range from 0.55 to 0.9, depending on the nature of the dissolved constituents. For a comparatively constant water source, the factor can be determined by dividing TDS by conductivity.

As with TDS, conductivity varies widely in natural waters. Values may range from 50 μmhos/cm in mountain streams to 50,000 μmhos/cm in seawater. Conductivity measurements are quick and reliable indicators of relative differences in water quality between different aquifers.

Major Cations

Calcium. Calcium is a major cation in most natural fresh waters. Together, calcium and magnesium are the primary elements contributing to water hardness. Calcium is abundant in many minerals and may be dissolved as water passes through or over them. The amount of calcium retained in solution is dependent on a variety of chemical equilibria. Equilibria involving carbonates, and to a lesser extent sulfates, are important for determining the calcium content in most natural waters.

In ground water, cation-exchange processes may affect the occurrence of calcium. The divalent calcium ions (Ca^{2+}) are held more strongly to the negatively charged exchange sites of soil minerals than are monovalent ions such as sodium (Na^+). Natural water softening may occur when calcium in ground water displaces sodium on the mineral-exchange sites. This would increase the ratio of sodium to calcium in the water. The reverse is also possible if a water high in sodium passes through a mineral formation with exchangeable calcium. However, the concentration of sodium in the water would have to be much higher than that of calcium to overcome the effects of calcium's greater affinity for the exchange sites.

Magnesium. Magnesium originates and behaves chemically in water much the same as calcium. It occurs in natural waters as a result of the weathering of minerals. Magnesium, however, is much less abundant in minerals than is calcium. In natural waters the magnesium concentration in milligrams per liter is usually lower than calcium. Due to the lower equivalent weight of magnesium, the concentrations of calcium and magnesium expressed in milliequivalents per liter are often similar.

Sodium. Sodium is the sixth most abundant element in the earth's crust. It is present in most natural waters in concentrations ranging from 1 to 500 mg/L. Once sodium is dissolved in water it tends to stay in solution. There are no major precipitation equilibria that control the amount of sodium dissolved in natural waters. For this reason levels can become very high in waters exposed to soluble sodium-bearing minerals, such as halite (sodium chloride). The sodium levels of sodium chloride brines, such as connate waters, may exceed 100,000 mg/L.

Connate waters, also called "oil-well brines," sometimes contaminate fresh ground water. These brines are ancient bodies of water that were formed at the same time as the sediments in which they reside. They are often saltier than present-day ocean water and usually contain relatively high concentrations of iodide and bromide, as well as sodium. Contamination of fresh ground water can occur via improperly sealed and abandoned oil wells. There are also instances where natural barriers between fresh ground water and connate water are breached.

Sodium can also be present as a result of natural ion-exchange phenomena (see the section on calcium), or as a result of pollution. Seawater intrusion is a major source of sodium in freshwater aquifers of coastal areas. The average concentration of sodium in seawater is 25,000 mg/L. Domestic wastewater can also contribute to the occurrence of sodium. Municipal use typically results in the addition of 20 to 50 mg/L of sodium ion (Na^+), primarily from urine and washing products (16). Certain industrial wastes and the use of sodium chloride as a deicing agent on roads are other sources of sodium related to human activities.

The ratio of sodium to total cations is important to agriculture. In irrigation water it is desirable to have a low ratio of sodium (Na^+) to calcium (Ca^{2+}) and magnesium (Mg^{2+}). When water containing relatively high amounts of sodium is applied to soil, sodium tends to displace exchangeable calcium and magnesium held on clay mineral surfaces. High levels of sodium cause certain soils and clay minerals to swell when irrigated. This results in soils with poor permeability.

Since the divalent calcium and magnesium ions are usually held more strongly on clay minerals than the monovalent sodium ions, sodium will displace calcium and magnesium only if the sodium is present in much greater amounts. In 1954 the U.S. Department of Agriculture Salinity Laboratory developed an empirical index to relate the chemical composition of irrigation water to the likelihood that the water will enter into cation-exchange reactions in the soil. The index, called the "sodium-adsorption ratio" (SAR), is defined as

$$SAR = \frac{(Na^+)}{\sqrt{\frac{(Ca^{2+}) + (Mg^{2+})}{2}}} \qquad (18.5)$$

where the sodium (Na^+), calcium (Ca^{2+}), and magnesium (Mg^{2+}) concentrations are expressed in milliequivalents[2] per liter (meq/L). In general, the use of waters with SAR values greater than 18 is expected to result in sodium problems. If the SAR is less than 10, it is unlikely that sodium problems will develop. The application of the SAR must be interpreted in relation to specific soil conditions and irrigation practices.

Sodium is an essential element for human health. However, the contribution of sodium from drinking water is small relative to normal dietary intake. Thus there is no health benefit to be derived from its presence in drinking water. Conversely, there is some evidence linking excessive intake of sodium with hypertension and associated cardiac disease. For this reason it is recommended that sodium be maintained at the "lowest practical levels" in drinking water (16).

Although there are no national drinking water standards for sodium, the National Research Council reports that sodium levels below 100 mg/L are desirable for general human consumption. Persons on severely restricted diets, who must limit sodium intake to less than 500 mg/day, may require water that contains less than 20 mg/L sodium. On a large

---

2. To convert milligrams per liter (mg/L) to milliequivalents per liter (meq/L), divide mg/L by the atomic weight and then multiply by the charge.

scale sodium is difficult to remove from water. It can be accomplished by demineralization techniques such as distillation, reverse osmosis, or hydrogen–ion exchange.

Potassium. Potassium is the seventh most abundant element in the earth's crust. Although it is only slightly less abundant than sodium, its levels in natural waters are typically much lower, rarely approaching 20 mg/L. This is due in part to the greater resistance to chemical weathering of the potassium minerals and the preferential incorporation of potassium ions into clay or mica mineral structures. The incorporation of potassium into mineral structures, termed "potassium fixation," effectively removes potassium from solution.

Major Anions

Bicarbonate/Carbonate. The primary ions that contribute to alkalinity in water are bicarbonate (HCO_3^-) and carbonate (CO_3^{2-}). There are two major sources. First, bicarbonate and carbonate can arise when carbon dioxide gas (CO_2) is dissolved in water to form carbonic acid (H_2CO_3). Carbon dioxide in ground water originates from the atmosphere and also from the oxidation of organic matter by bacteria. The chemical equilibria governing the proportions of bicarbonate, carbonate, and carbonic acid can be written

$$CO_2(g) + H_2O(l) = H_2CO_3(aq) , \quad (18.6)$$

$$H_2CO_3(aq) = H^+ + HCO_3^- , \quad (18.7)$$

$$HCO_3^- = H^+ + CO_3^{2-} , \quad (18.8)$$

where CO_2 is a gas (g), H_2O is a liquid (l), and H_2CO_3 is carbonic acid in aqueous solution (aq). Equation 18.6 describes the formation of carbonic acid (H_2CO_3), Equation 18.7 describes the dissociation of carbonic acid to bicarbonate (HCO_3^-), and Equation 18.8 describes the further dissociation of bicarbonate to carbonate (CO_3^{2-}). As can be seen, the equilibria in 18.7 and 18.8 are strongly pH dependent, with the addition of acid (H^+) forcing the reaction to the left in each case. The percentage of each of the constituents present in solution as a function of pH can be represented graphically (Figure 18.8). For example, at pH 6.4, about 50% of the dissolved carbon dioxide would exist as carbonic acid and 50% as bicarbonate. Very little carbonate will exist below pH 8.0.

The second major source of carbonate and bicarbonate in natural waters is dissolved limestone ($CaCO_3$) and other carbonate minerals such as dolomite ($CaCO_3 \cdot MgCO_3$). When limestone, a solid(s), is dissolved in water the net reaction may be written

$$CaCO_3(s) = Ca^{2+} + CO_3^{2-} . \quad (18.9)$$

The above equilibrium is linked to those defined in Equations 18.6, 18.7, and 18.8, in that the species of dissolved carbon dioxide is still very dependent on pH. Since the amount of hydrogen ion in water is largely controlled by the dissociation

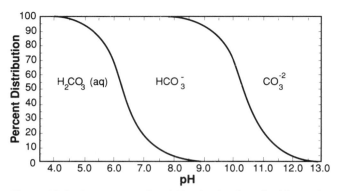

Figure 18.8. Percentages of total dissolved carbon dioxide species in solution as a function of pH at 25°C; pressure 1 atmosphere (from Hem, J. D., "Study of Interpretation of the Chemical Characteristics of Natural Water," *Geological Survey Water Supply Paper*, 1970).

of dissolved carbon dioxide, Equations 18.8 and 18.9 can be combined to illustrate this pH dependence. The combined equation is written

$$CaCO_3(s) + H^+ = Ca^{2+} + HCO_3^- . \quad (18.10)$$

As the hydrogen ion (H^+) increases, or as the pH decreases, the reaction is forced to the right, and more limestone is dissolved. Conversely, as H^+ decreases, or as the pH increases, limestone or calcium carbonate ($CaCO_3$) will tend to come out of solution.

The tendency of a particular water to either dissolve or precipitate $CaCO_3$ can be calculated. The calculation, resulting in the Langelier index (LI), is based on the measured pH of the water and the pH at which the water becomes saturated with respect to $CaCO_3$. Calculation of the LI and its relevance to corrosion are described in Chapter 10. A positive LI indicates that the water has scale-forming tendencies and precipitates $CaCO_3$. A negative LI indicates that the water will tend to dissolve $CaCO_3$. For most beneficial uses of water, it is desirable to avoid a substantial negative LI so that a protective mineral scale is preserved on pipe surfaces.

Alkalinity resulting from bicarbonate and carbonate is not considered a health hazard in drinking water supplies. However, alkalinity in excess of 400 mg/L (as $CaCO_3$) may result in corrosion of metal pipe. High alkalinity may also result in problems for certain industries. This is particularly true for food-processing operations that require acidity to flavor or stabilize products such as carbonated beverages. Table 18.12 lists maximum alkalinities desirable for some industrial uses.

In agriculture, high alkalinity may have adverse effects on crops by affecting the availability of iron. Alkalinity in excess of 600 mg/L (as $CaCO_3$) may cause iron to precipitate as a hydroxide, creating an iron deficiency in plants. High alkalinity in irrigation water may also lead to the precipitation of calcium and magnesium carbonates in soil as the soil dries out. This may increase the percentage of sodium in

TABLE 18.12 Maximum Alkalinity in Waters Used as a Source of Supply for Industry

Industry	Alkalinity mg/L as $CaCO_3$
Steam generation boiler makeup	350
Steam generation cooling	500
Textile mill products	50–200
Paper and allied products	75–150
Chemical and allied products	500
Petroleum refining	500
Primary metals industries	200
Food-canning industries	300
Bottled and canned soft drinks	85

Source: USEPA, "Quality Criteria for Water," July, 1976, Table 1 on p. 8. U.S. Environmental Protection Agency.

the soil, resulting in associated soil cultivation problems and plant damage (see the section on sodium).

Sulfate. Sulfate ion (SO_4^{2-}) is the most common form of sulfur found in water. Natural waters may contain sulfate in concentrations ranging from a few tenths of a milligram per liter up to several grams per liter. Rainwater may contain sulfate in concentrations greater than 10 mg/L.

The primary mineral sources of sulfate include evaporite sediments that contain calcium sulfate, such as gypsum ($CaSO_4 \cdot 2H_2O$), anhydrite ($CaSO_4$), or the sulfates of magnesium and sodium. The weathering of metallic sulfides, such as iron pyrites, also produce sulfate in oxygenated waters. Therefore mine drainage wastes are a potential localized source of high sulfate in surface and ground waters. Iron and hydrogen ions are produced along with sulfate when pyrite is oxidized.

There are few elements that will form insoluble salts with sulfate. Thus, once dissolved, sulfate largely remains in solution unless it is anaerobically reduced to sulfide. Sulfate is not easily removed from water by conventional treatment methods. Demineralization methods are effective but costly.

The major health effect of sulfate is laxative action. No adverse health effects are noted for concentrations less than 500 mg/L. The taste threshold for sulfate in drinking water is between 300 and 400 mg/L for most persons. A National Secondary Drinking Water Standard for sulfate has been set at 250 mg/L.

Chloride. Chloride ion (Cl^-) is the predominant natural form of the element chlorine. The major mineral source of chloride in natural water is sedimentary rock, particularly the evaporites. Seawater contains chloride at levels near 19,000 mg/L and is a significant source of both chloride and sodium in ground waters subject to seawater intrusion. Industrial and domestic wastewater also contribute chloride to water. High chloride levels in ground water may indicate infiltration by sewage, industrial waste, or brackish water.

Chloride ions are extremely stable in water. They do not form insoluble salts or stable solute complexes with other ions. In addition they are not significantly adsorbed on mineral surfaces and are not involved in major biogeochemical cycles.

Chloride may produce salty tastes in water, but the taste threshold is dependent on the types and quantities of other ions present. For example, waters containing 250 mg/L chloride may have a detectable salty taste when the major cation is sodium. However, if the major cations are calcium and magnesium, a water may have a chloride content approaching 1000 mg/L and still not have a salty taste. The National Secondary Drinking Water Standard of 250 mg/L chloride is based on the potential for producing taste.

Most industrial uses of water are adversely affected by high levels of chloride. Food-processing industries usually require less than 250 mg/L. Industries that manufacture paper, textiles, or rubber may require levels less than 100 mg/L. Chloride is removed by demineralization techniques.

Nitrate. Nitrate ion (NO_3^-) is the predominant form of inorganic nitrogen found in oxygenated natural waters. Smaller quantities of other inorganic nitrogen compounds may also exist in solution. These include nitrite (NO_2^-) and ammonium ion (NH_4^+). Nitrogen is also present in a wide variety of organic compounds such as protein and the biological waste product, urea (NH_2CONH_2).

Nitrogen is rapidly cycled through the biosphere and only a small amount is contained in minerals. Most of the earth's nitrogen is found in the atmosphere and in soils or biological materials. Certain microorganisms and leguminous plants convert atmospheric nitrogen gas (N_2) to other nitrogenous compounds. In the presence of oxygen, there is a tendency for nitrogenous materials to be converted to nitrate.

Nitrate enters ground water primarily by leaching from soils. It may originate in soil from natural processes (nitrogen fixation, decomposition of biological material) or be deposited as a result of human activities. In agricultural areas nitrates in ground water have resulted from the application and leaching of nitrogen-rich fertilizers. Similarly, nitrate can be released from urea and other organic nitrogen compounds found in domestic and industrial wastewaters. Other sources include animal wastes, such as feedlot discharges, and leachate from waste disposal dumps, sanitary landfills, sewage lagoons, and septic tanks. Consequently the occurrence of high levels of nitrate in ground water may signal the presence of other contaminants, such as pathogenic organisms.

The National Primary Drinking Water Regulations establish a maximum contaminant level (MCL) for nitrate (NO_3^-) at 45 mg/L. This is equivalent to 10 mg/L of elemental nitrogen (N). Healthy adults are unaffected by high nitrate levels. The standard is based on the susceptibility of human infants to a condition known as methemoglobinemia (blue baby syndrome). It is caused by the reduction of nitrate (NO_3^-) to nitrite (NO_2^-) in the gastrointestinal tract. Nitrite binds to hemoglobin in the bloodstream and reduces the oxygen transport capacity of the blood. Methemoglobinemia is especially hazardous for infants under six months of age.

In oxygenated waters, nitrate is very stable and difficult to remove. Conventional water treatment has no significant impact on nitrate levels. Biological denitrification to nitrogen gas or removal of nitrate by ion exchange or reverse osmosis are technically feasible.

Minor Constituents

Because the chemical composition of natural waters varies greatly, it is difficult to classify individual drinking water constituents as either "minor" or "major." For the purpose of this chapter, the constituents described in this section are considered to be "minor" because they typically occur in water at concentrations less than 1.0 mg/L or are considered of minor importance in most applications.

Arsenic. Minerals containing arsenic are ubiquitous in nature but are relatively insoluble. In natural waters arsenic exists as arsenate (AsO_4^{3-} or As V) or arsenite (AsO_3^{3-} or As III). In oxygenated waters arsenate is predominant. The forms of arsenate that exist are Eh and pH dependent, with $H_2AsO_4^-$ predominant below pH 7 and $HAsO_4^{2-}$ predominant above pH 7. Under mildly reducing conditions, ground water may also contain arsenite.

The concentration of soluble arsenic in water depends on the types and quantities of other constituents present because arsenate readily forms precipitates with a variety of cations. In ground water the predominant cations are calcium, magnesium, and sodium. Calcium, magnesium, and sodium arsenates are sufficiently soluble for some arsenate to remain in solution (14). Although the concentration of arsenic found in most natural waters is less than 0.01 mg/L, it has been reported that some ground waters contain greater than 1.0 mg/L (16).

Arsenic is toxic to humans, and there is some evidence that certain forms may be carcinogenic. The maximum allowable concentration in domestic water supplies is 0.05 mg/L as required by the National Primary Drinking Water Standards and the WHO International and European Drinking Water Standards. Arsenic is also toxic to sensitive food crops. The tolerance of crops to arsenic is dependent on soil type and irrigation practices. The EPA has established a criterion of 0.1 mg/L arsenic for irrigation water. Arsenic may be removed by adsorption on activated alumina or by standard demineralization techniques such as ion exchange or reverse osmosis.

Boron. High levels of boron are usually found in water associated with evaporite minerals containing sodium or calcium borate salts. When these salts are dissolved in water, orthoboric acid (H_3BO_3) is the primary dissolved species of boron. Orthoboric acid will dissociate

$$H_3BO_3(aq) = H^+ + H_2BO_3^-, \qquad (18.11)$$

but only at high pH. At pH less that 8.2, the ratio of H_3BO_3 to $H_2BO_3^-$ is greater than 10 to 1.

Boron usually occurs naturally in water but may result from pollution by industrial waste or sewage. Boron salts are common ingredients of laundry detergents. Natural waters rarely contain more than 1 mg/L boron and typically contain less than 0.1 mg/L. These concentrations are innocuous for human consumption.

Boron is an essential element for plant growth, but there is no evidence that it is required by animals. In excess of 2.0 mg/L, boron is toxic to many species of plants. The EPA criteria for protection of crops during long-term irrigation is 0.75 mg/L.

Fluoride. In natural waters the element fluorine exists as fluoride ion (F^-) or silicofluoride ion (SiF_6^{2-}). This depends on the solubility of the fluorine minerals with which the water has been in contact. Fluoride may also be present in water as a result of volcanic or fumarolic gases.

Fluoride concentrations in natural waters vary widely. It is not uncommon to find ground water with levels greater than 1.0 mg/L. Its beneficial effect in drinking water, mitigating tooth decay, is well documented. However, excessive fluoride can result in cosmetic damage to tooth enamel (dental fluorosis) and a crippling bone disease (skeletal fluorosis). The current EPA primary drinking water standard of 4 mg/L is designed to protect against skeletal fluorosis. The secondary drinking water standard of 2 mg/L is intended to protect against dental fluorosis.

Fluoride is removed from water by standard demineralization techniques, usually reverse osmosis and deionization. It can also be removed by adsorption on activated alumina or bone char. Lime softening incidentally removes some fluoride along with calcium and magnesium.

Iron. Iron is the fourth most abundant element in the earth's crust and is a common component of rocks and soils. Iron-bearing minerals include sulfides such as pyrite or marcasite (FeS_2), carbonates such as siderite ($FeCO_3$), hydroxides such as limonite ($Fe(OH)_3$), and oxides such as magnetite (Fe_3O_4). Although some minerals contain iron in the reduced ferrous (Fe^{2+}) state, weathering of these minerals results in its oxidation to the ferric (Fe^{3+}) state. Nearly all of the iron found in sedimentary or alluvial materials is in the ferric state. However, sediments deposited in lakes or stream beds may become a source of ferrous iron in local ground water under reducing conditions.

Iron can exist in a variety of forms in water. The ferrous (reduced) form of iron occurs only in waters devoid of oxygen. The ferric (oxidized) form is found in oxygenated waters and usually exists as a colloidal precipitate. Iron may also combine with natural organic substances or human-made wastes to form soluble organic complexes. Infiltration of organic waste into an aquifer can indirectly affect the forms and concentrations of iron species by contributing to the creation of reducing conditions. Iron bacteria may accelerate the oxidation of ferrous to ferric iron.

When excessive concentrations of dissolved iron are found in ground water it is always in the reduced ferrous (Fe^{2+}) form. Ferrous salts are generally quite soluble, whereas the oxidized ferric (Fe^{3+}) salts are quite insoluble. Thus, in the absence of oxygen, dissolved ferrous iron may occur in relatively high concentrations. In the presence of oxygen, it will be oxidized to the ferric form, and very little iron will remain in solution. The solubility of ferric salts is significant only when the pH is abnormally low (<3.5).

The presence of carbonate will also limit the concentration of dissolved iron in ground water. One iron salt that is quite insoluble is ferrous carbonate, $FeCO_3$ ($K_{sp} = 3.13 \times 10^{-11}$).[3] The solubility of ferrous carbonate is less than the solubility of calcium carbonate, $CaCO_3$ ($K_{sp} = 4.95 \times 10^{-9}$). In the presence of carbonate, ferrous iron may precipitate as ferrous carbonate (Equation 18.13). Thus, even in the absence of dissolved oxygen, the amount of iron in ground water may be limited by carbonate alkalinity. It should be noted, however, that most of the alkalinity found in ground water is due to bicarbonate rather than carbonate.

The principal equilibria that control the amounts of iron in solution involve the oxidation and reduction of iron and the solution and precipitation of hydroxides, carbonates, and sulfides. Thus oxidation-reduction potential (Eh) and pH are important variables in determining the species and quantities of iron in water. The importance of the various iron species in corrosion reactions is discussed in Chapter 10. The primary equilibria governing the solubility of iron in natural waters are

$$Fe(OH)_2(s) = Fe^{2+} + 2OH^- , \qquad (18.12)$$

$$FeCO_3(s) = Fe^{2+} + CO_3^{2-} , \qquad (18.13)$$

$$Fe^{2+} + H_2O = FeOH^+ + H^+ . \qquad (18.14)$$

It is not uncommon to find ground waters with ferrous iron concentrations between 1 and 10 mg/L. When such a water is exposed to air, however, the ferrous iron is oxidized to the ferric form. The resulting ferric iron precipitates (iron oxides or hydroxides) will cause the water to appear cloudy and colored. The oxidation of ferrous iron by dissolved oxygen is not instantaneous. The rate increases rapidly with pH, but in most wells (pH 6.5 to 8.5) the time required for 90% of the iron to be oxidized is many seconds (17). Thus ground water containing reduced iron may appear clear when first pumped from a well but may become cloudy or colored after standing in contact with air.

3. K_{sp} denotes the solubility product constant at 25°C.

The EPA's secondary drinking water standard for iron is 0.3 mg/L. This level corresponds to concentrations at which iron will cause aesthetic problems such as colored water, turbidity, staining, or taste. Its presence in water distribution systems may accentuate biological growths, resulting in additional aesthetic problems. Ferrous iron in excess of the standard may impart an astringent taste. Ferric iron in excess may cause a perceptible color or cloudiness. Either form in excess of the standard will cause a brownish stain on white fabric when laundered with hypochlorite bleach. Some industrial processes, such as refining sugar, require iron concentrations less than 0.1 mg/L.

Iron can be removed from water by softening (lime-soda process or ion exchange). When iron is removed by ion exchange, it becomes so tightly bound to the exchange resin that it cannot be removed by regeneration with brine. Ion-exchange plants designed to remove iron must occasionally regenerate the resin with strong acid. Iron is also removed by aeration or chemical oxidation followed by filtration to remove precipitated ferric iron (18). Alternatively, if the water contains relatively low concentrations of iron (less than 1.0 mg/L), it may be made acceptable to consumers by sequestering (chelating) the iron with polyphosphates. In this case the iron is not removed but is kept in solution to avoid the problems associated with iron deposition. For this reason polyphosphates should be added before final chlorination. Sequestration may also be accomplished by the nearly simultaneous addition of sodium silicate and chlorine (19).

Manganese. The chemistry of manganese is similar to that of iron, although there are some important differences. Manganic ion (Mn^{3+}) is unstable in water and decomposes:

$$2Mn^{3+} + 2H_2O = Mn^{2+} + MnO_2(s) + 4H^+ . \qquad (18.15)$$

Soluble manganese exists in ground water as the reduced manganous ion (Mn^{2+}). Oxidation of manganous ion results in the precipitation of manganese dioxide (MnO_2). Manganese dioxide has an intense black color in contrast to the reddish brown color of ferric oxide.

Although the reduced forms of both iron and manganese react with dissolved oxygen to form relatively insoluble precipitates, the oxidation reaction with manganese proceeds much more slowly than with iron. Thus manganese in the reduced form (Mn^{2+}) exists for longer periods of time in oxygenated waters than does ferrous iron. This has two important consequences. First, conditions are created that may be favorable for the growth of iron bacteria. Experience suggests that the reaction between manganous ion (Mn^{2+}) and dissolved oxygen is at least as good an energy source for iron bacteria as is the corresponding reaction with ferrous ion (Fe^{2+}). Second, the slower oxidation of manganous ion makes impractical the removal of manganese by aeration/filtration. In order to remove manganese by filtration, oxidation of the soluble reduced forms would have to occur more quickly than is practical by aeration alone. This could be accomplished by raising the pH of the water to a value approaching 10, or by using a strong chemical oxidant such as chlorine.

The nuisance effects of manganese in water are similar to those of iron, except that manganese is troublesome in much lower concentrations. The secondary drinking water standard for manganese is 0.05 mg/L. In acid soils (pH less than 6) manganese may be toxic to some crops. The maximum level of manganese recommended by EPA water quality criteria for protection of crops is 0.2 mg/L.

Like iron, manganese can be removed by lime-soda softening, cation-exchange softening, or oxidation followed by filtration. It may also be removed by a special type of greensand that is regenerated with potassium permanganate (18). The nuisance effects of manganese are sometimes controlled by sequestering the manganous ion with polyphosphates.

Silica. The element silicon ranks second to oxygen as the most abundant element in the earth's crust. It is the major component of most rocks and soils. The term silica (SiO_2) is commonly used to refer to silicon in water; however, the actual form is hydrated and should be written H_4SiO_4 (silicic acid). This is equivalent to SiO_2 bound to two water molecules.

The concentration of silica (SiO_2) in natural waters is usually in the range of 1 to 30 mg/L, although it is not uncommon to find some ground waters with concentrations above 100 mg/L. The hydrated form of silica (silicic acid) may dissociate to silicate ion ($H_3SiO_4^-$):

$$H_4SiO_4(aq) = H_3SiO_4^- + H^+ , \qquad (18.16)$$

but below pH 8.0 the dissociation is insignificant.

Silica is undesirable in many industrial process waters because it forms scale in boilers and on steam turbines. It may volatilize from boiler water in very high-pressure steam and then precipitate on turbine blades. These scales are extremely difficult to remove. Dissolved silica may be removed from boiler feedwater by deionization using strongly basic anion-exchange resins. The addition of magnesium salts during lime-soda softening also reduces the silica content of a water.

Other Inorganic Constituents. There are many other inorganic constituents that occur in natural waters at concentrations less than 1 mg/L. Many of these are either rare in nature or do not exist in soluble forms. They are not discussed in detail here because they normally do not occur in ground water in high enough concentrations to affect use. However, local geochemical phenomena as well as pollution may result in unusual and significant levels of a variety of

inorganic constituents in selected ground water basins. For example, pits containing tailings from mining operations or abandoned mines that have been allowed to flood may become highly concentrated sources of metal ions. Local sources of contamination should always be considered when evaluating ground water resources.

Radioactive Substances

Radioactivity is the release of energy by spontaneous disintegration of atomic or nuclear structures. Energy is released from radioactive substances (radionuclides) in the form of alpha particles, beta particles, or gamma rays. Minute quantities of radionuclides are generally found in all natural waters. The concentrations and types depend on the radiochemistry of the soil and mineral formations through which the water passes.

Radioactivity in water originates from both natural and human-made sources. Natural radioactive substances have been present in the earth's crust since its formation and are being produced continuously by cosmic ray bombardment. These include potassium-40, rubidium-87, and carbon-14. Most of the naturally occurring radionuclides in ground water are decay products of uranium-238 and thorium-232. These are primarily short-lived alpha emitters and include isotopes of radon, radium, and polonium.

All radium isotopes are strongly radioactive and can be detected in small amounts. Radium-226 and -228 have the greatest potential to impact the beneficial uses of ground water. The decay series of these compounds may result in alpha doses that increase the risk of bone cancer in humans. Usually, drinking water contributes only a small percentage of the total human dietary intake of radium-226 and -228. However, in areas where the concentration is high, radium in ground water supplies can add significantly to the alpha dose. Radium exists in water as the divalent cation Ra^{2+}, and its chemical behavior is similar to that of calcium and magnesium, the principal components of water hardness. Processes used to remove hardness will also remove radium. These include lime-soda softening, ion exchange, and reverse osmosis (20). Methods for removing other radionuclides are contaminant specific and depend on chemical or physical separation processes. An overview of treatment and analytical methods is provided by Aieta (21).

The National Primary Drinking Water Regulations specify maximum contaminant levels for gross alpha particle activity as well as for combined radium-226 and -228. Standards are also established for gross beta particle activity, as well as for the individual radionuclides tritium and strontium-90 (see Table 18.7). The concentration of radionuclides in water is usually expressed in picocuries per liter (pCi/L). This is an expression of activity rather than mass. One curie equals 3.7×10^{10} disintegrations per second. A picocurie is 10^{-12} curies or 2.22 disintegrations per minute.

Radon-222 is a decay product of radium-226. Radon is a radioactive gas and is found in varying degrees in almost every part of the United States. It exists in ground water as a dissolved gas. The highest concentrations generally occur in areas with granitic rock, uranium deposits, or phosphate deposits. The major health concern of radon is inhaling the gas rather than consuming drinking water. There is some evidence that inhalation may lead to the development of lung cancer. Radon in drinking water accounts for only a small percentage of human exposures. There is currently no drinking water standard for radon. However, radon is on the list of 83 substances to be regulated by the EPA as required by the 1986 amendments to the Safe Drinking Water Act. Radon concentrations can be reduced in water by aeration or carbon adsorption.

Human-made sources of radioactivity include nuclear waste, radiopharmaceuticals, and nuclear fission products released to the environment as a result of nuclear weapons testing. Many of the fission products are strong beta emitters, such as strontium-90 and cesium-137. In the 1950s the amount of tritium in the atmosphere was greatly increased as a result of nuclear weapons testing. Some of this tritium became incorporated into precipitation. Since the half-life of tritium is 12.4 years and the approximate date of the tritium rainfall is known, measurements of tritium remaining in ground water may indicate the age of that water. In general, surface water sources are more likely to contain fission products and other radionuclides produced by humans than are ground water sources.

Dissolved Gases

Dissolved gases are important to a variety of processes. Nitrogen fixation requires dissolved nitrogen gas (N_2), photosynthesis requires a source of dissolved carbon dioxide gas (CO_2), and respiration requires dissolved oxygen (O_2). Dissolved gases also play an important role in certain chemical equilibria, a fundamental example being the solution of carbon dioxide and subsequent formation of carbonic acid (H_2CO_3). The relationship of this reaction to the solubility of carbonate and silicate minerals has been described previously. Similarly, as described in Chapter 10, dissolved gases such as oxygen, carbon dioxide, and hydrogen sulfide play a role in metallic corrosion. The gases most likely to impact ground water quality are discussed here.

Oxygen. The solubility of oxygen in water is a function of temperature and pressure. In ground water dissolved oxygen concentrations are usually less than 10 mg/L. These levels frequently decrease with depth due to the consumption of oxygen by microorganisms.

Dissolved oxygen is considered an important constituent of natural waters because of its influence on the aesthetic quality of water and its requirement by fish and other aquatic life. Its absence may lead to anaerobic decomposition of organic matter and formation of noxious or malodorous gases such as methane (CH_4) and hydrogen sulfide (H_2S). The presence of dissolved oxygen may increase the rate of corrosion of metal surfaces.

The magnitude of the problems caused by too much or too little oxygen are entirely dependent on the other constituents and conditions found in the particular environment. Thus there are no national standards for dissolved oxygen in drinking water. EPA water quality criteria for surface waters recommend that "water should contain sufficient dissolved oxygen to maintain aerobic conditions" and that the minimum dissolved oxygen content required for good fish populations is 5.0 mg/L (5).

Oxygen can be added to water by simple aeration. However, some industrial uses may require reduction of dissolved oxygen levels. Both mechanical and chemical methods are available to remove dissolved oxygen from boiler feed water. Mechanical methods employ either vacuum deaerators or deaerating heaters. Chemical methods include reactions with oxygen-scavenging chemicals such as sodium sulfite or hydrazine.

Carbon Dioxide. Carbon dioxide (CO_2) in ground water originates from the solution of atmospheric carbon dioxide and from microbial respiration in the soil. Most of the carbon dioxide dissolved in natural waters is converted to bicarbonate (HCO_3^-) or carbonate (CO_3^{2-}), depending on the solution pH and availability of other reactive ions. Ground waters that effervesce under atmospheric pressure and contain significant amounts of dissolved carbon dioxide gas along with calcium and bicarbonate ions sometimes become supersaturated with respect to calcium carbonate.

Carbon dioxide may be removed from water by several methods. Excessively high concentrations may be reduced by simple aeration. Further reductions, such as may be necessary for industrial heating or cooling waters, can be accomplished by deionization processes or by addition of alkali.

Hydrogen Sulfide. Hydrogen sulfide gas (H_2S) is generated by the anaerobic decomposition of organic matter or by the action of sulfate-reducing bacteria. It may exist naturally or as a component of pollution from industrial wastes or sewage sludge. It is a common by-product of petroleum refining, tanning, coking, natural gas purification, and food processing. The sulfate-reducing process is a source of energy for certain anaerobic bacteria. The reaction may be written

$$SO_4^{2-} + CH_4 \rightarrow HS^- + HCO_3^- + H_2O + \text{energy} .$$

$$(18.17)$$

Within the pH range of most natural waters, the reduced sulphur species exists both as the undissociated hydrogen sulfide gas (H_2S) and as the hydrosulfide ion (HS^-). Hydrogen sulfide in solution is a weak acid and dissociates

$$H_2S \text{ (aq)} = HS^- + H^+ . \qquad (18.18)$$

At pH 7, equal amounts of H_2S and HS^- are expected. At pH 9, about 99% is in the form HS^-. At pH 5, 99% is in the form H_2S.

The presence of hydrogen sulfide in water can easily be detected by its characteristic "rotten egg" odor. This odor is detectable by humans at levels below 0.5 micrograms per liter ($\mu g/L$). The gas is soluble in water in 4000 mg/L at 20°C and 1 atm of pressure. Although hydrogen sulfide gas is toxic to humans and animals in high concentrations, sulfide standards (or criteria) have not been established for drinking water because the unpleasant taste and odor occur at much lower concentrations than are considered toxic.

Hydrogen sulfide can be partially removed by aeration. The primary mechanism is volatilization caused by the scrubbing action of air moving through water. In the oxygenated water additional sulfide is removed by oxidation to sulfur or sulfate. Chemical oxidation methods, such as chlorination, are also effective. An overview of the chemistry of reduced sulphur species and their removal from ground water is provided by Dohnalek (22).

Organic Constituents

Organic compounds are substances that contain the element carbon. Natural organic compounds are produced by biological activity. Synthetic organic compounds (SOCs) are manufactured chemicals. Both natural and synthetic organic compounds encompass a wide range of substances. Unlike the inorganic constituents, only a small percentage of organic compounds found in water have been chemically characterized. This is partly due to the size and complex nature of many naturally occurring organic molecules and the lack of analytical instrumentation with the capability to separate and identify individual compounds from complex mixtures.

Prior to 1975 most ground waters were believed to be relatively free from contamination by synthetic organic compounds. Since that time, however, increased sensitivity of analytical instrumentation and improved monitoring programs have demonstrated that such contaminants have found their way into many ground water supplies. Several federal surveys, including the National Organics Reconnaissance Survey (23) and the National Organics Monitoring Survey (24), have identified organic contamination in many U.S. water systems supplied by ground water. On a national scale the magnitude of contamination is small. It is estimated that organic contaminants are present in less than 1% of the usable aquifers (25). However, removal of contaminants is difficult, and any degradation of ground water resources must be considered serious. The following section describes the general clas-

sifications of organic constituents and their importance to the development of ground water resources.

Total Organic Carbon. Total organic carbon (TOC) is an expression of the concentration of all the organic substances found in a water. The value does not provide any information on the types of organic substances that make up the total. As such, TOC measurements are useful primarily for monitoring gross changes in organic content, such as may be caused by pollution, or for providing a general description of the organic content of a water.

Natural Organic Constituents. It is estimated that more than 10^{10} different kinds of organic molecules exist in the biosphere (26). Naturally occurring organic compounds range in size from simple monomers, like formaldehyde (CH_2O), to high molecular weight polymers, like proteins. Although many of these individual compounds are essential to biological functions within living organisms, they are rarely present in natural waters at levels that would affect its beneficial use. However, the accumulation and subsequent degradation of organic matter may affect the quality of water by contributing individual chemical contaminants or aggregate organic chemical materials, such as humic substances.

Humic substances are mixtures of high molecular weight, naturally occurring organic compounds. They are normally formed in soils as a result of microbial decomposition of plant or animal matter. The chemical structure of these substances cannot individually be identified. Rather, they are described as three groups of organics defined by their solubilities in acid and alkali. When extracted from soil, the acid-insoluble, alkali-soluble fraction is termed "humic acid." The acid- and alkali-soluble fraction is termed "fulvic acid." The third group, termed "humins," is insoluble in both acid and alkali. The composition and chemical nature of humic substances extracted from different soils may vary appreciably.

The color, yellowish to brownish, of some waters may be due to the presence of humic or fulvic acids. They have also been identified as trihalomethane precursors. Carbon adsorption is used to remove them. Fulvic acids may be removed incidentally during lime-soda softening (28).

Synthetic Organic Compounds. Synthetic organic compounds (SOCs) encompass a wide range of commercially produced chemicals, including pesticides, detergents, petroleum products, and industrial solvents. Their occasional presence in ground water is due to application, storage, or disposal on or beneath the land surface. Many of these substances, if present in high enough concentrations, can produce adverse effects on human health or affect the aesthetics of drinking water. The current National Primary Drinking Water Regulations set standards for fourteen SOCs. The 1986 amendments to the Safe Drinking Water Act mandate the EPA to establish standards for additional organic compounds.

Removal of SOCs from contaminated aquifers is difficult and time consuming. Many are degraded into different compounds or move at different rates with respect to ground water flow. An overview of the movement of SOCs in ground water is provided by Roberts (28). The methods most frequently used to remove SOCs from well water are aeration and adsorption. Reverse osmosis may also be effective in removing certain types of organic compounds.

Pesticides and Agricultural Chemicals. Pesticides and other agricultural chemicals, such as fertilizers, can enter the ground water via several routes. Spills and leaks of concentrated pesticide products are a source in agricultural areas. Other common routes are by infiltration of irrigation return flow or the leaching of chemicals at the point of application. In both cases the degree of contamination is dependent on irrigation practices and the physical characteristics of the soil as well as the nature of the chemical. Many pesticides, for example, are strongly adsorbed to soil particles and are relatively immobile from the point of application. Other pesticides are highly mobile and are readily leached if sufficient water is applied.

The stability of the chemical is also important in evaluating its potential for ground water contamination. Many pesticides and other agricultural chemicals are volatile or degrade by microbial metabolism, hydrolysis, or photooxidation. All of these processes reduce the amounts of chemicals available for leaching. Still other pesticides, like the chlorinated hydrocarbons, are quite stable and may exist in the environment for many years. If these chemicals are also relatively mobile in soil, they may be leached into ground water.

As with the other classes of organic compounds, agricultural chemicals are highly variable in their impacts on water quality. In sufficient quantities many of them are toxic to humans and animals or produce other adverse effects on the beneficial uses of water. Six pesticides are currently regulated by the EPA in the National Primary Drinking Water Regulations (see Table 18.7). These include four chlorinated hydrocarbon insecticides and two chlorophenoxy herbicides. Organic compounds to be regulated in the future are listed in Table 18.9.

Petroleum Products. Gasoline is but one of many petroleum products that could adversely affect ground water resources. Others include natural gas, kerosene, fuel oils, and lubricating oils. These deserve special attention because they are so widely used and are frequently stored underground.

The accidental release of these products may occur when storage tanks leak or distribution piping fails. The hazard posed depends on the nature of the compound as well as the amount released. Heavy oil spilled on dry soil frequently results in a hard immobile mass resembling asphaltic cement. Small quantities of biodegradable hydrocarbons may decompose so quickly that there is no threat to ground water.

A common source of contamination in wells is the oil used for lubricating enclosed line-shaft turbine pumps. The lubricating oil that leaks from these pumps frequently accumulates on the surface of the water. If too much oil accumulates, it may be pumped into the water system. The depth of the oil accumulated in the well casing can be accurately measured using a home-made instrument that senses changes in thermal conductivity among oil, water, and air (29).

Detergents. Household detergents include synthetic organic compounds that act as surface-active agents (surfactants). Surfactants are long-chain hydrocarbons with a polar group at one end and nonpolar group at the other end. Many of these compounds are not readily broken down by microorganisms or wastewater treatment practices and may find their way back to surface waters and some ground waters. The surfactants in detergents marketed in the United States after 1965 were formulated to be "biodegradable." However, even biodegradable detergents contain polyphosphates or other chemical builders that may contribute specific constituents to receiving waters. For example, polyphosphates are hydrolyzed to orthophosphate, a nutrient contributing to nuisance algal growths in surface waters. This problem has caused many states to ban or limit the amount of phosphates in detergents.

Surfactants in water can produce aesthetic problems. Concentrations less than 1 mg/L often result in noticeable froth or foam. Surfactants are defined and measured in water as methylene blue active substances (MBAS). The EPA has established a secondary drinking water standard of 0.5 mg/L MBAS.

Volatile Organic Chemicals. The term "volatile organic chemical" (VOC) designates a group of synthetic organic compounds that evaporate or volatilize easily. These chemicals do not occur naturally but are manufactured and widely used in industry as solvents or cleaning agents.

The VOCs most commonly encountered in ground waters are organohalide solvents, such as trichloroethylene (TCE), tetrachloroethylene or perchloroethylene (PCE), and trichloroethane (TCA). Some of these compounds are believed to be carcinogenic or produce other adverse effects on human health. In June 1987 the EPA established primary drinking water standards for the eight VOCs listed in Table 18.7. Other VOCs will be similarly regulated in the future. In the absence of formal federal standards, several states, including California, Arizona, and Florida, have established their own standards or health guidance levels for VOCs found in local ground water.

Due to their volatility VOCs are seldom found in surface waters at concentrations greater than a few micrograms per liter (μg/L). However, leaks from underground storage facilities or improper disposal on or below the ground surface may result in significant contamination of local ground water.

Several nationwide surveys have shown the occurrence of VOCs in aquifers (16).

The behavior of a particular VOC in soil and ground water depends primarily on the physical properties of the soil and density, solubility, and amount of chemical discharged. When released due to a spill or leak from an underground storage tank, some of the chemical will be adsorbed onto or absorbed into the soil. Other forces affecting its fate include degradation by microorganisms. However, if a sufficient quantity is released, it will tend to percolate toward the underlying ground water.

Chemicals percolating to ground water behave according to their density and solubility in water. Those that are fairly insoluble and heavier than water, such as 1,1,1,-trichloroethane (density = 1.3376), will continue moving downward through the water table. The ground water that it contacts may become contaminated with the chemical up to its solubility. If the chemical encounters an aquitard, it may collect at the bottom of the aquifer.

A chemical that is fairly insoluble and lighter than water, such as gasoline, will tend to float on top of the water table. Since flow in aquifers is laminar, there is little opportunity for the floating chemicals to mix with the ground water. This may account for the observation that relatively few wells are contaminated by gasoline even though gasoline is reported to be the chemical most frequently released from leaking underground storage tanks. Figure 18.9 illustrates the effects of density on the migration of contaminants.

A well may become grossly contaminated by gasoline, or other light and insoluble liquid, if the water table and overlying chemicals are drawn below the top of the screen or perforations. Figure 18.10 illustrates how such light liquids may cascade down the surface of the cone of depression and enter the well.

VOCs can be effectively removed from contaminated water supplies using either granular activated carbon (GAC) adsorption or aeration (air-stripping). In general, the more volatile the compound, the more effective the aeration. For compounds with lower volatility, the relative effectiveness of GAC treatment increases.

Trihalomethanes. Some organic contaminants occur in water as byproducts of disinfection rather than as a result of pollution. When a water is chlorinated, chlorine reacts with naturally occurring organic matter and certain halide salts to produce volatile, halogenated organic compounds known as trihalomethanes (THMs). THMs encompass the group of volatile organic chemicals consisting of halogen-substituted methanes (containing three halogens). In most natural waters, the most common THM formed is chloroform ($CHCl_3$). However, if bromide ion (Br^-) is present, the haloform reaction favors substitution with bromine to produce brominated compounds such as bromoform ($CHBr_3$), bromodichloromethane ($CHBrCl_2$), or chlorodibromomethane ($CHBr_2Cl$). Bromide ion is a constituent of seawater and is present in trace amounts in nearly all natural waters. In the

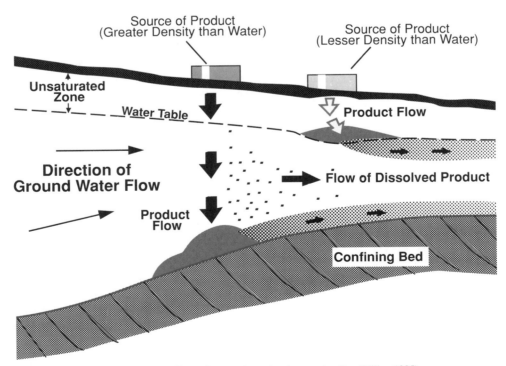

Figure 18.9. Effects of contaminant density on migration (Miller, 1985).

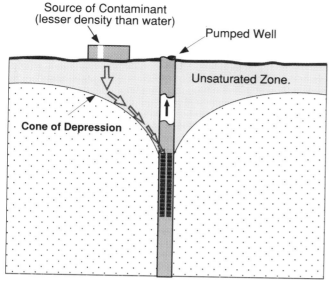

Figure 18.10. Contaminants with lower densities than water may cascade down the cone of depression during well pumping.

presence of chlorine, it is oxidized to bromine (Br_2). Thus seawater intrusion may influence the types of THMs formed upon chlorination.

Chloroform is a suspected human carcinogen and is regulated in domestic water supplies along with the other trihalomethanes. The National Primary Drinking Water Regulations limit the total concentration of trihalomethanes to 100 micrograms per liter ($\mu g/L$). This standard is currently limited to public water systems serving more than 10,000 people and using chlorine for disinfection.

There are two basic strategies that may be employed by water systems in controlling THMs. First, the formation of THMs can be avoided by the use of alternate disinfectants, such as ozone, chloramines, or chlorine dioxide. Second, concentrations of THMs that have already been formed can be reduced by treatment using aeration or GAC adsorption.

18.4 TREATMENT OF GROUND WATER

Alternatives to Treatment

A major portion of this chapter is devoted to identifying impurities in ground water. If a water is unsuitable for a particular use because of the types and amounts of impurities, there are two possible courses of action:

1. Develop an alternate source.
2. Modify the original source by blending with another source, or providing treatment.

The alternatives are evaluated based on a variety of issues, including costs and availability of treatment technologies or

alternate sources. Before abandoning any source, it is important that the alternate source be thoroughly evaluated. A possible problem is abandoning a well, only to drill a new one that has the same water quality.

Where a well is known to be contaminated and a better-quality source exists, it may be possible to produce acceptable water by blending. Blending is frequently practiced to dilute high nitrate levels in domestic water supply systems with multiple wells.

If the chosen alternative is to treat a contaminated ground water, there are a number of issues to consider. Very important is to evaluate the effectiveness and cost of point-of-use treatment devices versus a central treatment plant, and the need for trained operators and laboratory support. It may be necessary to plan for the disposal of waste products from the treatment process. For community water systems it may also be desirable to address public reaction to the planned treatment activity. This issue has received much attention in relation to air-stripping of volatile organic chemicals from ground water.

Treatment Methods

The treatment methods available for removing chemical contaminants from ground water are summarized in Table 18.13. In combination with conventional water treatment (filtration and disinfection) the technology currently exists to produce high-quality water from even the most contaminated source. The degree of treatment necessary for a particular water is determined by the types and amounts of contaminants and the desired quality of the finished water. Following are descriptions of some of the treatment methods that may be applied to ground water.

Filtration. Filtration is the process of removing suspended particles by the passage of water through a porous medium (filter). The particles to be removed may include sediments, precipitated metals (iron, manganese), or clay minerals that

TABLE 18.13 Treatment Technologies for Removal of Chemical Contaminants from Ground Water Supplies

Inorganics Removal Processes	Organics Removal Processes
Coagulation/filtration	Aeration
Lime softening	Packed column
Ion exchange	Diffused air
Reverse osmosis	Multiple tray
Activated alumina	Adsorption
Adsorption	GAC
Electrodialysis	PAC
Oxidation/filtration	Synthetic resins
Distillation	Biodegradation
	Oxidation
	Reverse osmosis
	Distillation

contribute to turbidity, or microorganisms such as bacteria and protozoans. Filters used for treatment of ground water generally employ sand media or fibrous cartridges, and are operated under pressure. Filtration is frequently preceded by coagulation, a process by which contaminants are incorporated into a precipitating floc.

Disinfection/Chlorination. Disinfection is the process of destroying microorganisms in water. Heat (boiling) and ultraviolet radiation are effective in killing microorganisms, but neither process is widely used in treating ground water supplies. Chemical disinfectants are much more common and economical. The most common chemical disinfectant used is chlorine. Alternate disinfectants include ozone, chlorine dioxide, and chloramines.

When chlorine gas is added to "pure" water, a mixture of hypochlorous acid (HOCl) and hydrochloric acid (HCl) is formed:

$$Cl_2(g) + H_2O = HOCl + HCl \ . \qquad (18.19)$$

Hypochlorous acid further dissociates to hypochlorite ion (OCl$^-$) and hydrogen ion (H$^+$):

$$HOCl = H^+ + OCL^- \ . \qquad (18.20)$$

The distribution of hypochlorous acid (HOCl) and hypochlorite ion (OCl$^-$) is dependent on pH. HOCl predominates below pH 7.5, and OCl$^-$ predominates above pH 7.5 (see Figure 18.11). Since hypochlorous acid is a more potent disinfectant

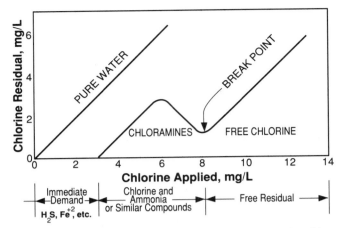

Figure 18.12. Graphical representation of the breakpoint chlorination reaction. The straight line at the left shows that chlorine residual is proportional to dosage in pure water. When impurities are present, they exert an initial chlorine demand (from *USEPA Report No. 570/9-83-012*, "Microorganism Removal for Small Water Systems," 1983).

than hypochlorite ion, chlorine is more effective at lower pH.

When chlorine is added to ground water, it oxidizes organic materials and any reduced minerals such as iron. It also combines with nitrogenous compounds, such as ammonia, to form chloramines. Chloramines (NH$_2$Cl, NHCl$_2$) are also oxidizing agents but are much less potent disinfectants than hypochlorous acid (HOCl). The chlorine in chloramines is termed "combined chlorine." The chlorine present as hypochlorous acid, hypochlorite ion (OCl$^-$) or molecular chlorine (Cl$_2$) is termed "free chlorine." Together, free and combined chlorine compose the "total available chlorine."

Figure 18.12 illustrates how the total chlorine residual in solution changes as chlorine is added to a ground water. If impurities such as ferrous iron or hydrogen sulfide are present, they exert an initial chlorine demand. If ammonia or organic nitrogenous compounds are present, further addition of chlorine results in the formation of chloramines. At this point the measured residual will be predominantly combined chlorine. A sufficient increase in dosage will result in the destruction of chloramines, and the total chlorine residual will drop. At the "breakpoint," all the combined chlorine species will be destroyed. The addition of more chlorine beyond the breakpoint results in a total chlorine residual composed of free chlorine. "Breakpoint chlorination" is the term applied to the dosing of a water with sufficient chlorine to produce a free chlorine residual.

In addition to its value as a disinfectant, chlorine is capable of oxidizing a variety of inorganic and organic constituents in water. It is effective in oxidizing the pungent gas hydrogen sulfide (H$_2$S) to less noxious compounds. Similarly, many naturally occurring organic compounds are destroyed by chlorine. Some of these may otherwise result in tastes or odors in water. It should be noted that chlorine by itself or

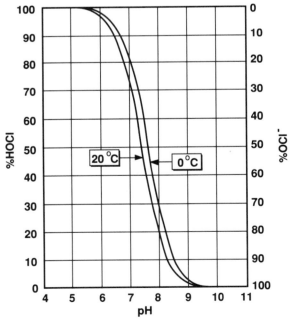

Figure 18.11. Distribution of HOCl and OCl$^-$ in water at indicated pH (from Laubush, E. J. and AWWA, *Water Quality and Treatment*, "Chlorination and Other Disinfection Processes"). © 1971. By permission of McGraw-Hill.

in combination with some organic substances, such as phenols, may also result in strong tastes or odors. Soluble ferrous iron (Fe^{2+}) is also oxidized by chlorine to the insoluble ferric (Fe^{3+}) form. Chlorine and other oxidants (potassium permanganate) are sometimes used in iron and manganese removal processes.

Softening. Softening refers to the removal or reduction of the dissolved constituents that contribute to water hardness (primarily calcium and magnesium ions). This may be accomplished by lowering the total mineral content of a water or by exchanging hardness ions for sodium.

The zeolite (ion-exchange process) is capable of complete removal of water hardness. Zeolites, including greensands, are naturally occurring aluminosilicate minerals that have cation-exchange capabilities. When hard water flows through a bed of zeolite, dissolved calcium and magnesium ions (Ca^{2+}, Mg^{2+}) displace sodium ions (Na^+) held by the ion-exchange sites. This is because the divalent calcium and magnesium cations have a greater attraction for the zeolite-exchange sites (anions) than do the monovalent sodium cations. The process may be defined by the reactions

$$Ca^{2+} + 2Na\text{—zeolite} = Ca\text{—zeolite} + 2Na^+ , \tag{18.21}$$

$$Mg^{2+} + 2Na\text{—zeolite} = Mg\text{—zeolite} + 2Na^+ . \tag{18.22}$$

When the cation-exchange sites become saturated with calcium or magnesium, the zeolite material may be recharged by a solution containing a relatively high concentration of sodium. A 10% solution of table salt (NaCl) is commonly used for recharge. Greensands and other natural zeolites have been largely replaced by synthetic organic resins that are more effective. Synthetic zeolites have rated capacities of more than 14,000 grains[4] per cubic foot compared to about 2800 grains per cubic foot for natural zeolites (30).

The lime-soda process softens water by lowering its calcium and magnesium content through a series of chemical precipitations. These are induced by the addition of lime ($Ca(OH)_2$) and either soda ash (Na_2CO_3) or caustic soda (NaOH). The lime-soda process requires relatively large treatment plants and skilled operators. The process is not capable of complete removal of hardness. However, most treatment plants using this method are able to reduce hardness to below 80 mg/L as $CaCO_3$. A secondary benefit of the lime-soda process is the precipitation (and removal) of iron and manganese along with the hardness minerals. Other undesirable substances, such as heavy metals and some dissolved organic compounds, may also be removed in the process.

Aeration. Aeration is the process of bringing water in contact with air to allow the transfer of volatile substances (gases).

4. The unit "grain" is traditionally used when discussing treatment to remove hardness. One grain = 64.8 mg; 1 grain per gallon = 17.12 mg/L.

The process has historically been used in water treatment to add oxygen to water or to remove hydrogen sulfide, carbon dioxide, or volatile taste and odor-causing organic substances.

There are two basic types of aeration equipment. Diffused aerators create bubbles of air that are passed up through a water basin. Waterfall aerators, such as cascades, multiple trays, spray nozzles, and packed columns, cause water to fall through air and form small drops or thin films that facilitate the transfer of gases. Diffused aerators provide an optimum treatment system for dissolving a costly gas such as oxygen or ozone into water. Packed column aerators are more efficient in removing volatile contaminants such as hydrogen sulfide or VOCs from water (31). Both types are shown schematically in Figure 18.13.

Iron has traditionally been removed from ground water by means of aeration in combination with filtration. In this

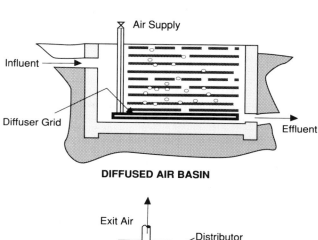

DIFFUSED AIR BASIN

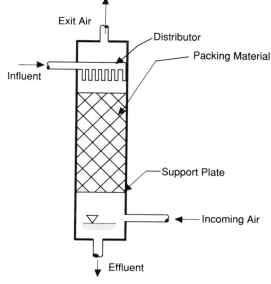

PACKED COLUMN

Figure 18.13. Diagram of aeration equipment (from Dyksen, J. E. and A. F. Hess III, "Alternatives for Controlling Organics in Groundwater Supplies"). Reprinted from *Journal* American Water Works Association, Vol. 74, No. 8, by permission. Copyright © 1982 American Water Works Association.

process oxygen transferred to the water results in the oxidation and precipitation of dissolved iron. The precipitated metals must then be removed by filtration. Because the oxidation of reduced manganese by air is much slower than the similar oxidation of iron, manganese removal often requires the additional use of a chemical oxidant such as chlorine.

A more recent application of the aeration process is the removal of volatile organic chemicals (VOCs) from water. This process, also called air-stripping, is generally more cost effective than adsorption techniques. In particular, industrial solvents with low molecular weight such as trichloroethane (TCA), trichloroethylene (TCE), and chloroform are removed. The effectiveness of the method is reduced for contaminants with lower volatility. For example, most pesticides are not sufficiently volatile to be removed by aeration.

Carbon Adsorption.

In water treatment, adsorption is the accumulation of dissolved substances onto the surface of a solid (adsorbent). The most commonly used adsorbent is activated carbon. There are two types of activated carbon. Powdered activated carbon (PAC) consists of finely ground particles. PAC is normally applied to water in a slurry and must be removed by filtration. Granular activated carbon (GAC) consists of larger particles that are packed in a column or placed in a bed through which the water will pass. The columns or beds are called carbon contactors. Water may be applied to the carbon contactors by gravity or pressure according to the system design. Figure 18.14 is a schematic drawing of a typical pressure contactor. When the carbon is exhausted, it must be replaced to prevent the breakthrough of absorbed contaminants.

GAC is effective in removing a variety of volatile and nonvolatile organic contaminants. It is commonly employed to remove taste or odor-causing organic compounds. Many of these compounds are relatively nonvolatile and are not removed by aeration. GAC is also used to remove many synthetic organic chemicals such as pesticides, aromatic solvents, phenols, and high molecular weight hydrocarbons.

Demineralization.

Highly mineralized waters may require treatment to reduce the total mineral content (demineralization). Demineralization reduces the levels of constituents for which specific treatment methods are not available. For example, it is employed to remove arsenic and fluoride, but other minerals are removed as well. Except for distillation processes, most of the methods require water that is relatively free of turbidity or of inorganic or organic matter that may foul equipment. Appropriate pretreatment using conventional methodologies may be necessary.

Distillation is the process of generating water vapor (steam) by heating the source water. The steam is then cooled and condensed to produce water that is relatively free of dissolved contaminants.

Reverse osmosis is a method of removing dissolved constituents by a filtrationlike process. This is possible because of the development of semipermeable membranes that allow the passage of water molecules but retain a high percentage of larger solute molecules. Water must be applied to the membranes under pressure not only to increase throughput but to "reverse" the driving force of osmosis. Osmosis is the process by which water moves across a semipermeable membrane from a solution containing low solute concentrations to one containing high concentrations.

Ion-exchange processes have previously been discussed in reference to softening methods. Zeolite resins are primarily cation exchangers that are effective in exchanging sodium for calcium and magnesium. Manufactured resins are capable of either cation- or anion-exchange processes, exchanging hydrogen ion (H^+) or hydroxyl ion (OH^-) for respective cations or anions. Mixed beds of these resins are capable of producing demineralized water of high purity.

Electrodialysis removes dissolved ions by passing water through a series of membrane stacks that are electrically charged. The stacks consist of alternating positive- and negative-charged membranes that selectively retain cations and anions. The final product is a partially demineralized solution.

18.5 WATER-QUALITY TESTING AND RECORDS

The data from analyses performed on samples of ground water should become part of a permanent water-quality record for that source. When the source is being developed for domestic water supply, much of the data collected are required by regulatory agencies. In addition to submitting this information to the appropriate agencies, well owners should maintain their own historical records of water quality. They are useful in tracking trends or changes in source water quality.

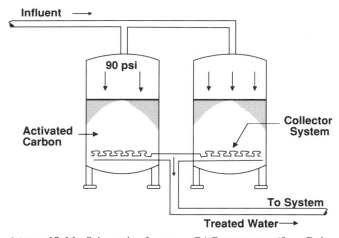

Figure 18.14. Schematic of pressure GAC contractors (from Dyksen, J. E., "Organics Treatment Techniques Overview," USEPA Workshop on Emerging Technologies for Drinking Water Treatment, September, 1987).

Records Required by Regulatory Agencies

In the United States, public water supply systems are regulated by the EPA or by state or local government. Requirements for water-quality monitoring may vary from region to region and with the source, size, and type of water system. For example, EPA regulations require that community water systems using ground water regularly test their sources for a wide range of inorganic constituents (see Table 18.7). Noncommunity systems, which basically serve only transient or intermittent users, are required to test for only one inorganic constituent—nitrate.

States that have accepted primary enforcement responsibility under the Federal Safe Drinking Water Act may impose additional or more frequent monitoring than is required by the EPA. Private wells are not regulated by the EPA but may be subject to local regulation. The appropriate regulatory agencies should be consulted to determine specific monitoring requirements and water-quality records that must be maintained for compliance.

Baseline Data and Historical Records

The term "baseline data" refers to the initial collection of information. For ground water resources, the initial water-quality test results provide an important baseline for evaluating future results. A comparison of current water-quality data with baseline and other historical data may provide useful information about changing source quality, the structural integrity of the well, or the validity of the test results. For example, nitrate levels in an aquifer may gradually increase as a result of contamination by agricultural chemicals or domestic wastewater. Baseline data should characterize the source in terms of the standards or criteria that apply to the intended use.

Sometimes a change in quality will reflect physical changes in the well or aquifer rather than source quality. If a well is not properly sealed, surface water or shallow ground water may flow down the well casing to the level of extraction. The intermittent intrusion of contaminating sources may produce unusual water-quality test results. Similarly, a well completed in more than one aquifer may yield water of varying quality. This may occur when the relative quantities of water extracted from each aquifer changes. This results from fluctuating water tables, plugging in the filter zone, or structural problems in the well such as incrustation of the well screens.

Sampling

Unusual or changing water-quality test results should not always be considered an indication of significant changes in the aquifer or well structure. There are many possible sources of error in laboratory tests. These include, but are not limited to, improper sampling procedures, field or lab-oratory contamination of samples, instrument failure, and inaccurate recording or reporting of test results.

Sampling is a critical step in obtaining valid water-quality data. A sample must be representative of the water residing in an aquifer (or produced from a well), and its integrity must be maintained until the laboratory tests are completed. Water standing in a well casing is probably not representative of the overall ground water quality. This can be due to the presence of drilling contaminants, biological growths, and corrosion by-products, or changes in environmental conditions, such as redox potential.

For these reasons it is necessary to pump or bail a well before collecting water samples. The recommended time of pumping depends on several factors, including the hydrogeology of the aquifer, the constituents or parameters to be tested, and the characteristics of the well. For small monitoring wells that are not easily bailed, a common practice is to pump or bail the well until a minimum of 4 to 10 bore volumes have been removed (32). If possible, it is desirable to pump a production well for one to two hours before collecting samples.

Newly completed wells sometimes require extended periods of pumping before a truly representative sample can be obtained. Samples collected during the first few hours of operation may be of a different quality than samples collected after several days. This phenomenon has been observed in wells that penetrate more than one aquifer. A plausible explanation is that during construction there may be exchanges of water between different aquifers due to pressure differences. Depending on the difference in quality and the amount of intruding water, the intended source may become temporarily contaminated. Over time, pumping of the new well would remove the intruded waters from the source and result in more representative samples.

Many of the chemical, physical, and biological parameters in ground water are unstable, so care must be taken in maintaining sample integrity. When a ground water sample is brought to the surface, it may be necessary to use special containers and preservatives to prevent deterioration. An obvious example is temperature, which must be measured in the field. Changes in temperature can alter chemical reaction rates, reverse cationic and anionic exchanges on solids, and affect microbial growth rates. Many types of analyses require that samples be chilled on ice for delivery to the laboratory. Volatile constituents, such as hydrogen sulfide and VOCs, may escape if the sample is not tightly contained. Conversely, atmospheric gases, such as carbon dioxide and oxygen, may diffuse into an open sample, affecting pH and alkalinity or resulting in the oxidation and precipitation of dissolved metals. Proper sample containers should be used to avoid these and other problems associated with the adsorption of constituents or release of contaminants into the sample.

A complete discussion of proper sampling procedures and laboratory quality assurance is beyond the scope of this book. Well owners should consult qualified laboratory personnel for instructions in obtaining valid samples for specific water-quality tests. An excellent reference is *Standard Methods for the Examination of Water and Wastewater* (9).

REFERENCES

1. Downes, C. J. 1985. "Redox Reactions, Mineral Equilibria, and Ground Water Quality in New Zealand Aquifers." *Ground Water Quality*. J. Wiley and Sons, New York.

2. Miller, D. W. 1980. *Waste Disposal Effects on Ground Water.* Premier Press, Berkeley, CA.

3. Asano, T., and G. S. Pettygrove. 1987. "Using Reclaimed Municipal Wastewater for Irrigation." *California Agriculture* (March–April).

4. McKee, J. E., and H. W. Wolf. 1963. *Water Quality Criteria.* State Water Quality Control Board Publication 3-A. Sacramento, CA.

5. USEPA. 1976. *Quality Criteria for Water.* U.S. Environmental Protection Agency. Washington, DC. July.

6. World Health Organization. 1970. *European Standards for Drinking Water.* 2d ed. Geneva, Switzerland.

7. World Health Organization. 1971. *International Standards for Drinking Water.* Geneva, Switzerland.

8. World Health Organization. 1984. *Guidelines for Drinking Water Quality.* Vols. 1 and 2. Geneva, Switzerland.

9. American Public Health Association. 1985. *Standard Methods for the Examination of Water and Wastewater.* 16th ed. Washington, DC.

10. Rossum, J. R. 1954. "Control of Sand in Water Systems." *JAWWA* (February).

11. Alexander, H. C., W. M. McCarty, E. A. Bartlett, and A. N. Syverud. 1982. "Aqueous Odor and Taste Threshold Values of Industrial Chemicals." *JAWWA* (November).

12. McGinnis, J. A., and F. DeWalle. 1983. "The Movement of Typhoid Organisms in Saturated, Permeable Soil." *JAWWA* (June).

13. Rossum, J. R. 1985. Unpublished notes.

14. Hem, J. D. 1970. *Study and Interpretation of the Chemical Characteristics of Natural Water.* Geological Survey–Water-Supply Paper 1473, U.S. Government Printing Office, Washington, DC.

15. Freeze, R. A., and J. A. Cherry. 1979. *Groundwater.* Prentice-Hall, Englewood Cliffs, NJ.

16. NAS. 1977. *Drinking Water and Health*, National Academy of Sciences, Washington, DC.

17. Olsen, L. L., and C. J. Twardowski. 1975. "$FeCO_3$ vs $Fe(OH)_3$ Precipitation in Water Treatment Plants." *JAWWA* (March).

18. O'Connor, J. J. 1971. "Iron and Manganese." *Water Quality and Treatment.* AWWA, McGraw-Hill, New York.

19. Dart, F. J. 1981. "Iron and Manganese Removal." Vol. 2. AWWA, Denver, CO.

20. Snoeyink, V. L., C. Cairns-Chambers, and J. L. Pfeffer. 1987. "Strong-Acid Ion Exchange for Removing Barium, Radium, and Hardness." *JAWWA* (August).

21. Aieta, E. M., J. E. Singley, A. R. Trussell, K. W. Thorbjarnarson, and M. J. McGuire. 1987. "Radionuclides in Drinking Water: An Overview." *JAWWA* (April).

22. Dohnalek, D. A., and J. A. Fitzpatrick. 1983. "The Chemistry of Reduced Sulfur Species and Their Removal from Groundwater Supplies." *JAWWA* (June).

23. Symons, J., et al. 1975. "National Organics Reconnaissance Survey for Halogenated Organics." *JAWWA* (November).

24. USEPA. 1978. *National Organics Monitoring Survey (NOMS).* Technical Support Division, Office of Drinking Water, U.S. Environmental Protection Agency, Cincinnati, OH.

25. Lehr, J. H. 1985. "Calming the Restless Native: How Ground Water Quality Will Ultimately Answer the Questions of Ground Water Pollution." *Ground Water Quality*. J. Wiley and Sons, New York.

26. Moore, J. W., and E. A. Moore. 1976. *Environmental Chemistry*. Academic Press, New York.

27. Liao, M. Y., and S. J. Randtke. 1985. "Removing Fulvic Acid by Lime Softening." *JAWWA* (August).

28. Roberts, P. V., M. Reinhard and A. J. Velocchi. 1982. "Movement of Organic Contaminants in Groundwater": Implications for Water Supply." *JAWWA* (August).

29. Rossum, J. R. 1970, "A Method of Measuring and Removing Oil from Wells." *JAWWA* (July).

30. Bowers, E. 1971. "Ion-Exchange Softening." *Water Quality and Treatment*. AWWA, McGraw-Hill, New York.

31. Dyksen, J. E., and A. F. Hess III. 1982. "Alternatives for Controlling Organics in Groundwater Supplies." *JAWWA* (August).

32. USEPA. 1982. *Handbook for Sampling and Sample Preservation of Water and Wastewater*. Environmental Monitoring and Support Laboratory, U.S. Environmental Protection Agency, EPA-600/4-82-029, Cincinnati, OH. September.

33. Hess, A. F., and J. E. Dyksen. 1984. "Utility Experiences Related to Existing and Proposed Drinking Water Regulations." *Proceedings AWWA Seminar on Experiences with Groundwater Contamination*. AWWA, Denver, CO.

34. Camp, T. R., and R. L. Meserve. 1974. *Water and Its Impurities*. 2d ed. Dowden, Hutchinson & Ross, Inc., Stroudsburg, PA.

35. Montgomery, J. M., Consulting Engineers, 1985. *Water Treatment Principles and Design*. J. Wiley and Sons, New York.

36. Sayre, I. M. 1988. "International Standards for Drinking Water." *JAWWA* (January).

37. AWWARF. 1989. *A Preliminary Assessment of Water Utility Monitoring Needs Under the Safe Drinking Water Act*. American Water Works Association, Denver, CO.

38. USEPA. 1976. *Quality Criteria for Water*. U.S. Environmental Protection Agency, Washington, DC. July.

READING LIST

AWWA. 1973. *Ground Water*. Manual M21. American Water Works Association, Denver, CO.

AWWA. 1987. *New Dimensions in Safe Drinking Water*. American Water Works Association, Denver, CO.

Bowen, H. J. M. 1979. *Environmental Chemistry of the Elements.* Academic Press, London.

Brown, T. L. 1963. *General Chemistry.* Charles E. Merrill Co., Columbus, OH.

Dyksen, J. E. 1987. "Organics Treatment Techniques Overview." Presented at the USEPA Workshop on Emerging Technologies for Drinking Water Treatment, Philadelphia, PA. September.

Houck, D. C., R. G. Rice, G. W. Miller, and C. M. Robson. 1985. *Contaminant Removal from Public Water Systems.* Noyes Publications, Park Ridge, NJ.

Johnsen, R. H., and E. Grunwald. 1965. *Atoms, Molecules and Chemical Change.* Prentice-Hall, Englewood Cliffs, NJ.

Keswick, B. H., and C. P. Gerba. 1980. "Viruses in Groundwater." *Environmental Sci. Technol.* (November).

Laubusch, E. J. 1971. "Chlorination and Other Disinfection Processes." *Water Quality and Treatment.* AWWA, McGraw-Hill, New York.

LeGrand, H. E. 1965. "Patterns of Contaminated Zones of Water in the Ground." *Water Resources Research* 1, 1.

Mackay, D. M., P. V. Roberts, and J. A. Cherry. 1985. "Transport of Organic Contaminants in Groundwater." *Environmental Science and Technology* 19, 5.

Miller, D. W. 1985. "Chemical Contamination of Ground Water." *Ground Water Quality.* J. Wiley and Sons, New York.

Mortimer, C. E. 1967. *Chemistry A Conceptual Approach.* Van Nostrand Reinhold, New York.

Rosen, A. A., and R. L. Booth. 1971. "Taste and Odor Control." *Water Quality and Treatment.* AWWA, McGraw-Hill, New York.

USEPA. 1983. "Microorganism Removal for Small Water Systems." Report No. EPA 570/9-83-012. U.S. Environmental Protection Agency, Office of Drinking Water, Washington, DC.

USEPA. 1984. *EPA Journal* 10, 6. Washington, DC.

CHAPTER 19

Artificial Recharge

19.1 INTRODUCTION

Artificial recharge is the process of replenishing ground water reservoirs by adding water to aquifer storage. It is planned, as in deliberate impounding or stream-bed infiltration, but may be unplanned in the case of leakage from canals, pipelines, or other conduits. Planned recharge which stores water underground for future use, involves procuring or collecting water to be recharged, selecting recharge method, and constructing the facilities to be employed. Facilities include spreading basins, stream channels, pits, and injection wells.

Major objectives of an artificial recharge program are:

1. Conservation of water resources.
2. Better use of ground water reservoirs by recharging close to points of demand.
3. Elimination of evaporation loss and other undesirable effects associated with aboveground reservoirs.
4. Increasing the ground water supply.

In addition to these general objectives, there are several special problems that can be mitigated or overcome by use of artificial recharge. For example, artificial recharge is used to block encroachment (landward movement) of seawater caused by inland pumping. Freshwater is injected into wells, forming a "pressure ridge" barrier parallel to the coast (1). Here the primary purpose is to prevent inland ground water contamination and recharge is only a minor side benefit. In cases where heavy withdrawals of ground water have resulted in land surface subsidence, the condition has been stabilized by injecting water into the depleted aquifers (2).

19.2 CONDITIONS FAVORABLE FOR RECHARGE

A comprehensive understanding of geologic, hydrological, and operational aspects of the project is necessary to select the appropriate artificial recharge method. These include aquifer boundaries, surface and ground water inflow and outflow, storage capacity, porosity, hydraulic conductivity, and available water sources.

Aquifers best suited for artificial recharge absorb large quantities of water but do not quickly release it. This implies a high vertical hydraulic conductivity with a moderate horizontal value. Successful artificial recharge requires the underlying aquifer to be unconfined with a sufficiently low water table to allow additional storage.

Many ground water basins contain an abundance of artesian aquifers. The inability to recharge them by direct methods precludes their use for large-scale artificial recharge programs.

Most large-scale recharge areas are located in the foothills or forebays[1] of alluvial basins. They are favorable to ground water storage because of the unconfined nature of the aquifers coupled with high infiltration rates. Unconsolidated alluvial basins containing coarse sediments and buried-river channels also represent favorable conditions for artificial recharge.

19.3 METHODS OF ARTIFICIAL RECHARGE

Water Spreading

The most common method of artificial recharge is to spread water over a large surface area, allowing it to sink into underlying aquifers. The terms "infiltration" or "percolation" describe the vertical movement of water through the non-saturated zone. The infiltration rate is defined as the volume of water that moves downward during a unit-time through a unit-area, and is expressed in cubic feet per day per square foot or more commonly, ft/day. Two primary methods of large-scale water spreading commonly in use are off-channel spreading and on-channel spreading.

Off-channel Spreading. Off-channel spreading consists of transferring water from a stream or river channel to areas where soil conditions favor high-infiltration rates. Spreading basins generally are constructed by building dikes or levees. The shape and size of the basins depend primarily on land availability, slope, and soil conditions. If storm runoff is diverted for recharge, in-line desilting basins may be required. In a typical case water is diverted into an abandoned gravel pit located near a stream or river. This basin fills and overflows back to the river (Figure 19.1).

1. The term forebay denotes areas of recharge composed of materials of high hydraulic conductivity, usually found near mountains.

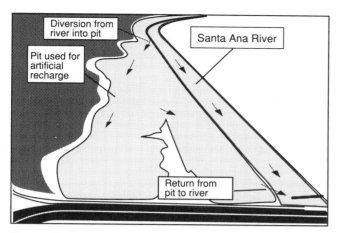

Figure 19.1. Off-channel spreading (after Orange County Water District).

Land flooding may be appropriate where relatively flat land requiring little preparation is available. Water is allowed to spread evenly in a very thin layer, which prevents erosion. To control depth and velocity, berms are often constructed on the perimeter of the area.

A less frequently used means of off-channel artificial recharge is the ditch-and-furrow method (1). A series of closely placed shallow and flat-bottomed ditches or furrows are dug, and water is diverted into them. Three layouts may be used:

1. Contour. The ditch follows the ground contour.
2. Tree-shaped. The main ditch successively branches into smaller ditches.
3. Lateral. A series of smaller ditches extend laterally from the main ditch.

The ditch-and-furrow method is particularly susceptible to clogging because the shallowness of the ditches encourages silt buildup. To prevent this, the gradient of the ditches should be steep enough to carry suspended material through the system.

On-channel Spreading. In the on-channel method of water spreading, a natural stream channel is improved to increase infiltration capacity. This is done by scraping, leveling, widening, ditching, and building sand levees (dikes) in the stream bed. The surface area over which the water infiltrates is increased, reducing water velocity and allowing more time for infiltration. Figure 19.2 shows an example of on-channel spreading facilities in the Santa Ana River forebay of Southern California.

Factors Affecting Infiltration. The downward movement of recharge water is governed by several factors: the average vertical hydraulic conductivity of the soil, the presence of

soil gases in the nonsaturated zone, and the changes of soil structure during infiltration. When new spreading grounds are put into operation, infiltration rates decrease initially and then increase after a few hours. Later the rates become less predictable and in most cases show a declining trend that eventually stabilizes (1). Sometimes a second, temporary, and smaller increase in infiltration rate may occur. The cycle repeats itself after the spreading ground dries and is refurbished (through removal of superficial fines), although at generally lower infiltration rates.

The first transient decrease of the infiltration rate is caused by swelling of clay particles in the formations. The first temporary increase results from gradual elimination of soil gas from the interstices. Later decline in recharge rate is caused by the physical-chemical alteration of the natural soil profile and by bacterial growth and accumulation of metabolic products at shallow depths. Since bacterial growth is effectively destroyed by oxidation, spreading grounds must be

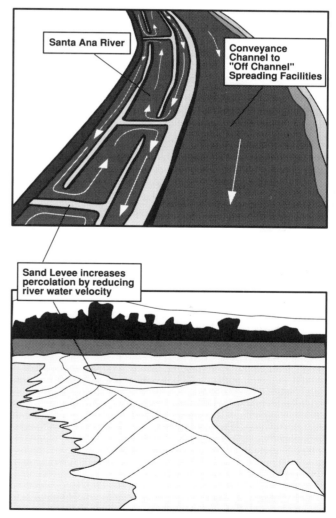

Figure 19.2. On-channel spreading (after Orange County Water District).

dried and cleaned periodically to restore their infiltration capacity. Consequently allowance must be made for the area that will be out of service during rehabilitation. Use of chlorinated water also improves infiltration rates by inhibiting the clogging effect of bacteria.

Clogging also occurs in spreading operations when the water contains suspended solids. The soil may clog with silt or mud near the surface, or fine particles may penetrate deeper into the soil and clog by joining with native material. Particles may penetrate as deep as 60 ft in porous soils. Very turbid water is therefore retained in settling or desilting ponds, with flocculents sometimes added. Water containing less than 1000 mg/L of suspended solids is generally acceptable for spreading.

Recharge Rates. Todd has shown that a reasonable estimate of recharge rates in alluvial soils may be calculated based on ground slope (1). For slopes in the range of 0.1% to 10%, long-term percolation rates may be calculated empirically from

$$W = 0.65 + 0.56i$$

where

i = ground slope [%],
W = long-term infiltration rate [vertical rate of infiltration in m/day].

Experience has shown that typical recharge rates in spreading basins range from 1 to 4 ft/day (see Table 19.1) (3).

Clayey soils may have infiltration rates of 0.5 to 1 ft/day. Recharge rates generally decrease with a decrease in soil particle size. Soil conditioners that aggregate or clump particles together have been used with some success.

It has been observed that higher percolation rates occur when natural vegetation is undisturbed (although extensive phreatophyte growth inhibits percolation). Other measures that increase effective porosity of the soil include alternating wet and dry periods and scarifying the soil.

Maintaining Recharge Rates. The following operations have proved successful in maintaining recharge rates in spreading basins (1, 2, 3):

1. The periodic removal of the fine material deposited on the surface, by scraping. Scraping is more effective in coarse-grained spreading grounds.
2. In some cases it may be neccessary to install a surface filter layer whose hydraulic conductivity is lower than that of the natural strata. This filter, however, must be periodically replaced.
3. Turbid water should be prevented from entering basins.
4. Recharge water should be introduced into a basin at its lowest point to prevent erosion and clogging of the soil surface.
5. Aquatic vegetation should be minimized to prevent biological clogging.
6. Infiltration-increasing organic matter or chemicals may be added to the ground surface.
7. Waters containing a high mineral content should not be used for recharge.

Measuring Infiltration, Full-Scale Field Tests. Several methods are commonly used to measure infiltration rates. Full-scale testing of the entire spreading basin gives the most reliable results. However, if transmission of water to the spreading basins is costly, or if the land itself has to be acquired, less-than-full-scale testing may be acceptable.

The "test-pond method" is a long-duration test of a representative part of the contemplated spreading grounds. It measures changes in infiltration rates over time, detects possible creation of perched saturated zones, and estimates the influence of recharge on the water table. Data are gathered on potential evaporation and underlying formation geology, including the presence of any semipervious layers to ensure a meaningful test-pond investigation. A test hole is then drilled, and samples taken.

The testing procedure consists of filling the test pond with water and measuring water-level declines over time. The amount of evaporated water is estimated and subtracted from the volume of decline to obtain the volume of water infiltrated during the period. If a shallow confining or semi-confining layer exists, the effective saturated area may be larger than the bottom of the pond and water levels might

TABLE 19.1 Representative Spreading Basin Recharge Rates

Location	Rate (ft/day)
Santa Cruz River, AZ	1.0–3.9
Los Angeles County, CA	2.3–6.2
Madera, CA	1.0–3.9
San Gabriel River, CA	2.0–5.2
San Joaquin Valley, CA	0.3–1.6
Santa Ana River, CA	1.6–9.5
Santa Clara Valley, CA	1.3–7.2
Tulare County, CA	0.3
Ventura County, CA	1.3–1.6
Des Moines, IA	1.6
Newton, MA	4.3
East Orange, NJ	0.3
Princeton, NJ	<0.3
Long Island, NY	0.7–3.0
Richland, WA	7.5

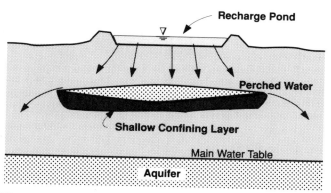

Figure 19.3. Increase in areal extent of recharge due to presence of a shallow confining layer.

be influenced by adjacent ponds. Figure 19.3 shows the effect of a shallow confining layer beneath a test pond.

Infiltrometer Testing.
In the infiltrometer method of measuring infiltration rate, metal casings 20 to 40 in. in diameter are installed into the first saturated layer. A measured volume of water is put into the casing, and downward seepage is observed. This method is best suited in regions where the upper formation is less pervious than the lower one. Concentric double casings are used to limit lateral movement near the surface. Correct interpretation of infiltrometer data requires considerable experience.

Laboratory analysis of undisturbed samples or use of pumping test data do not accurately reflect spreading basin parameters because they assume fully saturated conditions without bacterial clogging. Good representative undisturbed samples are difficult and costly to obtain.

Recharge through Pits and Shafts

Artificial recharging through pits or shafts requires less surface area than spreading basins since infiltration takes place primarily through the walls of the pit, due to horizontal hydraulic conductivity being greater than vertical conductivity. The pit method avoids evaporation problems associated with spreading. Its use is also advantageous where suitable land for spreading is limited and/or where clay or other impermeable layers prevent downward percolation. A pit is also less susceptible to clogging because the silt settles to the bottom, leaving the sides free for infiltration. Eventually, however, the pit must be dried and silt removed. Frequently existing pits or excavations must be renovated and cleared of debris prior to use as a recharge facility. Figure 19.4 illustrates the principle of artificial recharge using a pit.

Injection Wells

Recharge through injection wells is generally not done on a large-scale basis because capital and operating costs are much higher compared with other methods. Injection wells, however, are often used for special recharge purposes such as seawater barriers, or containment of contaminant plumes.

The design of a recharge well in many respects is similar to that of a pumping well. Some recharge wells have multiple-zone completions with screens set into each aquifer. The screens are individually gravel packed, and the aquifers are isolated by cement seals. Figure 19.5 shows a typical multi-casing well.

Development and redevelopment of multiple-zone completion recharge wells is relatively difficult, since the small-diameter screens limit the amount of water that can be pumped during development. Since recharge water is injected under pressure, it is important that the conductor and recharge well casing be sealed to a depth adequate to prevent the water from returning to the surface outside the well.

Clogging in Recharge Wells.
Water quality is very important to the successful operation of recharge wells. Fine material in the injected water tends to clog the filter zone. Bacteria and algae may produce the same effects.

If air is carried into a recharge well, it reduces the hydraulic conductivity of the near-well zone (air locking). In some instances wells have been completely air-locked. In a different air-related problem a recharge well in limestone failed because of the large volume of entrained air that had been introduced (2). When injection was stopped, the air release momentarily evacuated the well, causing casing collapse. Air can be eliminated by designing the installation to maintain positive pressure (exceeding atmospheric pressure) in the injection system.

Experience with clogging problems in recharge wells has shown:

1. When the injection water used is on a well-to-well basis (ground water pumped from one aquifer is injected into another), biological clogging is minimized.

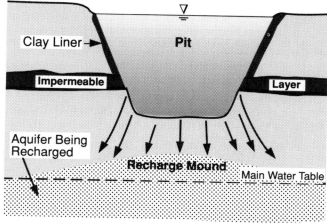

Figure 19.4. Use of a pit to recharge an aquifer overlain by an impermeable layer.

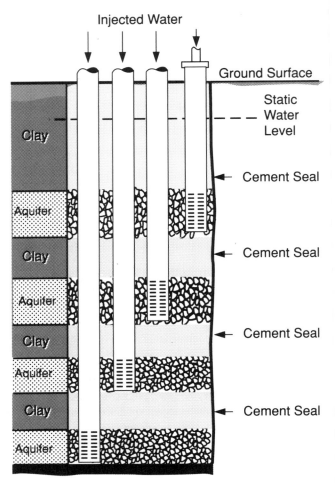

Figure 19.5. Multiple-zone recharge well (after Orange County Water District).

2. Surface water, although treated and brought up to drinking water standards, may cause clogging after comparatively short periods of injection. The phenomenon may be attributed to the higher temperatures, the relative abundance of nutrients, and the remnants of algae not completely destroyed by chlorination (2).

3. In most cases short periods of pumping quickly remove the clogging particles and improve recharge capacity. The best method of operation consists of short injection periods (about one month), alternating with one or two hours of pumping. Water pumped during redevelopment, however, should not be reinjected.

4. Chlorination at the well-head reduces bacterial clogging but does not entirely eliminate the need for pumping and/or mechanical redevelopment.

19.4   HYDRAULICS OF ARTIFICIAL RECHARGE

Two primary ground water hydraulic effects result from artificial recharge: the buildup or "mounding" effect due to

the applied head and an increased storage effect due to additional water in the aquifer. The increased head causes a rise in the water level in unconfined aquifers, or piezometric level in confined or semiconfined aquifers (injection wells). When water is introduced, the piezometric surface assumes a shape that is a function of aquifer parameters and boundaries. This mounding effect is related to aquifer diffusivity T/S (transmissivity/storativity). It is also related to the amount and duration of water recharged. Capillary forces, water temperature, quality, and presence of air in the aquifer also affect artificial recharge.

The increase in aquifer storage is related to storativity, amount of replenishment, transmissivity, and aquifer boundaries. Recharge water moves according to both hydrodynamic dispersion and regional ground water flow rates and direction.

Ground Water Mounds

Hantush has shown that the growth of a ground water mound in response to a uniform rate of deep percolation over a long recharge strip (river) may be expressed as (4)

$$h_1^2(x, t) = h_0^2 + \frac{2W\kappa t}{K}\left\{1 - \frac{1}{2}\left[4i^2\text{erfc}\left(\frac{L - x}{\sqrt{4\kappa t}}\right)\right.\right.$$
$$\left.\left. + 4i^2\text{erfc}\left(\frac{L + x}{\sqrt{4\kappa t}}\right)\right]\right\}$$

(19.1)

and

$$h_2^2(x, t) = h_0^2 + \frac{W\kappa t}{K}\left[4i^2\text{erfc}\left(\frac{x - L}{\sqrt{4\kappa t}}\right)\right.$$
$$\left. - 4i^2\text{erfc}\left(\frac{x + L}{\sqrt{4\kappa t}}\right)\right],$$

(19.2)

where

$4i^2\text{erfc}(x)$	= second repeated integral of the error function (see Table 2.2 in Chapter 2),
h_0	= original height of the water table above the base of the unconfined aquifer [L],
h_1, h_2	= height of the water table above the base of the unconfined aquifer after percolation has started in regions 1 and 2, respectively [L],
K	= hydraulic conductivity of the aquifer [LT^{-1}],
L	= horizontal length from the origin to start of region 2 [L],
t	= time since the initial condition of flow [T],

W = uniform rate of percolation per unit area $[LT^{-1}]$,

κ = Kh_0/θ, diffusivity $[L^2T^{-1}]$ where θ is effective porosity.

Figure 19.6 shows an example of artificial recharge occurring at a uniform rate under a river. The height of the mound after 30 days of recharge can be calculated by applying Equations 19.1 and 19.2.

Calculation for x = 50 ft. Applying Equation 19.1 with values from Figure 19.6,

$$\kappa = 66{,}845 \text{ ft}^2/\text{day} ,$$

$$\frac{L - x}{\sqrt{4\kappa t}} = \frac{(100 - 50)\text{ft}}{\sqrt{(4)(66{,}845)(30)}\text{ft}} = 0.0177 ,$$

$$\frac{L + x}{\sqrt{4\kappa t}} = \frac{100 + 50}{2832.2} \text{ ft} = 0.0530 ,$$

$$\frac{2W\kappa t}{K} = \frac{(2)(1.5)(66845)(30)(7.48)}{1000} = 45{,}000 \text{ ft}^2 ,$$

$$h_1(50, 30) = \sqrt{50^2 + 45{,}000[1 - 0.5(0.9618 + 0.8856)]} = 77 \text{ ft} .$$

Calculation for x = 175 ft

$$\frac{x - L}{\sqrt{4\kappa t}} = \frac{175 - 50}{2832.2} = 0.0441 ,$$

$$\frac{x + L}{\sqrt{4\kappa t}} = \frac{175 + 50}{2832.2} = 0.0794 ,$$

$$h_2(175, 30) = \sqrt{50^2 + 22{,}500(0.9048 - 0.8267)} = 65 \text{ ft} .$$

Thus, after 30 days of recharging, the height of the mound rose 27 ft and 15 ft above the initial water level at distances from the center of the river of 50 ft and 175 ft, respectively.

Hydraulics of Recharge Wells

Injection Well Pressure Head Losses. Injection wells have head-loss components similar to those found in pumping wells, as described by the following general equation:

$$P = BQ + B'Q + CQ^2 , \qquad (19.3)$$

where

P = injection well pressure head [ft],

Q = injection rate [gpm],

B = formation loss coefficient [ft/gpm],

B' = loss coefficient associated with drilling damage or other reduced permeability effects in the near-well filter zone [ft/gpm],

C = well loss coefficient [ft/gpm^2].

Figure 19.7 illustrates these loss components graphically.

Specific Injection Capacity and Pressure. The ratio of the injection rate to the pressure head producing it is denoted by Q/P. Since this is similar to specific capacity in a pumping well, it is defined here as "specific injection capacity," or simply "specific injection."

The reciprocal of specific injection is similar to specific drawdown in a pumping well and therefore may be called "specific injection pressure," or simply "specific pressure" (P/Q). The specific pressure is obtained by dividing both sides of Equation 19.3 by the injection rate Q:

$$\frac{P}{Q} = B + B' + CQ .$$

It is seen that specific pressure (P/Q) versus injection rate (Q) is the equation of a straight line, with the $Q = 0$ intercept equal to $B + B'$ and a slope equal to C.

The constants B, B', and C may be obtained using a field method similar to a step-drawdown test. Different injection

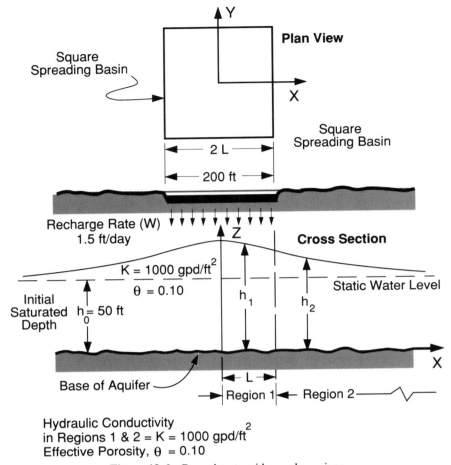

Figure 19.6. Ground water ridge under a river.

rates and pressures are measured, and the specific pressure is plotted against the injection rate. The value of the formation and well loss coefficients are then determined by methods similar to those used for pumping wells (see Chapter 15).

Injection Well Efficiency. The efficiency of an injection well may be stated as the ratio of the actual specific injection to the theoretical specific injection. In equation form this is written

$$E = \frac{Q/P_a}{Q/P_t} \times 100 = \frac{P_t}{P_a} \times 100 \ ,$$

or

$$E = \frac{100}{1 + (CQ/B)} \ ,$$

where

$$Q = \text{injection rate [cfs]},$$

P_a = actual injection pressure head (as measured at the well head) [ft],

P_t = theoretical injection pressure head (equal to aquifer loss [ft].

19.5 ARTIFICIAL RECHARGE EXAMPLES

Water Spreading—Montebello Forebay, Los Angeles County

The Montebello Forebay is a deposit of recent alluvium, beginning at the Whittier Narrows and extending approximately five miles west of the Rio Hondo River, five miles east of the San Gabriel River, and eight miles south of Whittier Narrows in Los Angeles County (see Figure 19.8). The ground surface in this area is in direct hydraulic continuity with the underlying older sediments containing the aquifers.

Over the years, when high surface water flows occurred through the Whittier Narrows, the relatively more ancient marine and fluvial deposits were eroded. When the flows receded, recent sediments from the San Gabriel Mountains and San Gabriel Valley were deposited. The fan of deposition is composed of variable-sized sediments, from highly permeable coarse-grained material near the apices to less permeable finer-grained material toward the peripheries. Lenses of clay and very fine silt dispersed intermittently throughout the Montebello Forebay affect permeability. The hydraulic continuity to the underlying aquifers through the highly permeable sediments makes this forebay suitable for artificial ground water recharge by surface spreading.

The Montebello Forebay is the major recharge area for the Central and West Coast Basins located near Los Angeles. Within the forebay are the Los Angeles County Department of Public Works' largest spreading facilities: the Rio Hondo Coastal Basin Spreading Grounds, the San Gabriel Coastal Basin Spreading Grounds, and the Lower San Gabriel River. The department has annually conserved in these recharge facilities an approximate average of 125,000 acre-ft of storm and dry-weather runoff, poststorm dam releases, and imported and reclaimed water. Additional water entering the forebay is supplied by underflow from the north and direct rainfall percolation.

Ground water levels in the Central and West Coast Basins have generally stabilized in recent years, as a result of adjudications limiting total annual extractions and artificial recharge. Increased artificial recharge would result in raising ground water levels, augmenting the water available for use in a sustained drought.

Injection Well Barrier—Dominguez Gap

Overproduction of ground water from the West Coast Basin during the first half of this century lowered water levels below sea level over the entire basin, reversing the normal gradient to the ocean (5). This resulted in seawater intrusion

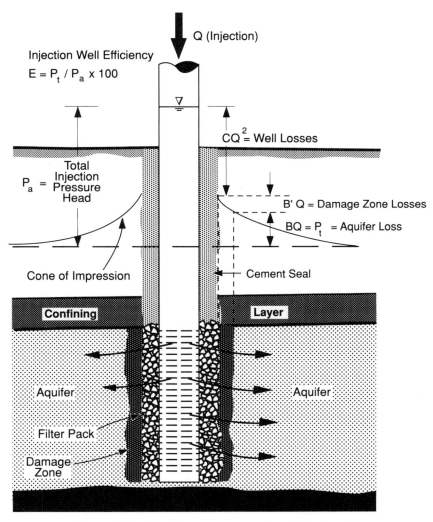

Figure 19.7. Head-Loss components in an injection well.

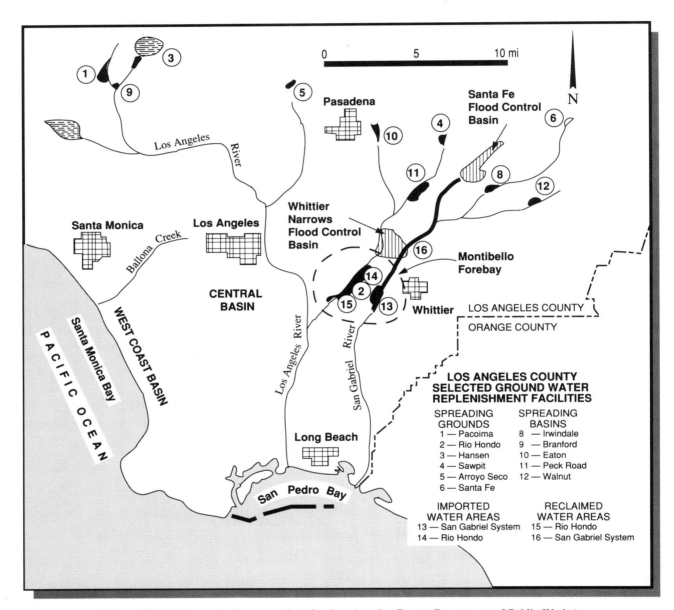

Figure 19.8. Water spreading example (after Los Angeles County Department of Public Works).

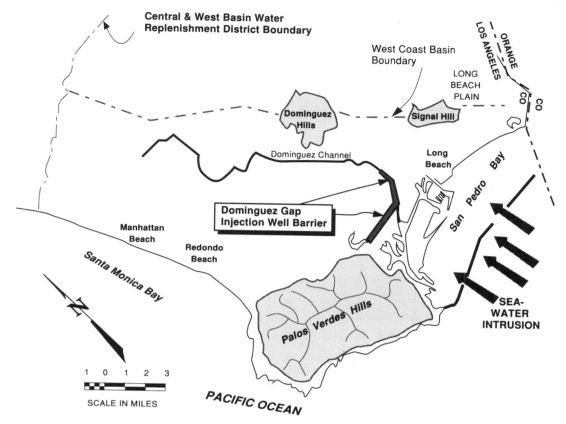

Figure 19.9. Dominguez gap injection well barrier.

from San Pedro Bay into the exposed aquifers. The two principal aquifers in the San Pedro Bay sector are known as the 200-ft sand (Upper Pleistocene) and the 400-ft gravel (Middle Pleistocene) aquifers. A third aquifer, the Gaspur Aquifer (Recent), is a channel deposit in hydraulic continuity with both the 200-ft sand and the 400-ft gravel aquifers.

In 1967 the Los Angeles County Flood Control District began construction of the Dominguez Gap Barrier Project. Figure 19.9 is the vicinity map of the project. The barrier, extending slightly over four miles and consisting of 29 injection wells and 181 observation wells, began operation in 1971. Of the 29 injection wells, 17 are single-zone wells injecting into the 200-ft sand. One well injects into the merged 200-ft sand and 400-ft gravel zone. The remaining 11 injection wells are dual-zone wells injecting into both aquifers.

REFERENCES

1. Todd, D. K. 1980. *Groundwater Hydrology*. J. Wiley and Sons, New York.

2. United Nations. 1975. "Ground-Water Storage and Artificial Recharge." Department of Economic and Social Affairs, Natural Resources/Water Series No. 2. United Nations, New York.

3. Todd, D. K. 1959. "Annotated Bibliography on Artificial Recharge of Ground Water through 1954." U.S. Geological Survey Water-Supply Paper 1477.

4. Maasland, D. E. L., and M. W. Bittinger. 1963. Proceedings of the Symposium on Transient Ground Water Hydraulics. Colorado State University, Fort Collins, CO.

5. PRC Engineering and Geoscience Support Service, Inc. 1987. "Dominguez Gap Barrier Project, Deficiency Study." Prepared for Los Angeles County Department of Public Works.

CHAPTER 20

Ground Water Management

20.1 INTRODUCTION

Ground water management involves the planning, implementation, and operation necessary to provide safe and reliable ground water supplies. Objectives typically focus on aquifer yield, recharge, and water quality and on legal, socioeconomic, and political factors. Formal ground water management, although generally more important in large-scale development, may also be applied to smaller-scale or even individual well projects.

Figure 20.1 illustrates some aspects of various ground water management programs.

20.2 SAFE YIELD

General Concept

Ground water management is concerned with renewability of the resource and its practical exploitation. Fundamental to this principle is a concept generally known as "safe yield," which is associated with the amount of supply that a water user can depend upon. Most sources state that the term was originated by Meinzer in 1920 (1). He defined safe yield of a ground water reservoir as "the practicable rate of perennially withdrawing water from it for human use." By today's standards this definition is vague and needs further clarification.

There are many other definitions of safe yield (2, 3). Most are based on the rate of withdrawal above which the quantity and quality of the ground water suffers. This chapter generalizes the definition and defines safe yield as the annual amount of withdrawal that does not exceed annual recharge, permanently lower the water table to an uneconomic level, or allow intrusion of poor-quality ground water. Figure 20.2 shows this safe yield concept.

In practice, safe yield should be less than average annual recharge to compensate for minor ground water losses. Water may be lost by underflow through leaky aquifers, or from natural ground water discharges (springs). Other losses may occur from rising ground water in unconfined aquifers, resulting in evapotranspiration and evaporation.

When safe yield is exceeded on a continuing basis, the aquifers are said to be mined. In some areas mining water for periods up to 40 or 50 years is not considered uncommon

(2). The reasons for long-term mining may include lack of other water sources or development of a economic base to finance importation of new supplies.

Equation of Hydrologic Equilibrium

Calculation of safe yield in an area involves relating geohydrologic and operational factors in a quantitative form, known as a "water balance" or "hydrologic budget" (see Chapter 3). The hydrologic budget may be written

$$\text{Inflow} = \text{outflow} \pm \text{change in storage} . \quad (20.1)$$

Equation 20.1 is commonly known as the "equation of hydrologic equilibrium," and is an example of the law of conservation of matter (see Chapter 2). Although the concept is invariant, the particular use may vary slightly in form and complexity depending on application. Inflow includes all precipitation infiltration and other recharge. Outflow includes exploitation, evapotranspiration, and seepage losses to the surface or adjacent ground water reservoirs. The following example employs an equation of the form

$$Q_i + Q_{cl} + Q_{ur} + EP + Q_{rfs} + Q_{rfg}$$
$$= Q_0 + Q_{ip} \pm \Delta V , \quad (20.2)$$

where

Q_i = ground water inflow $[L^3T^{-1}]$,

Q_{cl} = conveyance loss from unlined surface canals contributing to ground water recharge $[L^3T^{-1}]$,

Q_{ur} = recharge from unaccounted sources and inter-aquifer leakage $[L^3T^{-1}]$,

EP = effective precipitation recharging the ground water reservoir $[L^3T^{-1}]$,

Q_{rfs} = return flow from surface water irrigation $[L^3T^{-1}]$,

Q_{rfg} = return flow from ground water irrigation $[L^3T^{-1}]$,

Q_0 = ground water outflow $[L^3T^{-1}]$,

Q_{ip} = ground water pumping $[L^3T^{-1}]$,

ΔV = ground water storage change $[L^3T^{-1}]$.

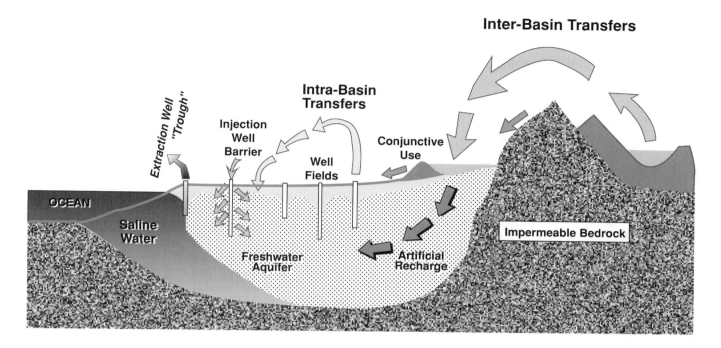

Figure 20.1. Aspects of ground water management.

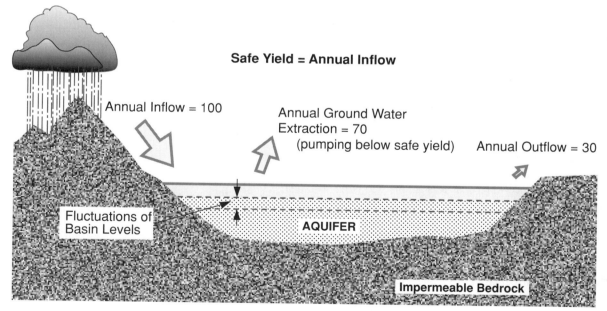

Figure 20.2. Safe yield concept.

Example Calculation

The following example applies this form of the hydrologic budget to calculate safe yield for a project in northern Iran (4) (Figure 20.3).

The time period chosen for the water balance was the four-year period of 1969 to 1972, and it was considered representative of mean geohydrologic conditions.

Ground Water Inflow, Q_i. The inflow term was calculated as

$$Q_i = (L)(Kb)\left(\frac{\Delta h}{\Delta x}\right) = 70,542 \text{ acre-ft/year },$$

where

Q_i = flow of ground water moving across a section L (normal to the flow path) having an average transmissivity $T(Kb = T)$ under a hydraulic gradient of $\Delta h/\Delta x$.

The recharge area originates at the southern boundary (foothill zone) of the area of investigation. Values of average transmissivity for the area were obtained from aquifer test results correlated with a geophysical resistivity survey. Hydraulic gradients were obtained from water-level contour maps, and cross-sectional areas were drawn parallel to equipotential lines.

Ground Water Outflow Q_0. Ground water outflow across the northern boundary (near the coast of the Caspian Sea) consists of the sum of the outflow from the water table and the deeper confined aquifers. However, as the upper aquifer water-level contour maps show little-to-no outflow, the estimate was based on the deeper aquifers only. The total subsurface ground water outflow across the northern boundary was estimated as 5676 acre-ft/year.

Conveyance Loss, Q_cl. Some water diverted for irrigation is lost during transit and percolates as recharge to the upper aquifer. The conveyance water loss was assumed to be 15% of the total volume of water diverted for irrigation (204,330 acre-ft/yr as measured at the main diversion located at the base of the mountains). Of this, 15% is the conveyance loss, or 30,649 acre-ft/yr.

Effective Precipitation (EP). In areas where the ground water table is reached by plant roots or is so shallow that the capillary fringe extends to the land surface, the amount of water discharged by transpiration through plants and evaporation from the land surface is the "potential evapotranspiration" of that area (5). The effective precipitation or infiltration is that portion of the precipitation that actually

TABLE 20.1   Potential Infiltration and Evapotranspiration Constants

Depth to Shallow Ground Water (ft)	Potential Evapotranspiration Constant, α	Potential Infiltration Constant, β
<3	0.4	0.1
3–15	0.1	0.5
>15	0	0.8

Note: α and β empirically determined for the area (14).

reaches the phreatic water surface. Its calculation involves subtraction of potential evapotranspiration (PE) from potential infiltration (PI). Potential infiltration is estimated by adjusting residual precipitation for soil moisture deficiency and depth to the water table.

In the study area monthly precipitation records were available at two stations. After adjusting for soil moisture deficiency and interception, the average value of residual precipitation was 8.67 in./yr.

Potential Infiltration was then calculated from

$$PI = P\beta A = 113,679 \text{ acre-ft/yr}$$

where

PI = potential infiltration [acre-ft/yr],
P = residual precipitation [ft/yr], 0.723 ft/yr,
β = empirical infiltration factor (function of depth to ground water; see Table 20.1),
A = total area, 338,567 acres.

Potential evapotranspiration was calculated in a similar manner:

$$PE = 3.445^1 \alpha A = 178,221 \text{ acre-ft/yr },$$

where

PE = potential evapotranspiration [acre-ft],
α = empirical evaporation constant (function of depth to water table, see Table 20.1).

Effective precipitation (EP) is the difference between potential infiltration (PI) and potential evapotranspiration (PE):

$$EP = PI - PE = -64,542 \text{ acre-ft/yr }.$$

The negative sign signifies a loss from the ground water reservoir.

---

1. The average of pan evaporation values in the area is 3.445 ft/yr.

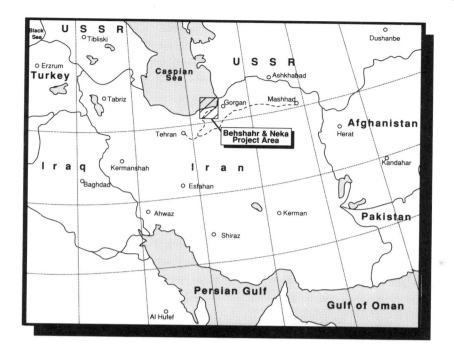

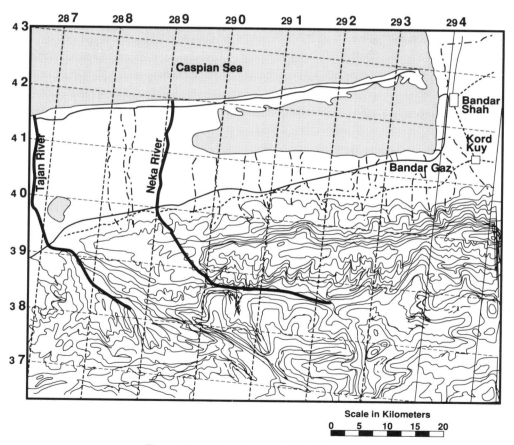

Figure 20.3. Location map of project area.

Irrigation Return Flow, Q_{rfs} and Q_{rfg}. Of the total water delivered for irrigation in the project area, the following disposition of use was assumed:

1. Consumptive use by plants (60% of total).
2. Evaporation of excess water (15%).
3. Percolation to the phreatic aquifer (25%).

Of the 204,330 acre-ft/yr of surface water diverted for irrigation, 173,680 acre-ft/yr reaches the fields (15% was lost during conveyance). Of this amount 25% or 43,420 acre-ft/yr infiltrates as irrigation return flow recharging the phreatic aquifer (Q_{rfs}).

A total amount of 103,786 acre-ft/yr was pumped from shallow and deep aquifers. Of this, 25% returns to the phreatic aquifer contributing 25,947 acre-ft/yr toward ground water recharge (Q_{rfg}).

Pumping, Q_{ip}. Total irrigation pumping was estimated from averages of the metered yearly discharge for all wells and ghanats in the area. Specifically,

- All shallow and deep wells and ghanats were located on 1:20,000 scale maps.
- Average pumping discharges were recorded from records or personal communication with the landowners in the field.

Average yearly discharge was obtained by weighting pumping days to the total days per year:

$$\overline{Q} = Q \times \frac{t}{365} \, ,$$

where

$\overline{Q}$ = average yearly discharge = 103,786 acre-ft/yr,
Q = metered or estimated discharge [acre-ft/yr],
t = days of pumping per year.

Ground Water Storage Change, ΔV. In the project area the ground water reservoir was approximated by a two-layered system:

1. Phreatic (less than 150-ft depth).
2. Deep (sum of all confined and semiconfined aquifers below 150 ft).

As a first approximation, it was assumed that:

$$\Delta V = \delta V_1 + \delta V_2 \, ,$$

where

ΔV = total ground water storage change [acre-ft/yr],
δV_1 = phreatic aquifer change [acre-ft/yr],
δV_2 = deep aquifer change [acre-ft/yr].

It was further assumed that $\delta V_2 \rightarrow 0$ (deep aquifer storage). This was a reasonable assumption as total deep aquifer exploitation during the study period was only 44,859 acre-ft/yr, whereas total ground water inflow alone was 70,542 acre-ft/yr.

Shallow aquifer storage change was computed from a plot of average ground water fluctuations (hydrographs) after subdividing the area into polygons (using Thiessen's method) with the shallow wells as the center of each polygon.

The average head (H) for each polygonal area for the time period in question was computed from

$$H = \frac{\sum\limits_{i=1}^{n} A_i h_i}{\sum\limits_{i=1}^{n} A_i}$$

where

A_i = polygonal area [acres],
h_i = water level elevation as measured from the individual well hydrograph [ft],
n = total number of wells measured (74).

The southern recharge area (Foothill Zone) was analyzed separately in order to compare its fluctuations with the fluctuations of the total area of the plain.

The phreatic aquifer storage change was computed from the following equation:

$$\Delta V = A\theta\delta \, ,$$

where

A = total aquifer surface area [acres],
θ = effective porosity,
δ = slope of hydrographs [ft/yr].

Foothill Zone

$$\Delta V = 87,237 \times 0.06 \times 0.0116 = 61 \text{ acre-ft/yr} \, .$$

Total Plain Area

$$\Delta V = 308,912 \times 0.04 \times (-0.00241)$$
$$= -30 \text{ acre-ft/yr} \, .$$

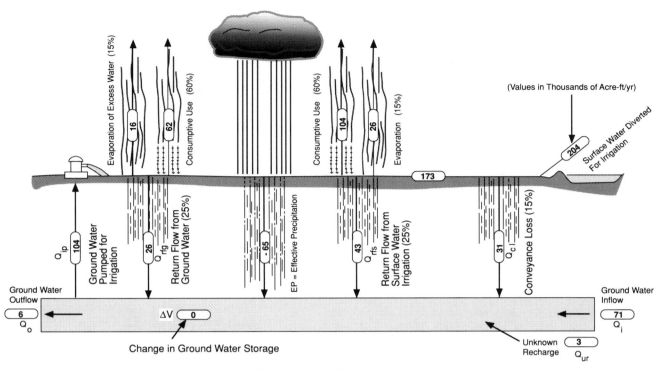

Figure 20.4. Hydrologic equilibrium equation components.

As can be see in the calculation, the low slopes indicated that in an average year no significant phreatic aquifer storage change occurred, and so ΔV was assumed to be zero and within the error of estimation of the regional water balance.

Unknown Recharge, Q_{ur}. The error term in the analysis is the sum of all hydrologic components not accounted for in Equation 20.2. The major portion probably consists of natural recharge components neglected in the subsurface inflow/outflow calculations.

Because unknown recharge is the most intangible item in the equation, a balance was forced at its expense. The resulting value of 3477 acre-ft/yr indicates a slight gain in recharge.

Safe Yield Estimate. The safe yield estimate for this example was subject to the following assumptions:

1. Pumping could lower the water table to 12 to 15 ft below land surface, thus eliminating losses due to evapotranspiration. (This would result in the effective precipitation EP = PI = 113,679 acre-ft/yr with a net savings of 178,221 acre-ft/yr.)
2. Ground water recharge from all sources could be utilized.
3. With the exception of the initial decline in the water table and piezometric surface, ground water would not be produced from storage ($\Delta V = 0$).

4. Ground water outflow from the basin is assumed to be zero.

The resulting annual safe yield (SY) was calculated as

$$
\begin{aligned}
SY &= Q_i + Q_{cl} + Q_{rfs} + Q_{rfg} + Q_{ur} + EP \\
&= 70{,}542 + 30{,}649 + 43{,}420 + 25{,}947 \\
&\quad + 3477 + 113{,}679 \\
&= 287{,}714 \text{ acre-ft/yr .}
\end{aligned}
$$

Figure 20.4 summarizes components in the hydrologic equilibrium equation example.

20.3 METHODS OF GROUND WATER MANAGEMENT

Developing a Plan

A well-organized plan is essential to any ground water management program. Because the plan should relate all necessary tasks, resources, and time, use of "critical path methods" (CPM) is encouraged.

Typically, a management program consists of three major phases:

1. Reconnaissance phase (general overall review).
2. Economic feasibility phase (costs, ranking, and recommendations).
3. Final design phase (preparation of detailed plans for implementation and operation).

Reconnaissance Phase. After project boundaries are identified, and special interests or problems reviewed, a reconnaissance study dealing with the general aspects of the area is initiated. Reconnaissance studies provide the necessary geologic, hydrologic, and economic data on which alternative development schemes are based. Methodology for performing a ground water reconnaissance study is found in Chapter 3.

Economic Feasibility Phase. Future water use in an area requires estimating water needs for the agricultural, municipal, and industrial components. Agricultural water demand is a function of irrigation methods as well as consumptive use of crops.

A ground water management plan must be compatible with and acceptable to the society in which it is to be implemented. An effective plan should also satisfy basic engineering and economic criteria ("Will it work and can we afford to do it?"). The result should be neither a continuing drain on the community's resources nor a laboratory research project. Finally, once hydrogeologic conditions are known and environmental constraints satisfied, cost/benefit analysis of alternatives are made and ranked in order of recommended implementation.

Conjunctive Use of Surface and Ground Water

Conjunctive use of surface and ground water is a management technique designed to maximize use of available water resources. It requires a coordinated operation plan for both surface and ground water designed to meet demands while ensuring maximum conservation. Plans vary from percolation of natural stream flows to complex programs involving inter- and intrabasin water transfers, with facilities for recharge, extraction, and distribution.

Some important benefits of conjunctive use are:

1. Reduced surface-water storage facilities.
2. Water conservation.
3. Smaller surface-water networks.
4. Less evaporation loss.

Interbasin Transfers of Water. In many areas of the world low precipitation rates, coupled with limited natural surface-water supplies, require importation of water from long distances. For example, in California aqueducts bring water hundreds of miles from areas where surface water is abundant to the southern semiarid region. This water is either consumed directly or stored in ground water reservoirs for later recovery (6). Figure 20.5 shows major California aqueducts.

Intrabasin Transfers of Water. Complex geologic and hydrologic conditions exist in most ground water development areas. For example, it may be possible to overdraft one area while excessively recharging another, and still not exceed safe-yield values predicted by regional ground water budget calculations. Detailed basin analysis is necessary therefore to delineate areas of excess or deficiency and design optimum pumping, distribution, and recharge programs.

Artificial Recharge and Seawater Barriers. Storing surface water underground for future use is an established practice in a conjunctive-use program. As discussed in the preceding chapter, ground water recharge is accomplished by inducing percolation of surface water, thereby replenishing underlying aquifers.

When near-coastal pumping creates depressions in water levels, seawater migrates inland, contaminating aquifers. Protection of coastal aquifers against seawater intrusion requires that either a ridge of "protective elevations" be constructed through use of a line of injection wells or a pumping trough be utilized to intercept intruding seawater (see Figure 20.1).

Indirect Recharge through In-Lieu Pumping. One ground water management technique practiced today makes use of an indirect method of recharge. This technique encourages or requires ground water pumpers to purchase imported water in lieu of pumping. In effect this is equivalent to recharging the basin by that quantity not pumped. In-lieu pumping programs are made effective by keeping costs of imported supplies equal to or below pumping costs. Such programs are implemented periodically by ground water basin managers to regulate water levels.

Control Well Fields

Another technique used to conserve ground water is through use of "control well fields." Control well fields are strategically placed to produce interference effects for the control of hydraulic gradients and induce desirable ground water flow directions. Control well fields typically control outflow from basins or contain contaminant plumes.

20.4 GROUND WATER MODELING

In this century the use of ground water has expanded worldwide, as have management techniques such as artificial recharge and conjunctive use for the exploitation of many basins. A serious threat to continued availability of this

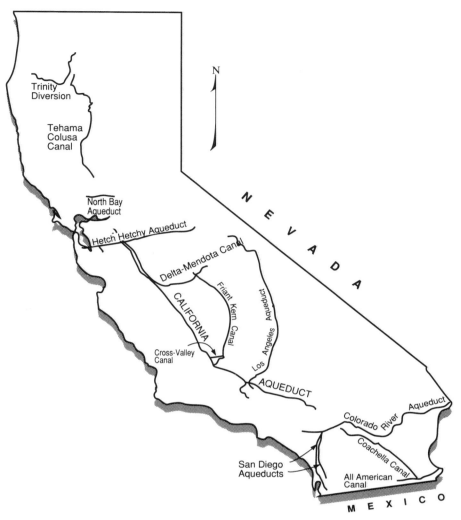

Figure 20.5. Major California aqueducts.

resource is contamination of human origin and from natural sources. Identifying the extent of contamination, restoring aquifers, and protecting them are serious problems facing ground water basin managers. Ground water modeling has emerged as a powerful tool to help managers optimize use of, as well as to protect, the ground water resource. Specific model uses include well field interference analysis, conjunctive use studies, prediction of the effects of artificial recharge, and tracing the movement of contaminants.

Figure 20.6 shows the evolution of models in ground water studies. Between the 1890s and 1950s several types of physical models were developed to solve specific problems. By the 1950s the growth of analytical theory coupled with electric analog/models expanded modeling to a basinwide level. Today computer simulation models and interactive software programs are commonplace and have become the primary tool employed in the industry for solution of small-and large-scale ground water problems and for the management of ground water.

Ground Water Basin Models

Ground water basin models can be divided into two main groups:

1. Lumped parameters (data are constant throughout the model, or vary in two directions at the most).
2. Distributed parameters (data vary throughout the model, usually on a discrete grid-wise basis).

Most ground water basin models employ the techniques of the distributed-parameter problem. Interactive data management thus can be applied to different complex field situations in conjunction with variable-grid multilayered models.

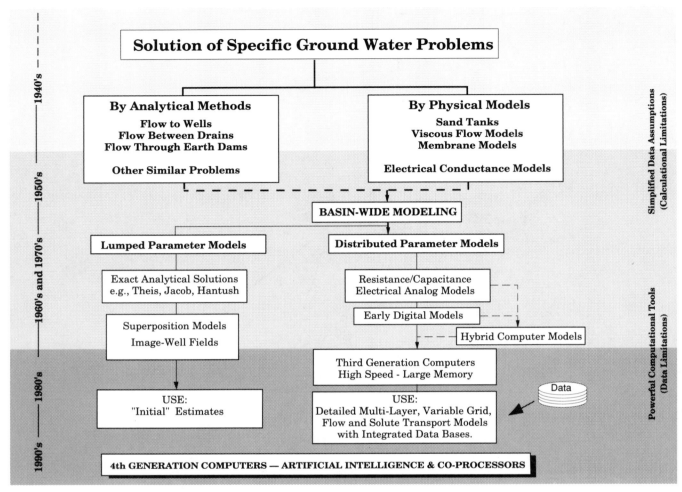

Figure 20.6. Evolution of ground water models.

The distributed-parameters approach requires that ground water flow and quality be simulated through numerical approximation methods (finite difference) which are then applied to the basic equation of ground water flow. The numerous calculations are solved by using a digital computer. Appendix K provides a complete description of the algorithms and methods used to simulate basinwide flow and solute transport.

Model Justification

The decision to model an area should be based on a real need for information, provided that useful and reliable data parameters are available. Sometimes it may seem fashionable to model an area for the sake of modeling alone when there is already sufficient information for decision making. This can be a detriment as there is a frightening willingness to blindly accept model results. An example of the thought processes involved in the modeling plan is illustrated in Figure 20.7.

Conceptualization

How well the model actually represents field conditions depends on how well it is conceptualized. For example, mathematical simulation of water-level responses in a large tank filled with spherical glass beads of a known diameter could be accurately modeled. In this environment the physical hydrologic system could be conceptualized as a single-layer problem with well-known and definable boundaries.

In a typical field problem the conceptualization process is not as easy. The job of the modeler is to define hydrogeologic conditions within the modeled area in the best possible way while staying within the economic and time constraints of the project. This includes not only estimates of the formation parameters (transmissivity, storativity, and leakance) but all hydrogeologic boundary conditions affecting the flow regime within the modeled area. In addition to hydrogeologic data, information must also be gathered on human activities affecting model results such as pumping and artificial recharge. The conceptualization process can be quite formidable, at times

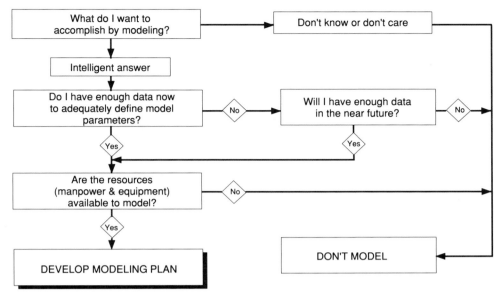

Figure 20.7. Model Planning.

requiring many months of study if model results are to be reliable. Figure 20.8 is an example of model conceptualization.

Model Grid Size, Number of Layers, and Time Steps

In a finite difference network, a node is the center of a rectangular (or square) finite difference grid block. In a finite element model, a node is the point of intersection of adjacent finite element grids. A model layer corresponds to a definable aquifer or group of aquifers combined to facilitate modeling.

The size of the model grid is selected according to the available data and model purpose. For example, if the model is for regional ground water planning of a 500-mi^2 or larger area, grid blocks of one square mile or more might be adequate. If, on the other hand, detailed interference calculations between wells in a multiple-well field is desired, grid spacing of a few hundred feet might be appropriate.

Sometimes there is a tendency is to "over-model" an area by creating many more nodes and layers then is necessary. In principle this is not harmful, provided that the conceptualization is correct and computer power adequate. However, the data and time required for calibration and operation are directly proportional to the number of nodes and layers.

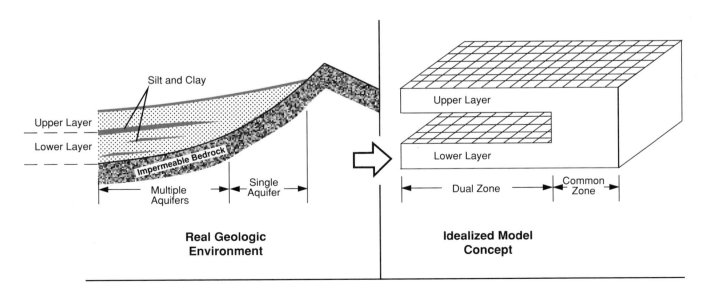

Figure 20.8. Model conceptualization.

Appropriate time intervals must be selected for the intended purpose of the model. Varied intervals are used for problems involving continuously changing pumping rates, with small increments used during pumping start-up or recovery.

Common and Multiple Zones

In practice, the ideal aquifer (pure confined or unconfined) rarely exists. Confined zones typically intermingle with unconfined or semiconfined zones, forming complex alluvial basins. Typically, coarse-grained deposits are found near mountain fronts where transport velocities of streams are high. Farther from the mountains, the streams lose their carrying power, and progressively finer materials are deposited. In addition changes in climatic and tectonic conditions result in interbedding of coarse- and fine-grained materials.

Coarse materials near mountains generally form unconfined aquifers with good recharge. Confinement may occur in these zones, but it is usually localized and all aquifers are considered interconnected or "common." Interbedded aquifers in the centers of valleys typically show confinement and semiconfinement depending upon the nature of the materials and the distance from the source of deposition. Recharge to these multiple-aquifer zones originates as percolation in the common zone and moves down-gradient, recharging the aquifers depending on their hydraulic gradient and conductivity.

Boundary Conditions

Two main boundary conditions are generally encountered in modeling: constant head (Dirichlet condition) and constant flow (Neumann condition).

Constant Head. Within the model area, regions may exist where surface water bodies (rivers, streams, lakes, etc.) are in hydraulic continuity with the aquifer system. In such regions ground water levels are "held" at the elevation of the surface-water body regardless of water-level fluctuations in the aquifer. These types of boundaries are known as "constant head boundaries."

Constant Flow. Constant-flux (flow) model boundaries occur where recharge or discharge is constant. Typically, these are found at impermeable barriers ("no-flow" boundaries) or near mountainous areas where a constant-recharge source enters the model area. Constant-flux boundaries are simulated by "sources" (recharge) or "sinks" (discharge) within the model. Figure 20.9 illustrates the common model boundary conditions.

In some cases the value of the source or sink term is dependent on the value of the head in a specified model area. This type of boundary is known as a head-dependent flow boundary and typically occurs in leaky aquifers or near streams with fluctuating water levels. Here influent or effluent

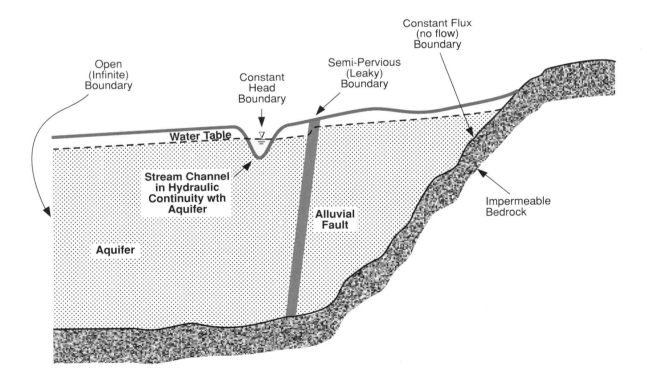

Figure 20.9. Model boundary conditions.

BASIC DATA FILES   **DATA INPUT**   **GEOSTATISTICAL ANALYSIS**   **DATA APPLICATIONS**

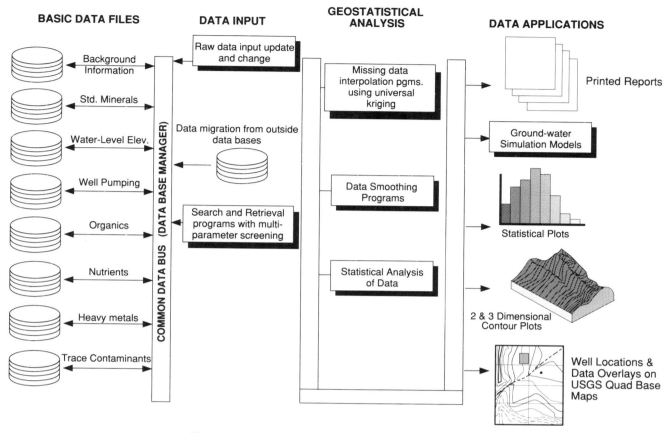

Figure 20.10. Interactive ground water modeling system.

seepage to or from the aquifer is dependent on head differentials existing between the stream and ground water levels.

Interactive Ground Water Modeling

In recent years emphasis on ground water modeling has shifted from the development of flexible physical and analytical models to the area of collection, recording, use, and upgrading of data (6). As more and more data regarding hydrologic and water-quality characteristics of ground water basins are collected, it is important that the data be assimilated into forms that are easily accessible by the models. Thus modern ground water computer models are constructed around the "integrated data-base concept." Computer systems are available that can model ground water reservoirs as well as provide this interactive data-base capability. These systems have the ability to store, retrieve, change, and display pertinent model data in graphic or tabular form. The ability to assimilate and categorize vast quantities of data in this manner results in highly accurate and versatile models. Figure 20.10 shows an overview of an interactive ground water modeling system.

Data Management Plan

A data management plan is required to assimilate and process raw data to a computer-acceptable form. Once this is done, monitoring and predictive analysis of the ground water reservoir is easier. Several steps are required; initial data screening, missing data interpolation, smoothing, and final model data formatting. Figure 20.11 is a simplified schematic of these steps.

In the development of the plan, several data management tools have proved useful:

- Use of digitizers and scanners for fast and accurate data input of water-level elevations, formation parameters, water-quality constituents, geophysical borehole logs, and other types of raw data.

- Migration and translation of existing computer data obtained from other sources such as governmental data bases.

- Categorization of the data based on area, time, chemical constituents, water levels, aquifer type, etc.

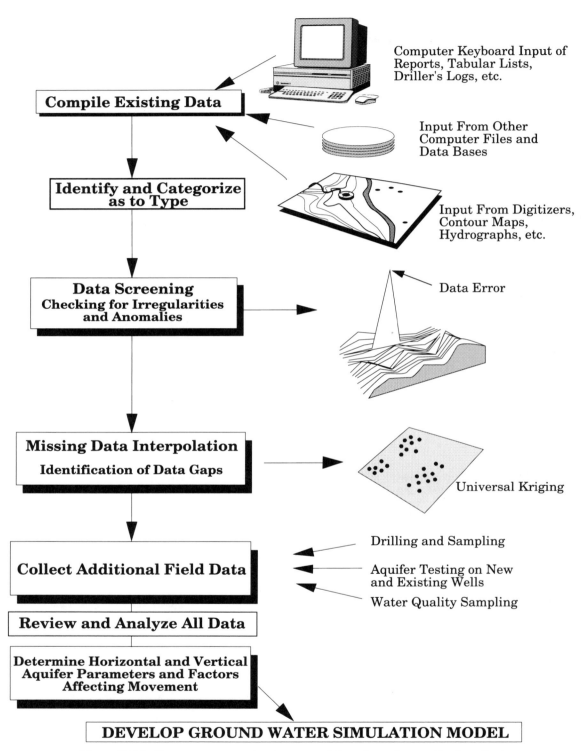

Figure 20.11. Data management plan (courtesy of Geoscience Support Services, Inc.).

- Checking for input and migration errors through use of graphical plotting and statistical checking for data irregularities and anomalies.
- Missing data interpolation, using the geostatistical technique of Universal Kriging or other grid-generating techniques.
- Translation of the final data to model formats.

Compilation of existing data should be oriented toward the intended final use. One primary purpose of the data is to establish reliable initial and boundary conditions for the model. Determination of the number of model layers and the nature of the aquifers, aquicludes, and aquitards is also required. This is usually accomplished by analyses of drillers logs, geophysical borehole logs, pumping test data, geologic maps, and geochemical data.

Identification of the model calibration period depends on the results of this phase, and specifically on the amount and quality of data available. Data gaps may be identified based on model or other requirements and the effort given to filling them.

Model Array Loading. Model simulation requires an orderly arrangement or "layering" of data. Data in finite difference models are easily indexed into a network of rows and columns called "arrays." Once the data necessary for model simulation are determined, they are entered into the model data arrays. One method for accomplishing this is to construct a contour map of the particular data parameter (e.g., water-level elevations or transmissivity) on a base map of the model area. Next, the model grid network is overlaid on the contour map and individual nodal values noted or interpolated. Entering these data into the model data arrays may also be done automatically using a digitizer and appropriate software.

The above technique is generally used for hydrogeologic data parameters. For pumping and recharge data, values are compiled (usually on separate maps) and tabulated by nodes for each model time period. Data is then entered only for those nodes containing pumping or recharge.

Kriging Techniques. In most cases a problem of insufficient data will exist in many areas of the model. The available data must then be interpolated to the missing areas using trends or other appropriate methods. A good approach to interpolation of missing data is through use of a stochastic approach called "Kriging." Kriging is a geostatistical method of interpolation involving relationships between spatial variations of the parameter to be interpolated. This spatial relationship is called a variogram. Theoretical models are then fit to the variograms and are used to estimate parameters in the unknown data areas (7, 8, 9).

Figure 20.12 gives an example of model data constructed from random control points using Universal Kriging.

Ground Water Model Data

Data for ground water models can be divided into three main categories:

1. Geohydrologic data.
2. Natural recharge data.
3. Operational data (pumping and artificial recharge).

Geohydrologic data are those model data that are necessary to physically represent the ground water system. They include:

1. Elevations of model layers.
2. Initial water-level elevations.
3. Hydraulic conductivity.
4. Specific storativity.
5. Effective porosity.
6. Leakance.
7. Initial water-quality concentrations.
8. Dispersion and retardation.

These data specify the model boundary conditions and initial flow, storage, and water quality in each model node.

Operational data impose dynamic conditions on the model. These data include:

1. Pumping.
2. Artificial recharge.
3. Ion concentration of recharge water.

Discussion of the more important model data follows.

Initial Water Levels. The initial water-level distribution is the starting point for model simulation runs. It generally is chosen at a time before the actual period of modeling interest begins, in order to accurately simulate "pre-model" trends that might affect early results. Starting water levels must be input for each node of each model layer.

Transmissivity. For model systems where the saturated thickness remains constant throughout the modeled operation, the transmissivity remains constant. For models of thin, unconfined aquifers where saturated thickness varies during model runs, the transmissivity must be calculated during each time step. This calculation is the product of hydraulic conductivity and the saturated thickness derived from the difference between the water table and the bottom of the aquifer (unconfined) or the layer thickness (confined).

Model transmissivity influences the flow between model nodes. Higher transmissivity results in more water moving into or out of a model node under a given hydraulic gradient.

Storativity. Storativity controls the ability of a model node to absorb or release water, and it varies with aquifer type.

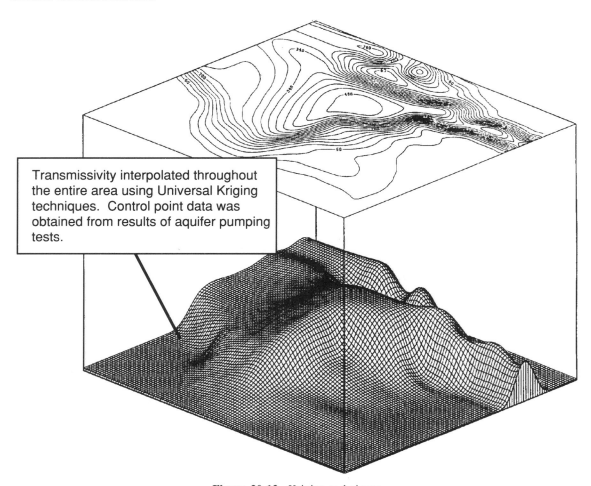

Transmissivity interpolated throughout the entire area using Universal Kriging techniques. Control point data was obtained from results of aquifer pumping tests.

Figure 20.12. Kriging techniques.

In unconfined aquifers, storativity is equal to the effective porosity, but in confined and semiconfined aquifers it is a function of aquifer elasticity and compressibility, as discussed in Chapter 2.

Leakance. The ability of the model to transfer water vertically between layers is governed by leakance. Leakance is a measure of the rate of flow crossing the interface between an aquifer and a semiconfining layer. Leakage may be either upward or downward depending on the difference in heads between the aquifers. In ground water models leakance is quite sensitive, and in large grid areas may result in abnormally large water-level fluctuations for small variations in leakance.

Natural Recharge. In modeling most ground water systems, the shape of the initial water-level elevations reflects a condition of natural recharge. The natural recharge can generally be assumed as a steady-state flow condition existing prior to the start of the simulation period. One method of estimating natural recharge is through the use of Poisson's equation which calculates individual flux (flow) elements for each node in the grid network (see Appendix K). This method eliminates problems of spurious anomalies developing during model simulation due to inaccurate estimates of natural recharge components (e.g., infiltration, influent seepage, etc.). Generally, natural recharge is considered a constant throughout the modeling period. However, relationships may be developed between ground water level elevations and precipitation, and natural recharge recalculated as necessary for more accurate simulations.

Pumping and Artificial Recharge Data. An important component of the model algorithm is the net vertical flux for each node. The net vertical flux is the sum of all pumping and recharge acting on that node for the particular model time step. Leakage is also assumed to act vertically but is treated separately as it is a function of water-level differences. The following are examples of various types of pumping and recharge (i.e., sinks and sources) which may act on the model:

- Pumping from wells (sink).
- Discharge from springs (sink).

- Return flow from irrigation (source).
- Return flow from municipal and industrial uses (source).
- Percolation of storm flow in river channels (source).
- Artificial recharge of natural, imported or reclaimed water (source).
- Percolation of precipitation (source).
- Effluent seepage from an aquifer to a stream or canal (sink).
- Influent seepage from a stream or canal to an aquifer (source).

Dispersion and Retardation. In addition to hydrogeologic data, water-quality models require additional parameters which affect movement of solute in the aquifers. Knowledge of these parameters requires an understanding of the mechanisms of both hydrodynamic dispersion and retardation.

Hydrodynamic dispersion (Figure 20.13) is a non-steady irreversible process which accounts for spreading (usually an elliptical shape) of a solute (e.g., contaminant) both in the direction of average flow (longitudinal dispersion) and in a direction perpendicular to it (transverse dispersion). The phenomenon is considered irreversible in the sense that the initial solute distribution cannot be recreated by simply reversing the direction of average flow. Jacob Bear (10) has shown that on a microscopic scale the process of molecular diffusion is caused by the random movement of fluid molecules from higher concentrations to lower ones. It is molecular diffusion which makes the phenomenon of hydrodynamic dispersion an irreversible process.

In addition to transport of solute at the average velocity (advection), and elliptical spreading due to hydrodynamic dispersion, other factors may affect the flow and concentration distribution. These include adsorption on the solid matrix, deposition, ion exchange, radioactive decay, and chemical reactions within the fluid itself (11).

Chemical reactions between the dissolved constituent and the porous medium tend to retard the movement of the constituents relative to the ground water velocity. The retardation coefficient may be written (11)

$$R_d = 1 + \rho_b K_d / \theta \, ,$$

where

R_d = retardation coefficient,
ρ_b = bulk mass density $[ML^{-3}]$,
θ = effective porosity,
K_d = distribution coefficient $[L^3 M^{-1}]$.

Model Calibration

Model efficacy is confirmed by duplicating a historical period of operation. One procedure used to accomplish this is known as the "history-matching method." The period selected is preferably one with major water-level fluctuations throughout the area modeled. The model will be more reliable with a longer and more varied calibration period. Records of all pumping and recharge are collected and coded into model data arrays and the model is forced to duplicate historic water levels.

Model-generated hydrographs, illustrated in Figure 20.14, or water-level contour maps are then checked against the selected well hydrographs for those nodes where historical data exist, showing where adjustments are necessary. Adjustments in geohydrologic formation parameters (transmissivity, storativity, and leakance) are made, and the run is repeated.

This trial-and-error procedure is continued until a satisfactory match is obtained. There is no hard-and-fast rule to decide when a model is fairly calibrated. The number of runs is a function of the intuitiveness of the modeler, required model accuracy, complexity of the flow system, and nature of the calibration period.

If geohydrologic parameters have been varied to allowable upper or lower limits with no acceptable match, a conceptual error might exist (e.g., leakage should be allowed where none was assumed) or recharge or discharge data might be in error. Where any of these are suspected, it is necessary to recheck the available data resources, to sort out reliable

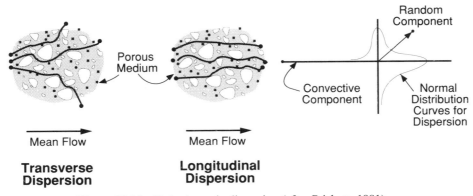

Figure 20.13. Hydrodynamic dispersion (after Prickett, 1981).

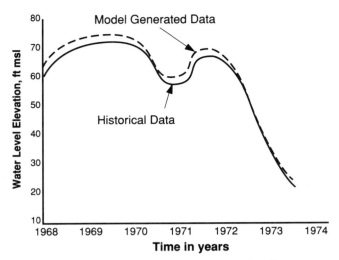

Figure 20.14. Hydrograph for model node.

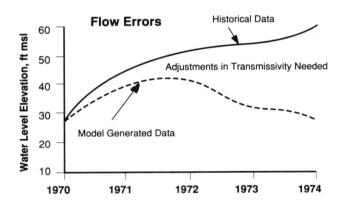

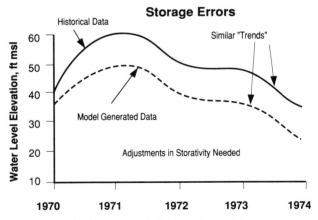

Figure 20.15. Transmissivity and storativity errors.

from unreliable data, and to modify formation parameters and boundary conditions as required. This process can be quite difficult, frequently requiring many months of careful analysis and study.

Sensitivity Analysis. The manner in which changes in data affect model results is known as sensitivity analysis. Models respond differently to changes in data parameters, and knowledge of this sensitivity is helpful during calibration. For example, one model may be insensitive to storativity changes, and estimation of this parameter in unknown areas may not be critical. On the other hand, in a model with high sensitivity to changes in storativity, estimation of unknown data could lead to significant calibration error.

Generally, changes in transmissivity affect the inflow and outflow between model nodes. Transmissivity errors can be identified by dissimilar hydrograph trends. Storativity errors, however, show similar hydrograph trends, but they are shifted above or below historical values, depending on whether storativity was under- or overestimated (see Figure 20.15).

Operational Runs—Predictive Analysis

Once model flow and quality calibrations are complete, different operational scenarios may be run for future periods of interest. The reliability of the results must be viewed with consideration of the degree of calibration performed. Long-calibration periods with good historic and model compatibility increase confidence. Predictions based on modeling must be intelligently qualified.

Development of the operational scenario is problem specific and requires analysis of present and future pumping and recharge patterns.

REFERENCES

1. Meinzer, O. E. 1920. "Quantitative Methods of Estimating Ground-Water Supplies." *Bulletin, Geological Society of America* 31.
2. ASCE. 1972. "Ground Water Management." ASCE Manual No. 40.
3. Todd, D. K. 1980. *Groundwater Hydrology*. J. Wiley and Sons, New York.
4. Louis Berger-Payab Consulting Engineers. 1974. "Behshahr-Neka Project." *Hydrology, Water Resources and Water Quality*. Vol. 3. Imperial Government of Iran, Ministry of Water and Power.
5. Hantush, M. S. 1959. "Potential Evapotranspiration in Areas along the Rivers of New Mexico." New Mexico Institute of Mining and Technology. Professional Paper 101.
6. Williams, D. E. 1986. "Groundwater Modeling in the Orange County Area." *Hydrogeology of Southern California*. Geological Society of America Guidebook. Cordilleran Section, 82nd Ann. Mtg.

7. Aboufirassi, M., and M. A. Marino. 1983. "Kriging of Water Levels in the Souss Aquifer, Morocco." *Mathematical Geology.* Vol. 15, No. 4.

8. Aboufirassi, M., and M. A. Marino. 1984. "Cokriging of Aquifer Transmissivities from Field Measurements of Transmissivity." *Mathematical Geology.* Vol. 16, No. 1.

9. Pucci, A. A., and J. A. Murashsige. 1987. "Applications of Universal Kriging to an Aquifer Study in New Jersey." *Ground Water* (November–December).

10. Bear, J. 1965. "Hydrodynamic Dispersion." Unpublished lectures delivered during a summer program on Hydrology and Flow through Porous Media. Princeton University. August.

11. Bear, J. 1987. *Modeling Ground Water Flow and Pollution.* D. Reidel. Dordrecht, Holland.

Sources of Geohydrological Information and Data

INTERNATIONAL ORGANIZATIONS

International Association of Scientific Hydrology
Rue de Ronces 61, Gentbrugge, Belgium
United Nations
  Food and Agricultural Organization
  Rome, Italy
  Water Resource Development Center
  New York
  UNESCO
  Paris, France
International Association for Hydraulic Research
Netherlands
International Association of Hydrogeologists
Paris, France
International Hydrographic Bureau
Monte Carlo, Monaco
International Water Supply Association
London, England

FEDERAL AGENCIES

Department of Agriculture

Agricultural Research Service
Forest Service
Soil Conservation Services
  Publications: Aerial photography, special reports
  Types of Data: Local geology and structure, areal data and methodology, especially vadose zones and shallow water table aquifers

Department of Commerce

National Oceanic and Atmospheric Administration
National Climatic Data Center

Publications: LANDSAT imagery, NASA aerial photography
Types of Data: Regional geology and structure

U.S. Army Corps of Engineers

Department of the Interior

United States Geologic Survey
Denver Federal Center
Denver, CO
  Publications: Topographic maps, geologic maps, water supply papers, open-file reports, water resource investigations
  Types of Data: Topography, location of natural and physical features, surficial extent of aquifer, geologic structures, areal surface and ground water data, areal surface and ground water data in projects
Tennessee Valley Authority
Bureau of Reclamation
Bureau of Land Management
EROS Data Center

Environmental Protection Agency

National Climatic Center

Publications: Climatological data
Types of Data: Precipitation, temperatures, evaporation

STATE AND LOCAL AGENCIES

Departments of Water Resources

State Geological Surveys

State Engineers

Publications: Areal geohydrological maps and reports
Types of Data: State and local geohydrologic data

Local Water and Flood Control Districts and Agencies

Publications: Varied

Types of Data: Stream flow, surface improvements

PRIVATE AGENCIES

Geological Society of America

Publications: *Bulletin*, special papers

Types of Data: Areal geology

National Water Well Association

Publications: *Ground Water, Ground Water Monitoring Review, Water Well Journal*

Types of Data: Geohydrologic methods and areal studies, investigational methods for ground water contamination, drilling methodologies

American Geophysical Union

Publications: *Water Resources Research, Journal of Geophysical Research*

Types of Data: Hydrological methodology and analysis, geophysical methodological applications

American Water Works Association

Publications: Journal

American Institute of Hydrology

Publications: *Bulletin*, special reports

Types of Data: Hydrologic methodology and data analysis

American Society of Civil Engineers

UNIVERSITIES AND COLLEGES

Geology, Engineering, Natural Resources, Environmental Departments

Publications: Master's theses, Ph.D. dissertations, faculty publications

Types of Data: Areal and methodological analyses

Geographic Ground Water Regions in the United States[1]

WESTERN MOUNTAIN REGION

Included in this region are the northern Coast Ranges, Sierra Nevadas, Cascades, northern Rocky Mountains, and portions of the Basin and Range Province. Mountains are covered with thin soil. Intermontane valleys contain alluvial and some glacial materials. Locally recent volcanics and tectonically fractured crystallines occur.

The dominant aquifer is confined and has fracture porosity, with fractures closing as depth increases. Rocks are largely insoluble. The confined aquifer is recharged by losing streams and by infiltration of precipitation from uplands. Discharge occurs via springs and seeps.

Representative values for aquifer parameters:

Effective porosity	0.01–0.2
Hydraulic conductivity [gpd/ft^2]	0.007–374
Recharge rate [in./yr]	0.1–2
Well yield [gpm]	10–100

THE ALLUVIAL BASINS

Tectonically controlled basins bordered by mountain ranges are partially filled with alluvial and occasionally with glacial materials that have eroded from the mountains. Rainfall is scanty and seasonal, falling during winter and early spring. The dominant aquifer system is a multiple aquifer–aquitard system that may be overlain by an interconnected unconfined aquifer, recharged by losing streams and by direct infiltration at the edges of the alluvial fans. Discharge is via seeps and springs, and by evaporation from playas.

Representative values for aquifer parameters:

Effective porosity	>0.2
Hydraulic conductivity [gpd/ft^2]	750–14,960
Recharge rate [in./yr]	0.001–1
Well yield [gpm]	100–5000

THE COLUMBIA LAVA PLATEAU

The Columbia Lava Plateau is composed of recent volcanics interspersed with alluvium and lake sediments. The dominant aquifer is confined with porosity due to cooling cracks and lava tubes. Recharge is via losing streams and by direct infiltration into exposed lavas. Discharge is by springs and seeps.

Representative values for aquifer parameters:

Effective porosity	<0.01
Hydraulic conductivity [gpd/ft^2]	3740–74,800
Recharge rate [in./yr]	0.2–10
Well yield [gpm]	100–20,000

THE COLORADO PLATEAU AND WYOMING BASIN

Thin soils overlie older (Permian to Triassic) fractured sedimentary sequences of sandstone, siltstone, shale, and limestones. The dominant aquifer is a complex multiaquifer system with sandstones, and occasionally limestones, functioning as aquifers separated by shale aquitards. Although these formations are not very permeable, they are areally extensive. The climate is semiarid; however, considerable rainfall may occur at higher elevations in the form of snow.

Porosity is actually a double porosity system, with major porosity due to fractures in rocks in which some intergranular porosity exists. Recharge is primarily from downward infiltration from uplands, as well as from losing streams. The recharge areas of aquifers that extend across the midcontinent, such as the Dakota Sandstone, occur in this region. Discharge is via springs and seeps, and to other aquifers via leakance through aquitards.

Representative values for aquifer parameters:

Effective porosity	<0.01
Hydraulic conductivity [gpd/ft^2]	0.07–37.4
Recharge rate [in./yr]	0.01–2
Well yield [gpm]	10–1000

---

1. In this appendix the aquifer parameters are from Heath (1982, p. 400). The regions are delineated in Figure B.1.

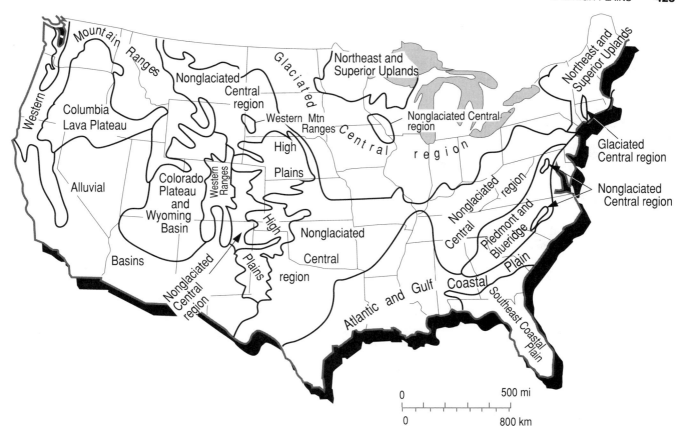

Figure B.1. Geographic ground water regions (from Heath, R.C., "Classifications of Ground Water Systems of the United States" in *Ground Water*, Vol 20, No. 4. July -Aug 1982, Copyright by the American Geophysical Union).

THE HIGH PLAINS

Thick alluvial deposits of the Ogallala Formation formed from erosion of the eastern front of the Rocky Mountains overlie fractured sedimentary rock units similar to those in the Colorado Plateau Region. Climate is semiarid throughout this region, with winter precipitation and brief but intense storms during spring and fall. In the north precipitation has been sufficient for recharge. In the south agricultural demand has caused major overdraft and quality deterioration.

The dominant aquifer, the Ogallala Formation, is unconfined, composed of insoluble, unconsolidated, or poorly consolidated alluvial materials. Porosity is intergranular. Recharge is by direct infiltration and by leakance from underlying confined aquifers. Discharge is via springs and surface seepage.

Representative values for aquifer parameters:

Effective porosity	>0.2
Hydraulic conductivity [gpd/ft^2]	750–7500
Recharge rate [in./yr]	0.2–3
Well yield [gpm]	100–3000

NONGLACIATED CENTRAL REGION

Thin, unconsolidated soils composed of weathering products and water- or wind-deposited materials overlie fractured sedimentary rocks.

Fluvial deposits form local aquifers. The dominant aquifer is a multiaquifer system composed of sedimentary rocks that in some areas is overlain by an unconfined aquifer. Total porosity is composed of intergranular and fracture porosity. Soluble karstic limestones provide additional porosity in the form of solution channels and caverns.

Recharge is from direct infiltration of precipitation and by upward leakance from deeper, confined, fractured aquifers and solution channels. Discharge is to the surface via springs and seeps, as well as by leakance into deeper aquifers.

Representative values for aquifer parameters:

Effective porosity	<0.01
Hydraulic conductivity [gpd/ft^2]	75–75,000
Recharge rate [in./yr]	0.2–20
Well yield [gpm]	100–5000

GLACIATED CENTRAL REGION

Thick glacial deposits overlie sedimentary sequences similar to those found in nonglaciated regions. The glacial drift occurs as outwash plains and bedrock valley fills. Although much of the glacial deposits are fine grained, outwash deposits occur as sand and gravel layers. Paleozoic sedimentary aquifers of marine origin occur at or near the surface, and are areally extensive, but may contain highly saline ground water.

The dominant aquifer is actually a complex, interbedded multiaquifer system composed of glacial outwash that is separated by thin, discontinuous aquitards, as well as by sedimentary rocks. Porosity is intergranular in the glacial deposits and secondary in the sedimentary rocks due to fractures and solution cavities.

Representative values for aquifer parameters:

Effective porosity	0.01–0.2
Hydraulic conductivity [gpd/ft^2]	37–7500
Recharge rate [in./yr]	0.2–10
Well yield [gpm]	50–500

PIEDMONT AND BLUE RIDGE

This region is composed of crystalline and metamorphosed sedimentary rock covered with a thick unconsolidated covering of weathering and alluvial sediments. Precipitation is high and occurs throughout the year. The dominant aquifer system is composed of an unconfined aquifer overlying a confined aquifer. Wells are located in the alluvium and in more permeable sandstone units.

Representative values for aquifer parameters:

Effective porosity	<0.01
Hydraulic conductivity [gpd/ft^2]	0.02–22
Recharge rate [in./yr]	1–10
Well yield [gpm]	50–500

NORTHEAST AND SUPERIOR UPLANDS

Crystalline and some metamorphosed sedimentary rocks are overlain by glacial drift. Precipitation is high and occurs throughout the year. The dominant aquifer system is confined and composed of the more permeable sandstone units. Locally, glacial outwash may form aquifers.

Porosity is composed of intergranular porosity and fractures in the crystalline and metamorphic rocks. Recharge is primarily due to leakance through confining layers from overlying formations, whereas discharge is via springs and seeps, or by leakance to other aquifers.

Representative values for aquifer parameters:

Effective porosity	<0.01
Hydraulic conductivity [gpd/ft^2]	37–750
Recharge rate [in./yr]	1–10
Well yield [gpm]	20–200

ATLANTIC AND GULF COASTAL PLAIN

This region includes the Atlantic Coast from Cape Cod southward through the Carolinas, as well as the Gulf Coast. Deposits are primarily unconsolidated and consolidated interbedded sands, silts, and clays. In the northeast, glacial outwash deposits form major aquifers, whereas in the southeast cavernous limestones occur. The climate is generally humid.

Porosity is primarily intergranular, with some contribution from solution cavities. Recharge occurs through direct infiltration of precipitation or by leakance from deeper formations. Discharge is primarily via seeps and springs. Overdraft in some areas has caused intrusion of ocean water into the aquifers.

Representative values for aquifer parameters:

Effective porosity	0.01–0.2
Hydraulic conductivity [gpd/ft^2]	75–2990
Recharge rate [in./yr]	2–20
Well yield [gpm]	100–5000

SOUTHEAST COASTAL PLAIN

This region is dominated by the Floridan Aquifer, a cavernous limestone formation that occurs throughout Florida and extends into Mississippi and Alabama. Unconsolidated sands and clays overlie the semiconsolidated carbonate rocks. The climate is humid, with precipitation occurring throughout the year.

Porosity is intergranular in the sands and due to solution cavities in the limestone. Recharge is from direct infiltration of precipitation and by leakance through adjacent confining layers. Discharge occurs via springs, some of which flow at many thousands of gallons per minute, forming streams and rivers. At some localities, overdraft has caused formation of sinkholes.

Representative values for aquifer parameters:

Effective porosity	>0.2
Hydraulic conductivity [gpd/ft^2]	750–75,000

Recharge rate [in./yr] 1–20
Well yield [gpm] 1000–20,000

ALLUVIAL VALLEYS

This is not a region in a geographical sense, but it is composed of all geohydrologically significant alluvial valleys, regardless of where they occur. Alluvial valleys throughout the coterminous United States provide significant amounts of ground water and are among the most intensively studied geohydrological environments. They are composed of water-deposited sand and gravel deposits beneath the flood plains and stream terraces.

The dominant aquifer system is an unconfined to semiconfined aquifer composed of insoluble sands and gravels with intergranular porosity. Recharge is by direct infiltration of precipitation, whereas discharge is via springs and seeps.

Representative values for aquifer parameters:

Effective porosity	>0.2
Hydraulic conductivity [gpd/ft^2]	750–37,400
Recharge rate [in./yr]	2–20
Well yield [gpm]	100–5000

HAWAIIAN ISLANDS

The islands of Hawaii are made up of thick sequences of recent lavas interbedded with less permeable ash deposits that form perching structures. The lava flows are segmented by dikes that may act as barriers to lateral ground water movement. In some areas the lavas are covered by alluvium. Precipitation is heavy and occurs throughout the year.

Porosity is principally due to cooling cracks and lava tubes. Recharge is by direct infiltration of precipitation and from losing streams. Discharge is via springs that form at the exposed ash flow contacts. They are often ephemeral (occurring during and shortly after rainfall events) because of the very high hydraulic conductivity of the lavas. Below sea level, the lavas are saturated with ocean water. Freshwater, because of its lower density, forms a double convex lens floating upon the saline water. This is known as a "Ghyben-Herzberg lens."

Representative values for aquifer parameters:

Effective porosity	>0.2
Hydraulic conductivity [gpd/ft^2]	3740–75,000
Recharge rate [in./yr]	1–40
Well yield [gpm]	100–5000

ALASKA

Alaska is treated as a single ground water region, although it is geohydrologically complex. Glacial and alluvial deposits form unconfined aquifers. In the upper zones ground water may remain frozen (permafrost) throughout the year. Crystalline, sedimentary, and metamorphic rocks lie beneath the alluvial and glacial deposits.

Porosity is chiefly intergranular in the unconfined aquifers, but may be due to fractures or even solution cavities in the ice. Recharge occurs from direct infiltration of precipitation or snow melt and also from losing streams. Discharge is via springs and seeps.

Representative values for aquifer parameters:

Effective porosity	0.01–0.2
Hydraulic conductivity [gpd/ft^2]	750–14,960
Recharge rate [in./yr]	0.1–10
Well yield [gpm]	10–1000

Methods of Estimation of Value of Parameters of the Hydrologic Budget

METHOD FOR DETERMINING HYDROLOGIC PARAMETERS

Parameter: Precipitation
  Point of Measurement or Estimate
    Precipitation gauge
  Volume Estimate
    Uniform—Point value times study area
    Nonuniform—Theisson polygonal isohyets[1]

Parameter: Surface-water inflow
  Point of Measurement or Estimate
    Stream hydrographs and channel geometry
  Volume Estimate
    Graphical analysis of hydrograph
  Simplifying Assumptions
    By including the head of the basin in the study area, $SW(i)$ may be considered zero

Parameter: Ground water inflow
  Point of Measurement or Estimate
    Flow net analysis along upper gradient boundary
  Volume Estimate
    Multiply estimated unit flow by study area
  Simplifying Assumptions
    By including the head of the basin in the study area, $GW(i)$ may be considered zero

Parameter: Other inflow sources
  Volume Estimate
    Injection or flow records

Parameter: Potential evapotranspiration
  Point of Measurement or Estimate
    Thornwaite, Blaney-Criddle, or other methods for which data are available[1]
  Volume Estimate
    Multiply point estimate by study area

[1] Equations and methods for performing these analyses may be found in most climate or hydrologic texts.

Parameter: Actual evapotranspiration
  Point of Measurement or Estimate
    Bimonthly summation of P − PE

Parameter: Surface-water outflow
  Point of Measurement or Estimate
    Stream hydrograph
  Volume Estimate
    Graphical analysis of hydrograph

Parameter: Ground water discharge
  Point of Measurement or Estimate
    Flow-net analysis along lower gradient boundary
  Volume Estimate
    Multiply estimated unit flow by study area
  Simplifying Assumptions
    Hydrograph analysis yields both surface-water outflow and base flow

Parameter: Withdrawals
  Volume Estimate
    From pumping and outflow records

Parameter: Change in storage
  Point of Measurement or Estimate
    Static water level elevation at beginning and end of time period considered in budget
  Volume Estimate
    Volume of aquifer materials represented by SWL difference multiplied by study area and average porosity
  Simplifying Assumptions
    Using long-term average data, change in storage may be considered zero

FLOW NETS

Flow nets are used in the estimation of ground water inflow (see Figure C.1). As such they are used to determine the

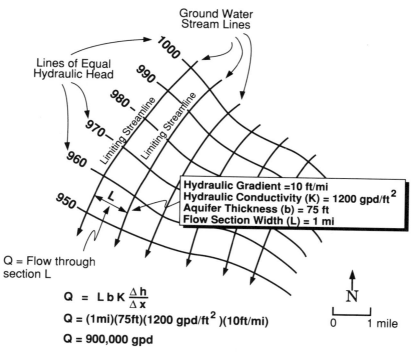

Figure C.1. Example of flow net.

parameters of inflow and outflow described in the preceding section. In addition the use of flow nets has been found useful to qualitatively evaluate multiple-aquifer systems, revealing aquifer relationships not easily envisaged and indicating direction of additional work. Changes in direction of flow vectors and widening or concentration of the vectors indicate influence by geohydrologic factors, such as barriers, and changes in the hydrologic parameters of the aquifers. This may alter the developing concepts of the geohydrology and suggest that further investigation is necessary in specific areas.

Figure C.2 shows such a multiaquifer system with numerous recharge and discharge zones and indicated leakage through an aquitard.

To prepare a flow net for a region, it is necessary to have sufficient ground water elevation and aquifer test data. An assumed one-unit-wide vertical section of the aquifer is usually modeled. Actual values are then obtained by extrapolation of data for the study area. The following assumptions are made:

1. Aquifer materials are homogeneous and isotropic.
2. Aquifer materials are saturated and incompressible.
3. Ground water flow is at steady state (there is no change in head over time and the velocity is relatively low, ensuring laminar flow).

Boundaries may be of three types:

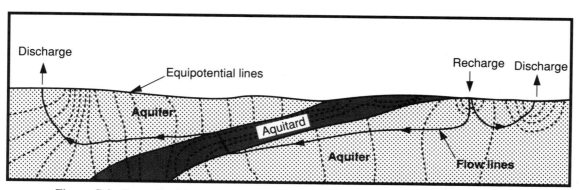

Figure C.2. Example of flow nets constructed from multi-aquifer systems data (after Freeze and Witherspoon, 1967).

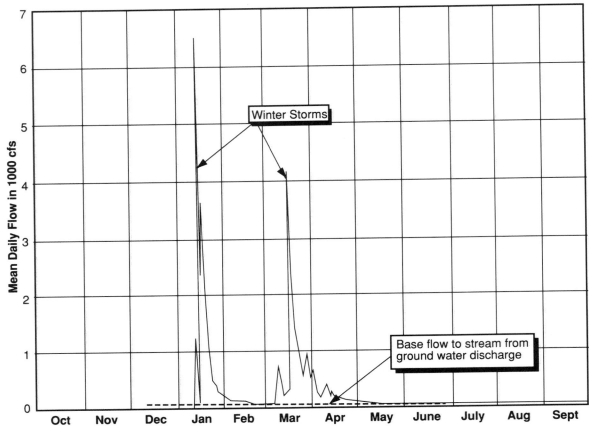

Figure C.3. Stream hydrograph of an area characterized by wet winters and dry summers.

1. Impermeable, or no-flow, boundaries. Equipotential lines meet impermeable boundaries at right angles. Flow lines are sketched parallel to impermeable boundaries.
2. Constant head boundaries, such as streams. Equipotential lines are parallel to such boundaries while flow lines cross them at right angles.
3. The water table, which is neither an equipotential nor a flow line.

Flow nets are also discussed in Chapter 2.

BASE FLOW RECESSION

One way to measure the amount of natural discharge under specific conditions is through base flow recession analysis.

Figure C.3 illustrates the hydrograph of a stream that drains an area characterized by wet winters and dry summers. The base flow of a stream occurs through ground water discharge. It is considered to commence whenever the elevation of the stream surface declines below the elevation of the water table adjacent to the stream. The characteristic hydrograph shows a logarithmic decline (which graphs as a straight line on semilogarithmic paper) with time on the arithmetic abscissa and stream flow on the logarithmic ordinate.

In the absence of extensive ground water extraction or rainfall (when the aquifer may be considered in a steady-state condition), average annual ground water recharge into an aquifer is equal to average annual ground water discharge so that recharge may be calculated based on stream discharge. In a small basin drained by a single stream, this method probably yields a satisfactory estimate. However, in more complex settings it may severely underestimate the portion of the recharge that enters deep, confined aquifers.

Grain Size Analysis to Estimate Hydraulic Conductivity

Hydraulic conductivity (K) may be defined as:

$$K = \frac{cd_{50}^2 g}{\nu}$$

where

ν = kinematic viscosity $[L^2 T^{-1}]$,

d_{50} = median grain diameter (50% passing) [L],

c = factors influencing hydraulic conductivity (grain shape, sorting, packing, etc.).

In grain size analysis the term $d_{(p\%)}$ may be used to denote the size of the sieve that will pass $p\%$ of the sample. For example, d_{90} indicates the diameter of the screen through which 90% of the sample passes, whereas d_{50}, which allows 50% of the sample to pass, is known as the median grain diameter (see Chapter 13).

Masch and Denny used the median d_{50} as a measure of average particle diameter with σ_I as a measure of dispersion (spread of grain size). $\phi(p)$ are grain size diameters retained at $p\%$ levels, expressed in phi units (Md_{50}), where ϕ (phi) is the negative of the logarithm to the base 2 of the particle diameter. The phi units listed below are from Masch and Denny (1966) (1).

d_{50} (mm)		ϕ
$\frac{1}{2}$	$= 2^{-1}$	1
$\frac{1}{4}$	$= 2^{-2}$	2
$\frac{1}{8}$	$= 2^{-3}$	3
$\frac{1}{16}$	$= 2^{-4}$	4

The grain size analysis on which this is based shows grain diameter expressed as phi units on the abscissa, while the ordinate expresses the cumulative percent coarser grain size. For well sorted samples, σ_I is close to zero, but for poorly sorted samples, it may rise to 4.

Masch and Denny were able to relate the hydraulic conductivity of samples measured in a laboratory with the median grain diameter and dispersion expressed as σ_I in families of curves as shown in Figure D.1. From these curves accurate

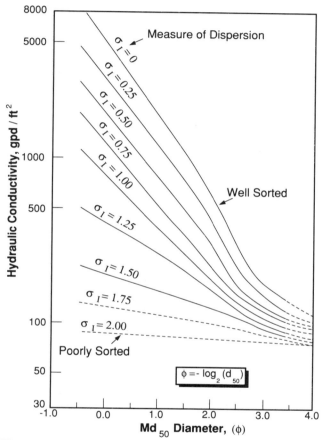

Figure D.1. Masch-Denny prediction curves (after Masch, F. D. and K. J. Denny, "Grain Size Distribution and its Effect on the Permeability of Unconsolidated Sands"). *Water Resources Research*, Vol. 2, No. 4, 4th Qrtr, 1966. Copyright © by the American Geophysical Union.

predictions of the hydraulic conductivity of other samples may be obtained by interpolation.

REFERENCE

1. Masch, F. D., and K. J. Denny. 1966. "Grain Size Distribution and Its Effect on the Permeability of Unconsolidated Sands." In *Water Resources Research*, 2 (Fourth Quarter), pp. 665–677.

Analysis of Factors Causing Borehole Deviations with Rotary Drilling Systems

The primary cause of failure to maintain proper alignment with rotary drilling systems is an improperly designed drill string combined with an inappropriate bit load. In any drilling situation there are formation effects or heterogeneities that will tend to cause a drill string to deviate out of alignment. However, the effect of such perturbations can be readily compensated.

The formation effects that cause alignment changes are well known and have long been recognized. However, sloping formations—the major difficulty in oil-well drilling—are generally not a problem in water-well drilling, since they are not encountered in unconsolidated alluvium. Rather, they are the heterogeneities engendered by boulders and cobbles. Once deflected, the bit will continue to drill in a direction off the original alignment, unless corrective action is taken.

In order to see how plumbness can be restored, it is necessary to understand how the drill bit load and "pendulum" effect of the drill collar weight interact to bend the bottom-hole assembly (bit, drill collars, stabilizers, reamers, etc.) and thereby to control the drilling direction. These two basic forces on the drill string are clearly not independent, and their interrelationship was first described in a formal and comprehensive way by Lubinski and coauthors in a series of papers in *API Drilling and Production Practice*, beginning in the early 1950s (1, 2, 3). The fundamental difficulty arises because a drill string is a long, slender column whose weight is supported by the rig and the bit. If the bit load is too high, the column will buckle under its own weight. As soon as buckling occurs, a lateral load is placed on the bit, causing it to drill in a direction other than along the center line of the hole. This situation is depicted schematically in Figure E.1.

One solution is to reduce the load on the bit so that the drill string remains straight. However, since penetration rate is dependent on bit load, reducing the load will lower the rate of penetration. A better solution is to stiffen the drill string by increasing its diameter at the bottom-hole assembly. This is done by adding drill collars and stabilizers as shown schematically in Figure E.2. This increases stiffness and the available weight on the bit, which augments both the bending

moment on the bottom-hole assembly and the rate of penetration.

The early papers by Lubinski et al. provide a detailed analysis of forces acting on the drill string and present mathematical methods for predicting the direction of drilling under any given set of circumstances (1, 2, 3). Although a complete mathematical analysis will not be presented here, discussion of the principles involved is useful.

There is a point along the drill string where it is tangent to the borehole wall and not curved, so there is no bending moment at the tangency point. With reference to Figure E.3, it can be seen that the forces acting on the drill string below this point are (1) the formation reaction, which has horizontal and vertical components $\mathbf{F}_h$ and $\mathbf{F}_v$, respectively, (2) the stresses in the drill column at point A, which result from both axial and shear forces, $\mathbf{T}$ and $\mathbf{S}$, respectively; (3) the weight of the drill pipe, collars and bit, and (4) the difference in hydrostatic pressure on the upper and lower surfaces of the collars and drill pipe (the Archimedian uplift from the drilling fluid in which the drill string is submerged).

The effect of these forces is to generate a shear stress and a bending moment in the drill string that are resisted by its rigidity. If y is the displacement at point x of the drill string from the line of borehole, then the bending moment, $\mathbf{M}$, in the drill string is

$$\mathbf{M} = EI\,\frac{\partial^2 y}{\partial x^2}\,, \qquad (\text{E.1})$$

where

E = drill string elastic modulus,

I = sectional moment of inertia of the drill pipe.

The bending moment is imposed by the reaction forces on the bit, and the submerged weight of the drill string.

Consider the equilibrium of the drill string element δx with submerged weight per unit length p, as shown in Figure E.3. If T is the axial force and S the shear force in the string, then force and moment equilibrium equations can be

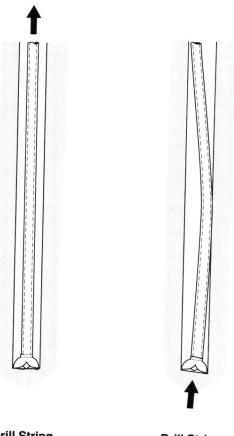

**Drill String
in Tension**   **Drill String
in Compression**

Figure E.1. Buckling of drill string.

to provide five equations for the five unknowns **S**, **T**, **M**, y, and θ. The solution of these equations requires the values of the five unknowns to be specified at the top and bottom of the section of drill string being considered. At the bottom of the string, the bending moment is zero, and the displacement is specified by the diameters of the borehole and the drill collar. The horizontal and vertical reactions at the bit specify the shear and axial forces in the string in terms of the drilling direction. At the upper end the bending moment and deflection are both zero, the drill string slope is defined by the borehole slope, and the shear force and axial load are specified by the overall equilibrium of the string, including

written such that, in the limit of δx becoming very small, the equations reduce to

$$\frac{\partial \mathbf{T}}{\partial x} + \mathbf{S}\frac{\partial \theta}{\partial x} + p \cos \theta = 0 \ ,$$

$$\frac{\partial \mathbf{S}}{\partial x} - \mathbf{T}\frac{\partial \theta}{\partial x} + p \sin \theta = 0 \ ,$$

$$\frac{\partial \mathbf{M}}{\partial x} - \mathbf{S} = 0 \ . \qquad \text{(E.2)}$$

These are coupled with the beam stiffness equation,

$$\frac{EI\partial^2 y}{dx^2} = \mathbf{M} \ , \qquad \text{(E.3)}$$

and the beam slope equation,

$$\frac{\partial y}{\partial x} \tan \theta = \theta \ , \qquad \text{(E.4)}$$

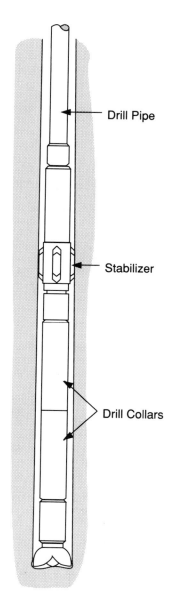

Figure E.2. Stiffening of drill string.

Drill Pipe

Stabilizer

Drill Collars

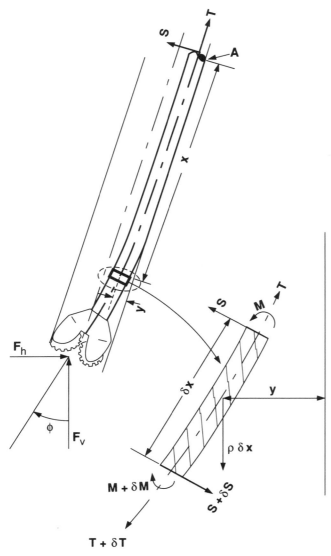

Figure E.3. Forces on the drill string.

deflection of the bit is reduced. However, if the collars are too large, they will interfere with circulation.

The approximate analysis by Lubinski and Woods makes clear that for any given set of drilling conditions, such as bit load, collar size and stiffness, and drilling fluid weight, there is an equilibrium borehole angle that is not vertical. Thus, although a straight hole may be drilled, it will not be vertical, unless there is continuous adjustment of the bit weight to maintain the desired plumbness.

More recent analyses of the problem have used finite element methods to take into account the varying stiffness of the different bottom-hole assembly components (4). Computer codes are now available to help design an optimum bottom-hole assembly. Recent comparisons by Sutko et al. between actual field measurements and predicted changes in hole angle show the inherent value of this analysis (5). The authors, in their comparisons of a finite element analysis method developed by Nicholson with a large number of field studies, conclude that the numerical methods are useful—although a number of other nonquantifiable factors, such as bit type, bit dullness, rotary speed, types of stabilizers, and formation properties and orientation, all act to generate field results that differ in some degree from the numerically predicted deviation angles (6). However, experience has shown that design of the bottom-hole assembly using these principles can minimize, and even eliminate, most problems of crooked boreholes in rotary drilling.

COMPARISON OF DIRECT AND REVERSE ROTARY SYSTEMS ON BOREHOLE ALIGNMENT

As the preceding analysis indicates, with direct rotary drilling placing sufficient weight on the drill bit to achieve satisfactory penetration rates may cause initial and continued borehole deviation if formation anomalies are encountered. Since typical direct rotary drilling begins with a pilot hole that is later enlarged, any deviation is continued into the completed well. This requires caution during pilot borehole drilling, particularly in the upper portion of the well that will house the pump. The deviation of the pilot hole can be checked during drilling operations by raising the bit from the bottom of the hole and measuring the movement of the drill pipe from fixed reference points. Otherwise, one of several types of properly centered survey instruments can be lowered down the drill pipe to measure the deflection from vertical.

Aside from the difference in the drilling-fluid flow path, there are several distinctions between direct and reverse rotary drilling procedures, in particular the drilling of the final large-diameter borehole in a single pass with reverse circulation. The high return velocity of the drilling fluid, which allows large drill cuttings to be removed, and the fact that reverse circulation drilling is usually only undertaken in unconsolidated formations, make rapid penetration rates

the total submerged weight. Thus, for a defined weight on the bit, borehole slope and diameter, and drill collar diameter, the solution defines the drilling angle as a function of the bit load and the distance between the bit and the point of tangency.

The first approximate numerical solution of these equations was given by Lubinski and Woods, who also published tables of the drilling angle as a function of the bit load and drill collar diameter (2, 3).

The results can be summarized in a qualitative way quite simply: increasing the weight on the bit increases bending in the drill string and therefore hole deviation. Increasing the drill collar diameter increases the stiffness of the bottom-hole assembly and thus reduces bending of the drill string. If clearance between the drill collars and the hole is decreased, with a larger-diameter borehole possible. Since reverse rotary equipment generally has less horsepower and derrick capacity than comparable direct rotary rigs, far less weight per unit bit diameter is available and applied. This lower weight with soft formations tends to make the borehole less apt to deviate. However, this is not always true, and plumbness and alignment must be checked as drilling proceeds.

REFERENCES

1. Lubinski, A. 1950. "A Study of the Buckling of Rotary Drilling Strings." *API Drilling and Production Practice*, pp. 178–210.
2. Lubinski, A., and H. B. Woods. 1953. "Factors Affecting the Angle of Inclination and Dog-legging in Rotary Boreholes." *API Drilling and Production Practice*, pp. 222–250.
3. Woods, H. B., and A. Lubinski. 1954. "Practical Charts for Solving Problems on Hole Deviation." *API Drilling and Production Practice*, pp. 56–71.
4. Millheim, J. S., and C. G. Ritter. 1978. "Bottom-Hole Assume Analysis Using Finite Element Method." *Journal of Petroleum Technology* (February), pp. 265–274.
5. Sutko, A. A., G. M. Myers, and J. D. Gaston. 1980. "Directional Drilling—A Comparison of Measured and Predicted Changes in Hole Angle." *Journal of Petroleum Technology* (December), pp. 2090–2096.
6. Nicholson, R. 1972. "Analysis of Constrained Directional Drilling Assemblies." Ph.D. thesis. University of Tulsa, OK.

Welding Casing in the Field

BUTT WELDING

Casing may be covered with a variety of coatings, which include primers, varnish, rust, scale, or moisture. Failure to properly clean the joint ends contributes to poor welds. Joint cleanliness is especially important in the root (first) welding pass. The cleaned area should extend at least 1 in. from the ends on both the inside and outside surfaces.

Use of a heavy-duty straight-shaft grinder with a rubber expanding wheel and a carbide-coated sleeve facilitates cleaning. The small shaft and reduced overall weight allow easy access to both the inside and outside pipe surfaces.

The two casing lengths to be joined are first aligned, and the upper joint is checked to be sure that it is vertical in two planes at right angles. If the casing ends are beveled, a root pass is made, typically using either a 1/8-in. or 5/32-in. welding rod. On larger diameters with all types of welded joints the root pass and all subsequent passes are best made by two welders working on opposite sides of the casing to speed up installation and minimize bowing at the joint from localized welding heat. The root pass should fully penetrate the flat surface of the ends without blowing through or creating slag inside the casing. There must be some buildup on the inside of the joint (inner bead). This should have an even surface and a minimum height of 1/16 in. When starting a new electrode while making the root pass, special care must be taken to maintain a smooth inside bead.

After the slag from the root pass is chipped off, the first filler pass is made. It should be made with the same diameter welding rod used in the root pass, also by two welders working on opposite sides of the casing for large diameters. After slag removal the joint is completed if necessary with a second filler pass sometimes made with a larger-diameter welding rod, depending upon casing thickness. A new bead should not begin at the same point a previous bead was started. The final passes should overlap the original groove by 1/16 to 1/8 in.

BUTT WELDING WITH BACK-UP RINGS

The use of back-up rings helps ensure complete penetration of the root pass and reduces the possibility of a crack on the inside of the casing at the root of the weld.

Back-up rings are typically fabricated from a 16 gauge × 1/2 in. wide steel strip, cut to the exact length to fit the casing's inside diameter. The strip is passed through rolls that form a small groove in the center. The cut length is rolled to a ring, with the groove side out, and is designed to expand against the inner surface of both joints.

A series of small holes are drilled or punched along the center line of the groove, and short lengths of wire (16 gauge) are threaded through the holes and extended outward. These wires act as spacers to separate the two casing ends during welding.

The welding procedure used is the same as for beveled-end welding, but the back-up rings physically support the molten weld metal and dissipate much of the heat, facilitating full penetration of the root pass. Penetration must be complete on both lips of the pipe and into the ring, and all three units are welded into one integral piece. Due to the thin metal used in the fabrication of back-up rings, there is minimum interference with tools passing through the casing.

WELDING COLLARS

The casings are joined using a fillet weld. After the casing joint to be added is seated and is confirmed to be vertical, a root pass is made at the fillet between the welding collar and the upper joint. It is also made with 1/8-in. or 5/32-in. welding rod, depending upon casing thickness.

After removal of slag, one or two filler passes are made on the fillet, again depending on the thickness of the casing and collar. The support for the molten weld metal by the collar allows a larger-diameter rod to be used for the filler pass. A properly made welded joint using a welding collar will have a joint strength equal to or exceeding the strength of the casing. A horizontal fillet weld is also an easier and more trouble-free welding procedure.

The alignment holes are filled by welding after the fillet weld is complete.

WELDING ROD

Welding rod conforming to AWS Class E6010 is commonly used for welding of all carbon-steel well casing, although

TABLE F.1 Physical Properties of Welding Rods Most Commonly Used for Well Casing

	Fleetweld 5P	Shield Arc 85	Stainweld 308-15	Stainweld 308L-16	Stainweld 316L-16
AS WELDED					
Tensile strength (psi)	62–69,000	70–78,000	87,000	79,000	85,000
Yield point (psi)	52–62,000	60–71,000			
Ductility (% elong. in 2 in.)	22–32	22–26	40	44	40
Charpy V-notch toughness (ft/lb)	20–60 @ −20°F	68 @ 70°F			
Hardness Rockwell B(avg)	76–82				
STRESS RELIEVED AT 1150°F					
Tensile strength (psi)	60–69,000	70–83,000			
Yield point (psi)	46–56,000	57–69,000			
Ductility (% elong. in 2 in.)	28–36	22–28			
Charpy V-notch toughness (ft/lb)	71 @ 70°F	64 @ 70°F			
Hardness Rockwell B(avg)		80–89			

Note: Fleetweld, Shield Arc, and Stainweld are trade names of products by the Lincoln Electric Company.
Source: "Weldirectory," the Lincoln Electric Company, 1986.

Class E7010 is used by some welders. Class E308-15 is used for welding type 304 stainless steel, E308L-16 for type 304L, and E316L-16 for types 316 and 316L.

AWS Class E309-16 rod is appropriate for welding carbon steel to stainless steel when a mechanical connection is not used. Technical data covering most commonly used rods for welding casing are given in Table F.1.

The information in this appendix is also applicable to the connection of screens, and casing to screen.

Double Well Casing

DOUBLE WELL CASING

Double well casing, commonly used in cable tool drilling in the alluvial valleys of the western United States, is designed to minimize the problems of formation friction while ensuring adequate joint strength. It consists of an equal number of inside and outside joints, fabricated in 4-ft lengths from a special high-tensile-strength, copper-bearing steel. This length accommodates the stroke of installation jacks (Chapter 7). The inside joints fit tightly (about 0.010 in. loose on the diameter) in the outside joints. The ends of the inside and outside joints lie at the midpoint of the other. Specifications for double well casing material are given in Table G.1.

This configuration inherently makes a very strong integral connection because of the 2-ft tight-fitting lap. Welded connections are made in the field by filling a ⅛-in. gap between the ends of the outside joints against the midpoint of the inside joint. Many wells completed with high-strength, corrosion-resistant double well casing have been operating more than 75 years, and some are over 2000 ft deep.

Frequently single thicknesses of this material are used to case shallow cable tool drilled wells. Advantages include corrosion resistance, the strength to withstand driving impact forces, and short lengths that are easier to handle when maintaining open hole is difficult.

TABLE G.1 Chemical and Physical Properties of Double Well Casing

Chemical Composition		
Element	Limiting Chemical Range (%)	Physical Properties
Carbon	0.20–0.30	Yield strength 55,000–70,000 psi
Manganese	0.85–1.30	Ultimate strength 80,000–95,000 psi
Phosphorous	0.05 Max.	Elongation 17–25% in 8 in.
Sulfur	0.05 Max.	Rockwell "B" hardness 80–90
Silicon	0.12 Max.	Elastic ratio 69–73
Copper	0.20 Min.	

The casing strength needed when driving or jacking is a function of the diameter and depth of the well, and the nature of the formations. In larger and deeper wells the strength requirement dictates the wall thickness. Collapse and crushing strengths for double-well casing are shown in Table G.2. Recommended minimum thicknesses are shown in Table G.3.

TABLE G.2 Collapse and Crushing Strengths of Double Well Casing

Nominal Diameter (in.)	U.S. Standard Gauge	Casing Wall Thickness (in.)	Outside Diameter (in.)	Inside Diameter (in.)	Weight (lb/ft)	Collapsing Strength (psi)	Collapsing Strength (ft/water)	Axial Compression Strength (tons)
8	12	0.218	8.436	8.000	19.13	302.65	698.96	76.36
8	10	0.282	8.504	7.940	24.76	662.74	1530.58	98.44
10	12	0.218	10.436	10.000	23.79	153.75	355.07	95.20
10	10	0.282	10.504	9.940	30.79	335.88	775.71	122.80
12	12	0.218	12.436	12.000	28.45	88.51	204.41	114.03
12	10	0.282	12.504	11.940	36.81	193.06	445.88	147.16
12	8	0.344	12.568	11.880	44.91	353.07	815.41	179.09
14	12	0.218	14.496	14.060	33.24	55.53	128.25	133.43
14	10	0.282	14.564	14.000	43.01	120.99	279.43	172.26
14	8	0.344	14.628	13.940	52.48	221.03	510.47	209.70
16	12	0.218	16.496	16.060	37.90	37.10	85.68	152.26
16	10	0.282	16.564	16.000	49.04	80.77	186.52	196.62
16	8	0.344	16.628	15.940	59.83	147.42	340.47	239.42
16	6	0.406	16.692	15.880	70.62	243.72	562.86	282.06
18	12	0.218	18.496	18.060	42.56	26.00	60.05	171.10
18	10	0.282	18.564	18.000	55.06	56.57	130.64	220.98
18	8	0.344	18.628	17.940	67.17	103.19	238.30	269.14
18	6	0.406	18.692	17.880	79.29	170.48	393.71	317.14
20	12	0.218	20.556	20.120	47.35	18.92	43.70	190.50
20	10	0.282	20.624	20.060	61.27	41.14	95.02	246.08
20	8	0.344	20.688	20.000	74.74	75.02	173.25	299.75
20	6	0.406	20.752	19.940	88.22	123.88	286.09	353.27
22	10	0.282	22.624	22.060	67.29	30.86	71.26	270.44
22	8	0.344	22.688	22.000	82.09	56.24	129.87	329.47
22	6	0.406	22.752	21.940	96.90	92.82	214.38	388.34
24	10	0.282	24.624	24.060	73.31	23.73	54.80	294.80
24	8	0.344	24.688	24.000	89.44	43.23	99.85	359.19
24	6	0.406	24.752	23.940	105.57	71.34	164.76	423.42
26	10	0.282	26.624	26.060	79.34	18.64	43.05	319.17
26	8	0.344	26.688	26.000	96.79	33.95	78.41	388.91
26	6	0.406	26.752	25.940	114.24	56.01	129.35	458.49

TABLE G.3 Recommended Minimum Thicknesses of Double Well Casing

Depth of Casing (ft)	Diameter (in.)								
	10	12	14	16	18	20	22	24	30
0–100	12	12	12	12	10	10	10	10	8
100–200	12	12	12	10	10	10	10	8	8
200–300	12	12	10	10	10	10	8	8	8
300–400	12	12	10	10	10	8	8	8	8
400–600	10	10	10	10	8	8	8	8	8
600–800	10	10	10	8	8	8	6	6	6
More than 800	10	8	8	8	8	6	6	6	6

Note: Values are U.S. standard steel thickness gauge.
Source: AWWA A100-84.

Properties and Standards for Water Well Casing

PHYSICAL PROPERTIES OF STEEL CASING

TABLE H.1 **Physical Properties of Steel Casing**

Outside Diameter (in.)	Inside Diameter (in.)	Wall Thickness (in.)	Weight (lb/ft)	Collapsing Strength		Axial Compression Strength (tons)	Tensile Strength (tons)
				(psi)	(ft/water)		
6.625	6.406	0.1094	7.61	199	459	39.2	67.2
6.625	6.344	0.1406	9.74	371	857	50.1	85.9
6.625	6.281	0.1719	11.85	593	1369	61.0	104.5
6.625	6.250	0.1875	12.81	720	1662	66.4	113.7
6.625	6.125	0.2500	17.02	1288	2974	87.6	150.1
8.625	8.406	0.1094	9.95	99	229	51.2	87.8
8.625	8.344	0.1406	12.74	192	444	65.6	112.4
8.625	8.281	0.1719	15.52	319	736	79.9	137.0
8.625	8.250	0.1875	16.79	393	908	87.0	149.1
8.625	8.125	0.2500	22.36	756	1745	115.1	197.3
10.750	10.375	0.1875	21.15	228	528	108.9	186.7
10.750	10.250	0.2500	28.04	461	1065	144.3	247.4
10.750	10.125	0.3125	34.71	760	1756	179.3	307.4
12.750	12.375	0.1875	25.16	147	340	129.5	222.0
12.750	12.250	0.2500	33.38	306	707	171.8	294.5
12.750	12.125	0.3125	41.52	521	1203	213.7	366.3
12.750	12.000	0.3750	49.57	795	1835	262.4	437.4
14.000	13.625	0.1875	27.73	115	265	142.4	244.1
14.000	13.500	0.2500	36.71	242	560	189.0	324.0
14.000	13.375	0.3125	45.68	419	967	235.2	403.1
14.000	13.250	0.3750	54.57	636	1469	280.9	481.6
14.500	14.125	0.1875	28.66	105	242	147.5	252.9
14.500	14.000	0.2500	38.05	222	512	195.9	335.8
14.500	13.875	0.3125	47.36	385	890	243.8	417.9
14.500	13.750	0.3750	56.57	588	1359	291.2	499.2
16.000	15.625	0.1875	31.67	80	185	163.0	279.4
16.000	15.500	0.2500	42.05	172	398	216.5	371.1
16.000	15.375	0.3125	52.36	303	700	269.5	462.0
16.000	15.250	0.3750	62.58	470	1085	322.1	552.2
16.625	16.250	0.1875	32.29	72	167	169.4	290.5
16.625	16.125	0.2500	43.73	156	360	225.1	385.8
16.625	16.000	0.3125	54.45	276	637	280.3	480.4
16.625	15.875	0.3750	65.09	429	991	335.0	574.3
18.000	17.625	0.1875	35.67	58	134	183.6	314.8
18.000	17.500	0.2500	47.39	126	292	244.0	418.2
18.000	17.375	0.3125	59.03	226	521	303.9	520.9

TABLE H.1 (*Continued*)

Outside Diameter (in.)	Inside Diameter (in.)	Wall Thickness (in.)	Weight (lb/ft)	Collapsing Strength (psi)	(ft/water)	Axial Compression Strength (tons)	Tensile Strength (tons)
18.000	17.250	0.3750	70.59	355	820	363.4	622.9
18.625	18.250	0.1875	36.92	53	122	190.1	325.8
18.625	18.125	0.2500	49.07	116	267	252.6	433.0
18.625	18.000	0.3125	61.33	207	478	314.6	539.4
18.625	17.875	0.3750	73.10	327	755	376.3	645.0
20.000	19.625	0.1875	39.68	43	100	204.2	350.1
20.000	19.500	0.2500	52.73	95	221	271.5	465.4
20.000	19.375	0.3125	65.71	172	398	338.2	579.8
20.000	19.250	0.3750	78.60	274	633	404.6	693.6
20.625	20.250	0.1875	40.93	40	92	210.7	361.2
20.625	20.125	0.2500	54.41	88	203	280.0	480.1
20.625	20.000	0.3125	67.80	159	367	349.0	598.3
20.625	19.875	0.3750	81.11	254	586	417.5	715.7
22.000	21.500	0.2500	58.07	74	170	298.9	512.5
22.000	21.375	0.3125	72.38	134	310	372.6	638.8
22.000	21.250	0.3750	86.61	215	498	445.8	764.3
22.500	22.000	0.2500	59.41	69	160	305.8	524.2
22.500	21.875	0.3125	74.05	126	292	381.2	653.5
22.500	21.750	0.3750	88.61	203	470	456.1	782.0
24.000	23.500	0.2500	63.41	58	134	326.4	559.6
24.000	23.375	0.3125	79.06	107	246	407.0	697.7
24.000	23.250	0.3750	94.62	172	398	487.1	835.0
24.500	24.000	0.2500	64.75	55	127	333.3	571.4
24.500	23.875	0.3125	80.73	101	233	415.6	712.4
24.500	23.750	0.3750	96.62	163	377	497.4	852.7
26.000	25.500	0.2500	68.75	47	108	353.9	606.7
26.000	25.375	0.3125	85.73	86	198	441.3	756.6
26.000	25.250	0.3750	102.63	140	323	528.3	905.7
26.500	26.000	0.2500	70.09	44	102	360.8	618.5
26.500	25.875	0.3125	87.40	82	188	449.9	771.3
26.500	25.750	0.3750	104.63	133	307	538.6	923.3
28.000	27.500	0.2500	74.09	38	88	381.4	653.8
28.000	27.375	0.3125	92.41	70	162	475.7	815.5
28.000	27.250	0.3750	110.64	115	265	569.5	976.4
28.500	28.000	0.2500	75.43	36	83	388.3	665.6
28.500	27.875	0.3125	94.08	67	155	484.3	830.2
28.500	27.750	0.3750	112.64	110	253	597.8	994.0
30.000	29.500	0.2500	79.43	31	72	408.9	701.0
30.000	29.375	0.3125	99.08	58	134	510.1	874.4
30.000	29.250	0.3750	118.65	95	221	610.8	1047.0
30.500	30.000	0.2500	80.77	30	69	415.8	712.8
30.500	29.875	0.3125	100.75	56	128	518.6	889.1
30.500	29.750	0.3750	120.65	91	211	621.1	1064.7

Note: Collapse strength values are based upon Timoshenko's formula with 1% as a value of ellipticity and 35,000 psi as the steel yield point. Strengths are based upon 35,000 psi yield strength and 60,000 psi ultimate strength steel.

MANUFACTURING STANDARDS APPLICABLE TO WATER WELL CASING

TABLE H.2   Manufacturing Standards Applicable to Water Well Casing

Specification	Title	Method of Manufacture	End Use	Grade	Yield (Minimum psi)
API 5L	API specification for line pipe	SAW, ERW and seamless	Line pipe	Grade A Grade B	30,000 35,000
AWWA C 200	Standard for steel water pipe 6 in. and larger	All	Water pipe	As specified by customer	To ASTM steel raw material standard or otherwise specified
ASTM A 53	Pipe, steel, black and hot-dipped, zinc-coated welded and seamless, ⅛–26 in.	CW, ERW, and seamless	Suitable for forming and bending	Grade A Grade B	30,000 35,000
ASTM A 134	Pipe, steel, electric-fusion (arc)-welded, 16 in. and over	Fabricated, press formed or spiral	Line pipe and various	According to ASTM A 283, 285, 570, 36 or other ASTM raw material as specified	As specified
ASTM A 135	Electric-resistance welded steel pipe	ERW	Line pipe	Grade A Grade B	30,000 35,000
ASTM A 139	Electric-fusion (arc)-welded steel pipe (sizes 4 in. and over)	SAW straight seam or spiral	Line pipe and various	Grade A Grade B Grade C Grade D Grade E	30,000 35,000 42,000 46,000 52,000
ASTM A 211	Spiral-welded steel or iron pipe	Spiral	Line pipe and various	According to ASTM A 570 or otherwise specified	As specified
ASTM A 252	Welded and seamless steel pipe piles	Seamless, ERW, SAW straight seam or spiral	Pipe piles	Grade 1 Grade 2 Grade 3	30,000 35,000 45,000
ASTM A 409	Welded large-diameter austenitic steel pipe for corrosive or high temperature service 14–30 in.	SAW straight seam or spiral	Corrosive or high temperature service	10 grades of chromium-nickel stainless steel are covered by this specification, including types 304 and 316	As specified according to grade
ASTM A 714	High-strength, low-alloy welded and seamless steel pipe ½– 26 in.	CW, ERW, and seamless	General purposes where saving in weight or added durability are important	Class 2 Class 4	50,000 36,000 to 50,000 depending on grade (8)
ASTM A 778	As Welded, unannealed austenitic stainless steel tubular products	Any method incorporating a shielded arc-welding process such as SAW or TIG	Corrosive service	5 grades of chromium-nickel stainless steel are covered by this specification, including types 304 L and 316 L	As specified according to grade

Note: Shown are frequently used specifications for steel pipe, their method of manufacture, end use, strength requirements of the finished product and chemistry as applicable. Titles and grades have been abbreviated in some cases for clarity.

Tensile (Minimum psi)	Carbon (%)	Manganese (%)	Copper (%)	Resistance to Atmospheric Corrosion
48,000			Residual	Carbon steel
60,000				
To ASTM steel raw material standard or otherwise specified	To ASTM steel raw material standard	To ASTM steel raw material standard	Residual	Carbon steel
			Residual	Carbon steel
			Residual unless specified	If copper is specified, 2 × carbon steel
48,000	0.25 maximum	0.95 maximum	Residual	Carbon steel
60,000	0.30 maximum	1.20		
As specified	As specified	As specified	Residual unless specified	If copper is specified, 2 × carbon steel
48,000	0.25 maximum	0.95	Residual	Carbon steel
60,000	0.30 maximum	1.20 maximum		
48,000	—	1.00 maximum	Residual unless specified	If copper is specified, 2 × carbon steel
60,000	0.30 maximum	1.00 maximum		
60,000	0.30 maximum	1.20 maximum		
60,000	0.30 maximum	1.30 maximum		
66,000	0.30 maximum	1.40 maximum		
As specified	As specified	As specified	Residual unless specified under table 1 of ASTM A 570	If copper is specified, 2 × carbon steel
50,000			Residual	Carbon steel
60,000				
66,000				
As specified according to grade	As specified according to grade	As specified according to grade	As specified according to grade	Corrosion rate is nil
According to grade	According to grade	According to grade	.20 minimum	2 × carbon steel
			.20 minimum	4 × carbon steel
As specified according to grade	As specified according to grade	As specified according to grade	As specified according to grade	Corrosion rate is nil

MATERIAL STANDARDS APPLICABLE TO WATER WELL CASING

TABLE H.3 Material Standards Applicable to Water Well Casing

Specification	Title	Grade	Yield (Minimum psi)	Tensile (Minimum psi)
ASTM A 36	Structural steel (plates)		36,000	58,000
ASTM A 167	Stainless chromium nickel steel plate, sheet, and strip	20 grades of chromium nickel stainless Steel are covered by this specification, including Types 304 and 316.	According to grade. The yield and tensile strengths of Types 304 and 316 are 30,000 psi and 75,000 psi	
ASTM A 242	High-strength, low alloy structural steel (plates)		50,000	70,000
ASTM A 283	Low and intermediate tensile strength, carbon steel plates, shapes, and bars	A	24,000	45,000–55,000
		B	27,000	50,000–60,000
		C	30,000	55,000–65,000
		D	33,000	60,000–72,000
ASTM A 569	Steel, carbon (0.15% max.), hot-rolled sheet and strip, commerical quality			
ASTM A 570	Hot-rolled carbon steel sheet and strip, structural quality	30	30,000	49,000
		33	33,000	52,000
		36	36,000	53,000
		40	40,000	55,000
		45	45,000	60,000
		50	50,000	65,000
		55	55,000	70,000
ASTM A 606	Steel sheet and strip, hot-rolled and cold-rolled, high-strength low-alloy, with improved corrosion resistance	Type 2	50,000	70,000
		Type 4	50,000	70,000
ASTM A 635	Hot-rolled carbon steel sheet and strip, commerical quality heavy-thickness coils (formerly plate)	15 grades		

Note: These specifications are associated with steel pipe and water well casing manufacture. Titles and grades have been abbreviated in some cases for clarity.

Although many pipe specifications cite manufacture from some of these raw material specifications, others do not, relying on testing of the finished pipe to ensure compliance with physical requirements. In such cases the manufacturer may order raw material according to one of the specifications above, another standard as agreed, or to a steel chemistry, which, taking in consideration physical changes occurring in the manufacturing process, results in the desired quality product.

The differences between plate, sheet, and strip are dimensional but are related to some extent to the mill process used to manufacture the product from the slab stage. This causes some confusion since pipe is manufactured from all three categories. ASTM A 635, a specification frequently used by manufacturers of ERW and spiral seam pipe, covers coils with plate thicknesses.

Carbon (%)	Manganese (%)	Copper (%)	Resistance to Atmospheric Corrosion
0.25 max.		0.20 minimum if specified under table 2 of ASTM A36	If copper is specified, 2 × carbon steel
			Corrosion rate is nil for types 304 and 316
Alloying elements to provide required strength and corrosion resistance vary from manufacturer to manufacturer.		0.20 minimum	4 × carbon steel
		0.20 min. if specified under table 2 of ASTM A 283	If copper is specified, 2 × carbon steel
0.15 max.	0.60 max.	0.20 min. if specified under table 1 of ASTM A 569	If copper is specified, 2 × carbon steel
0.25 max.	0.90 max.	0.20 min. if specified under table 1 of ASTM A 570	If copper is specified, 2 × carbon steel
0.25 max.	0.90 max.		
0.25 max.	0.90 max.		
0.25 max.	0.90 max.		
0.25 max.	1.35 max.		
0.25 max.	1.35 max.		
0.25 max.	1.35 max.		
0.22 max.	1.25 max.	Alloying elements vary from manufacturer to manufacturer but must produce strength and corrosion resistance for each type as specified	2 × carbon steel
0.22 max.	1.25 max.		4 × carbon steel
Carbon ranges from 0.08 max. to 0.25 max., and manganese from 0.45 max. to 1.65 max. according to grade		0.20 min. if specified under table 1 of ASTM A 635	If copper is specified, 2 × carbon steel

HYDRAULIC COLLAPSE PRESSURE AND UNIT WEIGHT OF PVC WELL CASING

TABLE H.4 Hydraulic Collapse Pressure and Unit Weight of PVC Well Casing

Nominal	Actual	SDR/SCH	Wall Thickness (Minimum in.)	DR	Weight in Air (lb/100 ft) PVC 12454	PVC 14333	Weight in Water (lb/100 ft) PVC 12454	PVC 14333	Hydraulic Collapse Pressure (psi) PVC 12454	PVC 14333
2	2.375	SCH-80	0.218	10.9	94	91	27	24	947	758
		SDR-13.5	0.176	13.5	78	75	22	19	470	376
		SCH-40	0.154	15.4	69	66	20	17	307	246
		SDR-17	0.140	17.0	63	61	18	16	224	179
		SDR-21	0.113	21.0	51	47	14	12	115	92
2½	2.875	SCH-80	0.276	10.4	144	139	41	36	1110	885
		SDR-13.5	0.213	13.5	114	110	32	28	470	376
		SCH-40	0.203	14.2	109	105	31	27	400	320
		SDR-17	0.169	17.0	92	89	26	23	224	179
		SDR-21	0.137	21.0	76	73	22	19	115	92
3	3.500	SCH-80	0.300	11.7	193	186	55	48	600
		SDR-13.5	0.259	13.5	169	163	48	42	470	376
		SCH-40	0.216	16.2	143	138	41	36	262	210
		SDR-17	0.206	17.0	136	132	39	34	224	179
		SDR-21	0.167	21.0	112	108	32	28	115	92
3½	4.000	SCH-80	0.318	12.6	235	227	67	59	589	471
		SDR-13.5	0.296	13.5	220	212	63	55	470	376
		SDR-17	0.235	17.0	178	172	51	44	224	179
		SCH-40	0.226	17.7	176	172	49	43	197	158
		SDR-21	0.190	21.0	146	141	42	36	115	92
4	4.500	SCH-80	0.337	13.3	282	272	80	70	494	395
		SDR-13.5	0.333	13.5	279	269	80	70	470	376
		SDR-17	0.265	17.0	226	218	64	56	224	179
		SCH-40	0.237	19.0	203	196	58	51	158	126
		SDR-21	0.214	21.0	185	178	53	46	115	92
		SDR-26	0.173	26.0	151	145	43	38	59	47
		SDR-32.5	0.138	32.5	121	117	34	30	29	23
		SDR-41	0.110	41.0	97	94	28	24	14	11
4½	4.950	—	0.248	20.0	235	226	67	58	134	107
		—	0.190	26.0	182	176	52	46	59	47
5	5.563	SDR-13.5	0.412	13.5	427	411	122	106	470	376
		SCH-80	0.375	14.8	391	377	112	98	350	280
		SDR-17	0.327	17.0	345	332	99	86	224	179
		SDR-21	0.265	21.0	283	273	81	71	115	92
		SCH-40	0.258	21.6	276	266	79	69	105	84
		SDR-26	0.214	26.0	231	222	66	58	59	47
5	5.563	—	0.190	29.3	206	198	59	51	40	32
		SDR-32.5	0.171	32.5	186	179	52	46	29	23
		SDR-41	0.136	41.0	149	144	43	37	14	11

Nominal size	Actual OD	SDR or SCH	Wall thickness	DR						
6	6.140	SDR-32.5	0.189	32.5	227	218	65	56	29	23
		SDR-41	0.150	41.0	181	175	52	45	14	11
6	6.625	SDR-13.5	0.491	13.5	605	584	173	151	470	376
		SCH-80	0.432	15.3	538	519	154	134	314	171
		SDR-17	0.390	17.0	489	472	140	122	224	179
		SDR-21	0.316	21.0	402	387	115	100	115	92
		SCH-40	0.280	23.7	358	345	102	89	78	62
		SDR-26	0.255	26.0	327	316	93	82	59	47
		SDR-32.5	0.204	32.5	264	255	75	66	29	23
		—	0.190	34.9	246	238	70	62	23	18
		SDR-41	0.162	41.0	211	204	60	53	14	11
7	7.000	—	0.300	23.3	405	390	116	101	83	66
8	8.160	SDR-32.5	0.251	32.5	400	386	114	100	29	23
		SDR-41	0.199	41.0	320	308	91	80	14	11
	8.625	SDR-17	0.508	17.0	830	800	237	207	224	179
		SCH-80	0.500	17.2	818	788	234	204	216	173
		SDR-21	0.410	21.0	678	654	194	170	115	92
		SDR-26	0.332	26.0	555	535	159	139	59	47
		SCH-40	0.322	26.8	539	520	154	135	54	43
		SDR-32.5	0.265	32.5	447	431	128	109	29	23
		SDR-41	0.210	41.0	356	343	102	89	14	11
10	10.200	SDR-32.5	0.314	32.5	626	604	179	156	29	23
		SDR-41	0.249	41.0	500	482	143	125	14	11
10	10.750	SDR-17	0.632	17.0	1290	1240	368	322	224	179
		SCH-80	0.593	18.1	1210	1170	346	303	184	147
		SDR-21	0.511	21.0	1050	1020	301	263	115	92
		SDR-26	0.413	26.0	860	830	246	215	59	47
		SCH-40	0.365	29.4	764	737	218	191	40	32
		SDR-32.5	0.331	32.5	695	670	199	173	29	23
		SDR-41	0.262	41.0	557	537	159	139	14	11
12	12.750	SDR-17	0.750	17.0	1810	1760	517	453	224	179
		SCH-80	0.687	18.6	1670	1610	476	417	168	134
		SDR-21	0.606	21.0	1480	1430	423	370	115	92
		SDR-26	0.490	26.0	1210	1170	346	302	59	47
		SCH-40	0.406	31.4	1010	974	288	252	33	26
		SDR-32.5	0.392	32.5	977	942	279	244	29	23
		SDR-41	0.311	41.0	780	752	223	195	14	11
14	14.000	SCH-80	0.750	18.7	2000	1930	572	500	166	133
		SDR-21	0.667	21.0	1790	1730	512	448	115	92
		SDR-26	0.539	26.0	1460	1410	418	366	59	47
		SCH-40	0.437	32.0	1200	1150	341	299	31	25
		SDR-32.5	0.430	32.5	1180	1140	336	294	29	23
		SDR-41	0.342	41.0	942	908	269	235	14	11

Note: SDR = standard dimension ratio; SCH = schedule, DR = dimension ratio (actual OD/wall thickness).

Sources: *Manual on the Selection and Installation of Thermoplastic Water Well Casing*, National Water Well Association, 500 West Wilson Bridge Road, Worthington, OH 43085. ASTM F 480-88a.

HYDRAULIC COLLAPSE PRESSURE AND UNIT WEIGHT OF ABS WELL CASING

TABLE H.5 Hydraulic Collapse Pressure and Unit Weight of ABS Well Casing

Outside Diameter (in.) Nominal	Actual	SDR/SCH	Wall Thickness (Minimum in.)	DR	Weight in Air (lb/100 ft) ABS 434	ABS 533	Weight in Water (lb/100 ft) ABS 434	ABS 533	Hydraulic Collapse Pressure (psi) ABS 434	ABS 533
2	2.375	SCH-80	0.218	10.9	71	70	3.4	2.7	829	592
		SDR-13.5	0.176	13.5	59	58	2.8	2.2	412	294
		SCH-40	0.154	15.4	52	51	2.5	2.0	269	192
		SDR-17	0.140	17.0	47	47	2.2	1.8	196	140
		SDR-21	0.113	21.0	39	38	1.8	1.5	100	71
2½	2.875	SCH-80	0.276	10.4	108	107	5.1	4.1	968	691
		SDR-13.5	0.213	13.5	85	85	4.0	3.3	412	294
		SCH-40	0.203	14.2	82	81	3.9	3.1	350	250
		SDR-17	0.169	17.0	69	68	3.3	2.6	196	140
		SDR-21	0.137	21.0	57	56	2.7	2.1	100	71
3	3.500	SCH-80	0.300	11.7	145	144	6.9	5.5	656	468
		SDR-13.5	0.259	13.5	126	125	6.0	4.8	412	294
		SCH-40	0.216	16.2	107	106	5.1	4.1	229	164
		SDR-17	0.206	17.0	103	102	4.9	3.9	196	140
		SDR-21	0.167	21.0	84	83	4.0	3.2	100	71
3½	4.000	SCH-80	0.318	12.6	176	175	8.4	6.7	515	368
		SDR-13.5	0.296	13.5	165	164	7.8	6.3	412	294
		SDR-17	0.235	17.0	134	133	6.4	5.1	196	140
		SCH-40	0.226	17.7	129	128	6.1	4.9	173	124
		SDR-21	0.190	21.0	109	108	5.2	4.1	100	71
4	4.500	SCH-80	0.337	13.3	211	209	10.0	8.0	432	308
		SDR-13.5	0.333	13.5	209	207	10.0	8.0	412	294
		SDR-17	0.265	17.0	169	168	8.0	6.5	196	140
		SCH-40	0.237	19.0	152	151	7.2	5.8	138	98
		SDR-21	0.214	21.0	138	137	6.6	5.3	100	71
		SDR-26	0.173	26.0	113	112	5.4	4.3	51	36
5	5.563	SDR-13.5	0.412	13.5	320	317	15.2	12.2	412	294

Nominal (in.)	OD (in.)		Wall (in.)	DR						
		SCH-80	0.375	14.8	294	291	14.0	11.2	306	218
		SDR-17	0.327	17.0	258	256	12.3	9.8	196	140
		SDR-21	0.265	21.0	212	210	10.0	8.1	100	71
		SCH-40	0.258	21.6	207	205	9.8	7.9	92	66
		SDR-26	0.214	26.0	173	171	8.2	6.6	51	36
6	6.625	SDR-13.5	0.491	13.5	454	450	21.7	17.3	412	294
		SCH-80	0.432	15.3	404	400	19.2	15.4	275	196
		SDR-17	0.390	17.0	367	364	17.4	14.0	196	140
		SDR-21	0.316	21.0	301	298	14.3	11.5	100	71
		SCH-40	0.280	23.7	268	266	12.8	10.2	69	49
		SDR-26	0.255	26.0	246	243	11.7	9.3	51	36
8	8.625	SDR-17	0.508	17.0	622	616	29.6	23.7	196	140
		SCH-80	0.500	17.2	613	607	29.2	23.3	189	135
		SDR-21	0.410	21.0	509	504	24.2	19.4	100	71
		SDR-26	0.332	26.0	416	412	19.8	15.8	51	36
		SCH-40	0.322	26.8	404	400	19.2	15.4	47	34
10	10.750	SDR-17	0.632	17.0	965	956	46.0	36.8	196	140
		SCH-80	0.593	18.1	909	901	43.3	34.6	161	115
		SDR-21	0.511	21.0	790	738	37.6	30.0	100	71
		SDR-26	0.413	26.0	645	639	30.7	24.6	51	36
		SCH-40	0.365	29.4	573	568	27.3	21.8	35	25
12	12.750	SDR-17	0.750	17.0	1360	1350	64.7	51.8	196	140
		SCH-80	0.687	18.6	1250	1240	59.6	47.6	148	106
		SDR-21	0.606	21.0	1110	1100	52.9	42.3	100	71
		SDR-26	0.490	26.0	908	899	43.2	34.6	51	36
		SCH-40	0.406	31.4	758	751	36.1	28.9	29	20
14	14.000	SDR-21	0.667	21.0	1340	1330	64.0	51.2	100	71
		SDR-26	0.539	26.0	1100	1090	52.2	41.8	51	36
16	16.000	SDR-21	0.762	21.0	1750	1740	83.5	66.8	100	71
		SDR-26	0.616	26.0	1430	1420	68.2	54.6	51	36

Note: SDR = standard dimension ratio; SCH = schedule, DR = dimension ratio (actual OD/wall thickness).

Sources: *Manual on the Selection and Installation of Thermoplastic Water Well Casing*, National Water Well Association, 500 West Wilson Bridge Road, Worthington, OH 43085. ASTM F 480-88a.

HYDRAULIC COLLAPSE PRESSURE AND UNIT WEIGHT OF SR WELL CASING

TABLE H.6 Hydraulic Collapse Pressure and Unit Weight of SR Well Casing

Outside Diameter (in.) Nominal	Actual	SDR	Wall Thickness (Minimum in.)	DR	Weight in Air (lb/100 ft)	Weight in Water (lb/100 ft)	Hydraulic Collapse Pressure (psi)
2	2.375	SDR-13.5	0.176	13.5	58	2.8	376
		SDR-17	0.140	17.0	47	2.2	180
		SDR-21	0.113	21.0	39	1.8	92
2½	2.875	SDR-13.5	0.213	13.5	85	4.0	376
		SDR-17	0.169	17.0	69	3.3	180
		SDR-21	0.137	21.0	57	2.7	92
3	3.500	SDR-13.5	0.259	13.5	126	6.0	376
		SDR-17	0.206	17.0	102	4.8	180
		SDR-21	0.167	21.0	84	4.0	92
3½	4.000	SDR-13.5	0.296	13.5	165	7.8	376
		SDR-17	0.234	17.0	134	6.4	180
		SDR-21	0.190	21.0	109	5.2	92
4	4.500	SDR-13.5	0.330	13.5	209	10.0	376
		SDR-17	0.265	17.0	169	8.0	180
		—	0.250	18.0	160	7.6	150
		SDR-21	0.214	21.0	138	6.6	92
		—	0.200	22.5	130	6.2	74
		—	0.175	25.7	114	5.4	49
4½	4.886	—	0.230	21.2	162	7.7	89
		—	0.200	24.4	142	6.8	57
5	5.300	—	0.320	16.6	240	11.4	194
		—	0.250	21.2	191	9.1	89
		—	0.200	26.5	154	7.3	44
		—	0.174	30.3	136	6.5	29
5	5.563	SDR-13.5	0.412	13.5	320	15.2	376

Nominal size	OD	SDR	Wall thickness	DR			
6	6.275	SDR-17	0.327	17.0	258	12.3	180
		SDR-21	0.265	21.0	212	10.0	92
		—	0.225	24.7	182	8.7	55
		SDR-26	0.214	26.0	173	8.2	47
		—	0.320	19.6	288	13.7	114
		—	0.250	25.1	228	10.9	52
		—	0.200	31.4	184	8.8	26
		—	0.175	35.8	161	7.7	17
6	6.625	SDR-13.5	0.491	13.5	454	21.6	376
		SDR-17	0.390	17.0	367	17.5	180
		—	0.320	20.7	305	14.5	86
		SDR-26	0.255	26.0	246	11.7	47
		—	0.250	26.5	241	11.5	44
7	7.000	—	0.250	28.0	255	12.1	37
8	8.625	SDR-17	0.508	17.0	622	29.6	180
		SDR-21	0.410	21.0	509	24.2	92
		SDR-26	0.332	26.0	416	19.8	47
		—	0.250	34.5	317	15.1	20
10	10.750	SDR-17	0.632	17.0	965	46.0	180
		SDR-21	0.511	21.0	790	37.6	92
		SDR-26	0.413	26.0	645	30.7	47
12	12.750	SDR-17	0.750	17.0	360	64.7	180
		SDR-21	0.606	21.0	1110	52.9	92
		SDR-26	0.490	26.0	908	43.2	47
14	14.000	SDR-21	0.667	21.0	1340	64.0	92
		SDR-26	0.539	26.0	1100	52.2	47
16	16.000	SDR-21	0.762	21.0	1750	83.5	92
		SDR-26	0.616	26.0	1432	68.2	47

Note: SDR = standard dimension ratio, DR = dimension ratio (actual OD/Wall Thickness).

Sources: *Manual on the Selection and Installation of Thermoplastic Water Well Casing*, written and produced by the National Water Well Association, 500 West Wilson Bridge Road, Worthington OH 43085. ASTM F 480-88a.

Design of an Airlift Pump

Proper design of an airlift pump requires knowledge of certain characteristics of the well:

S = static water level in well [ft],

D = estimated drawdown at pumping rate Q [ft],

L = pumping lift, or $S + D$ [ft],

Q = discharge rate [gpm],

A = total depth of air line below ground surface [ft].

The starting submergence of the pump is the total depth of the air line minus the static water level. Airlift pumps perform best if air is introduced at a point approximately twice as deep as the static water level, or when two-thirds of the air line is submerged. The operating submergence is the total depth of the air line minus the pumping lift. Table I.1 defines submergence, volume of free air, and start-up pressure.

The operating submergence is usually expressed in terms of a percentage of the total length of air line. This can be calculated as:

$$\text{Submergence} = \frac{100(A - L)}{A} \% .$$

The starting submergence dictates the air pressure needed to initiate pumping because the water in the air line must be displaced. Starting pressure can greatly exceed pumping pressure since the water level in the well declines during pumping.

The starting air pressure, P, can be calculated by

$$P = \frac{A - S}{2.31} \text{ psi} .$$

TABLE I.1   Submergence, Free Air Volume, and Start-Up Pressure

Lift (ft)	Optimum Submergence Range (%)	Approximate Volume of Free Air (V_a/gal)	Start-Up Pressure at Submergence (%)
20	55–70	0.12	25 psi, 65%
30	55–70	0.16	37 psi, 65%
40	50–70	0.27	49 psi, 65%
50	50–70	0.29	62 psi, 65%
60	50–70	0.32	74 psi, 65%
80	50–70	0.36	99 psi, 65%
100	45–70	0.47	108 psi, 60%
125	45–65	0.53	120 psi, 55%
150	40–65	0.69	144 psi, 55%
175	40–60	0.75	152 psi, 50%
200	40–60	0.81	173 psi, 50%
250	40–60	0.92	216 psi, 50%
300	37–55	1.2	236 psi, 45%
350	37–55	1.32	275 psi, 45%
400	37–50	1.72	289 psi, 40%
450	35–45	1.84	325 psi, 40%
500	35–45	1.97	361 psi, 40%
550	35–45	2.09	397 psi, 40%
600	35–45	2.2	433 psi, 40%
650	35–45	2.32	469 psi, 40%
700	35–40	2.99	505 psi, 40%

TABLE I.2   Table of Constants

Submergence (%)	75	70	65	60	55	50	45	40	35
C (OAL)[a]	366	358	348	335	318	296	272	246	216
C (IAL)[b]	330	322	306	285	262	238	214	185	162

[a] Constant for air line outside eductor pipe.
[b] Constant for air line inside eductor pipe.

Airlift pumps used for development do not require the design refinement of permanent airlift pump installations because operating efficiency is not a concern for the short term. However, satisfactory operation requires maintenance of certain relationships in diameters of eductor and air line pipes, and percentage operating submergence of the system.

The required size of the eductor pipe may be calculated from

$$Q_1 = av ,$$

where

Q_1 = the discharge rate of the mixture of air and water [ft^3/sec],

a = the cross-sectional area of the discharge pipe [ft^2],

v = the velocity of the mixture [ft/sec].

Satisfactory velocity of the mixture ranges from 1.2 to more than 30 ft/sec depending on submergence. Q_1 at any point in the discharge pipe is the volume of water pumped plus the volume of compressed air. Since the air expands as it passes up the discharge, Q_1 increases, and for a uniform-diameter discharge, the velocity increases.

The proper diameter of the air injection pipe will depend upon the volume and pressure of the air to be transmitted and its length. Common practice allows air injection velocities of 30 to 40 ft/sec, although this is often exceeded. It is possible to deliver too much air, which will result in excessive friction in the injection pipe and incomplete expansion in the discharge.

The following formula provides a reliable estimate of the quantity of air needed:

$$V_a = \frac{L}{C \log_{10}\{[(A - L) + 34]/34\}} ,$$

where

V_a = cubic feet of free air for each gallon of water pumped,

C = the constant from Table I.2.

Mathematical Analysis of Preliminary Development Methods for Gravel Envelope Wells

It is difficult to compare the effectiveness of various preliminary development methods. Except for jetting, they are difficult to physically model, and mathematical analysis is formidable. However, with the assumption that certain flow-velocity components are necessary for effective development and use of computer technology, it is possible to mathematically compare relative effectiveness of several procedures (1).

The following analysis provides a scientific basis for evaluation of some of the most commonly used preliminary development methods. It describes in general terms mathematical models of these methods. The results of a laboratory study of a test model of the jetting method will be included. Also included are techniques used in common well configurations. From the results presented, it will be concluded that development methods vary considerably in their effect and should be selected carefully.

The five basic methods to be analyzed will be described briefly. Figures illustrating these various development methods are found in Chapter 14. The first considered is jet development. As described in the *Johnson Driller's Journal*, January–February 1979, this method was developed in an attempt to provide high levels of flow energy to the filter cake on a borehole wall. Typical recommended jet velocities are 150 to 200 ft/sec with a jet orifice of ¼ to ½ in.

The second method, line swabbing, involves successively raising and lowering a rubber-flanged scow. Typical haul velocities will be on the order of 3 ft/sec. The scow is equipped with a flapper valve to facilitate lowering the swab.

The third method, a variation of the previous technique, uses a spudding beam to oscillate the swab as it is hauled. The swab is typically oscillated at 30 strokes/min with a 3-ft stroke.

The fourth method involves pumping below a single swab mounted on drill pipe so that the swab causes return flow to enter the gravel envelope and bypass the swab.[1] The swab may be hauled and dropped simultaneously with pumping. A typical fall velocity is 8 ft/sec.

The fifth method considered uses a double swab mounted on drill pipe. Fluid is pumped out between the swabs into the gravel envelope. In an alternative version, the swab is equipped with a bypass to allow flow from below the lower swab to pass to the well region above the upper swab.

The basic features of the mathematical models of each of these systems are similar. There is a completed well of radius a, a filter pack of radius b, and assumed hydraulic conductivities in the filter pack and formation of k_1 and k_2, respectively (Figure J.1). It is assumed that k_1 greatly exceeds k_2. Typical values of k_1 and k_2 are 10,000 gpd/ft^2 and 100 gpd/ft^2, respectively, giving a ratio k_2/k_1 of 0.01. This ratio may vary from 0.1 to 0.001, but as will be shown, the results obtained generally show a great insensitivity to the ratio k_2/k_1.

The basis for assessment of the development methods will be the magnitude of scouring velocity induced at the filter pack/formation interface by development flow circulation. This circulating fluid velocity has two components. One is radial to the well axis; the other tangential or parallel

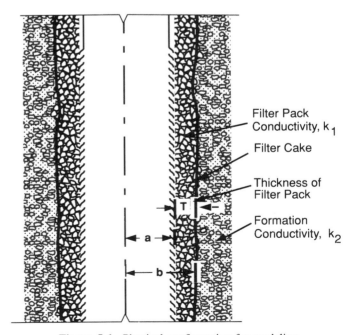

Figure J.1. Physical configuration for modeling.

---

1. In this appendix the terms gravel envelope and filter pack are used interchangeably.

to the well axis. Tangential fluid velocity at the filter pack/formation interface is primarily responsible for scouring the filter cake. The radial component of this velocity carries the material to the well.

J.1 WELL DEVELOPMENT MODELS

Basic models for each of the five well development techniques and computational results from the models are presented in this section.

Jetting

The purpose of jetting is to provide a high-energy flow through the gravel envelope to the filter cake at the formation. This is considered to occur either by flow through the motionless pack or by physical displacement of the pack material by the jet. Mathematical modeling of either mode of operation is difficult. In the first case, even though it is assumed that the pack will not move, the flow will not obey the Darcy equations for flow in porous media. In these equations velocity is directly proportional to the pressure gradient. However, for very high pressure gradients, the velocity produced is less than would be predicted by the Darcy equations.

In the second case, where filter pack is presumed to be displaced by the jet and move with the flow, equations describing the motion of the combined pack-fluid motion are extraordinarily difficult to solve. Furthermore it is not clear how to predict the boundary that will form between the pack material that moves and that which does not. It is apparent that any evaluation of this mode of jetting action must be performed through use of a laboratory test model.

In the following mathematical analysis it is assumed first that the pack material does not move and that any flushing of the filter cake must be caused solely by the jet flow. Subsequently, a laboratory test model will be described. This model has been used to determine when the jet does move the pack material, and it provides visual evidence of the mechanism involved.

Mathematical Model of Jetting. For the mathematical model, the Darcy equations are used to consider a flow field in a porous medium generated by injecting a flow Q over a circular surface of radius c. This is a favorable representation of a jet since, with any high speed jet, a significant fraction of flow would be deflected by the screen and filter pack.

The well wall is modeled as a filter pack of hydraulic conductivity k_1 and thickness T (which is equal to the difference in the screen and filter pack radii) and a formation of hydraulic conductivity k_2. In actuality the surface of the formation may well have a hydraulic conductivity much less than k_2 as a consequence of the filter cake. The physical configuration is as shown in Figure J.1.

Mathematically, this problem is identical to finding the flow of heat into a layered medium, with heat being applied over a circular area of radius c and the surrounding medium being kept at a constant reference temperature. The solution is given for a homogeneous medium (no layers) by Carslaw and Jaeger, and this serves as a starting point for the solution of the problem with two layers of different conductivity (2). The solution is in the form of an integral and involves the ratio of filter pack thickness to jet radius (T/c), velocity of the jet v_j, distance from the jet impact point on the screen, and the ratio of the hydraulic conductivities of the formation and filter pack. Numerical evaluation of the solution is possible, using a digital computer, and flow fields can be drawn in both the filter pack and formation.

Figure J.2 shows graphs of radial and tangential velocities into and along the formation/filter pack interface for the case when ratio of jet radius to filter pack thickness is 1 to 12 (e.g., ½-in. diameter jet into 3-in. filter pack). Velocities are given as a fraction of jet discharge velocity v_j, where v_j is the jet flow Q divided by jet area πc^2. Peak radial velocity into the formation is less than $\frac{1}{2000}$th of the original jet velocity v_j, and peak tangential velocity along the formation is only $\frac{1}{5000}$th of the original jet velocity. In other words, very little jet energy propagates into the formation. This is confirmed by the results appropriate to the problem when there is no filter pack. In this case an exact solution for the peak radial velocity can be found, and it indicates a magnitude of the order $(c/T)^3 v_j$ at depth T into the formation when c/T is small (1).

When the ratio of filter pack thickness to jet radius is 28 (7-in. filter pack and ½-in. diameter jet), velocities are even lower. Computations show a peak tangential velocity at the interface of the filter pack and formation of only $\frac{1}{59000}$th of the jet velocity.

These results make it clear that very little jet flow energy will penetrate more than a few jet diameters into the filter pack, and that very little flow will be generated at the filter pack/formation interface. Recall that this solution was based on the Darcy flow equations and a presumption that all jet flow would enter the filter pack. Given that at higher flow velocities friction will be greater and a fraction of the jet flow will be deflected by the well screen and filter pack, the induced velocities will be lower than those predicted in the preceding analysis. The preceding results are, of course, only valid when the integrity of the filter pack is maintained. In the event of disruption of the pack to a degree that the jet directly impacts the formation, these conclusions will not be valid, and recourse would be made to a laboratory test model for analysis.

Laboratory Model of Jetting. A laboratory test model of a section of a well with an artificial filter pack was constructed as shown in Figure J.3. The test section was filled with a selected pack, the top and diaphragm bolted

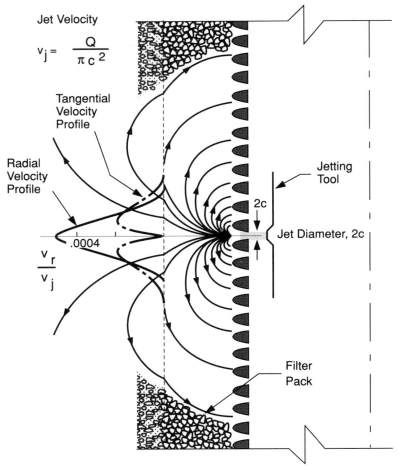

Jet Velocity

$$v_j = \frac{Q}{\pi c^2}$$

Tangential Velocity Profile

Radial Velocity Profile

.0004

$$\frac{v_r}{v_j}$$

Jetting Tool

2c

Jet Diameter, 2c

Filter Pack

Figure J.2. Radial and tangential velocities produced within filter pack by jetting development with no filter pack movement and $T/c = 12$.

down, and an overburden pressure as high as 60 psi imposed via the diaphragm. A test jet with internal diameter 0.167 in. was located in the model such that the flow from the jet directly impacted the section of well screen. The distance of the jet from the screen was adjustable, allowing simulation of a jet of larger diameter. The maximum jet flow possible

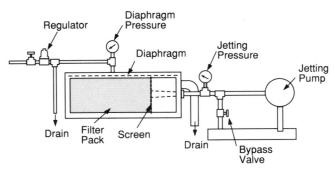

Regulator

Diaphragm Pressure

Diaphragm

Jetting Pressure

Jetting Pump

Drain

Filter Pack

Screen

Drain

Bypass Valve

Figure J.3. Laboratory jetting model.

was 12 gpm, corresponding to a jet velocity of close to 180 ft/sec and a stagnation pressure within the gravel of about 220 psi.

The purpose of this test facility was to determine the mode of jet operation and to evaluate the influences of pack material size distribution, jet flow rate, and overburden pressure on the jetting operation. Tests were performed with three pack material gradations commonly employed in gravel envelope wells. Details of the materials are given in Table J.1. The well screen used was the continuous wire wrap type, with 0.060 in. aperture size.

The fundamental result of the tests performed shows that pack material will not move under jet action unless there is sufficient free space in the filter pack. In other words, there must be "elbow room" for the particles. In the tests this could be provided either by release of the pressuring diaphragm or by waiting for the jet to wash enough fine particles through the screen to create the space. No motion occurred in the test using $1/8 \times 1/4$ gravel pack.

TABLE J.1 Gradation of Jetting Model Pack Materials

	Typical Analysis	
	U.S. Sieve Size	% Passing
6×14 filter pack is a	6	100
mechanically graded	8	70
gravel	10	30
	12	8
	14	2
$\frac{1}{8} \times \frac{1}{4}$ filter pack is a	3	100
mechanically graded	4	18
gravel. 100% passes	6	2
$\frac{1}{4}$ in. or No. 3	8	0
U.S. Sieve, and		
100% is retained on		
$\frac{1}{8}$-in. screen.		
6×10 filter pack is a	4	100
mechanically graded	6	95
smooth and rounded	8	18
gravel	10	1
	16	0

Operation of the test apparatus with finer pack material, such as a 6×14 gravel, disclosed that pack motion would develop only after a cavity was formed by elutriation of the finer size fraction of pack material through the screen. The progression of the pack motion was systematic. First, a few grains would move in the initial cavity formed. The motion of these grains in turn allowed the jet to penetrate deeper and wash out more fine material, thereby increasing the free space. The flow of jet fluid and pack material developed a vortex structure that advanced into the filter pack until an equilibrium penetration depth was attained in about two minutes.

At equilibrium it appears that total jet power is consumed in keeping a ball of fluid and pack material in motion, and none is available to generate new particle motion. It was to be expected that reducing the intergranular stress in the motionless pack should allow the depth of penetration of the jet to increase. This argument was confirmed in the test results.

Initial development of motion depends upon either the existence or creation of free space in which particles of pack material can move. Once particles are free to move, the scale of motion depends upon jet velocity, overburden pressure, the degree of compaction of the filter pack, and intergranular friction. The influence of each of these factors was apparent in the tests.

Considering only the effect of jet velocity, it was clear that reduction in flow velocity reduced the depth of penetration once the system was in motion. A second increase in jet velocity would not generate penetration depth equal to that

attained initially, a somewhat surprising result. The explanation lies in the greater degree of compaction attained in the pack material by the jetting action over what was initially present. This was also confirmed in another test, in which the overburden pressure was suddenly released to allow the filter pack material to move. The whirling mass of fluid and filter pack subsided to about half of its initial extent as the pack material compacted. In general, the penetration depth attained with a smooth rounded sand (6×10) appeared to be marginally greater than that with a sharp angular sand (6×14).

To summarize, the key element in getting pack material into motion is the ability of the jet to create moving space by washing out fine materials from the filter pack in the progressive fashion described. If the filter pack is compacted, and of a size distribution that none will pass through the well screen, it appears that no motion is possible. Once the pack material is in motion it can be sustained with lower velocities than are required to initiate motion. The size of the moving region is a function of the power available in the jet relative to the power necessary to keep a given volume of fluid and pack material in motion.

Development by Line Swabbing

Line swabbing is the term given to hauling a close-fitting rubber-flanged scow through the well. Its purpose is to develop a pressure differential across the section of well containing the swab. This pressure differential results in flow being forced into the well screen on the high-pressure side and out of the screen on the low-pressure side. Because the scow is equipped with a foot valve, it drops easily, and the process can be repeated.

Modeling the flows developed by this process must be done in two stages. First, there is radial production flow generated by placing the swab in motion. This is accompanied by a steady-state motion about the swab when seen from coordinates moving with the swab. We look at each of these separately, since analysis of the system over the complete range of motion is very difficult.

Consider the swab at rest with the well at equilibrium. Moving the swab upward causes the well to flow at a rate that matches the flow generated by the swab. If there is D feet of screen below the swab, then at the filter pack/formation interface radius b (and assuming no leakage past the swab), we must have a radial inflow velocity v_r, given by

$$2\pi b D v_r = \pi a^2 U$$

so that the inflow velocity is

$$v_r = \frac{a^2 U}{2bD}, \qquad (J.1)$$

where

a = well radius, [L]
U = swab velocity. $[LT^{-1}]$

This radial velocity will gradually decrease as D increases due to the rising swab. It is conceivable that the aquifer may not produce flow at the rate induced. If not, then pressure in the well below the swab will continue to drop and may even reach vapor pressure, forming a vapor cavity behind the swab. The magnitude of pressure drop is a function of the productive capability of the aquifer, leakage past the swab and bypass flow through the filter pack about the swab. This pressure drop will gradually decrease with time since more screen surface area capable of producing is exposed as the swab rises. Maximum well production clearly cannot exceed the equivalent flow produced by the swab motion, so the maximum value for the production velocity component is that given above in Equation J.1.

The pressure rise above the swab, as the swab lifts water in the well above the static point, will continue to increase until limited by one of two effects. The well will overflow as water reaches the surface, or a level will be attained where the head is sufficient to cause flow into the aquifer and bypass around the swab at a rate that exactly matches the flow produced by the swab motion. This bypass through the filter pack will flush drilling debris and filter cake from the filter pack/formation interface into the well behind the swab.

Swab-induced flow can be modeled mathematically by ignoring the top and bottom aquifer boundaries and considering the flow in a system of coordinates moving with the swab. Basic parameters controlling the solution are the ratio of filter pack to well radii b/a, ratio of hydraulic conductivities k_2/k_1, and the distance z up or down the well from the swab made dimensionless by the well radius a (z/a). Results are given in terms of multiples of a scaling velocity v_*, which is defined by

$$v_* = \frac{k_1 H}{\pi a} ,$$

where

H = head difference across the swab,
k_1 = hydraulic conductivity of the filter pack,
a = well radius.

The speed of the swab enters the problem in two ways. As noted earlier, it controls the magnitude of the head difference across the swab. It also enters in the form of a dimensionless parameter:

$$R = \frac{aSU}{2k_1} ,$$

where

U = swab velocity,
S = specific storativity of the filter pack.

In general, S is small so that the parameter R will usually be in the range of 0.1 to 0.001 (3). In the particular case when $R = 0.1$ and the ratio of filter pack radius to well radius is 1.5, distribution of tangential (along the well) velocity at the formation/filter pack interface is given in Figure J.4. Peak tangential velocity at the interface of the formation and pack is found to be $2.9v_*$.

In the case when $R = SaU/2k_1 = 0.001$, the magnitude of the peak tangential velocity at the interface increases slightly to $3.0v_*$. The solution appears to be relatively insensitive to R in the range of practical interest.

The preceding results were developed relative to a coordinate frame moving with the swab. To reduce these results to a fixed reference frame, the variable z, the distance from the swab, must be replaced by $z - Ut$. This means that at a fixed point in the well the velocity experienced will be similar to that shown in Figure J.4 as if the well axis were time. Tangential velocity at the formation/filter pack interface will be experienced for a time of approximately $2a/U$, which will be relatively short for common well sizes and swabbing speeds.

Radial velocity is also developed by the swab bypass flow. There will be flow into the formation ahead of the moving swab and flow out of the formation behind it. This flow velocity is also scaled by v_*, and a graph of radial velocity, v_r/v_*, as a function of distance ahead of and behind the swab is given in Figure J.4. The magnitude of this radial velocity will not be high relative to the peak tangential component of velocity along the well. Nevertheless, behind the swab it will form a steady inflow into the well which will carry drilling debris along with it.

In summary, line swabbing is a two-stage process involving production inflow generated by the swab displacement followed by high-velocity tangential motion in the filter pack coupled with a more uniform radial well inflow. Repeated applications of the process will be effective in clearing drilling debris and filter cake from the borehole.

Spudding Beam Swabbing

In the spudding beam swabbing method of well development the swab is oscillated up and down in the screen section. The exact modeling of this technique is extraordinarily difficult. The problems arise from having to satisfy a condition on the swab surface that is being accelerated and then decelerated continuously. In the previous case a simple change in coordinates was sufficient to enable a solution to be found. Here we look for a solution that has a periodic nature, with frequency f (circular frequency $w = 2\pi f$). The results involve

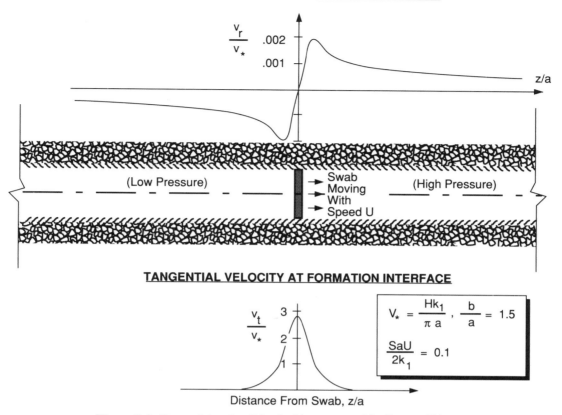

Figure J.4. Tangential and radial velocities computed for line swabbing.

numerical computations that are very time consuming, even with a digital computer. However, a simplification is possible by noting that the basic parameter of the solution is $a^2 Sw/k_1$. This has a very small magnitude, and the solution is equivalent to a motionless swab with a steady pressure drop across it. The oscillatory solution therefore is of the form of this steady-state solution multiplied by $\cos wt$.

Computation of the solution obtained in this way gives the tangential velocity distribution presented in Figure J.5. This figure includes a schematic of the flow field expected in the filter pack. It is interesting to note that, for a short swabbing stroke, results obtained for $k_2/k_1 = 0.001$ and $b/a = 1.5$ give a tangential velocity ratio $v_t/v_* = 3.0$, agreeing well with results obtained for line swabbing, even though separate methods were employed. This is a satisfying check on the computation.

For $b/a = 2.0$ the peak velocity is reduced to $1.5(v_*)$, or 50% of that for $b/a = 1.5$. The solution shown in Figure J.5 represents peak velocities to be expected from spudding beam swabbing. This would be repeated with each upstroke of the beam. Complete reversal will not occur because of the foot valve. Peak radial velocities induced by the swab motion are shown in Figure J.4. These are also cyclic with the swab motion.

Single Swab Mounted on Drill Pipe with Simultaneous Injection Pumping below the Swab

In this method of development, flow is injected into the well below a swab-equipped drill pipe. The injected flow must bypass the swab through the gravel envelope or enter the formation. The mathematical model for this operation is almost identical to that for line swabbing but with the pressure drop reversed. Head provided by the pump at the surface will control head difference across the swab when the swab is stationary.

There is a basic difference from line swabbing. Sections of the well above and below the swab will be subjected to a pressure in excess of the original static pressure in the well. Consequently the well will act as a recharge well.

Recharge potential of the well will be controlled by the hydraulic conductivity of the filter cake and increase in static pressure. For a given head drop across the swab, the effect of this recharge will not modify the magnitude of the tangential velocity profile because recharge flow is always radial. There remains the question of how much head is used in radial flow and how much is used to flush the filter pack. It seems reasonable to assume that the head difference between the pumping head and overflow is used for flushing and that

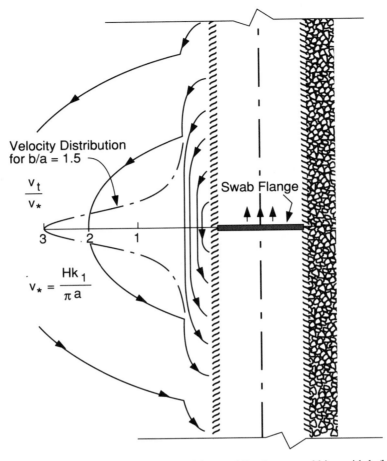

Figure J.5. Peak tangential velocities computed for spudding beam swabbing with $k_2/k_1 = .001$.

the head difference between the overflow and the original static is used for recharge. The line swab results in Figure J.5 utilizing the head difference between the overflow and pumping levels minus the head losses in the delivery pipe are used to determine the flushing velocities.

If the swab is placed in motion by hauling the drill pipe, then some modification of the previous analysis is required. On an up motion of the swab the tangential velocity in the filter pack will be reduced and may come to zero. The latter occurs when the speed of the haul, multiplied by the area of the swab, exactly matches the flow delivered by the pump. On the down stroke flushing effectiveness is enhanced since the effective flow rate will be that delivered by the pump added to that induced by the swab motion.

Double Swabs Mounted on Drill Pipe with Injection Pumping between Swabs (Flushing)

In this method of development, flow is injected into the filter pack from between two adjacent swabs (Figure J.6). A flow Q is forced into the filter pack and reenters the well above and below the swabs. There is a variation of this method that incorporates a bypass connecting the sections of well above and below the swabs.

Consider first the case without a bypass. Flow forced into the filter pack has two routes. It moves upward, reenters the well, and flows to the surface—or it moves downward to the lower section of well and must flow into the formation as recharge. In both cases it must pass through the filter pack.

If the formation has any resistance to recharge, the section below the swabs will be pressurized very quickly, and flow will pass upward through the filter pack to the surface. The net result will be similar to that described for a single swab. Peak tangential velocity will be given by the pumping head, corresponding to injection with a single swab, as shown in Figures J.4 and J.5.

If there is a flow bypass from the lower section of well to the upper section, the flow configuration is different. Flow will move in both directions within the filter pack, and even if significant recharge occurs in the well, all flow must exit between the swabs via the filter pack. Geometry for the mathematical model and numerical results obtained from the model described are depicted in Figure J.6.

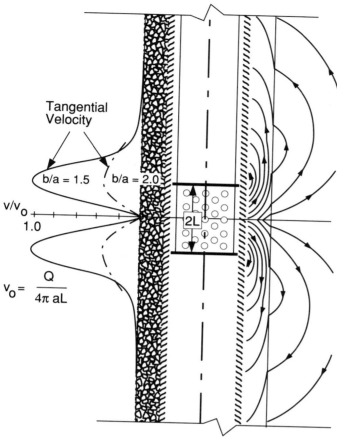

Figure J.6. Radial and tangential velocities obtained from mathematical model of swabbing with double swabs.

The scaling velocity in this case is

$$v_0 = \frac{Q}{4\pi aL} , \qquad (J.2)$$

which is the mean velocity for flow exiting between the swabs. For swabs spaced a distance apart equal to the well diameter, peak tangential velocity at the filter pack/formation interface is found to be equal to v_0 when $b/a = 1.5$, as shown in Figure J.6. When b/a is increased to 2.0, peak tangential velocity drops to $0.36(v_0)$, as shown by the dotted curves in Figure J.6.

Peak tangential velocity appears to depend rather weakly on the ratio of k_2/k_1 (the hydraulic conductivity ratio). Changing k_2/k_1 from 0.001 to 0.1 reduces peak tangential velocity by about 10% because more flow will leak into the formation.

Radial velocities are also induced by the pumping operation. There is a weak inflow into the formation, adjacent to the swabs, as depicted by the streamlines in Figure J.6. This flow reverses to become a flow out of the formation within a distance from the swabs about equal to half the

swab spacing. The magnitude of these velocities is small relative to the tangential velocities in the filter pack and probably does not contribute much toward flushing.

In this case it is possible to estimate the magnitude of the tangential velocity at the filter pack/formation interface. If we assume that the flow out between the swabs divides equally into an upflow and a downflow, and that vertical velocity distribution in the filter pack is uniform, then the estimated tangential velocity v_e is given by

$$\pi(b^2 - a^2)v_e = \frac{Q}{2} , \qquad (J.3)$$

since all flow must pass through the annular area composed of gravel envelope.

From Equations J.2 and J.3 it is possible to show that

$$\frac{av_e}{Lv_0} = \frac{2}{(b/a)^2 - 1} . \qquad (J.4)$$

This relationship is presented in Figure J.7 jointly with the tangential velocity to scaling velocity ratio as derived from the mathematical mode. Figure J.7 demonstrates that the dependence of the mathematically predicted tangential velocity on the ratios of b/a and L/a is very close to that

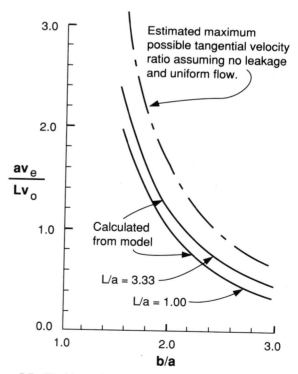

Figure J.7. Flushing velocities for swabbing with double swabs equipped with a bypass, given as a function of L/a and b/a as computed by the mathematical model.

estimated in Equation J.4. However, peak tangential velocities are found to be less than the estimated values. This can be ascribed to nonuniformity in the velocity distribution, as is apparent from the fact that the mathematically predicted value becomes a smaller fraction of the estimate as b/a increases, except for increasing L/a.

Results in Figure J.7 can therefore be used to evaluate peak tangential velocities for any practical ratio of b/a and L/a.

J.2 COMPARISON OF DEVELOPMENT METHODS

In this section we compare the methods as applied to a 14-in. diameter well with 7-in. filter pack, giving a dimensional ratio $b/a = 2.0$. Hydraulic conductivity of the filter pack is assumed to be 10,000 gpd/ft^2, and the formation 100 gpd/ft^2, giving a ratio $k_2/k_1 = 0.01$, which is within the range used in the computations.

Jetting Development

Consider first a well with filter pack particle size distribution that does not permit any filter pack material to pass through the screen, restricting pack motion. In this case assume that three jets, each ½ in. diameter producing 104 gpm at 250 psi, provide jet velocity of 190 ft/sec (4). The ratio $T/c = 28$. From mathematical modeling the peak tangential velocity at the filter pack/formation interface cannot exceed 0.003 ft/sec.

Jet development is essentially useless in this configuration. Results indicate that for any practical thickness of filter pack (>2 in.) in a situation where pack motion is impossible, jet development is inappropriate because velocities generated at the interface of the filter pack and formation are too small to be of much use.

Now consider the situation where gravel motion can be initiated by creation of a cavity through elutriation of finer material. Power in the jet is proportional to the product of the discharge rate Q and velocity v_j—or, writing this in terms of jet diameter $2c$ and jet velocity v_j, jet power is proportional to $c^2v_j^3$. Power dissipation in the motion of pack material will be proportional to v_j^3 and to the volume of material placed in motion. Thus, for a given jet flow velocity and pack material, it would be expected that the volume of gravel in motion will be proportional to c^2, everything else held constant. The linear dimension of volume in motion should then increase as $c^{2/3}$. Given a 0.167 in. jet diameter, delivering water at 180 ft/sec with a volume of moving pack material 5 in. thick, a 0.5 in. jet delivering water at about the same velocity should produce a moving pack volume about 10 in. thick, provided free volume for material to move either exists or can be created by flushing fine material

through the screen. The fact that this may not always occur helps explain mixed field results using this method of development.

If the filter pack can be placed in motion by the jet, effective flushing will develop. Down-well motion of the jetting tool must necessarily be slow, or the jet may be moved away from the moving pack material. In regions with high overburden pressures giving high intergranular stress, development of pack motion will be very slow and the technique less effective. With low intergranular stress and excessive fine filter material, slow jet motion may destroy effectiveness of the filter pack through the creation of cavities.

Line Swabbing

In the analysis of line swabbing it was explained that the process occurs in two stages. First, there is a radial flow velocity at the borehole of

$$v_r = \frac{a^2 U}{2bD} .$$

Second, a tangential velocity through the filter pack adjacent to the swab is set up. A scale velocity for this process was defined where

$$v_* = \frac{Hk_1}{\pi a} .$$

The peak tangential velocity v_t was found to be

$$v_t = 3.0(v_*) \quad \text{when} \quad \frac{b}{a} = 1.5 ,$$

$$= 1.5(v_*) \quad \text{when} \quad \frac{b}{a} = 2.0 .$$

Applying these results to the example of a 14-in. well and 7-in.-thick filter pack implies that for a swab speed of 3 ft/sec, the initial radial velocity at the filter pack/formation interface will be

$$v_r = \frac{0.44}{D} .$$

When the swab is moving near the bottom of the well screen, this velocity will be high. With 100 ft of screen below the swab, it will be 0.0044 ft/sec.

In the second stage, with $b/a = 2.0$, $k_1 = 10,000$ gpd/ft$^2 = 0.0154$ ft/sec, we have the tangential velocity

$$v_t = \frac{1.5 \times 0.0154 \times H}{\pi(7/12)} = 0.0126H .$$

Thus a head difference of 1 ft across the swab will produce four times the tangential velocity at the exterior of the filter pack, as would a ½-in. jet at 190 ft/sec. With a head difference of 10 ft the tangential velocity would be 1.5 in./sec. The head H, in effect, is controlled by swabbing speed since, as noted, if the swab is moving fast enough, the flow produced will exceed the capacity of the well to absorb it. Actual tangential velocity also depends upon the length of screen above and below the swab and how tightly the swab fits, since these control H, the head available.

The radial bypass flow produced by the swab motion is given in Figure J.4. Results show that at 50 well radii below the swab, there is an inflow velocity at the filter pack/formation interface of the order of $10^{-4}v_*$. This must be added to the radial production component induced by the swab, but as can be seen from the previous discussion, it is small in comparison and therefore probably does not play much of a role in the flushing process.

Speculation can arise as to how modifying the diameter of the swab to provide clearance will affect swabbing efficiency. Assuming swab clearance acts as an orifice, head loss across this orifice satisfies an equation of the form

$$h_1 = \frac{kV^2}{2g} ,$$

where V is the discharge velocity through the orifice. If V is large, then h_1 may exceed the head required to drive the flow through the filter pack and a reasonable velocity may still be attained within the filter pack. The key issue is whether head loss across the swab clearance is enough to divert flow through the filter pack. For the case considered here, with a 14-in. well and 7-in.-thick filter pack, a head of 10 ft would give peak tangential velocities in the pack of 1.5 in./sec. A swab with ¼-in. clearance provides a clearance area of 11 in.^{2}. Assuming a head loss coefficient 2 for the orifice, a velocity of 18 ft/sec is calculated. Thus a flow of about 600 gpm could bypass the swab at the head difference necessary to produce tangential velocities in the filter pack of 1.5 in./sec.

Spudding Beam Swabbing

The peak velocities obtained are, for the same head across the swab, equal to those for line swabbing. However, the effect of the valve in the swab, which allows it to fall back, results in a reduction in the head available to produce a flow through the filter pack. The effect of the spudding beam motion is to "pulsate" the flow in the filter pack. Although the head difference available may be reduced by valve action, accelerations associated with the oscillatory motion would provide an excess pressure difference equal to a modification of the gravitational acceleration. A 3-ft stroke and period of 2 sec results in a peak acceleration of $1.5\pi^2 = 14.8$ ft/

sec^2, or about g/2. Thus peak tangential velocity through the filter pack is increased by about 50%.

As with line swabbing, the radial effect can be split into a component representing production and a component representing swab bypass flow through the filter pack. Radial flow induced by the production component is governed by the depth of well screen below the swab and peak swab velocity. Peak swab velocity for a 3-ft stroke and period of 2 sec is 1.5π ft/sec, or about 4.7 ft/sec. Given equal heads, radial velocities are about 50% greater than for line swabbing. Flows are repeated every 2 sec, so there will be steady migration of particles through the well screen.

Given the pulsating character of flow fields produced by spudding beam swabbing, it is probably an effective development tool both for flushing of drilling debris and wall cake and consolidation of the filter pack.

Single Swab Mounted on Drill Pipe with Injection Pumping below the Swab

Peak tangential velocity at the formation/filter pack interface for a stationary swab is equal to three times the scaling velocity v_* defined by $Hk_1/\pi a$. The pumping head provides bypass flow, while the increase in the static head develops a recharge flow. Allowance must be made for the pumping head used to overcome pipe friction.

Suppose a tangential velocity of 1.5 in./sec is considered adequate at the filter pack/formation interface. For a 14-in. diameter well and 7-in. filter pack, this is accomplished with a head difference of 10 ft across the swab. Flow that must be delivered to attain this head difference depends on total screen length. Since the entire open well screen is available for recharge, flow to be delivered is controlled by total screen length and formation recharge capability. In any situation the operator can easily ascertain what is occurring from observation of pumping pressure and volume of makeup water required.

Swab motion produced by hauling and dropping the drill pipe creates significant velocities in the filter pack. A typical fall velocity is about 8 ft/sec. With a 14-in.-diameter well screen this produces a flow of 3800 gpm, which is generally in excess of injection pumping capacity. Without swab clearance this flow must either be forced into the formation or bypass the swab through the screen. Such flows forced into the filter pack will almost certainly fluidize it and be very effective in scouring wall cake from the borehole. However, if there is considerable screen below the swab, some flow could go into the formation as temporary recharge.

Double Swabs Mounted on Drill Pipe with Injection Pumping between Swabs

Without a bypass through the swabs, the system acts as a single swab, or it can be considered as one-half of the bypass-equipped double swab. The analysis is therefore subject to a degree of uncertainty in that some flow will enter the formations as recharge. Hauling and dropping the swab creates the same flows as a single swab. In an example illustrating this development method, 2200 gpm of flushing water were pumped into a newly constructed well with 18-in.-diameter casing and 30-in.-diameter filter pack. Water was pumped out between double swabs which in one case were spaced 5 ft apart and in another case 18 in. apart. Peak tangential velocity in both cases is approximately the same since, as we have seen, the ratio b/a is the predominantly controlling mechanism. If we assume that all flow passes upward through the filter pack and that there is essentially no recharge, tangential velocity cannot exceed

$$v = \frac{Q}{\pi(b^2 - a^2)}$$
$$= \frac{2200/449}{\pi[(1.25)^2 - (0.75)^2]} \quad (J.5)$$
$$= 1.56 \text{ ft/sec} .$$

As shown, actual tangential velocities are, for $b/a = 1.7$, about 70% of this estimate, giving tangential velocity of about 0.65 ft/sec. Swab spacing will not influence this significantly.

Where there is flow bypass and when $b/a = 1.5$, peak tangential velocity at the formation/filter pack interface is equal to the mean velocity of flow exiting between the swabs. When $b/a = 2.0$ peak tangential velocity drops to $0.36v_0$, as shown by the results in Figure J.6. Thus to obtain a peak tangential velocity of 1.5 in./sec in a 14-in. well with a 7-in. filter pack ($k_1 = 10,000$ gpd/ft^2),

$$Q = 1.5 \text{ cfs} = 666 \text{ gpm} .$$

The pressure difference necessary to produce this flow is

$$H = \frac{Q}{4\pi B_0 k_1} \quad \text{with } B_0 = 0.76 .^2$$
$$= 17.4 \text{ ft}.$$

For the actual case considered here with a flow of 2200 gpm and a swab bypass, then from Figure J.7, with $b/a = 1.67$ and 5-ft swab spacing ($L/a = 3.33$), peak tangential velocity is predicted to be 0.55 ft/sec. For $b/a = 1.67$ and $L/a = 1.0$ (18-in. swab spacing), tangential velocity is 0.45 ft/sec. The slight reduction that is predicted in the peak tangential flushing velocity at the formation/filter pack in-

2. See the last section on mathematical models for development methods, in particular the discussion of injection swabbing, at the end of this appendix.

terface is a consequence of a less uniform flow distribution in the filter pack as the swabs are moved closer together.

With a double swab the effect of undersized swabs must be compensated for by increasing pump capacity to offset leakage. As discussed, the effect of providing swab clearance is the same as introducing an orifice that permits flow to bypass the swab. With a ¼-in. clearance on the swab diameter in a 14-in. diameter well, it was estimated that 10 ft of head would produce a flow of about 600 gpm. Thus with two swabs an additional 1200 gpm is required. Adding this to the 666 gpm required for flushing the gravel gives a total of about 1900 gpm.

Performance Summary

We can summarize this discussion of the various methods in terms of the tangential velocities induced at the filter pack/formation interface to develop a typical 14-in.-diameter well with 28-in.-diameter filter pack.

Method	Injection Flow Required	Peak Tangential Velocity at the Borehole
Jetting (3 jets)	312 gpm	0.003 ft/sec
Line swab (10 ft of water above swab)		0.125 ft/sec transient
Spudding beam (10 ft of water above swab)		0.125 ft/sec pulsating
Single swab with injection	180 gpm minimum[a]	0.125 ft/sec
Falling single swab with injection	180 gpm minimum[a]	Very high
Double swab with no bypass with injection	180 gpm minimum[a]	0.125 ft/sec
Double swab with bypass with injection	666 gpm	0.125 ft/sec

[a] Assumes no loss to recharge.

J.3 CONCLUSIONS

The preceding analysis and computations permit some conclusions to be drawn regarding the efficacy of the various well development techniques:

1. Unless filter pack material is carefully chosen to provide some, but no excessive material loss through the well screen, high-velocity jetting, even when considered under the most favorable assumptions, is unlikely to provide effective well development. Without some

loss of particles to allow jet-induced motion of the pack material, the rate of velocity decay of the jet within the filter pack is extremely rapid. However, jetting can be an effective development technique, provided motion of the filter pack can be induced.

2. Line swabbing provides relatively high tangential velocities at the interface between the filter pack and formation. At any given location the velocities occur briefly during passage of the swab. Radial production flows out of the formation, in addition to tangential scouring velocities, are developed. The magnitude of the production flow is inversely proportional to the depth of screen below the swab. For these reasons many passes of the swab are likely to be necessary to develop the well.

3. Spudding beam swabbing effectiveness is limited by the head available to produce flow through the filter pack. This may be compensated for by the pulsating flow induced. The rocking mode, combined with slow hauling, is an effective method provided that sufficient power is available to encourage transport through the filter pack.

4. Single-swab development with injection pumping below the swab produces significant tangential velocities at the filter pack/formation interface. Since the entire well is under pressure causing recharge flow, makeup water is required. Hauling and dropping the drill pipe induces high tangential velocities in the filter pack. It is not obvious how this flow will be divided between recharging the formation and flushing the filter pack.

5. Operation of double-swab injection development depends markedly on whether a bypass conduit is available to permit return flow from below the swabs. Without this bypass the operation is virtually identical to a single swab, and flow losses to recharge may be significant from pressurization of the lower well section. With flow bypass the following comments are made:

   a. Efficiency is high since, depending on well depth, a reasonable fraction of energy expended is within the filter pack.

   b. Tangential velocities within the filter pack can be controlled easily by adjusting pumping rates.

   c. Leakage past swab clearances can be offset by increased pumping rates to maintain the same flow through the filter pack.

J.4   MATHEMATICAL MODELS FOR DEVELOPMENT METHODS

All the mathematical models are based on the well-known potential theory of flow in porous media. The basic equation for the velocity potential $\phi(r, z, t)$ is

$$K\left(\frac{\partial^2 \phi}{\partial z^2} + \frac{1}{r}\frac{\partial \phi}{\partial r} + \frac{\partial^2 \phi}{\partial r^2}\right) = \frac{\partial \phi}{\partial t} \ , \qquad (J.6)$$

where the velocity in the z-direction is

$$v_z = -k\frac{\partial \phi}{\partial z} \qquad (J.7)$$

and in the radial direction is

$$v_r = -k\frac{\partial \phi}{\partial r} \ , \qquad (J.8)$$

and where

k = hydraulic conductivity,

K = k/S, where S is the specific storativity for the medium.

In general, S is small ($10^{-4} - 10^{-6}$). For steady flow $\partial \phi/\partial t = 0$, and we have

$$\frac{\partial^2 \phi}{\partial z^2} + \frac{1}{r}\frac{\partial \phi}{\partial r} + \frac{\partial^2 \phi}{\partial r^2} = 0 \ . \qquad (J.9)$$

For layered media, we have two such potential functions, ϕ_1 and ϕ_2, and these satisfy conditions

$$\phi_1 = \phi_2 \ , \quad k_1\frac{\partial \phi_1}{\partial n} = k_2\frac{\partial \phi_2}{\partial n} \ , \qquad (J.10)$$

where n is the normal direction to the interface between region 1 and region 2. We consider each problem in turn.

Jetting

For steady jet flow we take coordinates (r, z) with $r = 0$ the axis of the jet, $z = 0$, the surface of the filter pack, and $z = T$, the filter pack/formation interface. It can be shown that

$$\phi_1(r, z) = \int_0^\infty f_1(\lambda)J_0(\lambda r)$$

$$\times \left[\frac{k_1 \cosh \lambda(T - z) + k_2 \sinh \lambda(T - z)}{k_1 \cosh \lambda T + k_2 \sinh \lambda T}\right]\partial \lambda \qquad (J.11)$$

and

$$\phi_2(r, z) = \int_0^\infty f_2(\lambda)J_0(\lambda r)e^{-0.112z} \, d\lambda$$

satisfy the differential Equation J.9, and conditions J.10. The solution then requires finding $f_1(\lambda)$ such that

$$\phi_1(r, 0) = 0 \quad \text{for } r > c, \tag{J.12}$$

$$2\pi \int_0^c k_1 \frac{\partial \phi_1}{\partial z} r\,dr = Q. \tag{J.13}$$

Condition J.12 specifies a constant potential surface outside the jet, and condition J.13 that the jet flow into the medium is correct. The correct function $f_1(\lambda)$ is found by analogy with the homogeneous medium heat flow problem given by Carslaw and Jaeger (1). It is

$$f_1(\lambda) = \frac{Q}{2\pi c k_1 A \lambda} \left(\frac{\sin \lambda c}{\lambda c} - \cos \lambda c \right), \tag{J.14}$$

where

$$A = \int_0^\infty \frac{J_1(t)}{t} \left(\frac{\sin t}{t} - \cos t \right) \\ \times \tanh\left(\frac{tT}{c} + \epsilon \right) dt. \tag{J.15}$$

In these formulas J_0 and J_1 are Bessel functions of order 0 and 1 and

$$\epsilon = \frac{1}{2} \log_e \left(\frac{1 + k_2/k_1}{1 - k_2/k_1} \right).$$

From formulas J.11, J.14, and J.15 and formulas J.7 and J.8, it is possible to evaluate the velocities in the filter pack. The integrals were computed using polynomial approximations for $J_0(x)$ and $J_1(x)$, Abramowitz and Stegun, and Simpson's ⅜-rule numerical integration formula (5). A step length of 0.01 appeared adequate with all computations performed in double-precision arithmetic (17 significant figures). Evaluation of formula J.15 gave a value of $A = 0.7648$ for $T/c = 28$ and $\epsilon = 0.001$, and integrating to $t = 750$. This agrees reasonably well with the maximum possible value for the integral, which occurs when T/c is very large, and the integral can be evaluated exactly to give $\pi/4 = 0.7854$. The integrals for velocity converge very quickly on $z = T$ due to the presence of a $\cosh(\lambda[T/c] + \epsilon)$ term in the denominator.

Swabbing

For the swabbing problems, coordinates were taken with z on the axis of the well and r radially from the well axis so that the well screen was $r = a$ and the filter pack/formation interface, $r = b$.

The moving swab problem was solved by changing to coordinates moving with the swab to give a general solution for ϕ_1 that satisfies conditions J.10 and Equation J.9 plus a term $-U \partial \phi / \partial z$. Since the solution must be odd in z, and the eigenfunctions of the equation are $\sin \lambda z$, $\cos \lambda z$, and $I_0(\zeta r)$, $K_0(\zeta r)$, where $\zeta = (\lambda^2 + U/2k)^{1/2}$ and I_0 and K_0 are modified Bessel functions of the first and second kind, the general solution for ϕ_1 is

$$\phi_1(r, z) = e^{\gamma z/a} \int_0^\infty f(t) \sin(tz/a)$$

$$\times \left[\frac{\alpha I_0 \left(\zeta \frac{r}{a} \right) + \beta K_0 \left(\zeta \frac{r}{a} \right)}{\alpha I_0(\zeta) + \beta K_0(\zeta)} \right] dt,$$

where

$$\zeta = (t^2 + \gamma^2)^{1/2},$$
$$\gamma = aU/2k,$$
$$\alpha = 1 - k_2/k_1,$$
$$\beta = I_1(\zeta b/a)/K_1(\zeta b/a) \\ + (k_2/k_1)I_0(\zeta b/a)/K_0(\zeta b/a),$$

and I_1 and K_1 are modified Bessel functions of order 1. $f(t)$ must be chosen so that

$$\phi_1 = \begin{cases} +H/2 & \text{on } z > 0, \ r = a, \\ \\ -H/2 & \text{on } z > 0, \ r = a, \end{cases}$$

This is achieved by choosing

$$f(t) = \frac{Ht}{\pi(t^2 + \gamma^2)}.$$

Although this expression appears somewhat formidable, the velocities can be evaluated using formulas J.7 and J.8. Numerical integration is performed as before, using polynomial approximations for the modified Bessel functions. The results for v_t are given in Figure J.4 for the case $\gamma = 0.1$. For $\gamma = 0.001$, the results appear independent of γ and agree identically with the results of a steady swab calculation to be described shortly.

The solution for ϕ_2 has not been evaluated numerically but must be of the form

$$\int_0^\infty f_2(t) \, \sin\left(t\frac{z}{a}\right) K_0\left(\zeta\frac{r}{a}\right) dt \ ,$$

since it is known that $\phi_2 \to 0$ as r becomes large.

Spudding Beam Swabbing

In the previous case where there is a steady velocity, a simple change in coordinates was sufficient to enable a solution. Here, the solution will be of the form

$$\phi(r \ , z \ , t) = \mathrm{Re} \left[e^{i\omega t} \, \psi(r \ , z)\right] \ ,$$

where Re stands for the real part and $i = \sqrt{-1}$, $\psi = 2\pi f$, where f is the the stroking frequency of the oscillation. For this problem the equation is J.6.

The solution for ϕ_1 in this case is given by

$$\phi_1 (r \ , z \ , t) = \frac{2H}{\pi} \, \mathrm{Re}\left\{e^{i\omega t} \int_0^\infty \frac{J_0(l\lambda) \, \sin \lambda z}{\lambda} \left[\frac{\alpha I_0(\zeta r) + \beta K_0(\zeta r)}{\alpha I_0(\zeta a) + \beta K_0(\zeta r)}\right] d\zeta\right\} \ ,$$

with α and β as before but with

$$\zeta = \left(\lambda^2 + \frac{i\omega}{k}\right)^{1/2}$$

and l = length of swab stroke.

Evaluation of this would be very time consuming, and it is fortunate that ω/k is very small. The solution can therefore be approximated closely by putting $\zeta = \lambda$ in the integral. The solution is then a function of the ratio of l/a. However, evaluation of v_t/v_* at $z = 0$ for $l/a = 0.1$ gives a value of 3.01 and for $l/a = 0.01$ a value of 3.01. It seems reasonable to conclude that for l/a small, the solution is then identical with that found for the line-hauling swab (a satisfying check!). In other words, the oscillatory solution is a succession of steady-state solutions, since the medium responds faster than the period of oscillation. The numerical computations for v_t/v_* at $r = b$ are given in Figure J.5.

Injection Swabbing

For this case the flow is steady, so the equation is J.9 and the boundary conditions at $r = a$ are that $\phi_1 = 0$ for $|z| > 2L$ and

$$-2\pi a \int_0^L k_1 \frac{\partial \phi_1}{\partial r} \, dz = Q \ ,$$

with condition J.10 at $r = b$.

The solution to this problem is given by

$$\phi_1(r \ , z) = \frac{Q}{4\pi aB_0} \int_0^\infty \frac{J_1(L\zeta) \, \cos z\zeta}{\zeta} \left(\frac{\alpha I_0(\zeta r) + \beta K_0(\zeta r)}{\alpha I_0(\zeta a) + \beta K_0(\zeta r)}\right) d\zeta \ ,$$

where α and β are as previously given and

$$B_0 = \int_0^\infty \frac{J_1(tL/a) \, \sin \, (tL/a)}{t} \left[\frac{\alpha I_1(t) - \beta K_1(t)}{\alpha I_0(t) + \beta K_0(t)}\right] dt \ .$$

The following values have been found for B_0 when $b/a = 1.5$, and $k_2/k_1 = 0.001$:

L/a	0.5	1.0	2.0
B_0	0.913	0.760	0.573

For $k_2/k_1 = 0.1$, $L/a = 1.0$, and $b/a = 1.5$, the value of B_0 changes to 0.844; thus it is not very sensitive to the ratio of hydraulic conductivities. For $b/a = 2$ and $k_2/k_1 = 0.001$, the value of $B_0 = 1.001$.

Numerically computed results for the tangential velocity for the cases $L/a = 1$ and $b/a = 1.5$ and 2.0 are given in Figures J.6 and J.7. The streamline patterns are not computed but are schematic, based on an evaluation of the ratio of v_t/v_* at the exterior of the gravel pack.

The flow exterior to the gravel pack in the formation was not evaluated numerically. ϕ_2 is of the form

$$C \int_0^\infty \frac{J_1(L\zeta) \, \cos \, (z\zeta)}{\zeta} K_0(\zeta r) d\zeta \ ,$$

where C is obtained by matching to ϕ_1 according to formula J.10.

REFERENCES

1. List, E. J. 1983. *Analysis of Development Methods for Gravel Envelope Wells. Roscoe Moss Company* publication, no. 26-FS/383, 16 pp.
2. Carslaw, H. S., and J. C. Jaeger. 1959. *Conduction of Heat in Solids*. Oxford University Press, Oxford.
3. Bear, J. 1979. *Hydraulics of Groundwater*. McGraw-Hill, New York.
4. Zdener, F. F., and R. E. Allred. 1979. "Correct Methods are Essential in Well Development." *Johnson Driller's Journal* (January-February).
5. Abramowitz, M., and I. A. Stegun. 1965. *Handbook of Mathematical Functions*. Dober, New York.

APPENDIX K

Ground Water Models

K.1 PHYSICAL GROUND WATER MODELS

Sand-Tank Models

Ground water modeling began as a simplified physical representation of ground water flow with observation and measurement of water levels and flow rates. Darcy's original apparatus (Chapter 1) was used to "model" changes in flow rate for varying water-level differences across a typical filter-bed sand.

Physical models consist of a tank filled with porous material (usually sand). Water is introduced into the model, and water-level distributions are measured. Models of this type are generally referred to as "sand-tank" models. Sand-tank models are still used to study complex hydraulic interrelationships not easily expressed in mathematical terms. Typical problems range from simple aquifer flow relationships to complex laminar/turbulent-flow regimes.

The world's largest sand-tank model, illustrated in Figure K.1, was built in 1979 by Roscoe Moss Company of Los Angeles, California, to study interrelationships between aquifers, gravel envelopes, and well screens (1, 2). It has been used to physically verify theoretical equations on well efficiency and development, and to measure the effect of entrance velocity and well screen open area on well efficiency.

In a sand-tank model, field conditions are approximated by tank shapes ranging from rectangular boxes representing an aquifer strip to pie-shaped models simulating a radial flow sector of a well. Water flows through porous media similar to aquifer material (sand or gravel). Both confined and unconfined aquifers can be modeled once simulation problems are overcome. In unconfined models excessive capillary rise is minimized by increasing the porosity of the sand. In confined models problems of channeling against the upper confining tank surface have been successfully eliminated using an inflatable membrane (1).

Scaling Relations. Ratios between model and field lengths, velocities, discharges, storativities, and time are known as scaling relationships. Model dimensions are determined by field dimensions and hydraulic and geometric similarity. Scaling relationships for sand-tank models are based on Darcy's law and are as follows:

Length Ratio (Lr)

$$L_r = \frac{L_m}{L_p},$$

where r is a subscript denoting the ratio of the model to the prototype.

Velocity Ratio (Vr)

$$V_r = \frac{V_m}{V_p} = \frac{K_m(\partial h/\partial x)_m}{K_p(\partial h/\partial x)_p},$$

where

K = hydraulic conductivity [LT^{-1}],
$\partial h/\partial x$ = hydraulic gradient.

Flow-Rate Ratio (Qr)

$$Q_r = K_r L_r^2 .$$

Time Ratio (tr)

$$t_r = \frac{S_r L_r}{K_r},$$

where

S_r = storativity ratio,
L_r = length ratio,
K_r = hydraulic conductivity ratio.

Viscous Flow Models

A physical model that has been used to simulate laminar flow through porous media is the viscous-flow or Hele-Shaw model. This model was developed by Professor Hele-Shaw in 1896 to study fluid flow around varying-shaped obstructions. Subsequently, other researchers (Dackler, Gunther, and Todd) have used viscous-flow models to simulate a vertical or

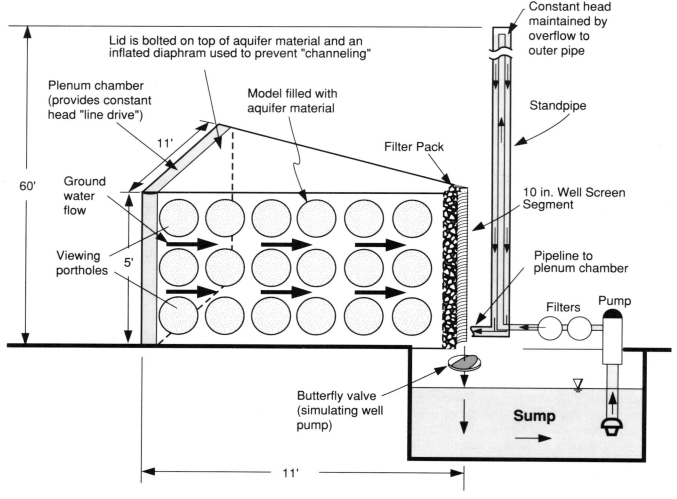

Figure K.1. Roscoe Moss Company well/aquifer model.

horizontal segment through an aquifer (3). As in other physical models, observation of flow rates and fluid levels are used to calculate field values via the scaling equations.

A viscous-flow model consists of two parallel plates (usually plastic or glass), separated by a capillary interspace, typically 1/16 inch. Fluid flowing between the plates is laminar, representing a "slice" through an aquifer, with analogy based on the similarity between the differential form of Darcy's law and Poiseuille's equation (the similarity between laminar flow in porous media and laminar flow of viscous fluids in the capillary spacing between two parallel plates).

Darcy's law states the velocity of flow is directly proportional to the hydraulic gradient. The differential form of Darcy's law is

$$v = -K \frac{\partial h}{\partial x} ,$$

where v is the bulk velocity at any point in the field of flow, h is the hydraulic head, and x is the direction of flow.

The flow between two parallel plates obeys Poiseuille's equation which is

$$v_m = -\frac{1}{12} \left[g \left(\frac{b_m^2}{\nu} \right) \right] \frac{\partial h}{\partial x} ,$$

where

v_m = average velocity of the fluid at a point (x) in the model $[LT^{-1}]$,

ν = kinematic viscosity of the fluid $[L^2 T^{-1}]$,

b_m = spacing between the plates $[L]$,

g = acceleration due to gravity $[LT^{-2}]$,

x = length along the direction of flow $[L]$.

Hele-Shaw models built to simulate ground water flow systems have assumed many aquifer shapes, dimensions, and types. Williams in 1965 built the first confined viscous-flow model to study ground water flow in a wedge-shaped aquifer, verifying Hantush's analytical equation for this flow (4). In this model confined aquifers were simulated by storage tubes placed along the back of the model. Inside each tube was a rod used to compensate for varying aquifer storage. The compensating rods were used to decrease storativity in the direction of convergence (see Figure K.2).

Scaling Relations. Design of a viscous-flow model involves an iterative calculation between model length, height, spacing, and fluid viscosity.

The scaling relationships between the model and field condition are as follows:

$$K_{xp} = \frac{K_{xm}}{R_{K_x}}, \quad K_{zp} = \frac{K_{zm}}{R_{K_z}},$$

$$x_p = \frac{x_m}{R_x}, \qquad z_p = \frac{z_m}{R_z},$$

$$h_p = \frac{h_m}{R_h}, \qquad S_{sp} = \frac{S_{sm}}{R_{S_s}},$$

$$t_p = \frac{t_m}{R_t},$$

(K.1)

where

K = hydraulic conductivity $[LT^{-1}]$,
h = hydraulic head $[L]$,
S_s = specific storativity $[L^{-1}]$,
t = time $[T]$,
p = prototype (field condition),
m = model,
x = horizontal (length) direction $[L]$,
z = vertical direction $[L]$,
R = model-prototype ratio.

Other physical models include simulations based on membrane, electrolytic, and electrical conduction.

K.2 MATHEMATICAL MODELING

Exact Analytical Solutions

An example of a simple mathematical model can be seen considering a well penetrating a homogeneous and isotropic confined aquifer of infinite areal extent. The particular solution of this boundary value problem is the Theis equation (see Chapter 5) which is

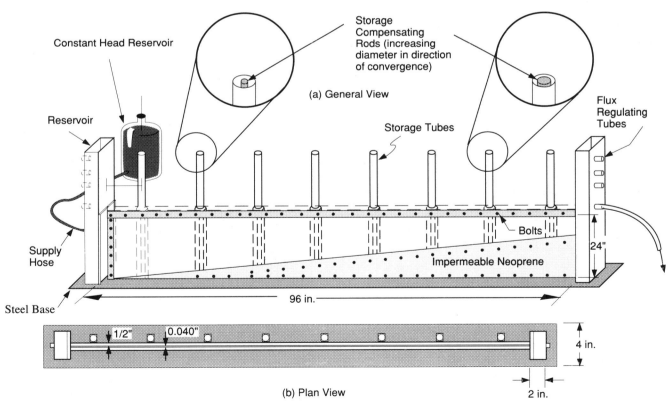

Figure K.2. Viscous flow model (after Williams, D.E., "Viscous-Model Study of Ground-Water Flow in a Wedge-Shaped Aquifer"). *Water Resources Research*, Vol. 2, 3rd Qrtr., 1966. Copyright © by the American Geophysical Union.

$$s(r, t) = \frac{114.6\ Q}{T} W(u)\ ,$$

where

$s(r, t) =$ drawdown at distance r and time t after pumping began [ft],

Q = well-pumping rate [gpm],

T = aquifer transmissivity [gpd/ft],

$W(u)$ = well function of Theis (Table 15.2 with r/B = 0),

u = $\dfrac{1.87\ r^2 S}{Tt}$,

S = aquifer storativity.

r = distance from well [ft],

t = time since start of pumping [days].

This exact solution of an idealized field case is easily applied. Tables of the well function $W(u)$ for different values of u are shown in Table 15.2. However, because of the assumptions leading to this solution, the model is of limited use and is primarily employed for single-well rather than basinwide studies.

Superposition Models

The superposition model is a more flexible analytical tool. In standard units it may be written

$$s(x, y, t) = \frac{114.6}{b\sqrt{K_x K_y}}$$

$$\times \sum_{i=1}^{n} Q_i W\left\{ \frac{1.87S}{t}\left[\frac{(x - x_i)^2}{T_x} + \frac{(y - y_i)^2}{T_y} \right] \right\}\ ,\quad (K.2)$$

where

$s(x, y, t) =$ drawdown [ft] at distances x and y [ft] after pumping time [days],

b = saturated thickness of aquifer [ft],

Q_i = discharge of ith well [gpm],

x_i, y_i = coordinate of the ith well [ft, ft]

T_x, T_y = transmissivity [gpd/ft] in x and y directions, respectively,

$$W\left\{ \frac{1.87S}{t}\left[\frac{(x - x_i)^2}{T_x} + \frac{(y - y_i)^2}{T_y} \right] \right\}$$

= well function of Theis

This equation is used to measure the regional well-field interference around a number of pumping and/or recharge

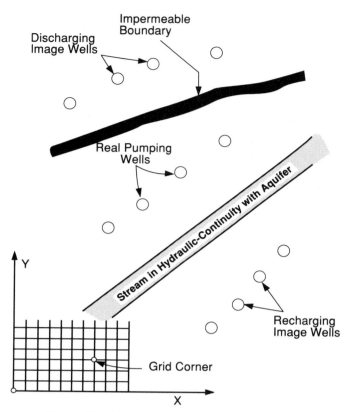

Figure K.3. Superposition model with image wells.

wells, with formation parameters varying anisotropically. It is sufficiently accurate in problems where seasonal recharge effects are not significant and geohydrologic properties do not vary significantly.

An example of a superposition model is shown in Figure K.3. Image wells are used to mathematically create the effect of hydrologic boundaries. After proper placement of the image wells (see Chapter 5), the problem reduces to solution of the superposition model for 12 wells. A grid network is overlaid on the model area, and Equation K.2 is repeatedly solved at each grid corner. With drawdowns calculated for each grid corner, contour maps may be constructed and the regional interference pattern studied. This iterative procedure is repeated for each time period required in the study.

K.3 NUMERICAL APPROXIMATION METHODS

Current ground water modeling involves approximation of the basic equation of ground water flow (as presented in Chapter 2) by a discrete (finite) set of parameters which may be used to calculate water-level distributions in the modeled area. The procedure involves numerial approximation of the terms in the equation using either: the method of finite differences (FDM); or the method of finite elements (FEM).

For example, in an isotropic artesian aquifer, the governing equation may be written

$$\frac{\partial^2 h}{\partial x^2} + \frac{\partial^2 h}{\partial y^2} + \frac{\partial^2 h}{\partial z^2} + Q = \frac{S}{T}\frac{\partial h}{\partial t} , \qquad \text{(K.3)}$$

where

h = hydraulic head [L],
S = storativity,
T = transmissivity [L^2T^{-1}],
Q = flux [L^3T^{-1}].

In operator form, Equation K.3 may be rewritten

$$\nabla \cdot T\nabla h + Q = \frac{S}{T}\frac{\partial h}{\partial t} .$$

The first term, $\nabla \cdot T\nabla h$, is determined by the water-level difference and aquifer transmissivity. The second term, the total flux Q, is composed of two major subterms: net pumping (pumping extraction less artificial recharge) and natural inflow (natural subsurface inflow less outflow plus leakage and infiltration). The third term, $S(\partial h/\partial t)$, the time-dependent change of storage, is governed by storativity and calculated heads at any time.

The first major subterm, net pumping, is usually found from historical operational records. The second subterm, natural inflow, is obtained by solving Poisson's equation for each model grid block. Poisson's equation for two-dimensional flow may be written as

$$\frac{\partial(T_x\partial h/\partial x)}{\partial x} + \frac{\partial(T_y\partial h/\partial y)}{\partial y} + QQ = 0 ,$$

where QQ represents the natural recharge term.

Finite Difference Approximation

Assuming that the function (h) and its derivatives are single valued, finite, and continuous functions of x and y, then Equation K.3 can be expanded in an infinite series. Using the differencing technique of Crank-Nicholson with both space- and time-centered differences (5), Equation K.3 can be discretized as follows:

$$\frac{1}{\Delta x_{i,j}} \left\{ T_{i+1/2,j} \left[\frac{h_{i+1,j}^{n+\theta} - h_{i,j}^{n+\theta}}{\Delta x_{i+1/2,j}} \right] \right.$$

$$\left. - T_{i-1/2,j} \left[\frac{h_{i,j}^{n+\theta} - h_{i-1,j}^{n+\theta}}{\Delta x_{i-1/2,j}} \right] \right\}$$

$$+ \frac{1}{\Delta y_{i,j}} \left\{ T_{i,j+1/2} \left[\frac{h_{i,j+1}^{n+\theta} - h_{i,j}^{n+\theta}}{\Delta y_{i,j+1/2}} \right] \right. \qquad \text{(K.4)}$$

$$\left. - T_{i,j-1/2} \left[\frac{h_{i,j}^{n+\theta} - h_{i,j-1}^{n+\theta}}{\Delta y_{i,j-1/2}} \right] \right\}$$

$$+ Q_{i,j} = S_{i,j} \left[\frac{h_{i,j}^{n+1} - h_{i,j}^{n}}{\Delta t^{n+1/2}} \right]$$

where

θ = ½ (Crank-Nicolson form),
h = head in aquifer [L],
$Q_{i,j}$ = flux per unit area [LT^{-1}].

Solution of Equation K.4 is generally performed by an implicit algorithm procedure. One method outlined by Halespaska employs a head extrapolation technique using the delsquared Aitken relation prior to each time step (5). The system of linear equations created forms a tridiagonal matrix that is then solved for head using the Thomas algorithm. The computed values of head are then compared to the extrapolated values. If the difference is greater than some chosen epsilon, the computed value is resubstituted and the linear equations solved again. This iterative procedure continues until two successive iterates fall below epsilon.

Use of the finite-difference methods has been shown to be unconditionally stable with respect to time and space and provides fast and accurate solutions requiring a minimum of computer time. Figure K.4a shows a typical model grid network and flow and storage terms associated with an individual model node.

Method of Finite Elements

The finite difference approximation used in ground water modeling employs a differential approach. The finite element method performs a similar task using an integral approach. In the method of finite elements the modeled region is divided into subareas (elements) whose geometry is determined by the grid network (see Figure K.4b). An integral expression is then derived using the method of weighted residuals or the variational method (6). The dependent variables (e.g., water or piezometric levels) are approximated in terms of interpolation functions (also called "basic" functions). For ease of comparison, these are usually chosen to be continuous, polynomial functions (generally linear-quadratic or cubic).

Once the basic functions are specified and the grid determined, an integral relationship for each element in the network is specified. The integrals are calculated for each element, yielding a system of first-order linear-differential equations in time. These are then solved using the method of finite differences and standard matrix techniques.

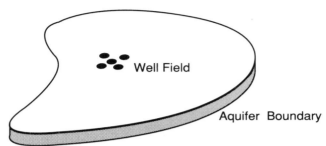

Well Field

Aquifer Boundary

Map view of aquifer showing well field and boundaries

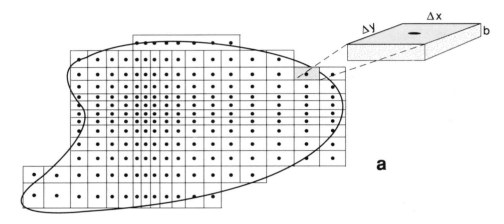

Δy Δx b

a

Finite difference grid for aquifer study, where Δx is the x-direction spacing, Δy the y-direction spacing, and b the aquifer thickness.

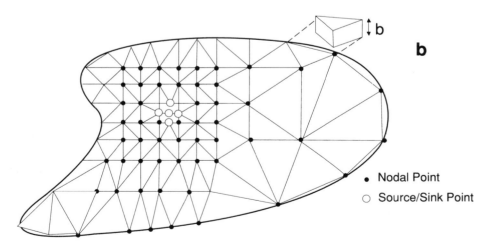

b

b

• Nodal Point

○ Source/Sink Point

Finite element configuration for aquifer study.

Figure K.4. Finite difference and finite element grid networks (after Mercer, J.W. and C.R. Faust, *Ground-Water Modeling*). Copyright © 1985, National Water Well Association, All Rights Reserved.

An advantage of the finite element method is the ability to readily define irregular model hydrogeologic boundaries. Disadvantages include lack of flexibility and cumbersome programming requirements.

K.4 WATER-QUALITY MODELING

Governing Equation

The governing equation for movements of solutes (in one direction) is (7, 8)

$$\frac{\partial}{\partial x}\left(\frac{D}{R_d}\frac{\partial C}{\partial x}\right) - \frac{v}{R_d}\frac{\partial C}{\partial x} \pm C_s Q = \frac{\partial C}{\partial t}\ ,\quad \text{(K.5)}$$

where

v = interstitial velocity [LT^{-1}],
D = coefficient of hydrodynamic dispersion [L^2T^{-1}],
x = space direction [L],
R_d = retardation factor,
C_sQ = source or sink function having a concentration C_s,
Q = discharge or recharge [L^3T^{-1}],
C = solute concentration.

The effects of mechanical dispersion as the fluid spreads through the pore spaces of the porous medium are described by the first and second terms of the left side of Equation K.5. The effects of dilution and mixing are expressed in the second and third terms on the left side.

Random Walk Concept

The random-walk technique as it relates to dispersion is illustrated by considering the progress of a unit slug or tracer-marked fluid, placed initially at $x = 0$, in an infinite column of porous medium with steady flow in the x-direction. With C_sQ equal to zero, Equation K.5 describes the concentration of the slug as it moves downstream. The solution is summarized by Prickett (9):

$$C(x, t) = \frac{1}{(4\pi d_L vt)^{1/2}} \exp\left[-\frac{(x - vt)^2}{4d_L vt}\right]\ ,\quad \text{(K.6)}$$

where

C = solute concentration,
d_L = longitudinal dispersivity [L],
v = interstitial velocity [LT^{-1}],
t = time [T],
x = distance along x-axis [L].

A random variable x is said to be normally distributed if its density function, $n(x)$, is

$$n(x) = \frac{1}{\sqrt{2\pi}\sigma} \exp\left[-\frac{(x - \mu)^2}{2\sigma^2}\right]\ ,\quad \text{(K.7)}$$

where

σ = standard deviation of the distribution,
μ = mean of distribution.

The following terms of Equations K.6 and K.7 are equated as

$$\sigma = \sqrt{2d_L vt}\ ,\quad \text{(K.8)}$$

$$\mu = vt\ ,\quad \text{(K.9)}$$

$$n(x) = C(x, t)\ .\quad \text{(K.10)}$$

Considering Equations K.8 through K.10, it can be seen that Equations K.6 and K.7 are mathematically similar expressions.

Particle in a Cell

Modeling the transport of dissolved constituents (solute transport model) uses the principle that the distribution of the concentration of chemical constituents of the water in an aquifer can be represented by the distribution of a finite number of discrete particles. Each particle is moved by ground water flow, known as "convective transport." The particles are assigned a mass that represents a fraction of the total mass of the chemical constituent involved. In this method of modeling, the number of particles used is many orders of magnitude less than the actual number of moles of constituent. However, experience has shown this approximation provides solutions adequate for most ground water model applications (9).

REFERENCES

1. Williams, D. E. 1981. *The Well/Aquifer Model, Initial Test Results*. Roscoe Moss Company, Los Angeles.

2. Williams, D. E. 1985. "Modern Techniques in Well Design." *J. Am. Water Works Assoc.* (September).

3. Todd, D. K. 1980. *Groundwater Hydrology*. J. Wiley and Sons, New York.

4. Williams, D. E. 1966. "Viscous-Model Study of Ground-Water Flow in a Wedge-Shaped Aquifer." *Water Resources Research* 2 (Third Quarter).

5. Halespaska, J. C., and F. W. Hartman. 1971. "Computer Program to Solve 3-Dimensional Equation of Heat Flow." Kansas Geological Survey Open File Report, September, p. 65.

6. Mercer, J. W., and C. R. Faust. 1981. "Ground-Water Modeling." *NWWA*.

7. Bear, J. 1965. "Hydrodynamic Dispersion." Unpublished lectures delivered during a summer program on Hydrology and Flow through Porous Media. Princeton University. August.

8. Bear, J., and A. Verruijt. 1987. Modeling Groundwater Flow and Pollution. D. Reidel, Dordrecht, Holland.

9. Prickett, T. A., T. G. Naymik, and C. G. Lonnquist. 1981. "A Random Walk Solute Transport Model for Selected Groundwater Quality Evaluations." Illinois State Water Survey, Bulletin No. 65.